AF321956

Hydrogen for Clean Energy Production: Combustion Fundamentals and Applications

Medhat A. Nemitallah · Mohamed A. Habib ·
Ahmed Abdelhafez

Hydrogen for Clean Energy Production: Combustion Fundamentals and Applications

 Springer

Medhat A. Nemitallah
Aerospace Engineering Department
Interdisciplinary Research Center for
Hydrogen Technologies and Carbon
Management (IRC-HTCM)
King Fahd University of Petroleum
and Minerals
Dhahran, Saudi Arabia

Mohamed A. Habib
Mechanical Engineering Department
Interdisciplinary Research Center for
Hydrogen Technologies and Carbon
Management (IRC-HTCM)
King Fahd University of Petroleum
and Minerals
Dhahran, Saudi Arabia

Ahmed Abdelhafez
Mechanical Engineering Department
Interdisciplinary Research Center for
Hydrogen Technologies and Carbon
Management (IRC-HTCM)
King Fahd University of Petroleum
and Minerals
Dhahran, Saudi Arabia

ISBN 978-981-97-7924-6   ISBN 978-981-97-7925-3   (eBook)
https://doi.org/10.1007/978-981-97-7925-3

This Springer imprint is published by the registered company Springer Nature Singapore Pte Ltd.
The registered company address is: 152 Beach Road, #21-01/04 Gateway East, Singapore 189721,
Singapore

If disposing of this product, please recycle the paper.

Preface

The demand for energy is expected to increase dramatically with the increase in population and the need for all energy sources, including renewable and fossil sources. The International Energy Agency (IEA) reported an expected surge in energy demand of as much as 30% by 2040 resulting in an additional 6700 GW to the existing power supply systems. It is also predicted that fossil fuels will remain as main source of energy for the next five decades. However, burning fossil fuels results in CO_2 emissions, as the main contributor to the global warming issue. This forces more research and development in the field of clean and efficient combustion of fossil fuels for a cleaner environment for the next generations. The motivation behind this work is to give more insight on clean combustion technologies along with the recent developments in hydrogen combustion systems for industrial applications.

The book is entitled *Hydrogen for Clean Energy Production: Combustion Fundamentals and Applications*. The book overviews combustion technologies and novel burner designs for reduced emissions and better fuel economy. Focus is made on the recently developed enhanced-design burners for higher turndown and lower NOx emissions, including the dual annular counter-rotating swirl (DACRS) and enhanced-vortex/environmental (EV) burners. The concept of flame stratification, i.e., heterogenization of the overall equivalence ratio, is introduced to widen the operability, control the emissions, and improve the turndown ratio of combustors. The recent developments and challenges toward the application of hydrogen combustion in the industrial combustion systems are introduced. As well, energy and hydrogen production in novel membrane reactors are reviewed in this book. With this, the book fills the existing gap in the literature on clean and hydrogen combustion technologies for industrial applications. This gas is created due to the absence of a comprehensive textbook that covers such kinds of developments. This book can be used as a textbook for graduate-level courses in the areas of clean and hydrogen combustion and as a reference book for short courses to be offered to mechanical and aerospace engineers and young researchers worldwide. The book chapters consider investigating clean and hydrogen combustion techniques for different applications based on experimental measurements along with detailed numerical simulations. Detailed descriptions of the different numerical models are presented for given applications to solve for the flow/flame fields, which are very important, especially for beginners and undergraduate students in the fields of clean and hydrogen combustion.

The book is composed of seven chapters, starting with an introductory chapter with a general description of combustion and emissions, global warming issues, and some proposed techniques for emission control with carbon capture. Chapter 2 presents some basic concepts and fundamentals of combustion, in order to help undergraduate students and early-career researchers understand the complicated technologies discussed in the preceding chapters of the book. This chapter discusses the different engine types and their classifications, basic fuel types and their classifications, in addition to basic thermodynamic analyses of combustion systems. Chapter 3 overviews the concept of fuel/oxidizer flexibility for controlling emissions and reserving flame stability in premixed combustion systems. This chapter presents the existing burner technologies for lean premixed combustion and the application of oxy-fuel combustion technology in such burners. Fuel flexibility, as well, is discussed considering hydrogen, syngas, ammonia, and fuel-blended combustion. Chapter 4 reviews in detail the technologies of stratified and hydrogen combustion for higher turndown and lower emissions of different combustion systems. Advances and recent developments in hydrogen gas turbines are introduced, as well. Chapter 5 presents the implementation of lean premixed combustion technology for emission control in different combustors. The application of fuel/oxidizer flexibility approach in premixed gas turbine combustors considering different burner designs is presented in Chap. 6. The last chapter discusses energy and hydrogen production in novel membrane reactors and oxy-syngas combustion considering membrane reactors for water splitting and partial oxidation of methane.

The authors appreciate the support received for the preparation of this book from the Deanship of Research Oversight and Coordination (DROC) at King Fahd University of Petroleum and Minerals (KFUPM). The support provided by the Interdisciplinary Research Center for Hydrogen Technologies and Carbon Management (IRC-HTCM) on projects numbered INHE2308 and INHT2409 is highly appreciated. Also, the support received through the KFUPM Consortium for Hydrogen Future through the projects numbered H2FC2309 and H2FC2315 is highly appreciated. Dr. Ahmed Abdelhafez would like to acknowledge the treasured support of his family during the compilation of this book. Dr. Medhat A. Nemitallah would like to dedicate this book to the spirit of his parents, may God forgive them and have mercy on them. Dr. Nemitallah also wishes to declare the continuous support and love he receives from his wonderful wife, Heba Abdelrahman Morshedy, as well as from his kids, Amira, Malak, and Mohammed.

Dhahran, Saudi Arabia   Medhat A. Nemitallah
                                                             Mohamed A. Habib
                                                             Ahmed Abdelhafez

Contents

Chapter 1
Introduction

1.1 Combustion and Emissions

Combustion is a set of exothermic chemical reactions, where fuel (typically a hydrocarbon material) reacts with an oxidizer (air, pure oxygen, or diluted oxygen) to oxidize the carbon and hydrogen in fuel and release copious amounts of heat energy. Carbon oxidizes in two steps, first to CO then to CO_2, as follows:

$$C + \frac{1}{2}O_2 \rightarrow CO + 110.53 \text{ MJ/kmol CO} \tag{1.1a}$$

$$CO + \frac{1}{2}O_2 \rightarrow CO_2 + 282.99 \text{ MJ/kmol } CO_2 \tag{1.1b}$$

It can be seen that the second step releases 72% of the total heat released when carbon (C) is fully oxidized to CO_2. This highlights the importance of ensuring complete combustion of carbon to utilize the maximum heat potential. Moreover, from a safety perspective, CO emissions from a combustion system must be minimized, because CO is a toxic gas that poses serious health hazards even in small concentrations. Emission regulations thus enforce strict limits on CO emissions from combustion systems, and the manufacturers of these systems must comply with those regulations. It should be noted that the second step of CO oxidation to CO_2 is inhibited if the local temperature within the reaction zone is low ($< \sim 1400$ °C). This occurs when combustion is lean or diluted excessively (e.g., with air), or when the flame is quenched by cold surfaces. The fuel-to-oxidizer ratio must thus be properly controlled to keep the reaction zone hot enough by avoiding excessively lean conditions, i.e., too much oxidizer.

Compared to carbon, hydrogen in the fuel is far more reactive and oxidizes readily to completion forming H_2O, as follows:

© The Author(s), under exclusive license to Springer Nature Singapore Pte Ltd. 2024
M. A. Nemitallah et al., *Hydrogen for Clean Energy Production: Combustion Fundamentals and Applications*, https://doi.org/10.1007/978-981-97-7925-3_1

$$H_2 + \frac{1}{2}O_2 \rightarrow H_2O + 241.82 \text{ MJ/kmol } H_2O \tag{1.2}$$

A perfect combustion system would thus burn hydrocarbon fuel to completion forming CO_2 and H_2O. But is CO the sole pollutant in combustion systems? The answer is no. Air, if used as oxidizer, consists of O_2 and N_2, which tend to react with each other to form nitrogen oxides (NO and NO_2), commonly referred to as NOx. From an environmental perspective, NOx is considered a pollutant because it reacts with moisture in the air to form acid rain, which is detrimental to the environment. Emission regulations thus enforce strict limits on NOx emissions as well, and the manufacturers of combustion systems must also comply with those regulations. NOx formation is governed by the Zeldovich mechanism [1] as follows:

$$O + N_2 \rightleftharpoons NO + N \tag{1.3a}$$

$$N + O_2 \rightleftharpoons NO + O \tag{1.3b}$$

$$N + OH \rightleftharpoons NO + H \tag{1.3c}$$

This mechanism is highly temperature dependent; NOx concentration increases exponentially with temperature. Figure 1.1 shows how NOx emissions peak near stoichiometric combustion, where the temperature is highest. The range of allowable flame temperatures in a combustion system that uses air as oxidizer is thus limited, typically 1650–1900 K, to minimize both NOx and CO emissions at excessively hot and cold temperatures, respectively, as seen in Fig. 1.2.

1.2 Lean Premixed Combustion for NOx Emission Reduction

1.2.1 Non-premixed Versus Premixed Combustion

The need to reduce NOx emissions drove gas-turbine manufacturers to abandon the conventional non-premixed combustor technologies and switch to the current lean-premixed (LPM) ones. In non-premixed combustors, fuel and oxidizer are introduced separately into the reaction zone and diffuse towards each other, so the flame has an unavoidable stoichiometric reaction sheet (Fig. 1.3). NOx formation peaks within this stoichiometric zone, as the local temperature there is highest. LPM flames, on the other hand, have no such stoichiometric zone, because the fuel and oxidizer are premixed in lean proportions upstream of the reaction zone. This reduces the NOx emissions significantly, because NOx formation becomes a sole function of the lean overall equivalence ratio.

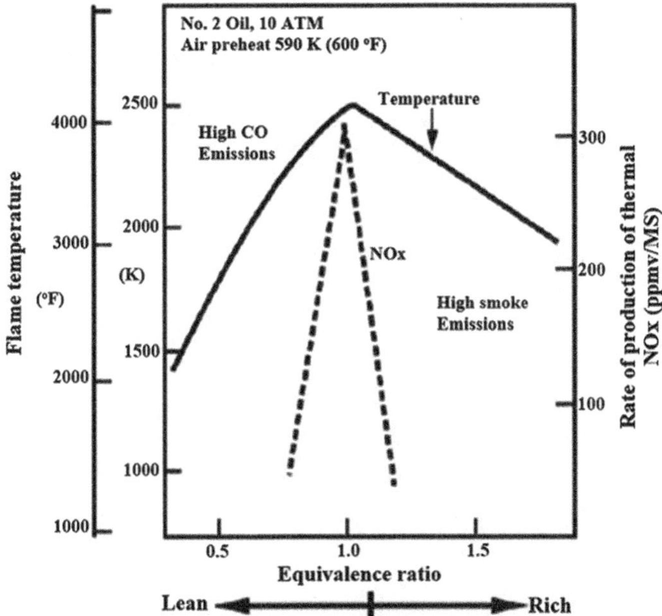

Fig. 1.1 Effect of equivalence ratio on NOx formation in a combustor of a gas turbine [2]

Fig. 1.2 Typical range of operation of LPM gas turbines to keep NOx and CO emissions below 15 and 25 ppm, respectively [3]

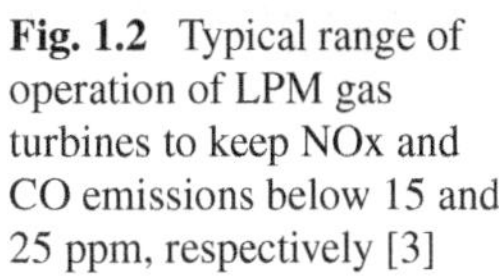

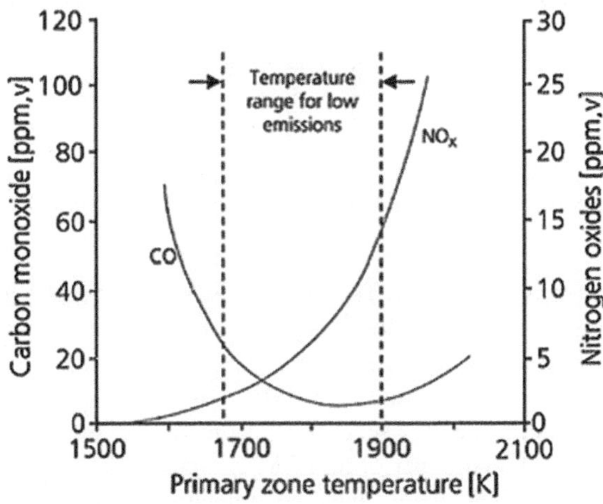

Figure 1.4 compares non-premixed and LPM gas-turbine combustors. It can be seen that the air entering the combustor is split significantly differently. Non-premixed combustors use a small portion of the inlet air in the primary flame zone, which makes it rich; most of the air is utilized as secondary air to cool the combustor liner, complete the combustion, and dilute the combustor discharge stream prior to

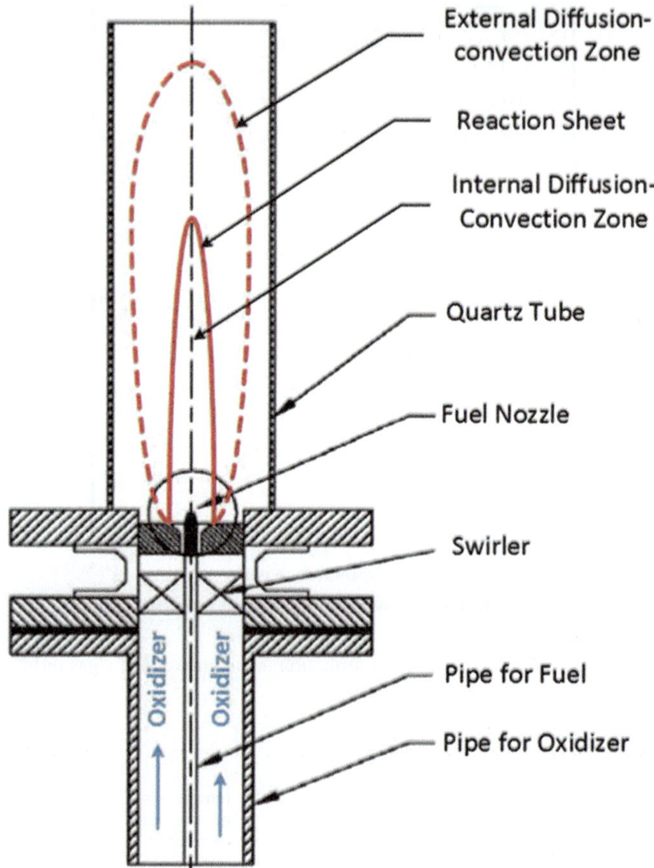

Fig. 1.3 Reaction zones in a non-premixed swirl-stabilized gas-turbine flame [6]

entering the turbine. The local equivalence ratio thus undergoes a gradient from rich near combustor headend to lean towards the exit. The high-temperature stoichiometric zones of non-premixed flames necessitate significant secondary air flow for liner cooling. In contrast, LPM combustors use a lesser amount of inlet air for liner cooling, since they send most of the air into the premixed primary flame to ensure it is lean. The liner of an LPM combustor can be effectively cooled using less air because LPM flames have no stoichiometric zones, and their temperature is much lower than the average of a non-premixed flame.

Non-premixed combustors have been used traditionally because of the superior flame stability they enjoy. The flame is resistant to blowout, because primary air is only a small part of the inlet air. No flashback occurs because air and fuel are not premixed upstream of the reaction zone. LPM combustors, on the other hand, are prone to both blowout and flashback, which limits the operability window significantly [4]. Blowout occurs at excessively low flame temperatures, because the

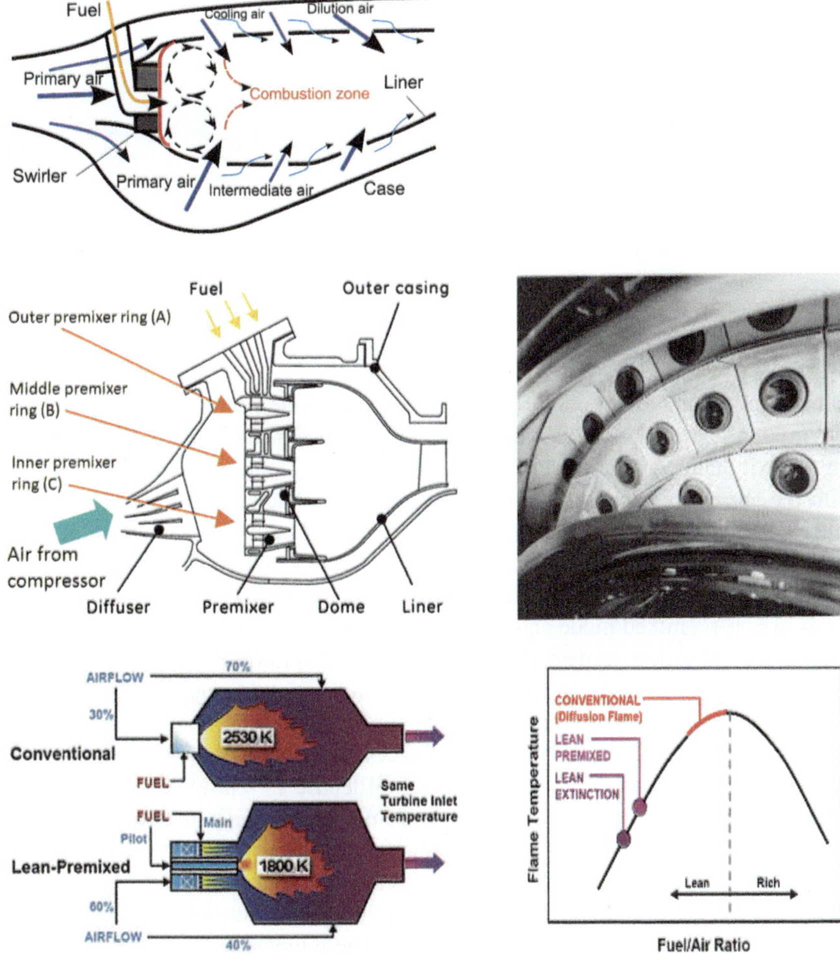

Fig. 1.4 Non-premixed (top) versus LPM (bottom) gas-turbine combustors [3, 7, 8]

combustor is already operated near its lean-extinction limit, see Fig. 1.4. Flashback occurs if the flame speed exceeds the incoming reactant velocity (at low pressure drop across combustor headend) and/or at excessively high flame temperatures [5].

Premixing of fuel and air is performed within the headend of LPM combustor by means of a premixer. Figure 1.5 shows a premixer example, namely the dual annular counter-rotating swirler (DACRS) premixer used in General Electric's LM2500 and LM6000 gas turbines [9]. The premix or primary component of compressor air (not used in liner cooling) is swirled using two counter-rotating swirlers to enhance flow turbulence (for air–fuel mixing) and stabilize the LPM flame effectively. Most of the fuel is supplied as premix fuel from within the outer-swirler vanes and is injected into the premix air through holes in these vanes (3 holes per vane, see Fig. 1.5). A

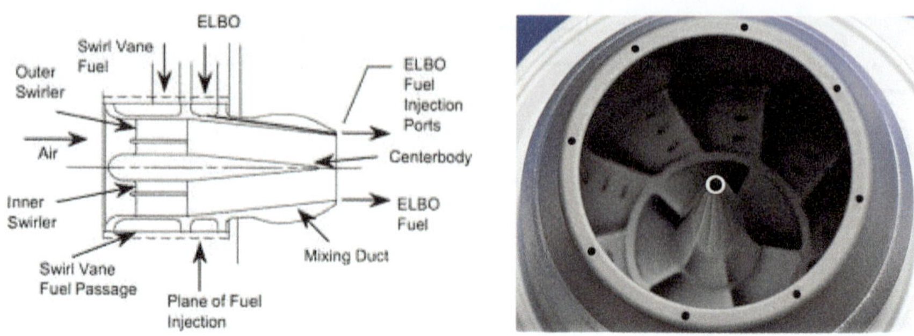

Fig. 1.5 DACRS premixer used in General Electric's LM2500 and LM6000 gas turbines [9]

converging mixing duct (aka shroud) provides the necessary residence time for air–fuel mixing and ignition delay, before the mixture exits the shroud and burns. The shroud exit diameter is critical in controlling the exit velocity, which, in conjunction with the shroud residence time, determines the flashback margin of the premixer. Since LPM flames are prone to blowout, as discussed earlier, a small portion of the fuel is injected directly into the flame at its base through holes at the shroud exit. This fuel component is called enhanced lean blowout (ELBO) fuel, see Fig. 1.5, because it burns in non-premixed mode and thus enjoys superior resistance to blowout of non-premixed combustion, as described earlier. The fuel split (premix vs. ELBO) has to be controlled precisely, of course, because NOx formation increases with ELBO fuel percentage due to the non-premixed combustion.

Micromixer (MM) burners are another successful premixer example that is already employed in utility gas turbines (e.g., General Electric's F-class) [10–12]. The term "micromixer" is based on the fact that these premixers mix fuel with the oxidizer at the micro-scale. A MM is made up of a two-dimensional array of straight, parallel, evenly spaced tubes (of a diameter of few millimeters) that are welded to two faceplates, one of which serves as the burner's intake and the other as its exit (see Fig. 1.6). The incoming fresh oxidizer flow is divided among these tubes. This geometry creates a plenum for gaseous fuel around the tubes between the faceplates. Fuel is supplied from this plenum and injected radially (crossflow) through side holes (each less than one millimeter in diameter) in the tube walls. A premixing zone is thus produced inside each tube, and the tube length is designed to ensure a fully developed and properly premixed flow at the exit section of each tube. To reduce the MM size axially, i.e., reduce the length needed to generate the premixed flow, each MM tube has multiple fuel holes (evenly dispersed along the tube circumference) that inject fuel at the same axial location close to the entrance faceplate. The diameter, number, and distribution of MM tubes on the faceplate are tuned to reduce emissions and broaden the operability limits. To stop flashback, the operational equivalence ratio and the pressure drop inside the tubes are regulated. Similar to the DACRS premixer, the MM injects a small amount of pilot (ELBO) fuel directly into the flame at its base to burn in non-premixed mode and enhance the resistance to blowout. A corresponding small amount of pilot air is used to burn the pilot fuel. The fuel split (premix vs.

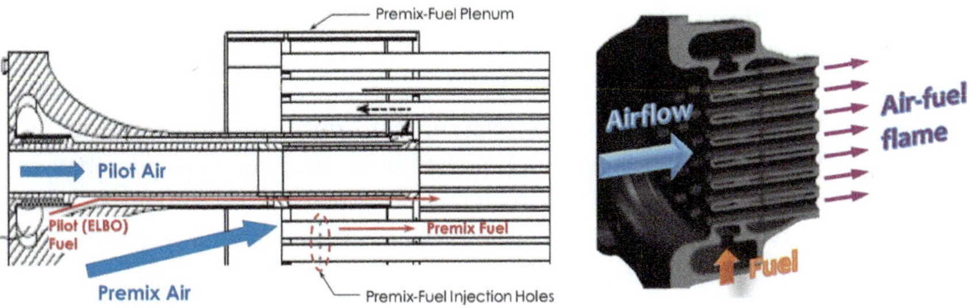

Fig. 1.6 Micromixer premixer used in the F-class gas turbines of general electric [12]

pilot) again has to be controlled precisely, because NOx formation increases with pilot fuel percentage due to the non-premixed combustion.

1.2.2 Stability Characteristics of Premixed Combustion

The term combustion instability describes the phenomena where the flame moves from the location it must be anchored at, which forces combustor shutdown and may cause hardware damage. There are two types of instabilities in premixed combustion, namely static and dynamic instabilities [13, 14]. Static instabilities are flame blowout and flashback, which have been discussed earlier, whereas the primary mechanism driving dynamic instabilities is thermoacoustic coupling. As the name suggests, dynamic instabilities are driven by the thermal (heat-release) field and the acoustic characteristics of combustor geometry [15]. The latter is simple to understand and is evident in many instances in our day-to-day lives. Any flow in a confined passage will produce noise or pressure waves, especially turbulent flows. The flute is a straight-forward example: blowing into it produces sound because the airflow generates low-amplitude broadband noise, but a specific frequency resonates with a larger audible amplitude. This frequency is determined by the passage geometry within the flute; even at constant airflow, covering or exposing each flute hole alters the geometry and, consequently, the resonance frequency. Consider gas-turbine combustors as massive flutes; turbulent air discharges from the compressor into the combustor, which is a confined passage with a specific shape and one or more resonance frequencies. Heat release in the flame is the second factor causing dynamic combustion insta-bilities. Due to the extreme turbulent nature of gas-turbine flames, the local heat release map is far from uniform. The release of heat causes combustion products to expand locally, which induces noise or pressure fluctuations in premixed combus-tion. Multiple pressure waves thus coexist within the combustor, some are caused by oscillations in heat release, while others are induced acoustically by turbulent flow through the confined combustor passage. Most broadband frequencies experience destructive interference when these pressure waves interact, while some frequencies interfere constructively, increasing the magnitude of pressure fluctuations. Large magnitudes cause the combustor hardware to shake, which may wear down or even break some parts. In severe circumstances, combustor fragments may detach and harm the turbine downstream [16–18]. It is interesting to note that even under stable working conditions, minor pressure oscillations are a consistent characteristic of all combustion systems, as long as the amplitude of these variations is smaller than ~5% of the mean combustor pressure [6]. Greater amplitudes are undesirable and deemed to be a sign of dynamic combustion instability [17–19]. Dynamic instabilities also change the location and shape of the flame [20–23], which poses a major operability risk.

Since LPM combustors are usually operated near their blow-off limits, dynamic instabilities are a major concern [24–29]. The temperature of a LPM flame is low enough to trigger local extinction and re-ignition at several locations, which induces oscillations in the heat-release map and, subsequently, pressure fluctuations that couple with the acoustic characteristics of the combustor [30, 31]. Furthermore, there are no stoichiometric zones in the premixed flame, as discussed earlier; that is, the local temperature map is nearly uniform. The resonant pressure waves are thus free to propagate within the quasi-uniform temperature field within the combustor in the absence of any major temperature (and density) gradients.

1.3 Oxy-fuel Combustion for Carbon Capture

A perfect combustion system would generate no CO or NOx emissions, just clean exhaust comprising CO_2, H_2O, O_2, and N_2. But is CO_2 really clean? Is it environmentally benign? Humans and animals exhale CO_2 in small quantities while breathing, and plants consume this CO_2 to produce sugar and oxygen within their photosynthesis process. The anthropogenic activities, however, release more than 30 gigatons of CO_2 annually into the atmosphere [32]. CO_2 is one of the major greenhouse gases; so, its presence in higher concentrations in the Earth's atmosphere increases the absorbance of solar infrared radiation, which increases the average atmospheric temperature, see Fig. 1.7. If this trend prevails, catastrophic climate changes are predicted, as seen in Fig. 1.8. The Paris Agreement was thus made in 2015 as a legally binding international agreement on climate change [33]. Its goal is to limit the rise in global temperature to less than 2 °C, preferably 1.5 °C, compared to pre-industrial levels. This can only be achieved by capturing the CO_2 and not emitting it into the atmosphere, aka carbon capture.

1.3.1 Carbon-Capture Technologies

To capture CO_2 from existing combustion systems, without modifying them, post-combustion carbon capture is implemented. The application is equipped with a carbon-capture unit that separates CO_2 from the exhaust stream. The vast majority of

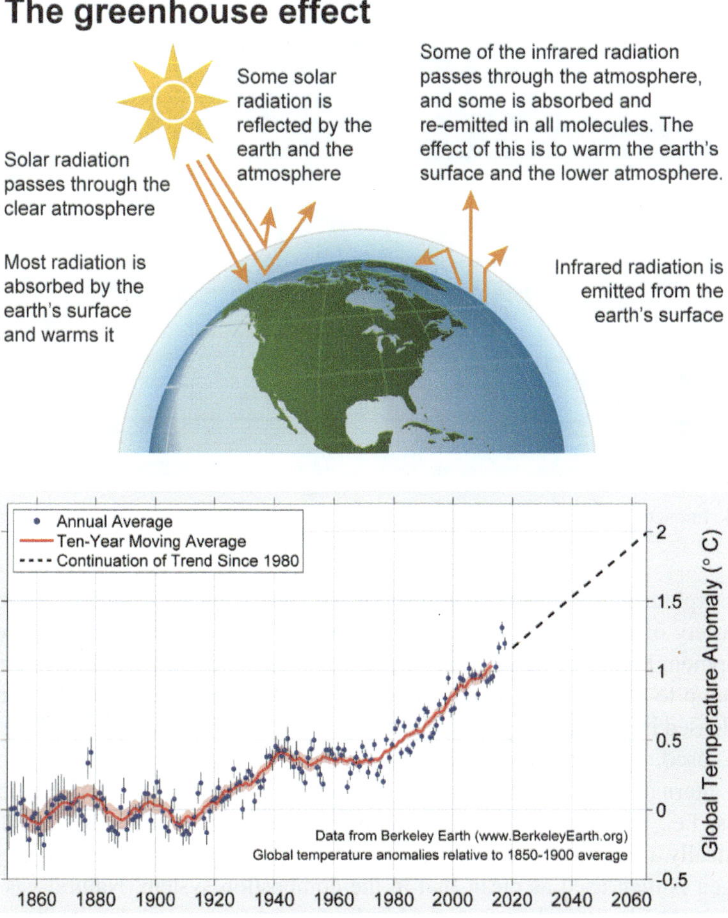

Fig. 1.7 Explanation of the greenhouse effect (top) [34] and plot of Earth's surface temperature (bottom) [35]

existing combustion systems utilize air as oxidizer, simply because it is an abundant, low-cost resource. The products of complete combustion thus consist of CO_2, H_2O, O_2, and N_2. Since air is ~79% N_2, nitrogen is the dominant constituent of the exhaust stream, and the concentration of CO_2 is rather small by comparison (~10%). This makes it challenging to capture all CO_2 cost-effectively.

CO_2 is removed by post-combustion capture methods through chemical or physical adsorption/ absorption processes. As seen in Fig. 1.9, the capture process can be classified as adsorption, absorption, and membrane separation, based on the principles of the process. The most mature process for post-combustion carbon capture is absorption [37]. It is implemented in 57% of existing applications, while 14% utilize adsorption, 8% employ membranes, and 21% use mineralization or bio-fixation.

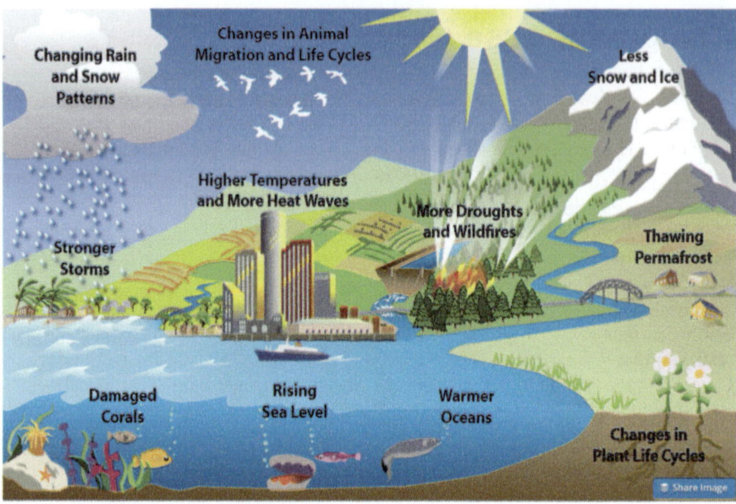

Fig. 1.8 Impacts of climate change [36]

This is explained by the fact that absorption gas separation has been used extensively in a variety of petrochemical sectors, whereas all other technologies require more development before being used on a large scale. Aqueous amine solutions can only remove up to 85% of the effluent CO_2 [38], meaning that the remaining portion is still released into the atmosphere. The capture effectiveness varies depending on the method used, but it has not yet achieved 100%.

An alternative approach to carbon capture is to do it prior to the combustion process, i.e., pre-combustion carbon capture. The hydrocarbon fuel is reformed catalytically to separate carbon from hydrogen, so that carbon is captured, and hydrogen is then used as clean fuel in the combustion system. Natural gas (CH_4) is the simplest hydrocarbon fuel and has been used for several decades to produce hydrogen [39]. The chemical reactions for steam methane reforming (SMR) can be simplified as:

$$CH_4 + H_2O + \text{heat} \rightarrow CO + 3H_2 \tag{1.4a}$$

$$CO + H_2O \rightarrow CO_2 + H_2 \tag{1.4b}$$

which can be summarized as:

$$CH_4 + 2H_2O + \text{heat} \rightarrow CO_2 + 4H_2 \tag{1.4c}$$

SMR is endothermic and energy-intensive, which poses an economic penalty on the combustion application. To alleviate this penalty, autothermal reforming (ATR) can be performed instead, where the hydrocarbon fuel is reformed with a mixture

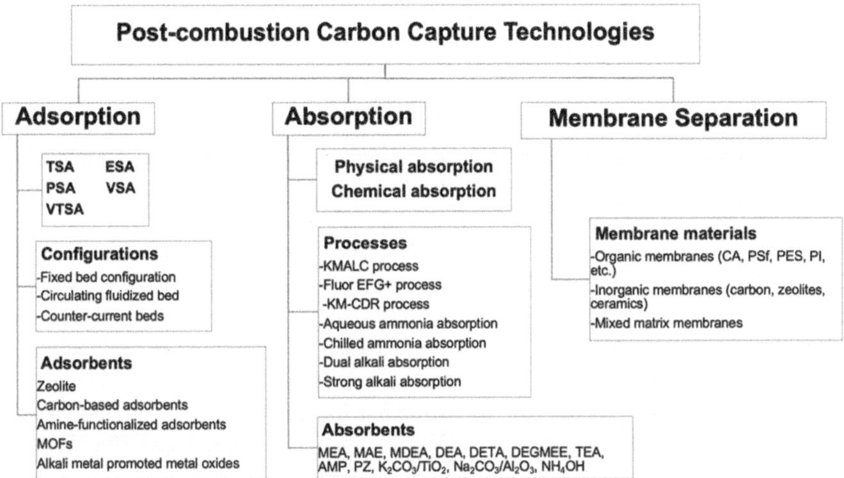

Fig. 1.9 Post-combustion carbon capture technologies [37]. TSA: Temperature swing adsorption; PSA: Pressure swing adsorption; ESA: Electrical swing adsorption; VSA: Vacuum swing adsorption; MOF: Metal organic framework; KMALC: Kerr-McGee/AGG Lummus Crest; Fluor EFG+ : Fluor Econamine FG Plus^{SM}; KM-CDR: Kansai Mitsubishi Carbon Dioxide Recovery; MEA: Monoethanolamine; MAE: Methylaminoethanol; MDEA: N-methyl-2,2-iminodiethanol; DEA: 2,2′-iminodiethanol; TEA: Triethanolamine; CA: Cellulose acetate; PSf: Polysulfone; PES: Polyethersulfone; PI: Polyimide

of steam and pure O_2 [40]. A portion of the fuel burns with oxygen to release the energy needed for reforming the remaining amount of fuel, so the overall demand for external energy is reduced drastically or possibly eliminated. However, the hydrogen yield of ATR is typically less than that of SMR.

Heavier fuels, such as biomass and solid waste, can also be used for hydrogen production. The fuel is gasified (i.e., partially oxidized) using pure oxygen with the aid of a catalyst to produce syngas ($H_2 + CO$), which is then treated to maximize the hydrogen yield using the water-shift reaction (1.4b) above [41]. Special treatment is also needed to desulfurize the fuel and/or syngas, neutralize any NOx emissions formed, and capture the fly-ash, which is inherent in solid fuels. The efficiency of producing carbon-free hydrogen by SMR, ATR, or gasification depends greatly on the carbon-capture section of the process [42]. Up to 95% of the CO_2 can be captured from the syngas using selexol solvent, because of the greater CO_2 concentration and higher CO_2/H_2 selectivity [43].

A revolutionary and cost-efficient way to capture all CO_2 is oxy-fuel combustion, where pure oxygen is used instead of air. Burning a hydrocarbon fuel in pure oxygen produces CO_2 and H_2O as the main combustion products, i.e., N_2 does not exist in the exhaust stream anymore. This facilitates capturing the CO_2 in high-purity form simply by condensing the H_2O. The next section is dedicated to discussing oxy-fuel combustion technology.

1.3.2   Oxy-fuel Combustion Technology

Unlike a conventional combustor, an oxy-fuel one uses nitrogen-free oxygen to burn the fuel. An air-separation unit (ASU) is thus needed upstream of the combustion system to separate air and provide oxygen for combustion; see Fig. 1.10. However, the use of pure undiluted oxygen raises the flame temperature. So, to dilute the reactants and regulate the flame temperature, some of the exhaust gases—mostly CO_2—must be recycled back into the combustor. This is known as flue-gas recirculation (FGR), see Fig. 1.10. The fuel is thus oxidized by an oxy-mixture of O_2 and CO_2, i.e., it can be thought that N_2 from air has been replaced by CO_2, except that the oxygen fraction (i.e., $\%O_2$) in this O_2/CO_2 oxidizer is variable (not fixed at 21%) and used as the primary parameter controlling flame temperature.

Replacing air with oxygen in combustion has its merits. The fundamental advantage, of course, is capturing all effluent CO_2 in high purity compressed form. Furthermore, in the absence of air-based nitrogen, NOx formation is nearly eliminated. If combustion is complete, i.e., no CO is emitted, the oxy-fuel system is virtually emission-free; oxy-fuel power plants are thus called zero-emission power plants (ZEPPs) [44]. Finally, the volume of oxy-fuel exhaust is considerably smaller in the absence of nitrogen, which used to be the largest constituent of air–fuel exhaust. This translates into a compact exhaust-treatment system in oxy-fuel applications.

There are major differences between the concepts of oxy-fuel and air–fuel combustion in terms of flame properties and combustor operability. Air–fuel combustors in gas turbines use the equivalence ratio as the primary control parameter that regulates flame temperature and prevents or alleviates combustion instabilities. Particularly in premixed combustors, the overall equivalence ratio is always maintained lean to minimize NOx emissions and adjust the combustor-exit temperature to the cycle specifications and the metallurgical constraints of the turbine first stage. Lean operation is made possible by the inexpensive plentiful nature of air. The gas turbine operates consistently with a fixed oxidizer composition, namely 21% O_2 +

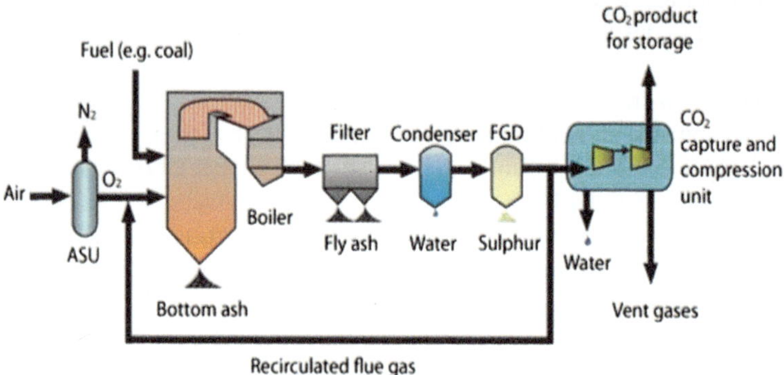

Fig. 1.10 Schematic diagram of an oxy-fuel combustion system

79% $N_2 \equiv$ air. Flue-gas recirculation (FGR) can be optionally used to reduce NOx emissions and alleviate combustion instabilities.

Oxy-fuel combustors, on the other hand, are significantly different. The ASU demands energy to produce the needed oxygen, which makes it a scarce and valuable resource (relative to air). For this reason, oxy-fuel combustors are usually operated close to stoichiometry to preserve oxygen. In this scenario, FGR takes the role of equivalence ratio as the primary control parameter. To regulate the flame temperature [45] and adjust the combustor-exit temperature to the cycle requirements and the metallurgical limitations of the turbine first stage [46], a portion of the effluent CO_2 is circulated in precisely the right amounts to dilute the primary oxy-reactants. Thus, the FGR ratio (or oxygen percentage) determines how the O_2/CO_2 oxidizer is composed. Air- and oxy-combustors are, consequently, very different in design and operability. However, present gas-turbine users will not be readily persuaded to transition to oxy-fuel combustion for carbon capture, unless their current air-based systems are modified to perform oxy-fuel combustion with minor hardware changes. A thorough examination of the operability and flexibility of the combustor is, therefore, necessary when converting a conventional air–fuel system to an oxy-fuel one.

Since oxy-fuel combustion uses CO_2 in place of air-based N_2, it is significantly different from air–fuel combustion in terms of physics and reaction kinetics. The density, volumetric heat capacity, and transport properties of the reacting mixture are all strongly impacted by the differences in the physical properties of CO_2 and N_2 [47–50]. Thermal conductivity, dynamic viscosity, and mass diffusivity make up the transport properties. Because CO_2 has greater density than N_2, it has an impact on the bulk velocity, flame shape, pressure drop across combustor headend, and total mass passing through the combustor and turbine. In addition, CO_2 has a larger volumetric heat capacity than N_2, which lowers the flame temperature and, consequently, the flame speed and stability.

The higher concentrations of CO_2 in oxy-fuel combustion have been reported to induce retarded kinetic rates in oxy-fuel flames [51, 52], as these rates are directly proportional to the laminar flame speed. CO_2 dissociation presents an additional challenge because it is an endothermic reaction that wastes energy and modifies the reaction kinetics [53, 54], particularly at greater CO_2 concentrations.

Therefore, it follows that a 21% oxygen fraction (by vol.) in the O_2/CO_2 oxidizer will not be sufficient to sustain a stable flame [55, 56] (as it used to do under air–fuel conditions), if an air–fuel combustor is converted to oxy-fuel operation without any geometry changes; that is, the minimum oxygen fraction in oxy-fuel systems is greater than 21%. Research has shown that this minimum is 30% to attain stable combustion, as compared to the 21% value of air–fuel combustion [57, 58]. More oxygen is needed to offset the greater volumetric heat capacity of CO_2. Burner designs must thus be modified to tackle these challenges and hold stable oxy-fuel flames.

1.3.3   Premixed Oxy-fuel Combustion

To motivate current users of air–fuel gas turbines to switch to oxy-fuel combustion for carbon capture, their current LPM air-based systems need to be retrofitted to employ oxy-fuel combustion with the least hardware changes. LPM combustion has its stability concerns, as discussed earlier. Oxy-fuel combustion brings its own challenges as well. It can thus be envisioned how challenging premixed oxy-fuel systems would be. So, what triggers the interest in premixed oxy-fuel combustion? Simply the fact that these systems need to operate close to stoichiometry to preserve the oxygen, which leaves very little excess oxygen to ensure complete combustion. The fuel and O_2/CO_2 oxidizer must, thus, be thoroughly premixed to ensure that all fuel finds its oxygen. Non-premixed oxy-fuel flames emit excessive levels of CO (incomplete combustion) and, therefore, cannot satisfy the requirement of stoichiometric operation for oxygen preservation [59].

Premixed oxy-fuel flames also have significantly narrower operability windows, in comparison to their air–fuel counterparts because of the adverse effect of adding CO_2 on the reaction kinetics [54, 60]. Moreover, the laminar flame speed of an oxy-fuel flame is about one-fifth that of the corresponding air–fuel one [61]. The inner recirculation zone in swirling oxy-fuel flames is located further downstream, compared to their air–fuel counterparts, which indicates inferior stability of the former [62]. The oxygen fraction has a key effect on flame characteristics. The 21% of air is unachievable in oxy-fuel flames; i.e., they have a greater minimum threshold, which limits the operational flexibility [62].

1.4   Stratified LPM Combustion for Ultra-Low Emissions

If you are asked to design a gas-turbine combustor, the typical vision you would have is a combustor hosting a single flame, where all input fuel is burned. You would go through a cumbersome exercise, attempting to ensure flame stability and acceptable emission levels at all required operating conditions, which is definitely not easy to achieve. So, why limit yourself to just one flame? Why not split the fuel and/or air to have multiple flames within the same combustor? The introduction of fuel and/or air staging, i.e., introducing fuel and/or air from multiple inlets into the combustor, creates gradients in the local equivalence ratio, where some zones are richer or leaner than the others, which is termed "stratification". The stratification concept intentionally seeks heterogenization of the overall equivalence ratio, which is achieved through staging or splitting of reactant flow. A controlled non-uniform spatial distribution of equivalence ratio is thus achieved within the combustion zone. Stratification adds more degrees of freedom to the combustor operability, widens the operability window, lowers the emissions, and allows for bypassing zones of dynamic instabilities [63–65], as will be discussed in the next sections. Stratified combustion is one of the attention-seeking contemporary technologies, employed in state-of-the-art gas turbines to achieve ultra-low emissions, enhanced stability, and superior turndown (ratio of maximum to minimum loads). The Beihang Axial Swirl Independently Stratified (BASIS) burner [66, 67], Cambridge Stratified Swirl Burner [68], Technology of Low Emission of Stratified Swirl (TeLESS) combustor [69], and non-separated stratified swirl burner [70] are a few examples of stratified-combustor designs.

1.4.1 Fuel-Staged Combustion

The simplest implementation of fuel splitting or staging is in the DACRS premixer discussed in Fig. 1.5. Fuel is split among two circuits to create a primary premix component and a smaller non-premixed one. The premix circuit receives most of the input fuel to the premixer; it is premixed with air within the shroud section to create a LPM flame downstream. The non-premixed circuit, on the other hand, receives fuel that is injected directly at the flame base to burn in non-premixed mode for benefiting from the superior stability of non-premixed combustion. This fuel component is of critical importance as the combustor approaches blowout, because it sustains the LPM flame, which would otherwise extinguish too early. The non-premixed fuel is thus termed enhanced lean blowout (ELBO) fuel. Logically, ELBO fuel is shut off or kept at its minimum level while the gas turbine is operating stably at high-to-full load, because non-premixed combustion generates excessive NOx emissions, as discussed earlier. The premix/ELBO split is thus a key control parameter to extend the operability window on the blowout side while keeping NOx emissions within acceptable limits. ELBO fuel also plays a vital role in mitigating dynamic combustion instabilities. Its injection creates a local zone of richer equivalence ratio, which stratifies the flowfield, disrupts the uniformity of local flame temperature, and induces density gradients. This disruption helps to prevent the pressure waves of dynamic instabilities from propagating freely and reciprocating within the flowfield. ELBO fuel is thus a powerful tool for suppressing dynamic instabilities. Other premixer technologies, such as micromixers (Fig. 1.6), also implement the same practice of fuel splitting or staging; ELBO fuel is also known as pilot fuel in this case. Pilot fuel serves to enhance lean blowout and mitigating dynamic instabilities.

So far, we are still discussing a single flame, created by either the DACRS or micromixer premixers. Does the combustor have a single premixer? The answer is no. Take a look at General Electric's LM6000 gas-turbine combustor (Fig. 1.11), which hosts 75 DACRS premixers, distributed among three rings: the outer ring (A) has 30 premixers, the middle ring (B) another 30, and the inner ring (C) 15 more. If all 75 premixers were lit during low loads (startup and idling), the flame temperature would have to be excessively low, and the combustor would have to operate beyond its lean blowout limit, which is not possible. Instead, the engine starts in the B-mode, firing the 30 B-ring premixers only, see Fig. 1.11. The A and C-rings are not lit, i.e., run "cold". Note that air passes through all three rings, but only the B-ring air is used for combustion, while the A- and C-ring air is used for dilution to reduce the

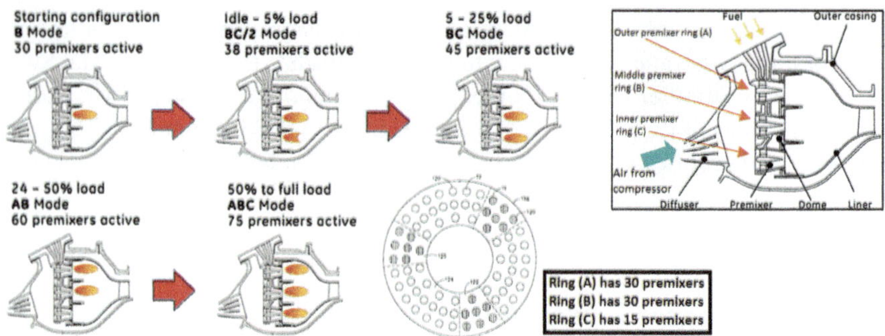

Fig. 1.11 Firing modes of General Electric's LM6000 gas turbine [71]

temperature at combustor exit to meet the cycle requirements at engine start. This allows the B-ring premixers to operate at reasonable equivalence ratio and flame temperature that facilitate stable operation with acceptable NOx and CO emissions. Interestingly, this B-mode of operation depicts a form of passive air staging; air is split to perform different functions in different parts of the combustor, but the air split is governed only passively, i.e., just by the area ratios of the three rings (without any means of active control).

As the load increases up to 5%, half of the C-ring is enabled (8 premixers), and the combustor is operating in the BC/2-mode with 38 active premixers. Increasing the load further (5–25%) enables the remaining C-ring premixers, and the combustor operates in the BC-mode with 45 lit premixers. When the engine transitions to medium loads (25–50%), the C-ring is disabled, and the A-ring is enabled instead, making the combustor operate in the AB-mode with 60 active premixers. Finally, high-to-full loads (50–100%) require all 75 premixers to be lit, which is the ABC mode of operation. It should be noted that the flame temperature (and load) is varied within each mode by controlling the fuel flow rate(s) to the active ring(s). A sophisticated control system monitors the load requirements and governs the transition from one mode to another safely and stably, without the combustor incurring blowout or dynamic instabilities.

1.4.2   Dual Lean-Premixed (DLPM) Combustion

An extension to the concept of LPM combustion is dual lean-premixed (DLPM) stratified combustion. Figure 1.12. shows an example of a stratified burner that is based on the DACRS design with modification to implement DLPM combustion. The DLPM DACRS premixer has two counter-rotating swirlers, just like the original DACRS design. The main difference, however, is that the DLPM concept has two shrouds, an outer one (like the original DACRS) and an inner shroud that separates the outer-swirler flow from the inner-swirler one. The incoming airflow is thus staged

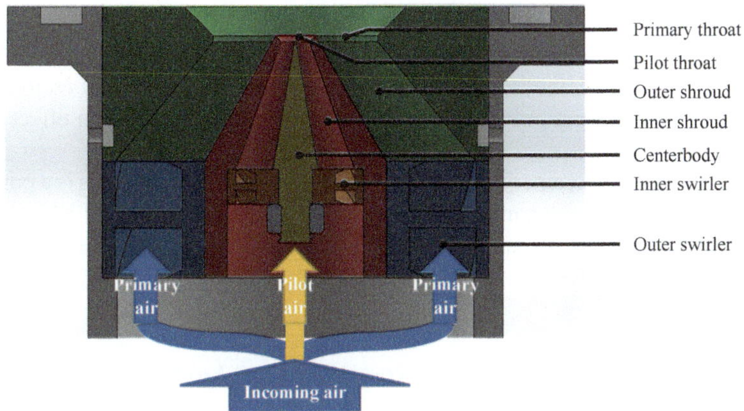

Fig. 1.12 DLPM DACRS premixer concept

into primary air (outer swirler) and pilot air (inner swirler). This facilitates having two distinct swirling airflows that interact downstream of the premixer throat, not before. This burner also implements fuel staging by having two fuel circuits: primary fuel that is injected and premixed with the primary air, and pilot fuel that is injected and premixed with the pilot air. Two concentric LPM flames (primary and pilot) are thus created downstream of the premixer throat, hence the name "dual" lean-premixed. The splits of air and fuel are designed and controlled to keep the pilot flame adequately hot (high enough equivalence ratio), while the primary flame is significantly colder (ultra-lean equivalence ratio). The pilot flame, thus, acts as a stabilizer and ignition source to keep the primary flame lit, which, otherwise, would have extinguished. The overall blowout limit of the combustor is, thus, significantly enhanced. The fact that the two flames are counter-rotating enhances their interaction and boosts the stabilization effect of the pilot flame.

The DLPM concept has several advantages over the original DACRS design:

1. The primary flame is stabilized by another LPM flame (pilot) that does not implement non-premixed combustion like the ELBO fuel component of the original DACRS. The absence of non-premixed combustion reduces the NOx emissions of the DLPM concept significantly.
2. Since both primary and pilot flames in the DLPM concept are premixed, the fuel split can be controlled more freely, i.e., pilot fuel can be increased safely, without the fear of generating too much NOx, as is the case with ELBO fuel in the original DACRS.
3. The area ratio of the outer and inner swirlers is utilized to create an extra degree of freedom (air split) during the design of the DLPM premixer. This degree of freedom does not exist in the original DACRS. For the same flow rate of incoming air, the original DACRS has to burn it all in one LPM flame, whereas the DLPM concept stages the incoming air to make a pilot portion of it assist

in the combustion of the primary component. This allows the primary flame to remain lit down to ultra-lean equivalence ratios.

4. The hotter pilot flame in the DLPM concept occupies the central recirculation zone of the swirling primary flame, which boosts the stabilization effect of the pilot flame.

5. The difference in temperature between the primary and pilot flames is a powerful tool for mitigating dynamic instabilities in the fully premixed DLPM concept, i.e., without having to use non-premixed fuel, as is the case with ELBO fuel in the original DACRS.

1.4.3   Fuel/Oxidizer-Flexible Stratified Combustion

Combustion stratification is also an effective approach to improve the adversely affected flame stability of oxy-fuel combustion systems due to the presence of high CO_2 concentrations, as discussed earlier. Consider, for example, a stratified oxy-fuel system with a central pilot flame and an annular primary one, similar to the system depicted in Fig. 1.12. If the primary flame has to be excessively diluted with CO_2 to reduce the overall combustor exit temperature, then it can be sustained by a hotter more-stable pilot flame that has less CO_2 dilution. Stratification thus allows for the staging of recirculated CO_2, i.e., stratifying the O_2/CO_2 oxidizer, which becomes a powerful tool for improving flame stability, especially close to the blowout limit. This is not possible without stratification.

Since combustion, emission, and flame-stability characteristics differ from one fuel to another, certain fuel compositions, or enrichment with highly reactive clean fuels, can extend flame operability to even lower loading conditions. Syngas, for example, despite being a substitute for natural gas, has different combustion behavior in comparison to natural gas. This is because the main constituents of syngas are carbon monoxide and hydrogen, which can produce higher adiabatic flame temperature and wider flame stability. Similarly, enriching hydrocarbon fuels with hydrogen can reduce some of the operability concerns of oxy-fuel combustion, by enhancing the burning velocity and extending the blowout stability limits [72]. In addition to using hydrogen for boosting the kinetics of conventional fuels, the utilization of hydrogen itself as a carbon-free fuel is gaining attention in recent times. The development of fuel/oxidizer-flexible stratified-combustion systems can lead to wider operability and ultra-low emissions in gas turbines.

1.5 Fuel-Flexible Combustion for Higher Turndown Ratio

1.5.1 Hydrogen-Enriched Combustion

Enriching hydrocarbon fuels with hydrogen improves the combustion characteristics because H_2 is a superior fuel, in terms of its faster reaction rate, higher burning velocity, and greater diffusivity [73]. Hydrogen is a clean source of energy, as it produces only H_2O when burned. The widespread use of H_2 for electricity production, however, is still constrained by its greater production costs, challenging transport and storage, and a safety concern with its high flammability. Research has demonstrated that blending H_2 with conventional fuels is beneficial for extending the static stability limits (i.e., improving the combustor turndown ratio) [74], reducing combustion instabilities, and decreasing NOx emissions [72, 75]. Therefore, hydrogen-enriched fuels are viewed as potential natural-gas alternatives because of their superior performance [76, 77]. Hydrogen enrichment can also be used in premixed oxy-fuel combustion to improve on its inferior characteristics [28, 50].

Studies reveal that adding H_2 to hydrocarbon fuels lowers the formation rates of NOx, CO, and soot [78, 79]. The addition of H_2 not only slows down the surface growth rate of soot [80, 81], but also decreases aldehyde emissions due to reduced formation rates of CH_2O and CH_3CHO [82]. Hydrogen enrichment can have a significant influence on the instantaneous and average characteristics of turbulent premixed flames. Even with little H_2, this influence can still be recognized [83]. The reaction rate [84] and unstretched laminar burning velocity [85] can both be enhanced by hydrogen enrichment. This is explained by the fact that H_2 has a higher molecular diffusivity and that O, H, and OH radicals are abundant in H_2-rich flames [86]. When 20% H_2 is added to CH_4, the mole fraction of OH increases by 20% [87]. The burning velocity of fuel-rich mixtures remains unchanged when H_2 is added, but it increases in lean mixtures [88]. Additionally, the combustion intensity improves, which is defined as the ratio of turbulent to laminar burning velocities (S_T/S_L), meaning that S_T rises faster than S_L [82, 89].

1.5.2 Syngas Combustion

Syngas, which stands for synthetic gas, is mainly made up of H_2 and CO with varying amounts of CO_2, H_2O, and N_2. Syngas can be generated by partially gasifying fossil fuels, most often coal, at high pressures and temperatures with the help of steam and oxygen. Many harmful contaminants such as sulfur dioxides (SOx), NOx, and H_2S are eliminated from the syngas after gasification. To capture the carbon prior to combustion, the fuel is reformed to generate syngas, which is then converted to H_2-rich syngas by removing the CO_2. Syngas is utilized in combined cycles with high efficiency and low emissions, such as the IGCC, which takes advantage of the associated higher combustion temperature. Moreover, pure H_2 and liquid fuels can be produced from syngas using the water–gas-shift technology [90, 91] and the Fischer–Tropsch conversion [92, 93], respectively.

Compared to natural gas, which is mostly CH_4, syngas has vastly different combustion characteristics because of its variable composition. Syngas-fueled gas-turbine combustors should thus be inherently fuel-flexible. Syngas has higher flame temperature, so LPM combustion is recommended for keeping the temperature under control and, thereby, lowering the NOx emissions [94]. Syngas is more reactive than natural gas because the former has more H_2, so the flashback risk is higher in premixed systems. Staging of syngas fuel (stratification) showed considerable improvement in the stability by minimizing flashback [95]. Therefore, the pressure fluctuations caused by flashback could also be lowered with syngas fuel staging. Staging is also useful in broadening the stability range and reducing the CO and NOx products in the emissions.

1.6 Concluding Remarks

The fact that CO_2, as a combustion product, is not environmentally benign if emitted in large quantities, has been discussed. Carbon-capture technologies have been presented, including post- and pre-combustion capture. The latter entails reforming the hydrocarbon fuel catalytically before combustion to separate carbon from hydrogen, so that carbon is captured, and hydrogen is then used as clean fuel in the combustion system. The inability of post- and pre-combustion capture technologies to capture all CO_2 has been discussed. The technology of oxy-fuel combustion has also been presented as a cost-efficient way for capturing all effluent CO_2 in high-purity compressed form. The differences between air–fuel and oxy-fuel combustion have been discussed in detail, enumerating the challenges associated with switching from current LPM air–fuel combustion to premixed oxy-fuel combustion with minimum hardware changes. Stratification, i.e., fuel and/or air staging has also been presented as a means of addressing most of the challenges of LPM air–fuel systems as well as premixed oxy-fuel ones. The stratification concept intentionally seeks heterogenization of the overall equivalence ratio, which is achieved through staging or splitting of reactant flow. Stratification adds more degrees of freedom to the combustor operability, widens the operability window, lowers the emissions, and allows for bypassing zones of dynamic instabilities. Stratified combustion is one of the attention-seeking contemporary technologies, utilized in state-of-the-art gas turbines to achieve ultra-low emissions, enhanced stability, and superior turndown (ratio of peak to minimum loads). Examples of stratified combustion systems have been enumerated, including the DACRS and micromixer premixers themselves, the use of several premixers within the same combustor, and the dual lean-premixed (DLPM) concept. The latter stages both fuel and oxidizer to create two concentric flames: a central pilot one, which is hotter and stable, and an annular primary one, which is colder and dependent on the pilot flame for stabilization. The overall blowout limit of the combustor is significantly enhanced with stratification. The advantages of DLPM concept over LPM combustion have been discussed. DLPM combustion does not implement a non-premixed fuel component for stabilization, like the LPM technologies, which reduces NOx emissions significantly in DLPM systems. Stratification also allows for the staging of recirculated CO_2 in oxy-fuel systems, i.e., stratifying the O_2/CO_2 oxidizer, which is a powerful tool for improving flame stability, especially close to the blowout limit. Finally, hydrogen-enriched and syngas combustion have been overviewed to present fuel flexibility as a viable solution to different challenges associated with LPM air–fuel and premixed oxy-fuel combustion. The overview presented in this chapter indicates that oxy-fuel combustion is the most promising option for cost-efficient carbon capture to make all existing power systems emission-free. Stratification allows for converting existing LPM air–fuel systems into premixed oxy-fuel ones with minimum hardware changes. The use of hydrogen (either by enriching hydrocarbon fuels or by syngas combustion) offers promising solutions to most of the incurred instability challenges in oxy-fuel combustion. Combustion systems in future gas turbines are thus recommended to be fuel/oxidizer-flexible stratified oxy-fuel systems for emission-free performance and cost-efficient carbon capture.

Acknowledgements The authors appreciate the support received for the preparation of this book from the Deanship of Research Oversight and Coordination (DROC) at King Fahd University of Petroleum & Minerals (KFUPM). The support provided by the Interdisciplinary Research Center for Hydrogen Technologies and Carbon Management (IRC-HTCM) on project number INHT2409 is highly appreciated. Also, the support received through the KFUPM Consortium for Hydrogen Future through the projects numbered H2FC2309 and H2FC2315 is highly appreciated.

References

1. Y.B. Zeldovich, *Oxidation of Nitrogen in Combustion and Explosions. Selected Works of Yakov Borisovich Zeldovich*, vol. I (Princeton University Press, 2014), pp. 404–410
2. W.R. Bender, Lean pre-mixed combustion, in *The Gas Turbine Handbook*, (U.S. Department of Energy, Morgantown, WV 26505, 2006), pp. 218–224
3. M.A. Darwish, H.K. Abdulrahim, A.A. Mabrouk, A.S. Hassan, *Cogeneration Power-Desalting Plants Using Gas Turbine Combined Cycle, Desalination Updates*, ed. by R.Y. Ning (IntechOpen). https://doi.org/10.5772/60209. Available from https://www.intechopen.com/books/desalination-updates/cogeneration-power-desalting-plants-using-gas-turbine-combined-cycle
4. A. Abdelhafez, S.S. Rashwan, M.A. Nemitallah, M.A. Habib, Stability map and shape of premixed $CH_4/O_2/CO_2$ flames in a model gas-turbine combustor. Appl. Energy 1(215), 63–74 (2018)
5. A. Abdelhafez, M.A. Nemitallah, S.S. Rashwan, M.A. Habib, Adiabatic flame temperature for controlling the macrostructures and stabilization modes of premixed methane flames in a model gas-turbine combustor. Energy Fuels **32**(7), 7868–7877 (2018)
6. M.A. Nemitallah, M.A. Habib, Experimental and numerical investigations of an atmospheric diffusion oxy-combustion flame in a gas turbine model combustor. Appl. Energy **111**, 401–415 (2013). https://doi.org/10.1016/j.apenergy.2013.05.027
7. https://en.wikipedia.org/wiki/Combustor
8. Y. Huang, V. Yang, Dynamics and stability of lean-premixed swirl-stabilized combustion. Prog. Energy Combust. Sci. **35**, 293–364 (2009)

9. J. Blouch, H. Li, M. Mueller, R. Hook, Fuel flexibility in LM2500 and LM6000 dry low emission engines. ASME J. Eng. Gas Turbines Power **134**(5), 051503 (2012). https://doi.org/10.1115/1.4004213

10. H.H.W. Funke, N. Beckmann, J. Keinz, S. Abanteriba, Numerical and experimental evaluation of a dual-fuel dry-low-NOx micromix combustor for industrial gas turbine applications, in *Proceedings of the 2017 ASME Turbo Expo*

11. A. Haj Ayed, K. Kusterer, H.H.W. Funke, J. Keinz, C. Striegan, D. Bohn, Experimental and numerical investigations of the dry-low-NOx hydrogen micromix combustion chamber of an industrial gas turbine. Propuls. Power Res. (2015)

12. T. Asai, S. Dodo, M. Karishuku, N. Yagi, Y. Akiyama, A. Hayashi, Performance of multiple injection dry low-NOx combustors on hydrogen-rich syngas fuel in an IGCC pilot plant. J. Eng. Gas Turbines Power (2015)

13. J.D. Mattingly, H. Von Ohain, *Elements of Propulsion: Gas Turbines and Rockets* (2006). https://doi.org/10.2514/4.861789

14. F.H. Verhoek, Thermodynamics and rocket propulsion. J. Chem. Educ. **46**, 140 (1969). https://doi.org/10.1021/ed046p140

15. S. Ducruix, T. Schuller, D. Durox, S. Candel, Combustion dynamics and instabilities: elementary coupling and driving mechanisms. J. Propuls. Power **19**, 722–734 (2003). https://doi.org/10.2514/2.6182

16. S. Taamallah, Z.A. Labry, S.J. Shanbhogue, A.F. Ghoniem, Thermo-acoustic instabilities in lean premixed swirl-stabilized combustion and their link to acoustically coupled and decoupled flame macrostructures. Proc. Combust. Inst. **35**, 3273–3282 (2015). https://doi.org/10.1016/j.proci.2014.07.002

17. T. Lieuwen, H. Torres, C. Johnson, B.T. Zinn, A mechanism of combustion instability in lean premixed gas turbine combustors. Trans. ASME **123**, 182–189 (2001). https://doi.org/10.1115/1.1339002

18. L. Crocco, Research on combustion instability in liquid propellant rockets. Symp. Combust. **12**, 85–99 (1969). https://doi.org/10.1016/S0082-0784(69)80394-2

19. A.M. Annaswamy, A.F. Ghoniem, Active control of combustion instability: theory and practice. Control Syst. IEEE **22**, 37–54 (2002). https://doi.org/10.1109/mcs.2002.1077784

20. M.Y.E. Selim, Sensitivity of dual fuel engine combustion and knocking limits to gaseous fuel composition. Energy Convers. Manage. **45**, 411–425 (2004). https://doi.org/10.1016/S0196-8904(03)00150-X

21. K. Krischer, H. Varela, Oscillations and other dynamic instabilities, in *Handbook of Fuel Cells: Fundamentals, Technology, Applications*, vol. 2 (2003), pp. 679–701. https://doi.org/10.1002/9780470974001.f206052

22. E.H. Robey, M.C. Janus, G.A. Richards, M.J. Yip, E. Robey, Effects of ambient conditions and fuel composition on combustion stability, in *ASME Conference Proceedings 97-GT-266*

23. A.G. Hendricks, U. Vandsburger, The effect of fuel composition on flame dynamics. Exp. Therm. Fluid Sci. **32**, 126–132 (2007). https://doi.org/10.1016/j.expthermflusci.2007.02.007

24. G. Sutton, O. Biblarz, *Rocket Propulsion Elements* (2001). https://doi.org/10.1017/CBO9781107415324.004

25. A. Lefebvre, D. Ballal, D. Bahr, *Gas Turbine Combustion* (2001). https://doi.org/10.1002/1521-3773(20010316)40:6%3c9823::aid-anie9823%3e3.3.co;2-c

26. A. Lefebvre, D. Ballal, D. Bahr, *Gas Turbine Combustion: Alternative Fuels and Emissions* (2010). https://doi.org/10.1002/1521-3773(20010316)40:6%3c9823::AID-ANIE9823%3e3.3.CO;2-C

27. A. Van Maaren, D.S. Thung, L.P.H. De Goey, Measurement of flame temperature and adiabatic burning velocity of methane/air mixtures. Combust. Sci. Technol. **96**, 327–344 (1994)

28. B.A. Imteyaz, M.A. Nemitallah, A. Abdelhafez, M.A. Habib, Combustion behavior and stability map of hydrogen-enriched oxy-methane premixed flames in a model gas turbine combustor. Int. J. Hydrogen Energy **43**, 16652–16666 (2018)

29. R.L. Speth, A.F. Ghoniem, Using a strained flame model to collapse dynamic mode data in a swirl-stabilized syngas combustor. Proc. Combust. Inst. **32**(2), 2993–3000 (2009). https://doi.org/10.1016/j.proci.2008.05.072

30. D.M. Wicksall, A.K. Agrawal, Acoustics measurements in a lean premixed combustor operated on hydrogen/hydrocarbon fuel mixtures. Int. J. Hydrogen Energy **32**, 1103–1112 (2007). https://doi.org/10.1016/j.ijhydene.2006.07.008

31. S. Li, Y. Zheng, D.M. Martinez, M. Zhu, X. Jiang, Large-eddy simulation of flow and combustion dynamics in a lean partially-premixed swirling combustor. Phys. Procedia **66**, 333–336 (2015). https://doi.org/10.1016/j.egypro.2015.02.080

32. https://yearbook.enerdata.net/co2/emissions-co2-data-from-fuel-combustion.html

33. https://unfccc.int/process-and-meetings/the-paris-agreement/the-paris-agreement

34. https://www.epa.gov/

35. http://berkeleyearth.org/

36. United Nations Framework Convention on Climate Change (2021) https://unfccc.int/

37. C. Chao, Y. Deng, R. Dewil, J. Baeyens, X. Fan, Post-combustion carbon capture. Renew. Sustain. Energy Rev. **138**, 110490 (2021)

38. A. Basile, S.P. Nunes, *Advanced Membrane Science and Technology for Sustainable Energy and Environmental Applications* (Woodhead Publishing, 2011). ISBN 978-1-84569-969-7

39. L. Barelli, G. Bidini, F. Gallorini, S. Servili, Hydrogen production through sorption-enhanced steam methane reforming and membrane technology: a review. Energy **33**(4), 554–570 (2008)

40. S. Cloete, M.N. Khan, S. Amini, Economic assessment of membrane-assisted autothermal reforming for cost effective hydrogen production with CO_2 capture. Int. J. Hydrogen Energy **44**(7), 3492–3510 (2019)

41. J. Ren, J.P. Cao, X.Y. Zhao, F.L. Yang, X.Y. Wei, Recent advances in syngas production from biomass catalytic gasification: a critical review on reactors, catalysts, catalytic mechanisms and mathematical models. Renew. Sustain. Energy Rev. **116**, 109426 (2019)

42. A. Giuliano, C. Freda, E. Catizzone, Techno-economic assessment of bio-syngas production for methanol synthesis: a focus on the water-gas shift and carbon capture sections. Bioengineering **7**(3), 70 (2020)

43. A.I. Papadopoulos, P. Seferlis, Process simulation of a dual-stage selexol process for pre-combustion carbon capture at an integrated gasification combined cycle power plant, in *Process Systems and Materials for CO_2 Capture: Modelling, Design, Control and Integration* (2017). https://doi.org/10.1002/9781119106418.ch24

44. A. Verma, A.D. Rao, G.S. Samuelsen, Sensitivity analysis of a Vision 21 coal based zero emission power plant. J. Power. Sources **158**(1), 417–427 (2006)

45. M.A. Habib, M.A. Nemitallah, P. Ahmed, M.H. Sharqawy, H.M. Badr, I. Muhammad, M. Yaqub, Experimental analysis of oxygen-methane combustion inside a gas turbine reactor under various operating conditions. Energy **86**, 105–114 (2015)

46. D.A. Granados, F. Chejne, J.M. Mejía, Oxy-fuel combustion as an alternative for increasing lime production in rotary kilns. Appl. Energy **158**, 107–117 (2015)

47. M.A. Nemitallah, A study of methane oxy-combustion characteristics inside a modified design button-cell membrane reactor utilizing a modified oxygen permeation model for reacting flows. J. Nat. Gas Sci. Eng. **28**, 61–73 (2016)

48. M.A. Nemitallah, M.A. Habib, K. Mezghani, Experimental and numerical study of oxygen separation and oxy-combustion characteristics inside a button-cell LNO-ITM reactor. Energy **84**, 600–611 (2015)

49. M.A. Habib, S.A. Salaudeen, M.A. Nemitallah, R. Ben-Mansour, E.M.A. Mokheimer, Numerical investigation of syngas oxy-combustion inside a LSCF-6428 oxygen transport membrane reactor. Energy **96**, 654–665 (2016)

50. S. Taamallah, K. Vogiatzaki, F.M. Alzahrani, E.M.A. Mokheimer, M.A. Habib, A.F. Ghoniem, Fuel flexibility, stability and emissions in premixed hydrogen-rich gas turbine combustion: technology, fundamentals, and numerical simulations. Appl. Energy **154**, 1020–1047 (2015)

51. D.M. Wicksall, A.K. Agrawal, R.W. Schefer, J.O. Keller, The interaction of flame and flow field in a lean premixed swirl-stabilized combustor operated on H_2/CH_4/air. Proc. Combust. Inst. **30**(2), 2875–2883 (2005)

52. T.C. Williams, C.R. Shaddix, R.W. Schefer, Effect of syngas composition and CO_2-diluted oxygen on performance of a premixed swirl-stabilized combustor. Combust. Sci. Technol. **180**(1), 64–88 (2007)

53. I.A. Ramadan, A.H. Ibrahim, T.W. Abou-Arab, S.S. Rashwan, M.A. Nemitallah, M.A. Habib, Effects of oxidizer flexibility and bluff-body blockage ratio on flammability limits of diffusion flames. Appl. Energy **178**, 19–28 (2016)

54. W. Jerzak, M. Kuznia, Experimental study of impact of swirl number as well as oxygen and carbon dioxide content in natural gas combustion air on flame flashback and blow-off. J. Nat. Gas Sci. Eng. **29**, 46–54 (2016)

55. Y. Xie et al., Experimental and numerical study on laminar flame characteristics of methane oxy-fuel mixtures highly diluted with CO_2. Energy Fuels **27**(10), 6231–6237 (2013)

56. P.H. Joo, M.R.J. Charest, C.P.T. Groth, Ö.L. Gülder, Comparison of structures of laminar methane–oxygen and methane–air diffusion flames from atmospheric to 60 atm. Combust. Flame **160**(10), 1990–1998 (2013)

57. M. Ditaranto, R. Anantharaman, T. Weydahl, Performance and NOx emissions of refinery fired heaters retrofitted to hydrogen combustion. Energy Proc. **37**, 7214–7220 (2013)

58. M. Ditaranto, J. Hals, Combustion instabilities in sudden expansion oxy–fuel flames. Combust. Flame **146**(3), 493–512 (2006)

59. Z. Abubakar, M.R. Shakeel, E.M.A. Mokheimer, Experimental and numerical analysis of non-premixed oxy-combustion of hydrogen-enriched propane in a swirl stabilized combustor. Energy **165**(B), 1401–1414 (2018)

60. S.S. Rashwan, A.H. Ibrahim, T.W. Abou-Arab, M.A. Nemitallah, M.A. Habib, Experimental study of atmospheric partially premixed oxy-combustion flames anchored over a perforated plate burner. Energy **122**, 159–167 (2017)

61. X. Hu, Q. Yu, J. Liu, N. Sun, Investigation of laminar flame speeds of $CH_4/O_2/CO_2$ mixtures at ordinary pressure and kinetic simulation. Energy **70**, 626–634 (2014)

62. P. Kutne, B.K. Kapadia, W. Meier, M. Aigner, Experimental analysis of the combustion behaviour of oxyfuel flames in a gas turbine model combustor. Proc. Combust. Inst. **33**(2), 3383–3390 (2011)

63. A. Mansour, *Gas Turbine Fuel Injection Technology. Proceedings of ASME Turbo Expo* (2005), pp. 141–149

64. D.A. Nickolaus, D. Scott Crocker, D.L. Black, C.E. Smith, The structure of premixed and stratified low turbulence flames. ASME Conf. Proc. 713–720 (2002)

65. H.J. Bauer, New low emission strategies and combustor designs for civil aeroengine applications. Prog. Comput. Fluid Dyn. **4**(3–5), 30 (2004)

66. X. Han, D. Laera, D. Yang, C. Zhang, J. Wang, X. Hui, Y. Lin, A.S. Morgans, C.-J. Sung, Flame interactions in a stratified swirl burner: flame stabilization, combustion instabilities and beating oscillations. Combust. Flame **212**, 500–509 (2020)

67. X. Han, D. Laera, A.S. Morgans, C.J. Sung, X. Hui, Y.Z. Lin, Flame macrostructures and thermoacoustic instabilities in stratified swirling flames. Proc. Combust. Inst. **37**, 5377–5384 (2019)

68. M.S. Sweeney, S. Hochgreb, M.J. Dunn, R.S. Barlow, The structure of turbulent stratified and premixed methane/air flames II: swirling flows. Combust. Flame **159**, 2912–2929 (2012)

69. L. Li, Y. Lin, Z. Fu, C. Zhang, Emission characteristics of a model combustor for aero gas turbine application. Exp. Therm. Fluid Sci. **72**, 235–248 (2016)

70. C.T. Chong, S.S. Lam, S. Hochgreb, Effect of mixture flow stratification on premixed flame structure and emissions under counter-rotating swirl burner configuration. Appl. Therm. Eng. **105**, 905–912 (2016)

71. https://www.ge.com/gas-power/products/gas-turbines/lm6000

72. H.S. Kim, V.K. Arghode, A.K. Gupta, Flame characteristics of hydrogen-enriched methane-air premixed swirling flames. Int. J. Hydrogen Energy **34**, 1063–1073 (2009)

73. A.A. Subash, R. Collin, M. Alden, A. Kundu, J. Klingmann, Investigation of hydrogen enriched methane flame in a dry low emission industrial prototype burner at atmospheric pressure conditions, in *Proceedings of ASME Turbo Expo 4A-2017* (2017)

74. A.E.E. Khalil, A.K. Gupta, Fuel flexible distributed combustion for efficient and clean gas turbine engines. Appl. Energy **109**, 267–274 (2013)

75. S. Abdelwahid, M. Nemitallah, B. Imteyaz, A. Abdelhafez, Effects of H_2 enrichment and inlet velocity on stability limits and shape of CH_4/H_2-O_2/CO_2 flames in a premixed swirl combustor. Energy Fuels **32**, 9916–9925 (2018)
76. M. Ball, M. Wietschel, The future of hydrogen—opportunities and challenges. Int. J. Hydrogen Energy **34**, 615–627 (2009)
77. P.P. Edwards, V.L. Kuznetsov, W.I.F. David, N.P. Brandon, Hydrogen and fuel cells: towards a sustainable energy future. Energy Pol **36**, 4356–4362 (2020)
78. F. Delattin, L.G. Di, S. Rizzo, S. Bram, R.J. De, Combustion of syngas in a pressurized microturbine-like combustor: experimental results. Appl. Energy **87**, 1441–1452 (2010)
79. A.E.E. Khalil, A.K. Gupta, Hydrogen addition effects on high intensity distributed combustion. Appl. Energy **104**, 71–78 (2013)
80. M. Gu, H. Chu, F. Liu, Effects of simultaneous hydrogen enrichment and carbon dioxide dilution of fuel on soot formation in an axisymmetric co-flow laminar ethylene/air diffusion flame. Combust. Flame **166**, 216–228 (2016)
81. S.H. Park, K.M. Lee, C.H. Hwang, Effects of hydrogen addition on soot formation and oxidation in laminar premixed C_2H_2/air flames. Int. J. Hydrogen Energy **36**, 9304–9311 (2011)
82. S. Daniele, P. Jansohn, J. Mantzaras, K. Boulouchos, Turbulent flame speed for syngas at gas turbine relevant conditions. Proc. Combust. Inst. **33**, 2937–2944 (2011)
83. F. Halter, C. Chauveau, I. Gökalp, Characterization of the effects of hydrogen addition in premixed methane/air flames. Int. J. Hydrogen Energy **32**, 2585–2592 (2007)
84. E. Hu, Z. Huang, J. He, C. Jin, J. Zheng, Experimental and numerical study on laminar burning characteristics of premixed methane-hydrogen-air flames. Int. J. Hydrogen Energy **34**, 4876–4888 (2009)
85. C. Tang, Z. Huang, C. Jin, J. He, J. Wang, X. Wang et al., Laminar burning velocities and combustion characteristics of propane-hydrogen-air premixed flames. Int. J. Hydrogen Energy **33**, 4906–4914 (2008)
86. R. Sankaran, H.G. Im, Effects of hydrogen addition on the Markstein length and flammability limit of stretched methane/air premixed flames. Combust. Sci. Technol. **178**, 1585–1611 (2006)
87. R.W. Schefer, Hydrogen enrichment for improved lean flame stability. Int. J. Hydrogen Energy **28**, 1131–1141 (2003)
88. M. Nakahara, H. Kido, Study on the turbulent burning velocity of hydrogen mixtures including hydrocarbons. AIAA J. **46**, 1569–1575 (2008)
89. J. Wang, Z. Huang, C. Tang, H. Miao, X. Wang, Numerical study of the effect of hydrogen addition on methane-air mixtures combustion. Int. J. Hydrogen Energy **34**, 1084–1096 (2009)
90. M. Chaos, F.L. Dryer, Syngas combustion kinetics and applications. Combust. Sci. Technol. **180**, 1053–1096 (2008)
91. N.F. Othman, M.H. Boosroh, Effect of H_2 and CO contents in syngas during combustion using micro gas turbine. IOP Conf. Ser. Earth Environ. Sci. **32** (2016)
92. M.E. Dry, The Fischer-Tropsch process: 1950–2000. Catal. Today **71**, 227–241 (2002)
93. D.J. Wilhelm, D.R. Simbeck, A.D. Karp, R.L. Dickenson, Syngas production for gas-to-liquids applications: technologies, issues and outlook. Fuel Process. Technol. **71**, 139–148 (2001)
94. P. Glarborg, Hidden interactions—trace species governing combustion and emissions. Proc. Combust. Inst. **31**(I), 77–98 (2007)
95. T. García-Armingol, Á. Sobrino, E. Luciano, J. Ballester, Impact of fuel staging on stability and pollutant emissions of premixed syngas flames. Fuel **185**, 122–132 (2016)

Chapter 2
Engines, Fuels, and Thermodynamics of Combustion

2.1 Engine Types and Parts

2.1.1 Combustion Engines

Combustion engines (see Fig. 2.1) are those engines where the chemical energy of the fuel is liberated by the combustion progression, and part of this energy is transformed to valuable mechanical work, while the remaining is vanished to the surroundings. Engines can be categorized into three types: internal combustion engines, external combustion engines, and jet and rocket engines.

The distinction of "internal" versus "external" is related to the conversion of heat to work. Most internal combustion engines are piston engines, where heat addition and work production happen within the same closed space inside the engine cylinders. The combustion process occurs inside the cylinders, as shown in Fig. 2.2. The heat energy resulting from combustion increases the temperature and pressure of the working substance, which exerts force on the moving piston to produce useful work.

In external combustion engines, on the other side, the combustion process occurs in one device, and work is produced in another device; see Fig. 2.3 for a gas-turbine example. The heat energy resulting from combustion increases the enthalpy of the working substance, which expands through a series of turbine stages, converting enthalpy to positive rotational work on turbine shaft.

In a jet or rocket engine, the engine is composed primarily of a combustion chamber and a nozzle. The working substance leaves the combustion chamber and enters the nozzle with high enthalpy, expanding and leaving with very high kinetic energy, see Fig. 2.4. Due to this change of momentum, a thrust force is produced, resulting in forward propulsion of the jet engine and any body attached to it, like an aircraft.

© The Author(s), under exclusive license to Springer Nature Singapore Pte Ltd. 2024
M. A. Nemitallah et al., *Hydrogen for Clean Energy Production: Combustion Fundamentals and Applications*, https://doi.org/10.1007/978-981-97-7925-3_2

Fig. 2.1 Combustion engine

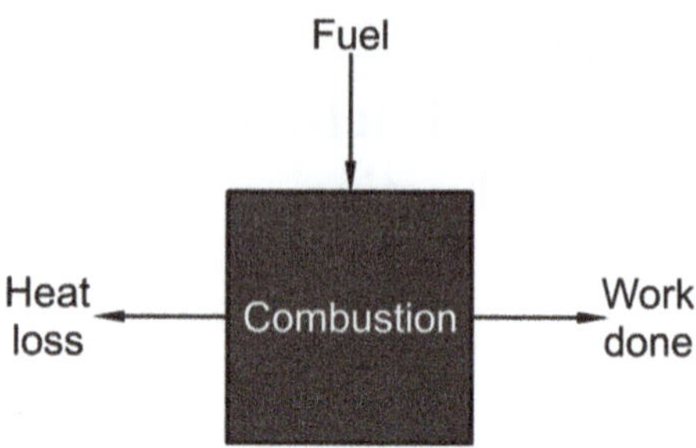

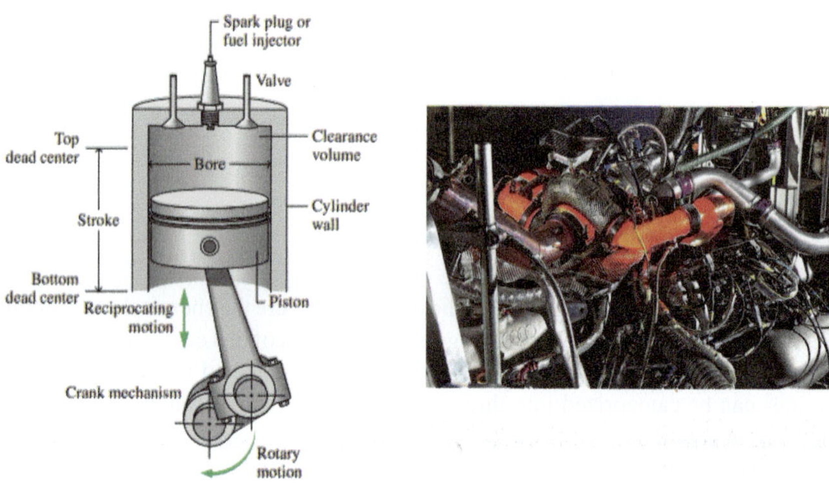

Fig. 2.2 Internal combustion engine

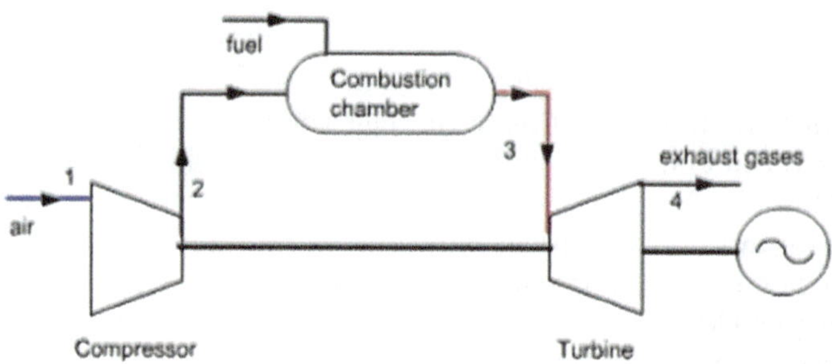

Fig. 2.3 External gas-turbine combustion engine

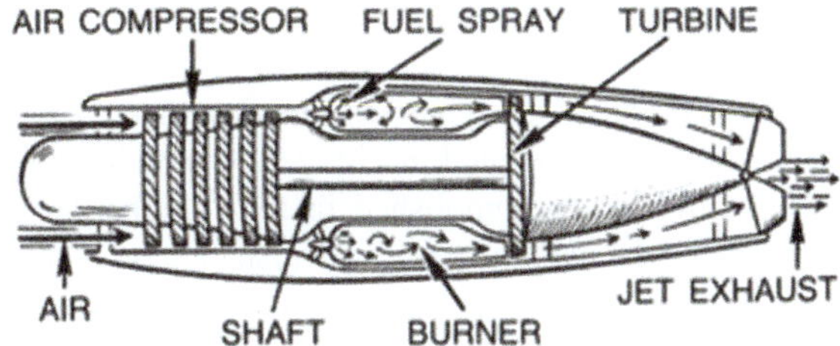

Fig. 2.4 Jet and rocket engines

2.1.2 *Classifications of Internal Combustion Engines*

Internal combustion engines can be categorized based on:

1. Application

Automobile, locomotive, truck, marine, light aircraft, power generation, or portable power systems.

2. Basic engine design

The reciprocating engines can be classified by the prearrangement of cylinders, as shown in Fig. 2.5, e.g., in-line, v-type, u-type, radial, or opposed. Another special kind of internal combustion engine is the rotary engine, like the Wankel rotary engine, see Fig. 2.6, and other geometries.

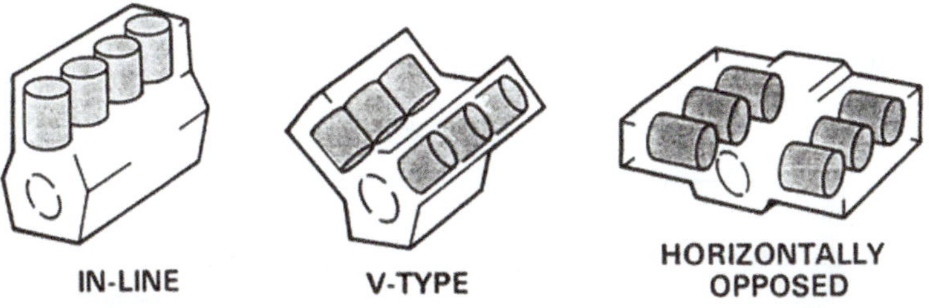

Fig. 2.5 Classification of internal combustion engines based on cylinder arrangement

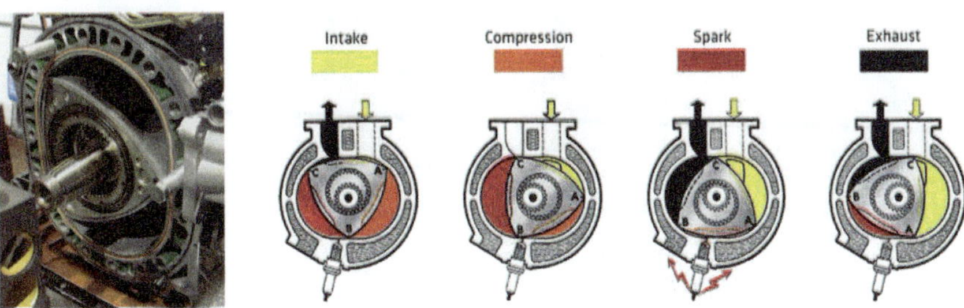

Fig. 2.6 Wankel rotary engine

3. Working cycle

Internal combustion engines can work either on a two-stroke cycle or on a four-stroke cycle, as presented in Fig. 2.7. The name "four-stroke" comes from the fact that the cycle is accomplished in four distinct strokes of the piston, performing four processes, namely, intake of fresh charge, compression of charge, combustion and expansion to produce power, and exhaust discharge. A four-stroke cycle can be naturally aspirated (receiving atmospheric air) or supercharged (receiving a pre-compressed fresh charge). The two-stroke cycle, on the other hand, is termed "two-stroke" because the same four processes of intake, compression, expansion, and exhaust are augmented in two strokes only of the piston.

4. Valve or port design and location

The internal combustion engines can be classified based on valve/port design and arrangement. Valves in four-stroke engines can be designed as under-head (or T-head) valves or over-head (or I-head) valves, or rotary valves. Scavenging ports in

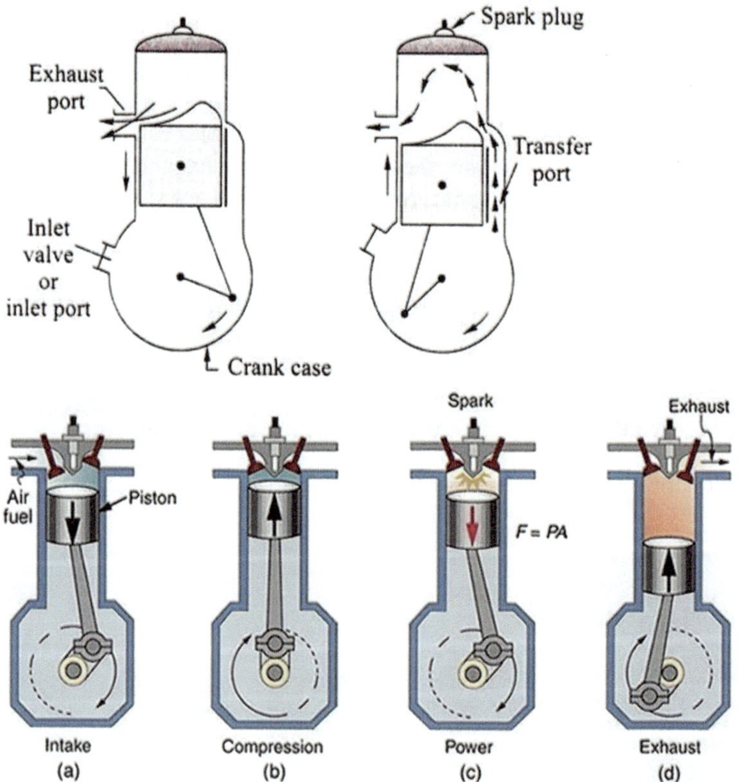

Fig. 2.7 Two-stroke (top drawing) versus four-stroke (bottom drawing) internal combustion engines

two-stroke engines can be classified as cross-scavenged ports (inlet and exhaust ports are existing on opposite sides of the cylinder at one end), loop-scavenged ports (inlet and exhaust ports exist on the same side of cylinder at one end), or uniflow scavenged (inlet and exhaust ports exist at different ends of cylinder).

5. Fuel

Internal combustion engines can operate using petrol (or gasoline), diesel fuel (or fuel–oil), liquified petroleum gas (LPG), natural gas, alcohols (methanol, ethanol, etc.), or hydrogen fuel.

6. Method of ignition

Engines are classified as either spark-ignition engines (SIEs), where a spark plug is utilized in igniting the premixed compressed fuel–air charge, or compression-ignition engines (CIEs), where fuel autoignites upon injection into the compressed air charge without the need for a spark. Spark-ignition fuels are gasoline, natural gas, LPG, alcohols, and hydrogen, while compression-ignition fuels are diesel and biodiesel.

7. Method of mixture preparation

The reacting mixture in SIEs can be prepared by carburetion, by fuel injection into the intake manifold at the inlet ports, or by direct fuel injection into the engine cylinders.

8. Combustion-chamber design

Combustion chambers can have an open or continuous-volume design (e.g., wedge, disc, hemisphere, and bowl-in-piston), or a separated design (comprising large and small auxiliary chambers, such as swirl-chamber, pre-chambers).

9. Method of load control

The load of an internal combustion engine can be controlled through throttling of the fuel–air mixture flow, controlling the fuel flow alone, or a combination of both.

10. Method of cooling

Internal combustion engines can be classified as water-cooled engines (by circulating water in a cooling jacket around the engine cylinders), or air-cooled engines (by combined heat convection and radiation to the surrounding air).

2.1.3 Internal Combustion Engine Parts

A four-stroke internal combustion engine is composed of the following parts, see Fig. 2.8:

1. Cylinder head

- Manifolds (intakes and exhaust)
- Valves (intake and exhaust)

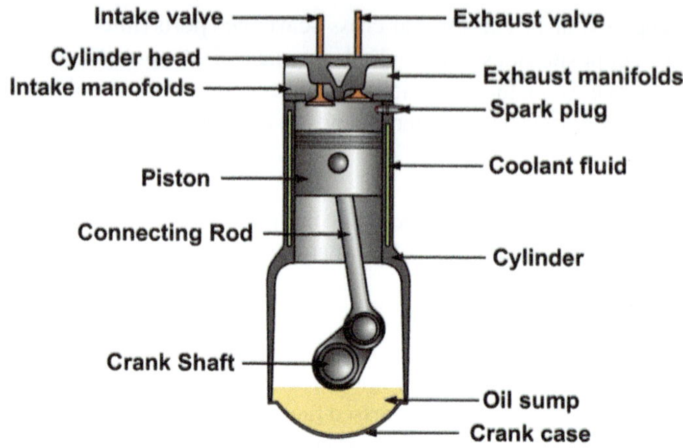

Fig. 2.8 Four-stroke engine components

- Spark plug or fuel injector

2. Cylinder block

   - Cylinders
   - Pistons
   - Connecting roods
   - Crank shaft(s)

3. Crank case

The components of an internal combustion engine, which house the lubrication oil and oil pump, are presented in Fig. 2.9.

4. Crankshaft and bearings

The components of the crankshaft and its bearings are presented in Fig. 2.10.

5. Piston and Connecting rod

The components of piston and connecting rod are presented in Fig. 2.11.

6. Camshaft and camshaft drives

The components of the camshaft and camshaft drives are presented in Fig. 2.12.

2.1.4 Operating Cycle and Valve Mechanism

Figure 2.13 shows the pressure–volume diagram inside an engine cylinder operating on the four-stroke cycle. Figure 2.14 shows the limits of reciprocating piston movement.

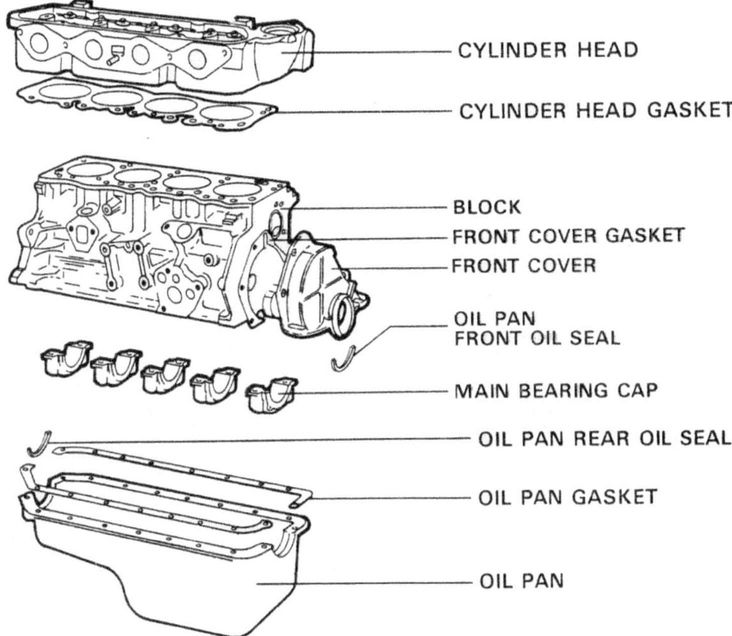

Fig. 2.9 Crankcase and other components for oil lubrication cycle

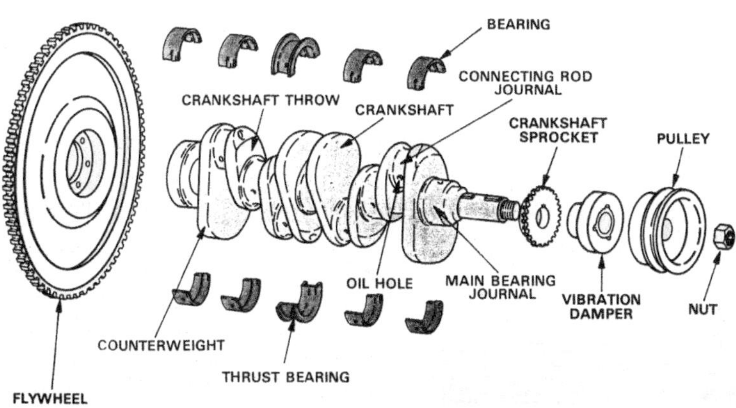

Fig. 2.10 Components of the crankshaft and its bearings

The limits of reciprocating piston movement are called the "dead centers", because the instantaneous piston speed is zero at each limit. When the piston confines the minimum volume inside the cylinder, it is at the *top dead center*, TDC; the minimum volume is also called the *clearance volume*. As the piston expands to confine the maximum volume, it is at the *bottom dead center*, BDC.

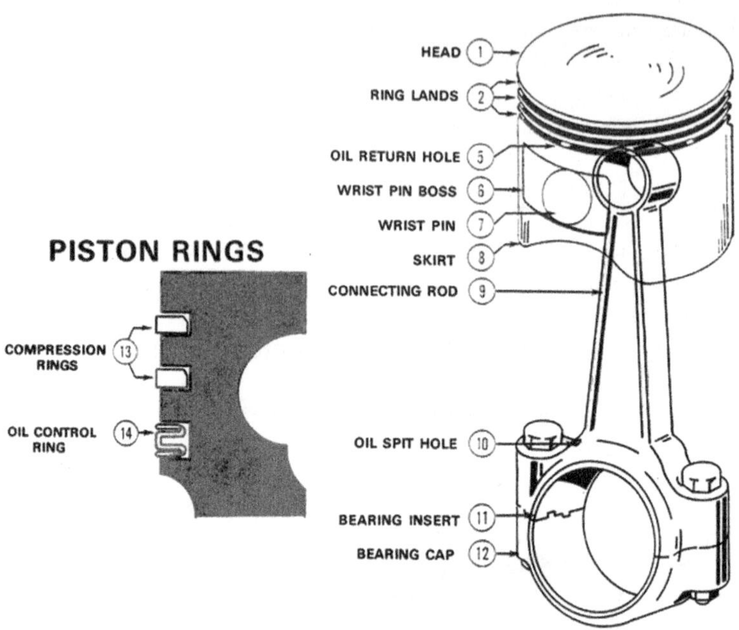

Fig. 2.11 Components of piston and connecting rod

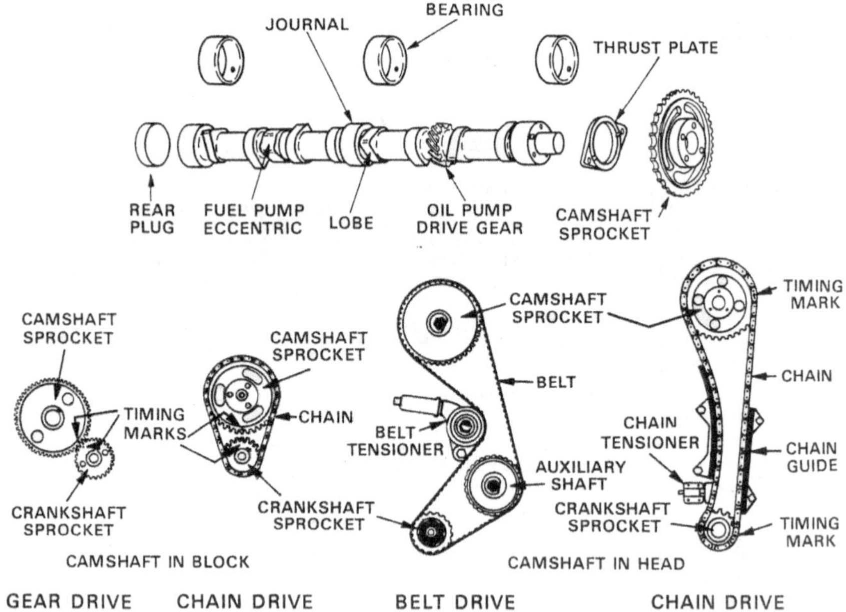

Fig. 2.12 Components of the camshaft and camshaft drives

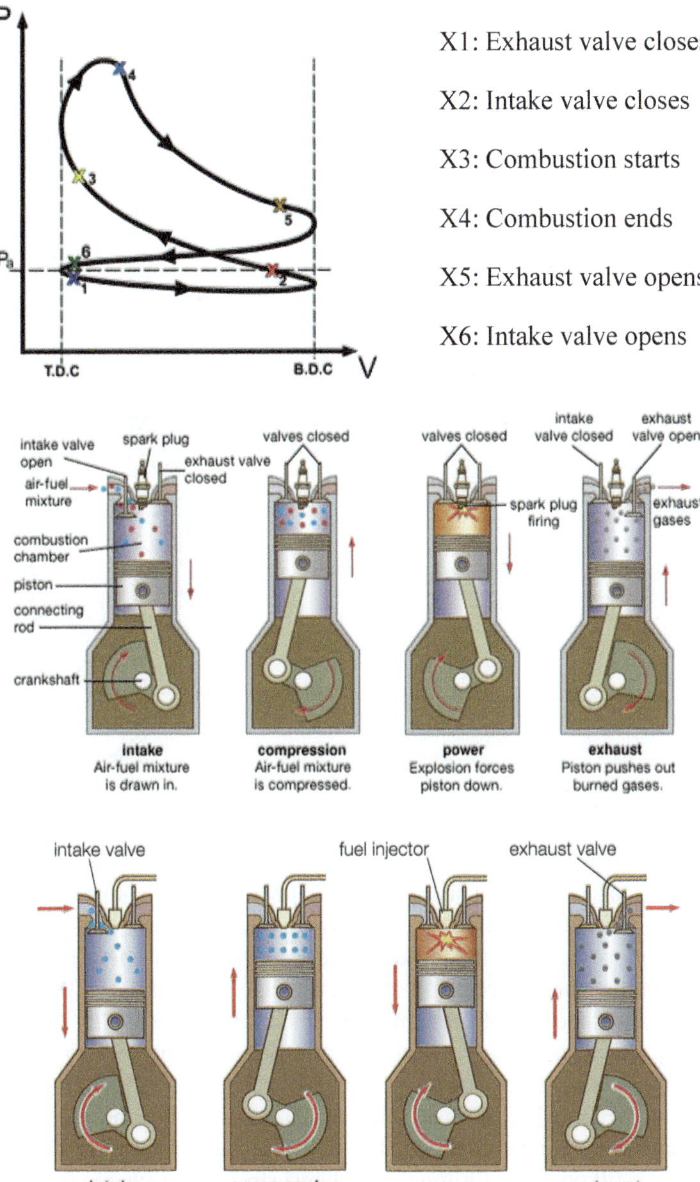

Fig. 2.13 Top: representation of a four-stroke engine cycle on the pressure–volume diagram; middle: four-stroke SIE; bottom: four-stroke CIE

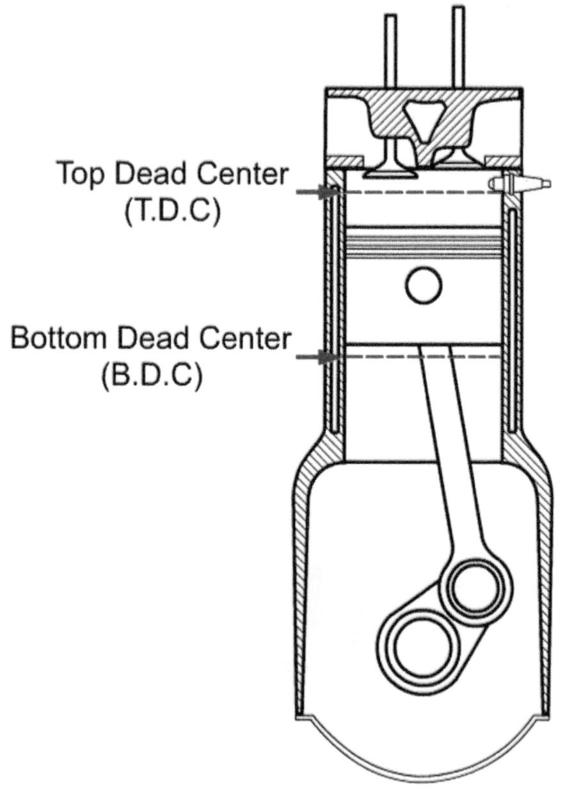

Fig. 2.14 Movement of the piston between top and bottom dead centers

The four-stroke cycle of a SIE starts with opening the intake valve while the piston is close to TDC. As the piston expands, it performs the intake stroke, in which the cylinder is filled with a fresh charge of fuel–air mixture. Close to BDC, the intake valve closes, ending the intake stroke, and the piston starts moving towards TDC, performing the compression stroke, in which the confined fresh charge is compressed. Near the end of the compression stroke, the spark plug fires, initiating combustion, which takes place until shortly after TDC. The peak energy content after combustion exerts positive work on the piston, pushing it down to perform the power stroke towards BDC. The exhaust valve opens, close to the end of power stroke, to blow down the cylinder contents. The piston moves up towards TDC, with the exhaust valve still open, performing the exhaust stroke to discharge the combustion products. Close to TDC, the exhaust valve closes, the intake valve opens, and the cycle is repeated. The four-stroke cycle completes in two crankshaft revolutions.

The corresponding four-stroke cycle of a CIE is very similar but with two major differences. First, the fresh charge is air only, so the piston compresses air only during compression stroke. Second, there is no spark plug; instead, there is a fuel injector that injects fuel near the end of compression stroke. Fuel autoignites release the desired energy.

The valve actuation mechanism, along with the components of a poppet valve, are presented in Fig. 2.15. With every piston stroke, the crankshaft turns half a revolution (180°), while the camshaft turns a quarter of a revolution only (90°). This means that the camshaft on a four-stroke engine revolves at half of the crankshaft speed, as the intake and exhaust valves open once (each) per cycle, i.e., once every two crankshaft revolutions. The operating cycle of a two-stroke engine is presented on the pressure–volume diagram shown in Fig. 2.16, along with the timings of inlet and exhaust ports.

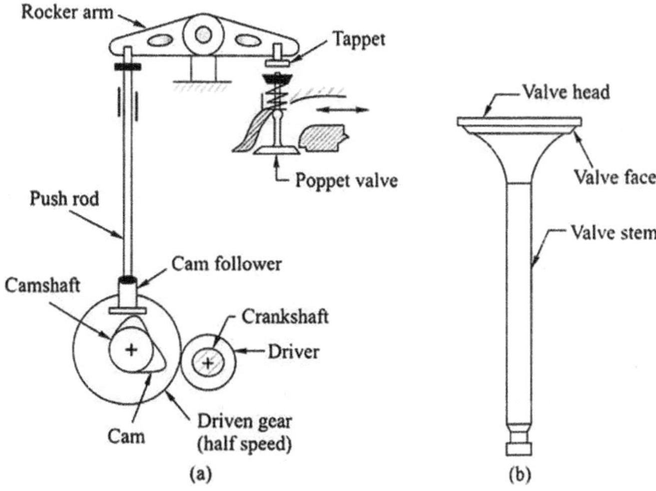

Fig. 2.15 Valve actuation mechanism and the components of a poppet valve

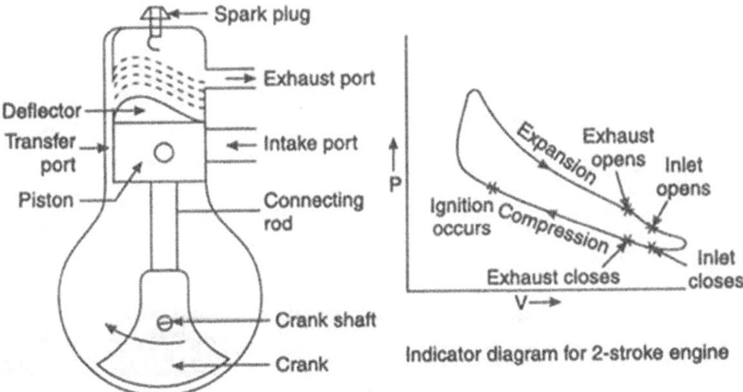

Fig. 2.16 Representation of a two-stroke engine cycle on the pressure–volume diagram with inlet and exhaust ports timings

2.2   Engine Design and Performance Parameters

In this section, definitions used to evaluate engine performance are presented as follows:

Cylinder Bore (D)

Another name for cylinder diameter.

Stroke Length (L)

Distance between TDC and BDC.

Stroke or Swept Volume (V_S)

The volume swept by the cylinder during its movement between the two dead centers.

$$V_S = \frac{\pi}{4}D^2L \tag{2.1}$$

Cylinder Clearance Volume (V_C)

Minimum volume the piston can confine when at TDC.

Compression Ratio (r)

The ratio of maximum confined volume (at BDC) to minimum volume (at TDC).

$$r = \frac{V_{max}}{V_{min}} = \frac{V_S + V_C}{V_C} = 1 + \frac{V_S}{V_C} \tag{2.2}$$

Number of Cylinders (Z)

Total number of cylinders in an engine.

Engine displacement Volume (V_D)

The combined swept volumes of all cylinders.

$$V_D = Z \times V_S \tag{2.3}$$

Engine Speed (N)

The speed of rotation of engine crankshaft; usually expressed in rpm $\equiv$ revolutions per minute.

$$\dot{V}_D = V_D(N/\xi) \tag{2.4}$$

where $\xi = 2$ in four-stroke engines, because the cycle completes in two crankshaft revolutions, and $\xi = 1$ in two-stroke engines, because the cycle completes every revolution.

Nominal Fresh Charge ($\dot{m}_{air,nom}$)

The ideal rate of air consumption of the engine, assuming no losses in the intake system. Note that even if the fresh charge is a mixture of fuel and air, it is typically approximated as air only, because fuel mass is significantly smaller than air mass.

$$\dot{m}_{air,nom} = \rho_{in}\dot{V}_D \tag{2.5}$$

where ρ_{in} is the fresh-charge density based on inlet pressure and temperature.

Volumetric Efficiency (η_v)

The ratio of actual to ideal rate of air consumption. η_v thus quantifies the effect of intake-system losses on air intake.

$$\eta_v = \frac{\dot{m}_{air}}{\dot{m}_{air,nom}} \tag{2.6}$$

Indicated Power (IP)

Recall the pressure–volume diagram in Fig. 2.13. Computing the closed-system work integral ($\oint pdV$) over the entire cycle gives the net positive power exerted by the cylinder contents on the piston surface. Since this p–V diagram is called the *indicator diagram*, the calculated power is termed *indicated power*.

Indicated Thermal Efficiency (η_i)

Evaluates the partial conversion of combustion heat (Q_{add}) to positive net power (IP) inside the cylinder.

$$\eta_i = \frac{IP}{\dot{Q}_{add}} = \frac{IP}{\dot{m}_{fuel} \times LHV} \tag{2.7}$$

Friction Power (FP)

The power lost to friction between all moving components inside the engine, including the power consumed in driving all engine auxiliary components, such as fuel, water, and oil pumps.

Brake or Clutch Power (BP)

The useful fraction of indicated power (IP) that can be utilized at the engine clutch as engine output, after losing the friction power (FP).

$$BP = IP - FP \tag{2.8}$$

Mechanical Efficiency (η_m)

$$\eta_m = \frac{BP}{IP} = 1 - \frac{FP}{IP} = \frac{1}{1 + FP/BP} \tag{2.9}$$

Brake Thermal Efficiency (η_b)

The overall efficiency that evaluates engine performance, as it relates the measurable useful output (BP) to the paid input (heat energy of fuel).

$$\eta_b = \frac{BP}{\dot{Q}_{add}} = \frac{BP}{\dot{m}_{fuel} \times LHV} \tag{2.10}$$

Torque (τ)

The torque developed by the engine.

$$\tau = \frac{BP}{2\pi N} \tag{2.11}$$

Specific Fuel Consumption (sfc)

Similar to thermal efficiency, the specific fuel consumption is another proportional metric for evaluating the conversion of fuel energy to work produced by the engine.

$$isfc = \frac{\dot{m}_{fuel}}{IP} \tag{2.12}$$

$$bsfc = \frac{\dot{m}_{fuel}}{BP} \tag{2.13}$$

Mean Effective Pressure (mep)

Work produced per unit displacement volume. mep is a very beneficial parameter, because it scales out the influence of engine size, thus, allows performance assessment of engines of different sizes.

$$imep = \frac{IP}{V_D(N/\xi)} \tag{2.14}$$

$$bmep = \frac{BP}{V_D(N/\xi)} \tag{2.15}$$

$$fmep = \frac{FP}{V_D(N/\xi)} \tag{2.16}$$

Mean Piston Speed (mps)

The average speed of the piston during one stroke. mps is an significant parameter in engine design, because stresses and further factors rule with piston speed somewhat different from engine speed.

$$mps = 2NL \qquad (2.17)$$

Maximum Rated Power

The highest power that an engine can change for brief periods of process.

Normal Rated Power

The engine highest power that can be developed continuously.

2.3 Fuels

Fuel is defined as any substance that can be burned to produce energy in the form of heat. Most familiar fuels are fossil fuels that consist principally of carbon and hydrogen; therefore, they are named *hydrocarbon fuels* and can be signified by the general formula C_mH_n. The name of a hydrocarbon molecule consists typically of a prefix and a suffix. The prefix denotes the number of contiguous carbon atoms forming the primary backbone of molecule. Table 2.1 lists the prefixes corresponding to 1–16 carbon atoms.

As for the name suffix, three concepts should be defined before discussing it. The first concept relates to the structure of the molecule. There are chain (aliphatic) and ring structures; the difference between the two is illustrated in Fig. 2.17. The same figure illustrates the second concept, which relates to the classification of unsaturated and saturated hydrocarbons. An unsaturated hydrocarbon can have two or extra adjacent carbon atoms mostly joined by a double or triple bond. However, in a

Table 2.1 Hydrocarbon prefix based on the number of carbon atoms

No. of carbon atoms	Name	No. of carbon atoms	Name
1	meth	9	non
2	eth	10	dec
3	prop	11	undec or hendec
4	but	12	dodec
5	pent	13	tridec
6	hex	14	tetradec
7	hept	15	pentadec
8	oct	16	hexadec or cet

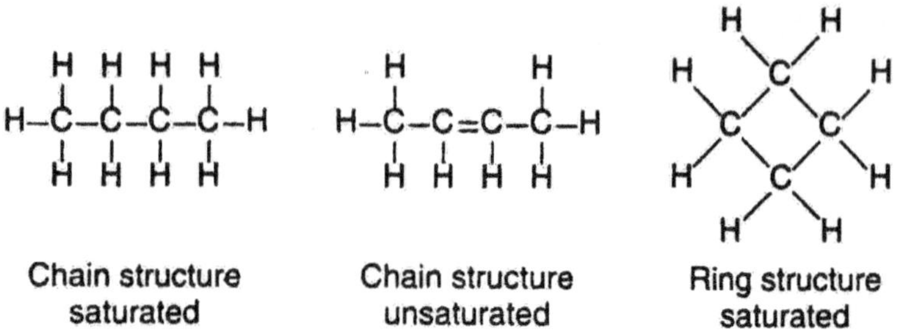

Chain structure
saturated

Chain structure
unsaturated

Ring structure
saturated

Fig. 2.17 Molecular structure of some hydrocarbon fuels

saturated hydrocarbon, all the carbon atoms are combined by a single bond. The third concept to be demarcated is an isomer, thus, two hydrocarbons with similar number of hydrogen and carbon atoms but having different molecular structures are called isomers. For example, there are several different isomers of C_8H_{18}, each one has 8 carbon atoms and 18 hydrogen atoms, but each of a different molecular structure.

2.3.1   *Molecular Structure of Hydrocarbon Fuels*

2.3.1.1   Paraffins

Paraffins (also known as alkanes) have saturated chain structure with the general formula C_nH_{2n+2}. The name "alkanes" implies that they have the suffix "-ane." The simplest alkane is methane, which is the principal component of natural gas. Figure 2.18 shows the structure of the methane molecule.

Higher members of the paraffin family are formed by joining more carbon atoms in a chain as shown in Fig. 2.19.

Fig. 2.18 Structure of
methane molecule

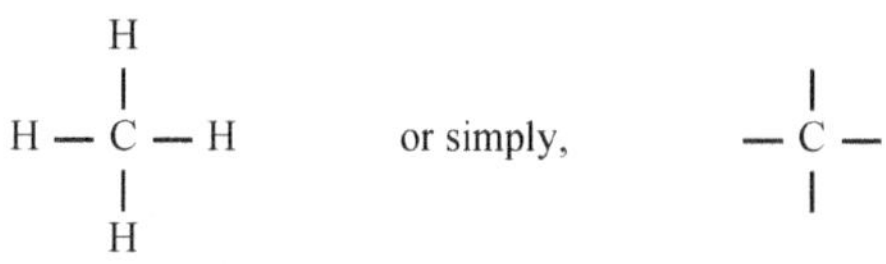

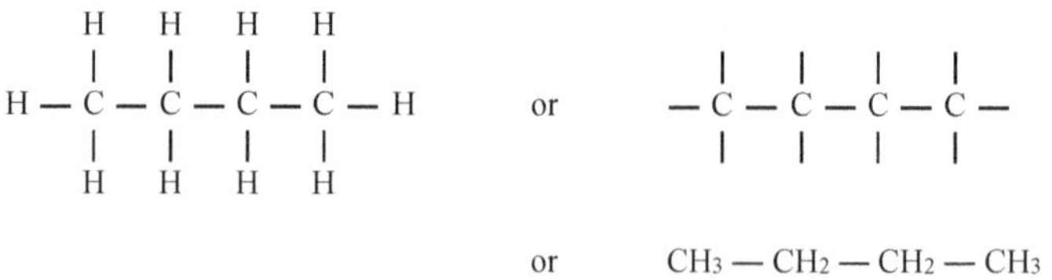

$$H - \overset{\displaystyle H}{\underset{\displaystyle H}{\overset{|}{\underset{|}{C}}}} - H \qquad \text{or simply,} \qquad -\overset{|}{\underset{|}{C}}-$$

$$H - \overset{\displaystyle H}{\underset{\displaystyle H}{\overset{|}{\underset{|}{C}}}} - \overset{\displaystyle H}{\underset{\displaystyle H}{\overset{|}{\underset{|}{C}}}} - \overset{\displaystyle H}{\underset{\displaystyle H}{\overset{|}{\underset{|}{C}}}} - \overset{\displaystyle H}{\underset{\displaystyle H}{\overset{|}{\underset{|}{C}}}} - H \qquad \text{or} \qquad -\overset{|}{\underset{|}{C}}-\overset{|}{\underset{|}{C}}-\overset{|}{\underset{|}{C}}-\overset{|}{\underset{|}{C}}-$$

or $CH_3 - CH_2 - CH_2 - CH_3$

Fig. 2.19   Structure of butane (C_4H_{10})

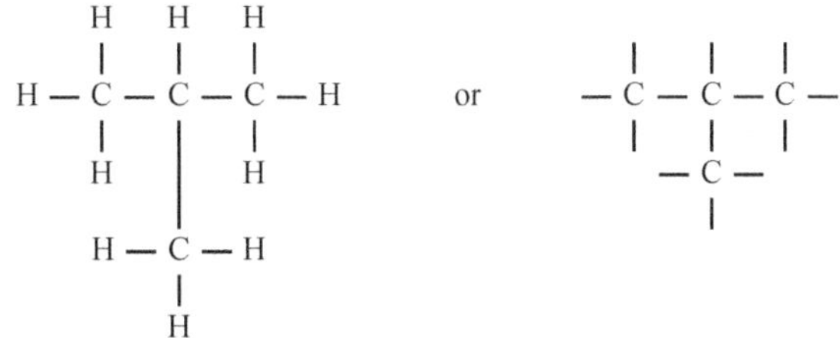

Fig. 2.20 Structure of isobutane (C_4H_{10})

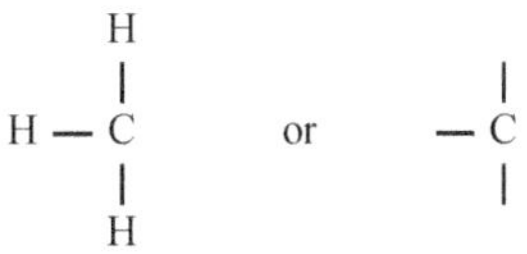

Fig. 2.21 Structure of methyl group

If all carbon atoms form one continuous chain, the alkane name is preceded with "n-", which stands for "normal"; the molecule depicted in Fig. 2.19 is thus called n-butane. Molecules containing four or more carbon atoms have isomers; the number of isomers increases substantially with the number of carbon atoms. Isomers are also saturated with the formula C_nH_{2n+2}, but they have a branched chain structure. Isobutane (C_4H_{10}), for example, is the single isomer of n-butane; see Fig. 2.20.

Branches in the isomer molecule are paraffin radicals or groups, with the formula C_nH_{2n+1}. They are also called alkyl radicals, where "alk-" is a prefix from Table 2.1 and "-yl" denotes the fact that they are radicals, i.e., incomplete molecules. The structure of methyl CH_3 group, for example, is shown in Fig. 2.21.

Since heavier hydrocarbons have a larger number of isomers, like octane C_8H_{18}, the simple name isooctane does not accurately describe which isomer is meant. A more sophisticated, explanatory name is thus required to distinguish the different isomers. If the molecule has a backbone chain of five carbon atoms, for example, with three methyl branches, as depicted in Fig. 2.22, then the name 2,2,4 tri-methyl pentane accurately identifies the isomer, because it has a pentane backbone with two methyl groups attached to the second carbon atom and one methyl group attached to the fourth one. Note that "tri" means 3, and that the backbone carbon atoms are numbered from left to right.

This particular isomer of octane is of great importance in the characterization of internal combustion engine fuels. Alkanes are stable in storage, clean burning, and do not chemically attack the usual gasket materials or metals, because they are saturated with hydrogen.

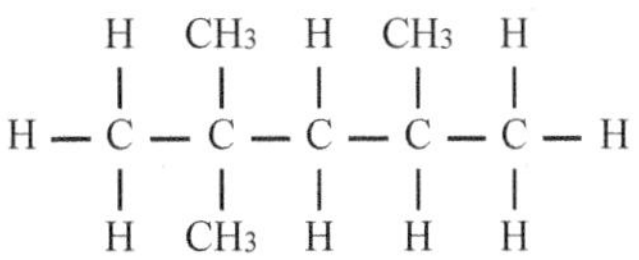

Fig. 2.22 Structure of 2,2,4 tri-methyl pentane (C_8H_{18})

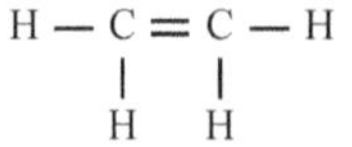

$$H - C = C - H$$

Ethene C_2H_4 (also known as ethylene)

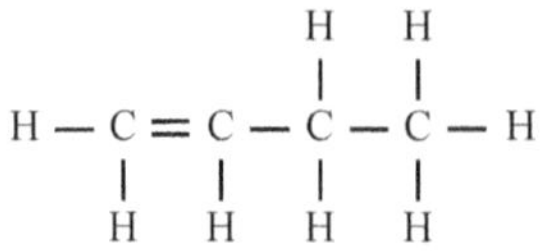

$$H - C = C - C - C - H$$

Butene-1 or 1-butene C_4H_8
(also known as butylene)

Fig. 2.23 Structure of some alkenes

Fig. 2.24 Structure of isobutene or methyl propene (C_4H_8)

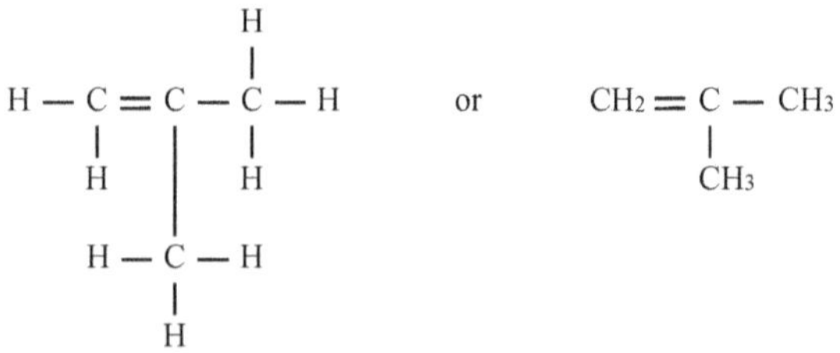

$$H - C = C - C - H \quad \text{or} \quad CH_2 = C - CH_3$$

2.3.1.2 Mono-olefins

Mono-olefins (also known as alkenes) have unsaturated chain structure with the general formula C_nH_{2n}, because the molecule has one double bond between two carbon atoms. The name "alkenes" implies that they have the suffix "-ene". Figure 2.23 shows examples of alkenes.

Since the double bond may occur at one of many locations, especially in larger molecules, the different isomers are distinguished by either a suffix or prefix number that denotes the order of the carbon atom that is double-bonded, as illustrated in Fig. 2.23 for butene-1 (or 1-butene). Butene has two normal isomers only, the second being butene-2, where the double-bond exists between carbon atoms 2 and 3. However, another isomer can be formed but with a branched chain, as shown in Fig. 2.24. The physical properties of olefins are closely similar to corresponding compounds in the paraffin family. Olefins are nearly as clean burning as paraffins and have higher octane rating.

2.3.1.3 Diolefins

Diolefins (also known as alkadienes) have unsaturated chain structure with the formula C_nH_{2n-2}, because the molecule has two double bonds or one triple bond. The name "alkadienes" implies that they have the suffix "-adienes". Figure 2.25 shows an example of diolefins.

Diolefins are not stable in storage, because cross-linking reactions take place that join the molecules at their double or triple bonds, leading to the formation of heavier

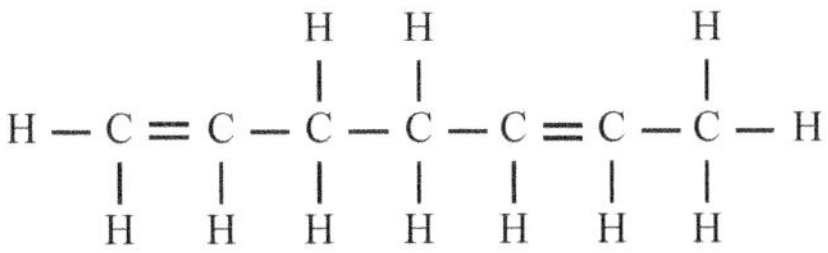

Fig. 2.25 Structure of 1,5 heptadiene (C_7H_{12})

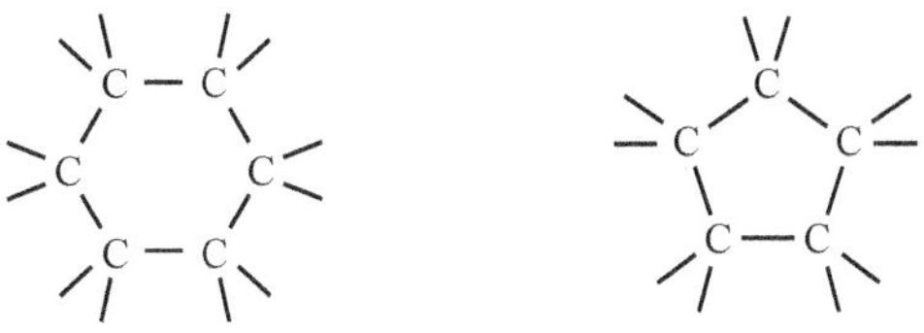

$$H-C=C-C-C-C=C-C-H$$

Fig. 2.26 Structure of cyclo-hexane (C_6H_{12}) and cyclo-pentane (C_5H_{10})

molecules that form a cloudy gum. This gum can, in turn, form engine deposits that adversely affect carburetion and valve operation. Diolefins are used primarily in the manufacturing of synthetic rubber.

2.3.1.4 Naphthenes or Cycloparaffins

Naphthenes have the same formula as mono-olefins (C_nH_{2n}), but naphthenes are saturated with a ring structure; the descriptive name cycloparaffins (or cyclanes) thus well describes them. A naphthene compound is typically named by the addition of the prefix "cyclo-" to the designation of the conforming straight-chain paraffin, see Fig. 2.26.

Although over 25% of crude oil is made up of naphthenic compounds, they are all resultant from any cyclo-hexane or cyclo-pentane with various side chains that are usually alkyl groups swapping one (or more) of the devoted hydrogen atoms. Naphthenes are required machineries for gasoline motors.

2.3.1.5 Aromatics (Benzene Derivatives)

Benzene has the formula C_6H_6 and is an unsaturated ring-structured molecule with alternating single and double bonds between the carbon atoms, see Fig. 2.27. Naphthalene has the formula, $C_{10}H_8$, and comprises two adjacent benzene rings. Both benzene and naphthalene and their derivatives are called aromatics. Some of the hydrogen atoms can be substituted by alkyl groups, as in toluene (methyl benzene) and ethyl benzene, see Fig. 2.28. The members of the aromatic family are admirable gasoline fuels and can be selectively shaped by catalytic cracking or by thermal cracking at high temperature (650 °C). Benzene, which is commercially known as benzol, is an exceptional combination agent to increase the octane ratings of the low-grade fuels. Table 2.2 summarizes the characteristics of the aforementioned hydrocarbon families.

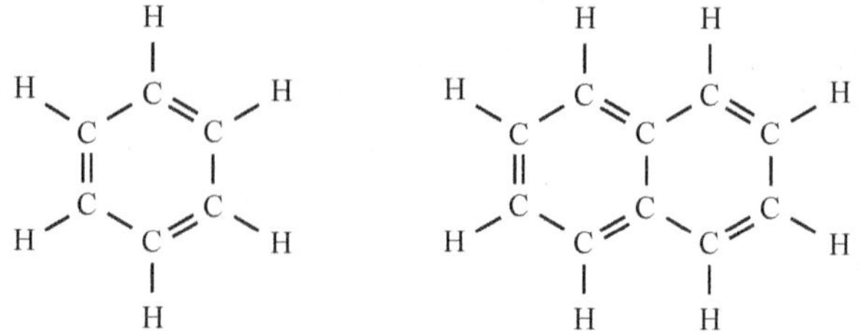

Fig. 2.27 Structure of benzene (C_6H_6) and naphthalene ($C_{10}H_8$)

Fig. 2.28 Structure of toluene (C_7H_8) and ethyl-benzene (C_8H_{10})

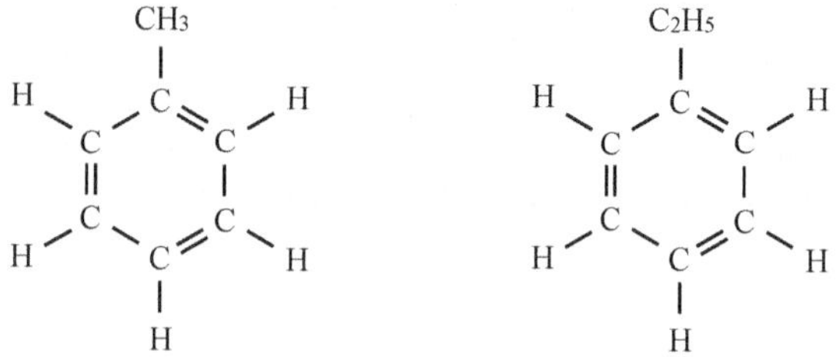

Table 2.2 Characteristics of hydrocarbon families

Family	Formula	Structure	Saturated
Paraffin (alkanes)	C_nH_{2n+2}	Chain	Yes
Naphthene (cyclo-alkanes)	C_nH_{2n}	Ring	Yes
Diolefin (alkadienes)	C_nH_{2n-2}	Chain	No
Olefin (alkenes)	C_nH_{2n}	Chain	No
Aromatic			
Naphthalene	C_nH_{2n-12}	Ring	No
Benzene	C_nH_{2n-6}	Ring	No

2.3.2 Fuel Types

Hydrocarbon fuels exist in all phases, namely solid, liquid, and gaseous fuels.

2.3.2.1 Solid Fuels

Solid fuels are primarily coal, coke, and wood. The most important solid fuel is coal, which exists in several grades.

1. Anthracite
2. Semi-anthracite
3. Bituminous
4. Semi-Bituminous

5. Lignite

The characteristics of coal vary considerably by location. Coal consists mostly of carbon, combined with some hydrogen, oxygen, nitrogen, sulfur, moisture, and non-combustible mineral salts (or ash), which remain as residue after combustion. The analysis of a solid fuel is either approximate or ultimate.

Approximate Analysis

On mass basis, the relative amount of moisture, volatile matter, carbon, and ash can be determined.

- **Moisture**

The moisture content is found by the reduction in the weight of a fuel sample after being heated at 100–110 °C for some time, until a minimum weight is obtained.

- **Volatile Matter**

The amount of volatile matter is determined by heating the dry sample at ~ 500 °C in a special container, allowing the volatile matter to escape without air coming into contact with the sample. Heating is continued until a minimum weight is reached.

- **Ash**

Air is next allowed to enter the container, and heat is applied to initiate the combustion of carbon. After all carbon is burned, the weight of remaining residue is the ash content.

- **Carbon**

Carbon content is obtained by subtracting the ash weight from the sample weight before combustion.

The ratio of carbon to volatile matter is called the fuel ratio.

Ultimate Analysis

This analysis specifies, on a mass basis, the percentages of carbon, sulfur, hydrogen, nitrogen, oxygen, and ash in the fuel. This is done by burning the fuel sample and analyzing the products of combustion. The ultimate analysis can be specified on an "as received" (wet) foundation or on a dry base. In the later case, the ultimate analysis would not comprise any moisture.

2.3.2.2 Liquid Fuels

Gasoline, kerosene, and fuel oils are examples of liquid fuels. Most commercial liquid fuels are combinations of diverse hydrocarbons attained from crude oil through distillation and cracking processes. Crude oil itself is a mixture of numerous hydrocarbon compounds. The percentage of carbon in crude oil, as it comes from the ground, varies generally from 83 to 87% by mass, and that of hydrogen from 11 to

14%. Oil also comprises various smaller quantities of sulfur, oxygen, nitrogen, water and sand. When crude oil is distilled or refined, the greatest volatile hydrocarbons vaporize first, to form what is known as gasoline. The less volatile fuels are diesel fuel, kerosene, and fuel oil. A sample of fuel is heated gradually until it vaporizes in order to produce the distillation curve. After that, the vapor condenses, and the volume is determined. The temperature of the non-vaporized fraction plotted against the amount of condensed vapor is known as the distillation curve, and it provides information about the fuel's volatility.

The conformation of a specific liquid fuel is subject to the basis of the crude oil and also on the refinery. Less volatile fuels, like fuel oils, typically have small percentages of sulfur, nitrogen, oxygen, moisture, and ash. Although sulfur combustion releases energy, sulfur presence in the fuel is undesired, because the resulting products are sulfur dioxide and trioxide, which combine with water vapor (combustion product of hydrogen) to form the highly corrosive sulfurous and sulfuric acids that attack and corrode metal surfaces in the combustion and exhaust systems.

For convenience, commercial liquid hydrocarbon fuels are sometimes approximated as a single hydrocarbon even though they are mixes of numerous distinct hydrocarbons. For instance, diesel fuel is referred to as dodecane (C12H26) and gasoline as octane (C8H18). Methyl alcohol, or CH3OH, sometimes known as methanol, is another popular liquid hydrocarbon fuel that is utilized in certain gasoline blends.

2.3.2.3 Gaseous Fuels

Natural gas, liquefied petroleum gas, producer gas (syngas), butagas, cock-oven gas, and blast-furnace gas are examples of gaseous fuels.

Gaseous fuels have a number of advantages over solid and liquid fuels:

1. They may be easily piped into combustion systems, so that no physical handling is necessary.
2. Gases are free of ash or other foreign matters.
3. The control of gas-fuel flame is easy, and complete smoke-free combustion can be obtained with a low percentage of excess air.

One of the disadvantages of gaseous fuels is the difficulty of storing them in large quantities.

Natural Gases (NG)

It can be found in the upper part of crude-oil or gas wells, and consists mainly of methane (CH_4) with varying smaller quantities of heavier hydrocarbons, CO_2, N_2, and H_2S. Natural gas is typically transported through pipelines after being compressed to above 10 bar. It can also be cooled and liquified for marine transport overseas. The heating value of natural gas is around 50 MJ/kg.

Liquefied Petroleum Gas (LPG)

Table 2.3 Volumetric analyses of some typical gaseous fuels [1]

Constituent	Various natural gases				Procedure gas from bituminous coal	Carbureted water gas	Coke-oven gas
	A	B	C	D			
Methane	93.9	60.1	67.4	54.3	3.0	10.2	32.1
Ethane	3.6	14.8	16.8	16.3			
Propane	1.2	13.2	15.8	16.2			
Butanes plus[a]	1.3	4.2		7.4			
Ethene						6.1	3.5
Benzene						2.8	0.5
Hydrogen					14.0	40.5	46.5
Nitrogen		7.5		5.8	5.9	2.9	8.1
Oxygen					0.6	0.5	0.8
Carbon monoxide					27.0	34.0	6.3
Carbon dioxide					4.5	3.0	2.2

In the petroleum refinery, considerable quantities of propane (C_3H_8) and butane (C_4H_{10}) are produced. These gases can be liquefied at normal temperature by subjecting them to a moderate pressure. LPG is typically 50% C_3H_8, 50% C_4H_{10} by volume. LPG is bottled in cylinders or shipped in large pressure tanks. The heating value of C_3H_8 is 46.3 MJ/kg, while that of C_4H_{10} is 45.7 MJ/kg.

Producer Gas (Syngas)

Synthetic gas, or syngas, is produced by reforming or gasifying solid fuels or heavy fuel oils/residues using a mixture of steam and air at high temperature. The resulting syngas is typically a mixture of CO, H_2, CO_2, and N_2. Syngas composition is widely variable and depends on the original fuel and the conditions of gasification process. The heating value of syngas is typically low (~ 5.8 MJ/kg) because of the diluting effect of CO_2 and N_2. Table 2.3 lists the volumetric analyses of some typical gaseous fuels.

2.3.3 *Fuel Properties*

Fuel properties are very important to determine which fuel is more suitable for a certain application and also to determine the types and amounts of fuel additives that enhance the fuel combustion characteristics. In this section, the general common properties of liquid fuels are presented, such as specific gravity, heating value, purity, and flash point. The properties of gasoline and diesel fuels for use in internal combustion engines are also presented.

Specific Gravity

Specific gravity is the ratio of fuel density (mass per unit volume) to water density at 15 °C.

Heating Value

Fuel heating value is the measure of energy content of the fuel. It is also called the heat of combustion and is determined by burning the fuel to completion with air or oxygen in a calorimeter, where the released heat increases the temperature of a cooling-water stream. The combustion products are cooled inside the calorimeter down to room temperature, and the flow rate and temperature rise of cooling water are used to calculate the fuel heating value. If all water vapor in the combustion products is condensed, the higher heating value is obtained, whereas if none of the water vapor is condensed, the lower heating value is obtained.

Purity and Combustion Residues

If the water content in fuel exceeds specification, it can cause corrosion and adversely affect all moving parts of the fuel-injection system, especially the injector nozzle. Water droplets do not pass through the multi-hole nozzle, which can damage it.

Sediments in the fuel cause sludge buildup within the fuel system and, consequently, wear of the injection components and filter clogging. Carbon residue is the deposit left in the combustion chambers and fuel-injection equipment because of the use of fuel made from residual blends. Ash originates from metal compounds in the fuel; it causes wear of fuel-injection components, engine pistons, and piston rings. The sulfur content of any fuel should be as low as possible to prevent corrosion damage and minimize the SOx content in the exhaust gases.

Flash Point

The flash point is the lowest temperature of a fluid that allows inflammable vapors to be formed on its surface. The fuel flash point is found by heating the fuel slowly while swiping a flame across the surface. A distinct flash is obtained at the flash point. This is an important property from a safety perspective and serves as a measure of fire hazard. The obtained temperature depends on the apparatus and procedure used in making the test; hence, the test method must be specified. Gasoline, being a volatile fuel, may flash at temperatures below 15 °C, whereas the flash point of the less volatile kerosene varies between 40 and 75 °C. Most heavier fuel oils have flash points between 70 and 140 °C. The common practice of adding kerosene or regular gasoline to diesel fuel for winter use can significantly lower its flash point. Just a few percent of gasoline can reduce the flash point of diesel to the room-temperature range, thus allowing for the formation of explosive fuel–air mixtures above the storage tank. The flash point of diesel fuel must not fall below 55 °C.

2.3.3.1 Gasoline-Fuel Properties

Volatility

It is the ability of a liquid to vaporize. Volatility is one of the main characteristic properties of fuel, which determines its suitability for use in a given engine. In spark-ignition engines, for example, gasoline has to be volatile enough, even in winter, to produce a combustible mixture of fuel vapor and air at the intake-system temperature and to have been completely vaporized and mixed with air inside the cylinder prior to ignition.

Boiling Curve (Distillation Curve)

The distillation curve is obtained by slowly heating a sample of fuel, so that its hydrocarbon constituents vaporize one at a time. The vapor is then condensed, and the distillate amount is measured, see Fig. 2.29. The distillation curve, which is a plot of the temperature of the non-vaporized fraction versus the amount of vapor condensed, is an indication of the volatility and purity of the fuel. Three ranges on the boiling curve are important for characterizing the behavior of fuel in an engine. These ranges can be distinguished by the percentages of fuel that vaporize at three distinct temperatures, see the minimum fuel requirements in Table 2.4. The fuel volume that vaporizes at temperatures equal and up to 70 °C should be large enough for a cold engine to start easily, see Fig. 2.30, yet not too large to origin vapor lock when the engine is hot. An engine is considered to be vapor locked when liquid fuel flow to the engine is either partially or completely blocked by the presence of fuel vapor, because vapor occupies a greater volume than liquid. The blocking of liquid fuel causes either loss of power or even complete stoppage of the engine. The volume of fuel that vaporizes at temperatures up to 180 °C, on the other hand, should not be too small, in order to prevent dilution of the engine oil, particularly when the engine is cold.

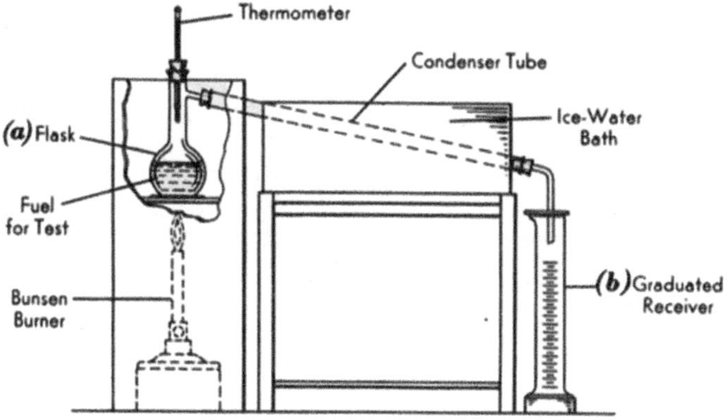

Fig. 2.29 Distillation test apparatus

Table 2.4 Minimum requirements for liquid spark-ignition-engine fuels [1]

Requirements		Premium gasoline		Regular gasoline		Test standard
		Summer	Winter	Summer	Winter	
Density at 15 °C	g/mL	0.730…0.780		0.715…0.765		DIN 51 757
Antiklock quality	min. RON	98.0		91.0		DIN 51 756
	min. MON	88.0		82.7		
Lead contents (lead Alkis)	max.g Pb/l	0.15		0.15		DIN 51 769
						DIN EN 13
Boiling curves: Total amounts of fuel vaporized						DIN 51 751
Up to 70 °C	% by vol	15…40	20…45	15…40	20…45	
Up to 100 °C	% by vol	42…65	45…70	42…65	45…70	
Up to 180 °C	% by vol	90	90	90	90	
Final boiling point	min max. °C	215		215		
Distillation residue	max. % by vol	2		2		DIN 51 751
Vapor pressure according to Raid method	bar	0.45…0.7	0.6…0.9	0.45…0.7	0.6…0.9	DIN 51 754
Evaporation residue	max. mg/ 100 mL	5		5		Din EN 5
Sulfur content	max. % by wt	0.10		0.01		DIN EN 41
						DIN 51 400

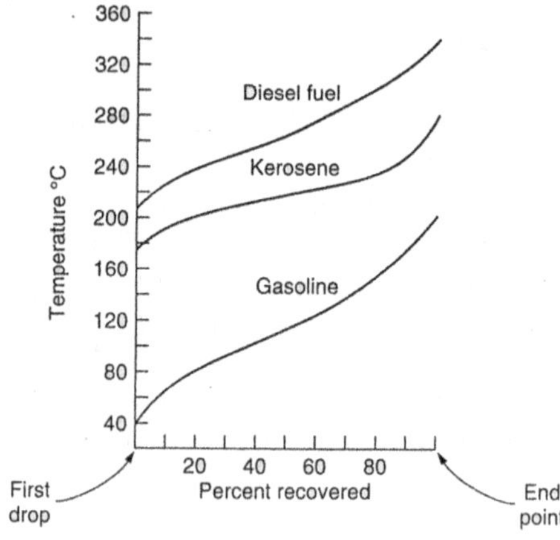

Fig. 2.30 Distillation curves of different fuels

Vapor Pressure

The vapor pressure of a fuel at 38 °C should not be greater than 0.7 bar for summer fuels and 0.9 bar for winter fuels. The vapor-pressure/temperature curves, however, are highly dependent on fuel composition. Fuels containing alcohol have considerably steeper curves than alcohol-free fuels. This means that fuels containing alcohol have a greater tendency to cause engine malfunction because of vapor lock at high temperature.

Antiknock Quality (Octane Number)

Normal combustion in spark-ignition engines takes place when a single flame front, initiated by the spark plug, expands to consume the unburned fuel–air mixture. Abnormal combustion, on the other hand, occurs when secondary ignition sources, other than the primary flame front, exist at the cylinder walls to ignite the fresh charge at more than one location, thus creating several flame fronts that interfere constructively, resulting in extreme heat-release rates and pressure spikes of substantial amplitude; see Fig. 2.31. These spikes resonate through the combustion chamber, and when they hit the piston surface, they (a) exert large and localized forces that induce detrimental piston vibration and an undesirable knocking noise, and (b) cut through the cooler thermal boundary layer that protects the piston surface from direct contact with the high-temperature cylinder contents. Abnormal combustion and the resulting knock are thus of great concern because they can induce major engine damage, and even in the case of minor knock, they are regarded as an objectionable source of noise by the engine or vehicle operator. Examples of secondary ignition sources are overheated valves or spark plugs, or glowing soot or residue deposits, which highlights the importance of fuel purity.

The locations furthest from the spark plug, also known as the "end gas", are most susceptible to inducing knock. Even in clean engines that are free of surface-ignition sources, the end gas gets inevitably compressed and heated by the approaching primary flame front, which inherently creates risk of end-gas autoignition before the arrival of primary flame front. End-gas autoignition is analogous to surface ignition, because both induce the knock scenario described above. Knock primarily arises

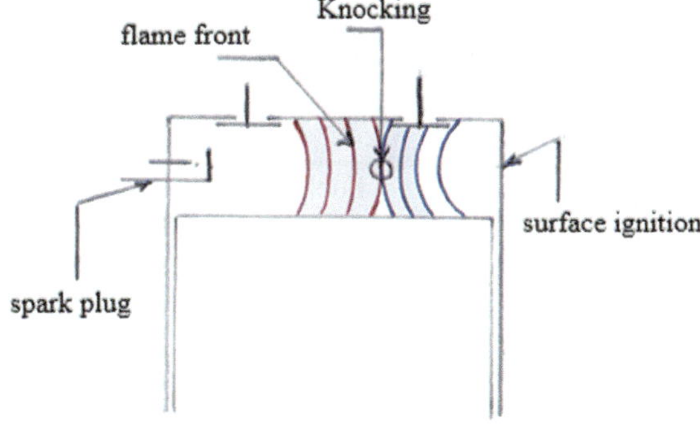

Fig. 2.31 A sketch showing the phenomenon of engine knock

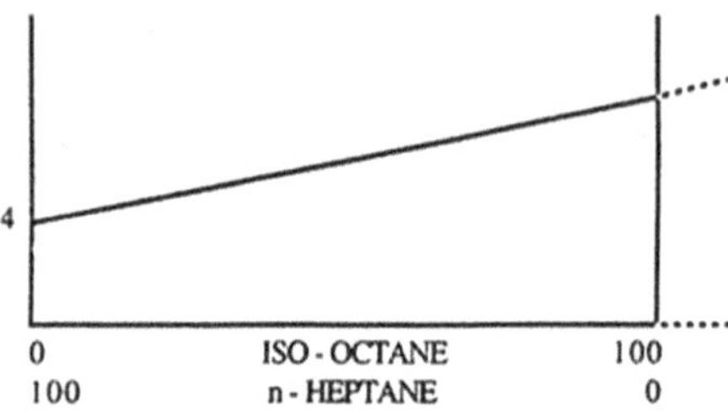

Fig. 2.32 Octane rating of fuels between zero and 100

under wide-open-throttle operational conditions, which directly impacts engine performance. Knock also constrains engine efficiency, because setting harsh limits on the pressure and temperature of the end gas means limiting the compression ratio of the engine. The occurrence as well as the severity of the knock phenomenon is thus strongly dependent on the ability of fuel in the end-gas region to resist autoignition and also on the antiknock characteristics of the combustion chamber. The ability of a fuel to resist knock is characterized by the fuel "octane rating"; higher octane ratings indicate greater resistance to knock.

The octane rating indicates the antiknock quality of fuel. A reference scale to characterize SI knock was established by arbitrarily selecting two reference fuels: Isooctane C_8H_{18} (2,2,4 tri-methyl pentane), with excellent knock resistance, was arbitrarily assigned an "octane rating" of 100, whereas n-heptane (C_7H_{16}), with poor knock resistance, was assigned an octane rating of zero, also quite arbitrarily; see Fig. 2.32. The octane rating of any fuel is found by comparing its tendency to knock in a standard test engine with various reference mixtures of isooctane and n-heptane. For example, if a fuel has an octane rating of 91, it means that the knock characteristics of this fuel are analogous to those of a reference mixture of 91% isooctane/9% n-heptane (by volume).

Two different procedures are used in the standard test engine to determine the octane number (ON), namely the research method and the motor method, see Table 2.5. Consequently, each fuel has two octane numbers, a Research one (RON) and a Motor one (MON). The motor method uses more severe operating conditions than the research method, e.g. higher inlet air temperature, higher engine speed, and greater spark advance, which represents more realistic operating conditions in most engines. Thus, the MON of a fuel is usually lower than its RON. Practically, all fuels exhibit a difference between their research and motor octane numbers. This difference is called the fuel sensitivity:

$$\text{Fuel sensitivity} = \text{RON} - \text{MON} \tag{2.18}$$

The arithmetic mean of RON and MON is termed the "antiknock index", and is used globally to accurately characterize the antiknock quality of a fuel:

$$\text{Antiknock index} = \frac{1}{2}(\text{RON} + \text{MON}) \tag{2.19}$$

Table 2.5 Operating conditions for research and motor methods to determine the octane number [2]

	Research method	Motor method
Inlet temperature	52 °C (125 °F)	149 °C (300 °F)
Inlet pressure	Atmospheric	
Humidity	0.0036–0.0072 kg/kg dry air 100 °C (212 °F)	
Engine speed	600 rev/min	900 rev/min
Spark advance	13° BTC (constant)	19–26° BTC (varies with compression ratio)
Air/fuel ratio	Adjusted for maximum knock	

These days, gasoline is enriched with several chemical additions intended to enhance its quality. In order to improve engine cold start, decrease deposits in the intake system, avoid valve sticking, resist gum formation, reduce spark-plug fouling, control surface ignition, and raise the overall octane number, these additives are utilized. Antiknock chemicals are often used to enhance the octane number of hydrocarbon fuels. Compared to changing the fuel's composition within the refinery, this is less expensive. Lead alkyls, tetraethyl lead (TEL), Pb(C2H5)4, and tetramethyl lead (TML), Pb(CH3)4, were the most successful antiknock agents. TEL was originally developed in 1923, while TML was introduced later in 1960. With TML's mid-range boiling point of 110 °C and TEL's high-end boiling point of 200 °C, the engine cylinders were able to distribute octane more evenly after TML was introduced. MMT, also referred to as methylcyclopentadienyl manganese tricarbonyl, was first offered as an additional antiknock agent for TEL in 1959.

In 1970, low-lead and unleaded gasoline were introduced in the United States. Two factors influenced the reduction of lead alkyl use: the concern about the toxicological aspects of lead in the urban environment and the poisoning of exhaust catalytic converters. In the US, unleaded and reduced-lead gasoline are now mandated. Japan has switched to unleaded gasoline nearly entirely. In Europe, in the late 1980s, laws were put into place to lower the lead concentration and introduce unleaded gasoline. MMT is occasionally added to unleaded gasoline as an antiknock ingredient. Its use is restricted to low quantities because it causes deposits that clog exhaust catalytic converters. The growing popularity of unleaded fuels has raised curiosity about other strategies for raising gasoline's octane rating.

The use of oxygenates (i.e., oxygen containing organic compounds) as octane boosters is becoming more common, because they have excellent antiknock qualities, thus allowing for meeting the octane-quality demands of modern gasoline while limiting the use of lead alkyls, which are banned by the increasingly stringent environmental regulations. Some oxygenates can also be produced from renewable non-petroleum sources (e.g., biomass), which offers strategic or economic benefits. Several oxygenates have been used as octane boosters; those of major interest are bio-methanol (CH_3OH), bio-ethanol (C_2H_5OH), tertiary butyl alcohol (TBA) (C_4H_9OH), and methyl tertiary butyl ether (MTBE), see Table 2.6.

Table 2.6 Oxygenates properties [3]

	Methanol	Ethanol	TBA	MTBE	Gasoline
Typical (R + M)/2 blending value	112	110	98	105	87–93
Weight percent oxygen	50	35	22	18	0
Stoichiometric (A/F)	6.5	9.0	11.2	11.7	14.5
Specific gravity	0.796	0.794	0.791	0.746	0.740
Lower heating value, MJ/kg	20.0	26.8	32.5	35.2	44.0
Latent heat of vaporization, MJ/kg	1.16	0.86	0.57	0.34	0.35
Boiling temperature, °C	65	78	83	55	27–227

In conclusion, it should be noted that, because of the great number of variables, it is impossible to tailor every characteristic of the fuel for the best operation of the engine. If each specification of the fuel were made too rigid, the resulting fuel would be too expensive for general use. Hence, the specifications of motor fuels must be flexible enough to make them economically affordable. Engine manufacturers design and build their engines to use commercially available fuels. If the engine compression ratio is low, which allows for using cheaper low-octane fuels, the use of high-octane fuels yields no better performance and is thus an economic waste. If the carburetion system has a large hot spot to evaporate high-boiling-point gasoline, a highly volatile gasoline might not be necessary; in fact, it may deteriorate the engine performance.

2.3.3.2   Diesel–Fuel Properties

Viscosity

In addition to its primary function as an energy source, diesel fuel also acts as a lubricant for its injection system. Fuel viscosity thus plays an important role in injection-system health and engine performance. If viscosity is lower than specified, the effectiveness of fuel as lubricant is reduced, which accelerates wear of injection components. Component wear can cause a reduction in the amount of metered fuel due to leakage in the injection pump or injector. High viscosity can also reduce the amount of metered fuel within the system due to higher flow resistance. Fuel viscosity also affects the injector spray pattern, see Fig. 2.33. Droplet size decreases with fuel viscosity, so lower viscosities induce smaller droplets that evaporate and burn too rapidly, resulting in large heat-release rate and excessive cylinder peak pressure and temperature. Higher viscosities, on the other hand, result in larger droplets that evaporate and burn too slowly, which increases the smoke emissions of the engine.

Ignitability

Unlike spark-induced ignition in gasoline engines, diesel fuel autoignites upon injection in diesel engines without the need for a spark. Fuel ignitability, i.e., the ease

Fig. 2.33 Effect of fuel viscosity on injector spray pattern

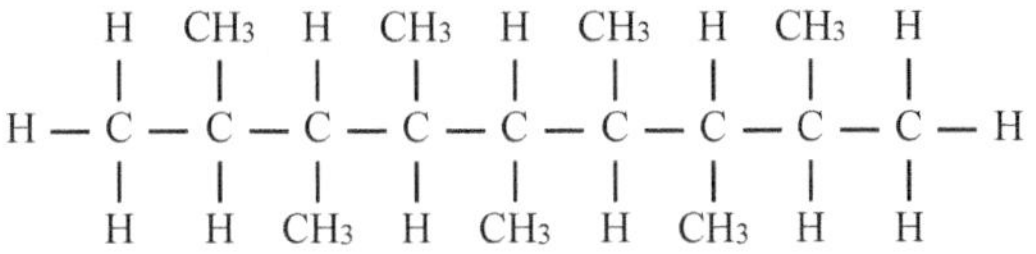

Fig. 2.34 Structure of hepta-methyl nonane (HMN) ($C_{16}H_{34}$)

of igniting fuel in the engine by autoignition, is thus a critically important property of diesel fuel. The Cetane Number (CN) scale was devised to characterize fuel ignitability. Again, two reference fuels were selected for comparison, namely n-cetane ($C_{16}H_{34}$) and hepta-methyl nonane (HMN) ($C_{16}H_{34}$); see Fig. 2.34. n-cetane, with excellent ignitability, was arbitrarily assigned a CN of 100, whereas HMN, with poor ignitability, was assigned a CN of 15, also quite arbitrarily. The CN of any fuel is found by comparing its ignition delay in a standard test engine with various reference mixtures of n-cetane and HMN. For example, if a fuel has a CN of 60, it means that its ignitability is analogous to that of a reference mixture containing 60% n-cetane (by volume). Higher CN means shorter ignition delay and, thus, better ignitability.

Middle distillates of crude oil generally demonstrate adequate ignitability in modern diesel engines. Fuels with a high percentage of cracked constituents have cetane numbers that come at the bottom of the standardized tolerance range. The more readily the fuel molecules break up, as is the case in unbranched chain structures like n-cetane, the better the fuel ignitability is. Heavy oils, e.g., heating oil, have large molecules that break up readily, so they also ignite easily. However, the use of heavy oils as fuels is limited due to their high viscosity and the presence of sulfur-containing constituents. Fuels with poor ignitability can be improved by mixing other fuels of good ignitability with them.

Additives are less widely used in diesel fuels, compared to spark-ignition fuels. Additives are sometimes used to improve certain characteristics of diesel fuel. For example, inhibitors and preventatives may be used to improve ignitability, to reduce oxidation and corrosion, and to keep the fuel-injection system clean. Sometimes, additives are also used to reduce smoke and odor. A problem that diesel fuel exhibits at lower temperatures is the precipitation of paraffin crystals that can grow to the extent of being stopped by the engine fuel filter, which accelerates filter clogging. Polymeric flow improvers are thus added, not to prevent the precipitation of paraffin crystals, but to impede their growth, so that they remain microscopically small to pass through the fuel filter. The effectiveness of these additives depends on the structure of the diesel fuel. The addition of kerosene helps, to some extent, re-dissolve the paraffin crystals, but at the expense of dangerously reducing the flash point of fuel. Normal medium distillates exhibit sufficient stability in storage; however, cracked constituents tend to age more rapidly. This tendency towards oxidation can be effectively counteracted through the use of additives.

Pour Point

The pour point is the minimum temperature at which the fuel can flow through filters or be transferred by the transfer pump to the fuel-injection pump. Fuel flow is affected only when the ambient temperature drops below the pour-point temperature. The fuel should have a pour point 5–8 °C below the operating temperature. In order to find the pour point, chill a fuel sample in a test jar until there is no longer any noticeable fuel movement (within 5 s) when the jar is moved from a vertical to a horizontal position (refer to Fig. 2.35).

2.4 Thermodynamics of Combustion

2.4.1 Composition of Atmospheric Air

Dry air is composed of 21% oxygen and 79% nitrogen on a mole or volume basis. Therefore, each kmol of oxygen entering a combustion chamber, or combustion process, will be accompanied by $0.79/0.21 = 3.76$ kmol of nitrogen. That is, 1 kmol $O_2 + 3.76$ kmol $N_2 = 4.76$ kmol air. Air is composed of 76.7% nitrogen and 23.3% oxygen by mass. That means that 3.29 kg of nitrogen will enter a combustion chamber or process for every kg of oxygen that enters $(0.767/0.233)$. Thus, 3.29 kg $N_2 + 1$ kg O_2 equals 4.29 kg air.

The triple bond between the two nitrogen atoms in the N_2 molecule is strong enough to make most of the nitrogen behave as an inert gas during the combustion process. However, in some cases, like extreme flame temperatures and/or the presence of fuel-bound nitrogen, a small fraction of nitrogen reacts with oxygen to form nitrogen oxides $(NO + NO_2 \equiv NOx)$. The mechanisms of NOx formation are beyond the scope of the present simplified analysis, so nitrogen will be treated as perfectly inert, i.e., all nitrogen entering the combustion system leaves as N_2.

2.4.2 Combustion Process

Combustion is a chemical reaction during which a fuel is oxidized (reacted with oxygen or air), and a large quantity of energy is released. The oxidizer most often used

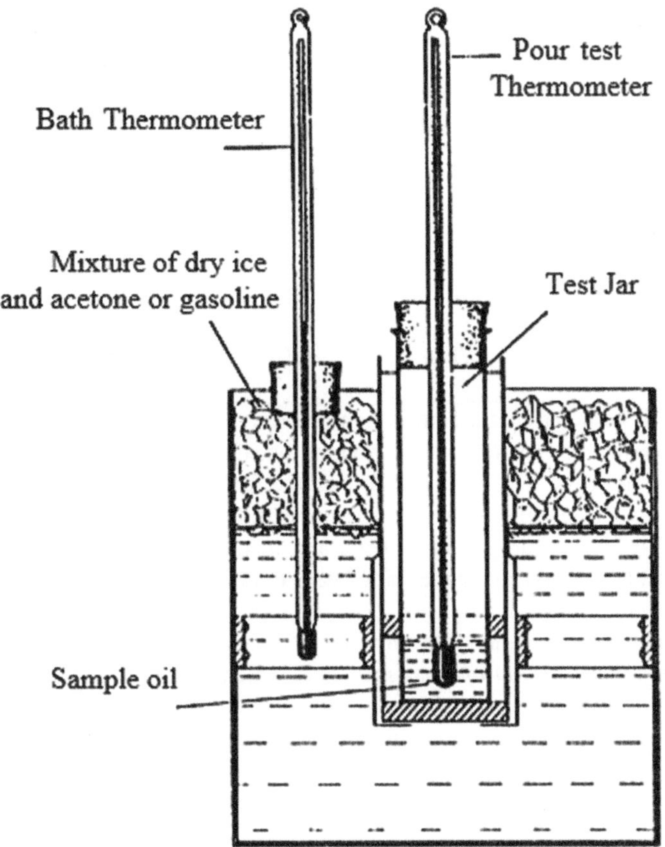

Fig. 2.35 Pour-point apparatus

in combustion processes is air, for obvious reasons; it is free and readily available. Pure oxygen, O_2, is used as oxidizer only in some specialized applications, like oxy-fuel combustion. For the most part of engineering applications, heat energy is the ultimate goal of the combustion process. However, in some other applications, light is required such as when using torches or candles. The components that enter a combustion process are referred to as reactants, and the components that remain after the reaction are referred to as products (see to Fig. 2.36).

Consider, for example, the combustion of one kmol of carbon with one kmol of pure oxygen, forming one kmol of carbon dioxide:

$$C + O_2 \rightarrow CO_2 \tag{2.20}$$

Here, C and O_2 are the reactants, since they enter the combustion process, and CO_2 is the product, since it exists after combustionWhen air is used in place of pure oxygen when burning carbon, N_2 is produced on both sides of the combustion equation. In

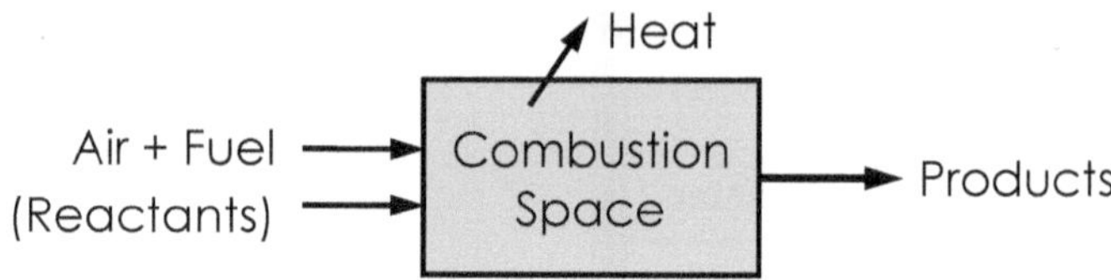

Fig. 2.36 Schematic of combustion process

other words, N_2 will manifest as both a reactant and a product.

$$C + O_2 + 3.76N_2 \rightarrow CO_2 + 3.76N_2 \tag{2.21}$$

It should be noted that a combustion process requires more than just putting a fuel into touch with oxygen (or air). An ignition source ought to be available for the combustion process. It is necessary to heat the fuel above the point of ignition. In atmospheric air, the minimum ignition temperature of gasoline is roughly 260 °C, carbon is 400 °C, hydrogen is 580 °C, and methane is 630 °C. In addition, for combustion to start, the air to fuel ratio needs to be within a certain range; this ratio will be covered later.

The combustion equation, like any chemical equation, should be balanced on the basis of the *conservation of mass principle*, which can be stated as follows: The total mass of each element is conserved during a chemical reaction. That is, Every element's total mass on the right side of the reaction equation (the products) must match every element's total mass on the left side (the reactants). Moreover, during a chemical reaction, the total number of atoms in each element is preserved. For example, consider the following combustion processes.

$$C + O_2 \rightarrow CO_2$$
$$12\,kg\,C + 32\,kg\,O_2 \rightarrow 44\,kg\,CO_2 \tag{2.22}$$
$$1\,kg\,C + (32/12)kg\,O_2 \rightarrow (44/12)kg\,CO_2$$

$$H_2 + 1/2O_2 \rightarrow H_2O$$
$$2\,kg\,H_2 + 16\,kg\,O_2 \rightarrow 18\,kg\,H_2O \tag{2.23}$$
$$1\,kg\,H_2 + 8\,kg\,O_2 \rightarrow 9\,kg\,H_2O$$

$$S + O_2 \rightarrow SO_2$$
$$32\,kg\,S + 32\,kg\,O_2 \rightarrow 64\,kg\,SO_2 \tag{2.24}$$
$$1\,kg\,S + 1\,kg\,O_2 \rightarrow 2\,kg\,SO_2$$

Note that the number of atoms of each element is balanced throughout the reaction. However, the total number of moles of the reactants is not necessarily equal to the total number of moles of the products.

Hence, given the composition by mass (or gravimetric composition) of fuel, e.g., x% of carbon, y% of hydrogen, and z% of sulfur, we may combine and apply for 1 kg of fuel as follows:

$$\left(\frac{x}{100} + \frac{y}{100} + \frac{z}{100}\right)\text{kg fuel} + \left[\left(\frac{x}{100} \times \frac{32}{12}\right) + \left(\frac{y}{100} \times 8\right) + \frac{z}{100}\right]\text{kg } O_2 \rightarrow$$

$$\left(\frac{x}{100} \times \frac{44}{12}\right)\text{kg } CO_2 + \left(\frac{y}{100} \times 9\right)\text{kg } H_2O + \left(\frac{z}{100} \times 2\right)\text{kg } SO_2 \qquad (2.25)$$

As discussed earlier, most fuels are mixtures of hydrocarbons, so the fuel can be described by the general formula $C_m H_n$. The combustion equation thus may be written as:

In molar form:

$$C_m H_n + \left(m + \frac{n}{4}\right)O_2 \rightarrow m\, CO_2 + \frac{n}{2}H_2O \qquad (2.26)$$

In mass form:

$$(12m + n)\text{ kg fuel} + 32\left(m + \frac{n}{4}\right)\text{kg } O_2 \rightarrow 44m\text{ kg } CO_2 + 9n\text{ kg } H_2O \qquad (2.27)$$

Alternatively,

$$1\text{ kg fuel} + \frac{32\left(m + \frac{n}{4}\right)}{(12m + n)}\text{kg } O_2 \rightarrow \frac{44m}{(12m + n)}\text{kg } CO_2 + \frac{9n}{(12m + n)}\text{kg } H_2O \qquad (2.28)$$

If the fuel is burned with air, the equation may be written as:

$$1\text{ kg fuel} + \frac{32(m + n/4)}{(12m + n)}\text{kg } O_2 + \frac{32(m + n/4) \times 3.29}{(12m + n)}\text{kg } N_2 \rightarrow$$

$$\frac{44m}{(12m + n)}\text{kg } CO_2 + \frac{9n}{(12m + n)}\text{kg } H_2O + \frac{32(m + n/4) \times 3.29}{(12m + n)}\text{kg } N_2$$

$$(2.29)$$

Recall that the multiplier 3.29 is the ratio of N_2 to O_2 by mass in the atmospheric air. The quantity $(m + n/4)$ is termed the minimum or theoretical oxygen demand of the fuel, because this is the minimum amount of oxygen needed to burn the fuel to completion, assuming that all carbon and all hydrogen will theoretically find their oxygen during combustion. In reality, it is impossible to mix fuel and oxidizer thoroughly enough to guarantee that each fuel molecule is in intimate contact with its entire oxygen demand. This will be discussed later in detail.

2.4.3   Air–Fuel Ratio

A very important term or quantity used in the analysis of combustion processes is the air–fuel ratio (A/F). It is the ratio of the amount of air to the amount of fuel.

On mass basis:

$$\frac{A}{F} = \frac{\text{Mass of air}}{\text{Mass of fuel}} = \frac{m_{air}}{m_{fuel}} \tag{2.30}$$

On volume basis:

$$\frac{A}{F} = \frac{\text{Volume of air}}{\text{Volume of fuel}} = \frac{\text{Number of moles of air}}{\text{Number of moles of fuel}} \tag{2.31}$$

The air–fuel ratio on mass basis is more often used.

Example 1 Calculate the theoretical air–fuel ratio for the combustion of octane, C_8H_{18}.

Solution

For C_8H_{18}, m = 8 and n = 18, so the theoretical O_2 is 8 + 18/4 = 12.5. The combustion equation is:

$$C_8H_{18} + 12.5O_2 + 12.5(3.76)N_2 \rightarrow 8CO_2 + 9H_2O + 47.0N_2$$

The air–fuel ratio on a mole basis is

$$\left.\frac{A}{F}\right)_{mole} = \frac{12.5 + 12.5 \times 3.76}{1} = 59.5 \, \text{kmol air/kmol fuel}$$

The air–fuel ratio on a mass basis is

$$\left.\frac{A}{F}\right)_{mass} = \frac{m_{air}}{m_{fuel}} = \frac{(NM)_{air}}{(NM)_{fuel}} = \frac{(NM)_{O_2} + (NM)_{N_2}}{(NM)_C + (NM)_{H_2}}$$

$$\left.\frac{A}{F}\right)_{mass} = \frac{12.5 \times 32 + 12.5 \times 3.76 \times 28}{8 \times 12 + 18 \times 1} = 15 \, \text{kg air/kg fuel}$$

Example 2 Air is burned to produce one kmol of octane (C_8H_{18}), which contains 20 kmol of O2. Calculate the kmol of each gas in the products and the air–fuel ratio for this combustion process, assuming that the products solely comprise CO_2, H_2O, O_2, and N_2.

Solution

The chemical equation for this combustion process can be written as

$$C_8H_{18} + 20O_2 + 20(3.76)N_2 \rightarrow \alpha CO_2 + \beta H_2O + \gamma O_2 + \delta N_2$$

where α, β, γ, and δ represent the unknown kmol of the gases in the products. These unknowns are determined by applying the conservation of mass principle to each of the elements.

C balance	$\alpha = 8$	$\rightarrow$	$\alpha = 8$
H balance	$2\beta = 18$	$\rightarrow$	$\beta = 9$
O balance	$2\alpha + \beta + 2\gamma = 40$	$\rightarrow$	$\gamma = 7.5$
N_2 balance	$\delta = 20*3.76$	$\rightarrow$	$\delta = 75.2$

Substituting yields:

$$C_8H_{18} + 20O_2 + 20(3.76)N_2 \rightarrow 8CO_2 + 9H_2O + 7.5O_2 + 75.2N_2$$

The air–fuel (A/F) ratio is determined by taking the ratio of the mass of air to the mass of fuel.

$$\frac{A}{F} = \frac{20 \times 32 + 20 \times 3.76 \times 28}{8 \times 12 + 1 \times 18} = 24.1 \text{ kg air/kg fuel}$$

2.4.4 Theoretical and Actual Combustion

If all of the fuel's carbon burns to CO_2, all of the hydrogen burns to H2O, and all of the sulfur, if any, burns to SO_2, the combustion process is finished. That is, during a full combustion process, all of a fuel's combustible components are consumed completely. However, if some unburned fuel or chemicals like C, H_2, or CO are present in the combustion products, the process of combustion is incomplete. Atomic carbon (C) is less reactive than atomic hydrogen (H). Thus, even in situations where there is insufficient oxygen to allow for the full combustion of the fuel, the hydrogen in the fuel typically burns completely to generate H_2O. However, some of the carbon ends up in the combustion products as CO or just as plain C particles, or soot. The primary cause of incomplete combustion is insufficient air, but it's not the only one. Another cause of incomplete combustion is inadequate air–fuel mixing in the combustion chamber.

Stoichiometric, or theoretical, air is the bare minimum, ideal or theoretical, required for the full combustion of fuel. Therefore, there won't be any oxygen in the product gases when the fuel burns entirely with theoretical air. The chemically right amount of air, or 100% theoretical air, is another name for the theoretical air. A combustion process that uses less air than theoretically possible is unavoidably imperfect. The stoichiometric or theoretical combustion of a fuel is the ideal combustion process in which the fuel burns entirely with theoretical air. As an illustration, the hypothetical burning of methane is

$$CH_4 + 2(O_2 + 3.76N_2) \rightarrow CO_2 + 2H_2O + 7.52N_2 \qquad (2.32)$$

Notice that the products of theoretical combustion do not contain unburned fuel (CH_4), C, H_2, CO, or free O_2. Since nothing is ideal in reality, and to account for fuel–air mixing deficiencies, the amount of actual air needed for complete combustion is logically greater than the stoichiometric amount.

2.4.5 Excess Air or Percent Theoretical Air

It is standard procedure to use more air than the stoichiometric amount in real combustion operations to boost the likelihood of full combustion. Excess air is the volume of air that is present over the stoichiometric volume. % surplus air or % theoretical air are common ways to describe the amount of excess air in terms of the stoichiometric air. 50% surplus air, for instance, is equal to 150 percent theoretical air, while 20% excess air is equal to 300% theoretical air. It follows that the stoichiometric air itself might be written as 100% theoretical air or as 0% extra air. The surplus-air factor, λ, can also be used to express the amount of excess air. It is defined as

$$\lambda = \frac{\text{Actual air}}{\text{Theoretical air}} = \frac{A/F)_{actual}}{A/F)_{theoretical}} \qquad (2.33)$$

The actual and theoretical amounts of air in Eq. 2.33 can be either by mass or by volume, resulting in the same value of excess-air factor. For a unit amount of fuel, the excess-air factor also relates to the air–fuel ratio, as seen in Eq. 2.33.

Another parameter used to describe the excess or deficiency of air with respect to fuel is the *equivalence ratio*, ϕ, which is defined as

$$\phi = \frac{F/A)_{actual}}{F/A)_{theoretical}} = \frac{A/F)_{theoretical}}{A/F)_{actual}} \qquad (2.34)$$

Note that, by definition, ϕ is the reciprocal of λ, i.e.,

$$\phi = 1/\lambda \qquad (2.35)$$

For instance, in case of 125% theoretical air, $\lambda = 1.25$, so $\phi = 0.8$.

2.4.6 Dew-Point Temperature

Since the main purpose of the combustion process is usually to extract heat energy for a particular application, the more heat is extracted, the lower the temperature of combustion products will be. The products contain CO_2 and H_2O, and in some cases, NOx and SO_2. If they are cooled enough, water vapor will start to condense to liquid and dissolve CO_2, NOx, and SO_2 to form diluted carbonic, nitrous/nitric, and sulfurous/sulfuric acids, which are corrosive by nature and can damage the hardware. It is thus of critical importance to monitor the *dew-point temperature*, which is defined as the temperature at which the water vapor in the combustion products starts to condense upon cooling. For this reason, the products of combustion are often kept above the dew point until discharged into the atmosphere.

Example 3 Determine the molar analysis of the products of combustion, when methane, CH_4, is burned with 110% theoretical air. Also determine the equivalence ratio and the dew point of the products, if the pressure is 0.1 MPa.

Solution

The combustion equation for methane with 110% theoretical air is:

$$CH_4 + 1.1 \times 2O_2 + 1.1 \times 2 \times 3.76N_2 \rightarrow CO_2 + 2H_2O + 0.1 \times 2O_2 + 8.272N_2$$

Total kmol of product $= 1 + 2 + 0.2 + 8.272 = 11.472$ kmol.
Molar analysis of products:

CO_2	$= 1/11.472$	$= 8.717\%$
H_2O	$= 2/11.472$	$= 17.434\%$
O_2	$= 0.2/11.472$	$= 1.743\%$
N_2	$= 8.272/11.472$	$= 72.106\%$
Total		$= 100.000\%$

The dew point of the products is the temperature at which the water vapor starts to condense as the products are cooled. *The dew-point temperature of a gas–vapor mixture is the saturation temperature of the water vapor corresponding to its partial pressure.* Therefore, the partial pressure of the water vapor (p_{H_2O}) in the products must be determined first.

The partial pressure of the water vapor is

$$p_{H_2O} = \left(\frac{N_{H_2O}}{N_{products}}\right)(p_{products}) = 0.17434 \times (100\,kPa) = 17.434\,kPa$$

From saturation steam tables, the dew-point temperature corresponding to 17.434 kPa is 57.13 °C.

Since 110% theoretical air means $\lambda = 1.1$, then the equivalence ratio $\phi = 1/1.1 = 0.909$.

Example 4 Ethane (C_2H_6) is burned with 20% excess air during a combustion process. Assuming complete combustion and a total pressure of 100 kPa, determine:

(a) The air–fuel ratio and
(b) The dew point temperature of the products.

Solution

Ethane is burned completely; therefore, the combustion products will contain only CO_2, H_2O, N_2, and the unused O_2.

The combustion equation is written as:

$$C_2H_6 + 1.2 \times 3.5(O_2 + 3.76\,N_2) \rightarrow 2\,O_2 + 3H_2O + 0.2 \times 3.5O_2 + 15.792N_2$$

$$C_2H_6 + 4.2O_2 + 15.792N_2 \rightarrow 2CO_2 + 3H_2O + 0.7O_2 + 15.792N_2$$

$$\frac{A}{F} = \frac{m_{air}}{m_{fuel}} = \frac{4.2 \times 32 + 15.792 \times 28}{2 \times 12 + 1 \times 6} = 19.22\,\text{kg air/kg fuel}$$

$$p_{H_2O} = \left(\frac{N_{H_2O}}{N_{prod}}\right)(p_{prod}) = \frac{3}{2 + 3 + 0.7 + 15.792}(100\,\text{kPa})$$

$$= \frac{3}{21.492} \times 100 = 13.96\,\text{kPa}$$

The dew point is the saturation temperature at 13.96 kPa, i.e., 52.3 °C.

Example 5 Natural gas has the volumetric analysis of methane (CH_4) 93.9%, ethane (C_2H_6) 3.6%, propane (C_3H_8) 1.2%, and butane (C_4H_{10}) 1.3%. This fuel is burned with 150% theoretical air. Assuming complete combustion and a total pressure of 100 kPa, determine the dew-point temperature of the products.

Solution

The combustion process is assumed to be complete; therefore, all the carbon in the fuel will burn to CO_2, and all the hydrogen to H_2O. Also, the fuel is burned with some excess air; therefore, there will be excess O_2 in the products. Considering 1 kmol of fuel, the individual combustion equations for the different constituents of the fuel are:

$$0.939CH_4 + 0.939 \times 1.5 \times 2O_2 + 0.939 \times 1.5 \times 2 \times 3.76N_2 \rightarrow$$
$$0.939CO_2 + 0.939 \times 2H_2O + 0.939 \times 0.5 \times 2O_2 + 10.59N_2$$

$$0.036C_2H_6 + 0.036 \times 1.5 \times 3.5O_2 + 0.036 \times 1.5 \times 3.5 \times 3.76N_2 \rightarrow$$
$$0.036 \times 2CO_2 + 0.036 \times 3H_2O + 0.036 \times 0.5 \times 3.5O_2 + 0.71N_2$$

$$0.012C_3H_8 + 0.012 \times 1.5 \times 5O_2 + 0.012 \times 1.5 \times 5 \times 3.76N_2 \rightarrow$$
$$0.012 \times 3\,CO_2 + 0.012 \times 4H_2O + 0.012 \times 0.5 \times 5\,O_2 + 0.338N_2$$

$$0.013C_4H_{10} + 0.013 \times 1.5 \times 6.5O_2 + 0.013 \times 1.5 \times 6.5 \times 3.76N_2 \rightarrow$$
$$0.013 \times 4CO_2 + 0.013 \times 5H_2O + 0.013 \times 0.5 \times 6.5O_2 + 0.477\,N_2$$

Products	No. of kmol	% in the products
CO_2	1.099	6.706
H_2O	2.099	12.810
O_2	1.07425	6.555
N_2	12.115	73.929
	$\Sigma = 16.38725$	100.000

$$p_{H_2O} = \left(\frac{N_{H_2O}}{N_{products}} \right) (p_{products}) = 0.1281 \times (100\,kPa) = 12.81\,kPa$$

The dew point is the saturation temperature at 12.81 kPa, namely 50.74 °C.

2.4.7 Gas Analysis

In theoretical calculations, the combustion process is assumed to be complete, i.e., only CO_2, H_2O, O_2, N_2, and SO_2 (if fuel contains sulfur) are produced. In practice, however, even when an excess amount of air is available, some other gases are emitted in minor concentrations, such as CO, C_mH_n, NO, and NO_2. The concentrations of these emissions are difficult to estimate through simple calculations. Rather, gas analyzers are used to measure these concentrations in ppm (part per million) or in mg/m^3. Electronic gas analyzers, such as infrared and electro-chemical-cell analyzers, are often used. They are reliable (if calibrated), portable, and can measure gases in %, mg/m^3, and ppm.

There are two types of gas analyses, namely dry and wet. The dry gas analysis is the volumetric (molar) analysis of the dry combustion products, i.e., excluding H_2O. Water vapor naturally exists in the products, but it is just not counted while conducting the molar analysis of the products. Inside the dry gas analyzer, H_2O is first separated from the sampled products. This is achieved by cooling (condensation), mechanical separation, and/or drying reagents such as silica gel. After the products have been dried, they are analyzed for their volumetric percentages. The wet gas analysis, on the other hand, is the volumetric (molar) analysis of all combustion products, i.e., including H_2O. So, water vapor is <u>counted</u> while conducting the molar analysis of the products. In practice, the majority of gas analyzers operate on dry basis.

Example 6 Octane (C_8H_{18}) is burned with dry air. The volumetric analysis of the products on a dry basis is CO_2 10.02%, O_2 5.62%, CO 0.88%, and N_2 83.48%. Determine:

(a) the air–fuel ratio
(b) the percent theoretical air used
(c) the fraction of H_2O that condensed as the products are cooled down to 25 °C at 100 kPa.

Solution

In this example, the dry analysis of the products is known, but we do not know how much fuel or air is used in the combustion process. The amounts of fuel and air can be determined from mass balances. If the combustion products are assumed to be ideal gases, the volume fractions given above become equivalent to mole fractions.

Considering 100 kmol of dry products for convenience, the combustion equation can be written as:

$$\alpha C_8H_{18} + \beta(O_2 + 3.76N_2) \rightarrow 10.02CO_2 + 0.88CO + 5.62O_2 + 83.48N_2 + \delta H_2O$$

The unknown coefficients α, β, and δ are determined from mass balances:

N_2 balance	$3.76\,\beta = 83.48$	$\rightarrow \beta = 22.2$
C balance	$8\,\alpha = 10.02 + 0.88$	$\rightarrow \alpha = 1.36$
H balance	$18\,\alpha = 2\,\delta$	$\rightarrow \delta = 12.24$
O_2 balance	$\beta = 10.02 + 0.88/2 + 5.62 + \delta/2$	$\rightarrow \beta = 22.2$

The O_2 balance is not necessary, as three unknowns need three equations only, but it can be used to double-check the values obtained from the other mass balances. Substituting,

$$1.36C_8H_{18} + 22.2(O_2 + 3.76\,N_2) \rightarrow 10.02\,CO_2 + 0.88\,CO$$
$$+ 5.62\,O_2 + 83.48\,N_2 + 12.24\,H_2O$$

$$\left(\frac{A}{F}\right)_{act} = \frac{m_{air}}{m_{fuel}} = \frac{22.2 \times 32 + 22.2 \times 3.76 \times 28}{1.36(8 \times 12 + 18 \times 1)} = 19.66\,\text{kg air/kg fuel}$$

To find the percentage of theoretical air used, we need first to know the theoretical amount of air, which is determined from the theoretical combustion equation of the fuel:

$$C_8H_{18} + 12.5(O_2 + 3.76N_2) \rightarrow 8\,CO_2 + 9H_2O + 47N_2$$

$$\left(\frac{A}{F}\right)_{st} = \frac{12.5 \times 32 + 12.5 \times 3.76 \times 28}{8 \times 12 + 18 \times 1} = 15.05\,\text{kg air/kg fuel}$$

Percent theoretical air $= \frac{(A/F)_{act}}{(A/F)_{st}} = \frac{19.66}{15.05} = 131\% \frac{(A/F)_{act}}{(A/F)_{st}} = \frac{19.66}{15.05} = 131\%$

That is, 31% excess air was used during this combustion process.

To obtain the dew point of the combustion products:

Products	No. of kmol	% in the products
CO_2	10.02	8.927
CO	0.88	0.784
H_2O	12.24	10.905
O_2	5.62	5.007
N_2	83.48	74.377
	$\Sigma = 112.24$	100.000

$$p_{H_2O} = \left(\frac{N_{H_2O}}{N_{products}} \right) (p_{products}) = 0.10905 \times (100\,kPa) = 10.905\,kPa$$

The dew point is the saturation temperature at 10.991 kPa, namely 47.51 °C.

Some of the water vapor will condense as the products are cooled to 25 °C. If the number of moles of H_2O that condensed to water is (N_{water}), then the remaining vapor is:

$$N_{vapor} = 12.24 - N_{water}$$

vSince the dry combustion products do not condensate upon cooling, and assuming that the total pressure (100 kPa) does not change because of condensation, the following equation can be defined for the remaining water vapor after condensation:

$$\frac{N_{vapor}}{N_{products}} = \frac{p_{H_2O}}{p_{products}}$$

where $N_{products} = (112.24 - N_{water})$ in this case, and $p_{H_2O} = 3.169$ kPa, which is the saturation pressure at 25 °C. Substituting,

$$\frac{12.24 - N_{water}}{112.24 - N_{water}} = \frac{3.169\,kPa}{100\,kPa} \qquad \rightarrow N_{water} = 8{:}967\,kmol.$$

The fraction of the H_2O that condensed is $(8.967 \div 12.24) \times 100\% = 73.3\%$

Thus, upon cooling the products to 25 °C, 73.3% of the water will be separated by condensation.

Example 7 Bagasse fuel is the fibers remaining after squeezing the sugar cane stalks in the sugar industry. After being dried, bagasse is used, as solid fuel, in firing the steam boilers in the same industry. Bagasse has the gravimetric composition: C 24.8%, H_2 3.1%, O_2 20.6%, H_2O 50.0%, and ash 1.5%. Calculate (a) the stoichiometric air–fuel ratio and (b) the dew-point temperature of the combustion products.

Solution

Assume 1 kg of fuel is burned with the stoichiometric air, the proportion of carbon is then 0.248 kg, the proportion of hydrogen is 0.031 kg, the proportion of oxygen is 0.206 kg, the proportion of moisture is 0.50 and the proportion of ash is 0.015 kg.

The only two combustible constituents are C and H_2.

The combustion equation of burning carbon with oxygen is:

$$C + O_2 \rightarrow CO_2$$
$$12\,kg\,C + 32\,kg\ O_2 \rightarrow 44\,kg\ CO_2$$

Multiplying through by (0.248 / 12)

$$0.248\,kg\,C + 0.6613\,kg\,O_2 \rightarrow 0.9093\,kg\,CO_2$$

The combustion equation of burning hydrogen with oxygen is:

$$H_2 + 1/2O_2 \rightarrow H_2O$$
$$2\,kg\,H_2 + 16\,kg\ O_2 \rightarrow 18\,kg\,H_2O$$

Multiplying through by (0.031 / 2)

$$0.031\,kg\,H_2 + 0.248\,kg\,O_2 \rightarrow 0.279\,kg\,H_2O$$

Therefore, the total mass of theoretical (stoichiometric) oxygen required to burn one kg of this fuel is the sum of the oxygen requirements of burning the carbon and the hydrogen.

$$\text{Mass of stoichiometric } O_2 = 0.6613 + 0.248 = 0.9093\,kg\,O_2.$$

Note that the fuel already has a proportion of oxygen (0.206 kg); this oxygen must be consumed first before demanding more oxygen from air. In other words, the oxygen content in the fuel must be deducted from the total required to burn the fuel. Therefore, mass of O_2 required from the atmospheric air $= 0.9093 - 0.206 = 0.7033$ kg.

Thus, mass of stoichiometric air $= 0.7033/0.233 = 3.0185$ kg, where 0.233 is the percentage of O_2 in air by mass.

Air–fuel ratio $\left(\frac{A}{F}\right)_{mass} = \frac{3.0185}{1} = 3.0185$ kg air/kg fuel.

To obtain the dew point, the partial pressure of H_2O must be determined first.

	Mass (m)	M	% mass = m/Σm	kmol = (% mass)/M	% volume kmol/Σkmol
CO_2	0.9093	44	22.74	0.5168	14.11
N_2	$0.7033 \times 3.29 =$ 2.31 (where 3.29 is N_2/O_2 mass ratio in air)	28	57.78	2.0636	56.34
H_2O	0.279 + 0.5 (from fuel) = 0.779	18	19.48	1.0822	29.55
Total mass = 3.9983				3.6626	100

The partial pressure of H_2O is then 29.55 kPa. The corresponding dew point is 68.76 °C.

2.5 Energy Balance of Combustion Process

Any system's molecules can hold energy in different forms, including perceptible energy (hot versus cold), latent energy (related to a change in state), chemical energy (related to the structure of the molecules), and nuclear energy (related to the structure of the atoms). Some of the chemical bonds that hold atoms together to form molecules are broken and new ones are generated during a chemical reaction. Generally speaking, the reactants and products have differing chemical energies connected to these linkages. Consequently, alterations in chemical energies will occur during a chemical reaction process, and this needs to be taken into consideration in the system's energy balance. assuming that there are no nuclear reactions and that the atoms of each reactant stay intact, while ignoring any shifts in the kinetic and potential energies, then:

$$\Delta E_{system} = \Delta E_{state} + \Delta E_{chemical} \tag{2.36}$$

If the products leave the combustor at the same inlet state of the reactants, then $\Delta E_{state} = 0$. So, the energy change of the system in this case is only due to the changes in its chemical composition.

$$\Delta E_{system} = \Delta E_{chemical} \tag{2.37}$$

To estimate the change in energy of a reacting system, a reference state must be designated. The reference state commonly chosen is 25 °C and 1 atm. This reference state is known as the *standard reference state*; the sensible enthalpy and internal energy of any substance are both zero at this state. Property values at the standard reference state are indicated by a superscript "°" (such as h° and u°). When analyzing reacting systems, we must use property values relative to the standard reference state. The ideal-gas property values tabulated in most textbooks are relative to 0

K; therefore, the sensible enthalpy relative to the standard reference state (298 K) has to be calculated as per the following example. The sensible enthalpy of N_2 at 500 K relative to the standard reference state is: $\left(\bar{h}_{500K} - \bar{h}^\circ\right) = 14581 - 8669 = 5912\,kJ/kmol$. Table 2.7 lists the sensible enthalpies of some substances relative to the standard reference state.

Table 2.7 Values of sensible enthalpy (in kJ/kmol) relative to 298 K for different species [3]

$\left[\bar{h}(T) - \bar{h}^\circ\right]$, kJ/kmol						
T(K)	C	CO	CO_2	H_2	H_2O	N_2	O_2
298	0	0	0	0	0	0	0
300	17	54	67	54	63	54	54
400	1046	2975	4008	2958	3452	2971	3029
500	2381	5929	8314	5883	6920	5912	6088
600	3962	8941	12,916	8812	10,498	8891	9247
700	5740	12,021	17,761	11,749	14,184	11,937	12,502
800	7661	15,175	22,815	14,703	17,991	15,046	15,841
900	9665	18,397	28,041	17,682	21,924	18,221	19,246
1000	11,816	21,686	33,405	20,686	27,702	21,460	22,707
1100	14,004	25,033	38,894	23,723	30,167	24,757	26,217
1200	16,246	28,426	44,484	26,794	34,476	28,108	29,765
1300	18,543	31,865	50,158	29,907	38,903	31,501	33,351
1400	20,870	35,338	55,907	33,062	43,447	34,936	36,966
1500	23,230	38,848	61,714	36,267	48,095	38,405	40,610
1600	25,614	42,384	67,580	39,522	52,844	41,903	44,279
1700	28,016	45,940	73,492	42,815	57,685	45,848	47,970
1800	30,439	49,522	79,442	46,150	62,609	48,982	51,689
1900	32,874	53,124	85,429	49,522	67,613	52,551	55,434
2000	35,321	56,739	91,450	52,932	72,689	56,141	59,199
2100	37,777	60,375	97,500	56,379	77,831	59,748	62,986
2200	40,250	64,019	103,575	59,860	83,036	63,371	66,802
2300	42,727	67,676	109,671	63,371	84,111	67,007	70,634
2400	45,216	71,346	115,788	66,915	93,604	70,651	74,492
2500	47,710	75,023	121,926	70,492	98,964	74,312	78,375
2600	50,216	78,714	128,085	74,090	104,370	77,981	82,274
2700	52,727	82,408	134,256	77,718	109,813	81,659	86,199
2800	55,241	86,115	140,444	81,370	115,294	85,345	90,144
2900	57,768	89,826	146,645	85,044	120,813	89,036	94,111
3000	60,300	93,542	152,862	88,743	126,361	92,738	98,098

2.5.1 *Enthalpy of Formation,* $\overline{h}_f^0$

Examine how carbon and oxygen combine to generate carbon dioxide (CO_2) in a steady-state steady-flow (SSSF) combustion process (Fig. 2.37). At 25 °C and 1 atm, both carbon and oxygen enter the combustion chamber. Assume that at 25 °C and 1 atm, the CO_2 produced during this process also exits the combustion chamber. An exothermic reaction, or one in which chemical energy is released as heat, is what happens when carbon burns. Since CO_2 is created during this process, some heat will be transported from the combustion chamber to the surrounding area. This heat transfer can be estimated at 393,520 kJ per kmol of CO_2. Applying the SSSF conservation of energy equation, with no work interactions, the total reactant enthalpy entering the system, H_R, plus the heat transfer, Q, in its positive sign convention (i.e., entering the system) must be equal to the total product enthalpy leaving the system, H_P.

Therefore:

$$H_R + Q = H_P \tag{2.38}$$

$$Q = H_P - H_R = -393520 \, \text{kJ/kmol} \tag{2.39}$$

Note that the negative sign is intentional to indicate that heat actually leaves the system in a combustion process. The heat transfer, Q, is also called the *enthalpy of reaction*. Since this reaction is a combustion reaction, the magnitude of Q (i.e., without the negative sign) is commonly termed the *enthalpy of combustion*, H_C.

Now, if the total enthalpy is discretized as the sum of chemical and sensible enthalpies, then:

$$(H_{chem} + H_{sens})_P - (H_{chem} + H_{sens})_R = -393520 \, \text{kJ/kmol} \tag{2.40}$$

Since both the reactants and products are at the reference state (25 °C) in this example, all sensible enthalpies are equal to zero, and the change of enthalpy during this process is due to the changes in chemical enthalpies only.

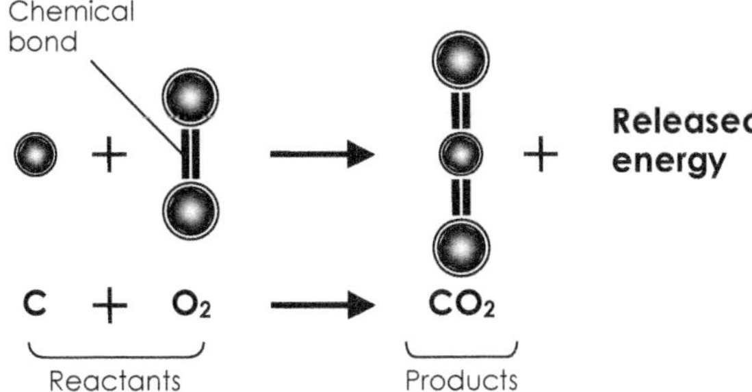

Fig. 2.37 Formation of CO_2 from its elements

$$(H_{chem})_P - (H_{chem})_R = -393520 \, \text{kJ/kmol} \qquad (2.41)$$

The reactants, C and O_2, are elements that have not been formed from any other substances, so the chemical enthalpy of each of them is assigned a value of zero. Thus,

$$(H_{chem})_{CO_2} = -393520 \, \text{kJ/kmol} \qquad (2.42)$$

The chemical energy of CO_2 has thus been deduced. This is a property of CO_2, termed the *enthalpy of formation*, $\overline{h}_f^o$, since it was calculated from the formation reaction of CO_2 from its elements. Note that the superscript "^{o}" indicates that both the products and reactants are at the reference state of (25 °C). Also note that, by definition, the *enthalpy of formation of any element is zero*. The enthalpies of formation of different substances are listed in Table 2.8.

Important remarks on Table 2.8:

1. A negative $\overline{h}_f^o$ means that the formation reaction of that substance from its elements is exothermic, i.e., releases heat, whereas a positive $\overline{h}_f^o$ means that the formation reaction of that substance from its elements is endothermic, i.e., consumes heat.
2. Some substances have two tabulated values of $\overline{h}_f^o$. Such substances have the ability to exist as either liquid or vapor in atmospheric conditions. The difference between the two values of $\overline{h}_f^o$ is thus the latent enthalpy of vaporization/

Table 2.8 Enthalpy of formation of some substances at 25 °C, 0.1 MPa [4]

Substance	Formula	M (kg/kmol)	State	$\overline{h}_f^o$ (kJ/kmol)
Carbon monoxide	CO	28	Gas	$-110{,}530$
Carbon dioxide	CO_2	44	Gas	$-393{,}520$
Water	H_2O	18	Gas	$-241{,}820$
Water	H_2O	18	Liquid	$-285{,}830$
Methane	CH_4	16	Gas	$-74{,}850$
Acetylene	C_2H_2	26	Gas	$+226{,}730$
Ethene	C_2H_4	28	Gas	$+52{,}280$
Ethane	C_2H_6	30	Gas	$-84{,}680$
Ethyl alcohol	C_2H_5OH	46	Gas	$-235{,}310$
Ethyl alcohol	C_2H_5OH	46	Liquid	$-277{,}690$
Propane	C_3H_8	44	Gas	$-103{,}850$
Propane	C_3H_8	44	Liquid	$-118{,}910$
Butane	C_4H_{10}	58	Gas	$-126{,}150$
Octane	C_8H_{18}	114	Gas	$-208{,}450$
Octane	C_8H_{18}	114	Liquid	$-249{,}950$

condensation at 25 °C. Consequently, latent energy is also accounted for in the energy balance of the system.

2.5.2 Heating Value

Consider the stoichiometric combustion reaction of CH_4 with air (Eq. 2.32).

$$CH_4 + 2(O_2 + 3.76N_2) \rightarrow CO_2 + 2H_2O + 7.52N_2 \qquad (2.32)$$

If both the reactants and products are again at the reference state, then the first law reduces to

$$Q = \sum_P \overline{H}_f^\circ - \sum_R \overline{H}_f^\circ \qquad (2.43)$$

where $\overline{H}_f^\circ = N \times \overline{h}_f^\circ$ and N is the number of moles. Therefore,

$$Q = 1 \times -393520 + 2 \times -285830 + 7.52 \times 0 - (1 \times -74850 + 2 \times 0 + 7.52 \times 0)$$

And $H_C = |Q| = 890330\,\text{kJ/kmol}\,CH_4 = 55646\,\text{kJ/kg}\,CH_4$.

This unique value is a property of CH_4, termed the *heating value*, which is defined as the amount of energy released, in form of heat, when a unit mass (or unit volume) of a fuel is burned stoichiometrically to completion in a SSSF process, and the products are returned to the state of the reactants.

Note that the above calculation assumed that all the H_2O in the products is in the liquid state. Since H_2O can exist as either liquid or vapor at 25 °C, another heating value can be calculated considering the H_2O in the products to be in the vapor state.

$$Q = 1 \times -393520 + 2 \times -241820 + 7.52 \times 0 - (1 \times -74850 + 2 \times 0 + 7.52 \times 0)$$
$$H_C = |Q| = 802310\,\text{kJ/kmol}\,CH_4 = 50144\,\text{kJ/kg}\,CH_4$$

Both heating values are properties of CH_4, termed *higher* and *lower heating values*, HHV and LHV, respectively. Since it is impossible to condensate all H_2O in the products to liquid upon cooling to 25 °C, as seen earlier in Example 6, the actual heating value of fuel is always somewhere between the HHV and LHV, which set easy-to-calculate upper and lower boundaries on the expectation of actual heating value.

The two heating values are related by:

$$HHV = LHV + \left(N \times \overline{h}_{fg}\right)_{H_2O} \text{kJ/unit fuel} \qquad (2.44)$$

where N is the molar amount of H_2O (per unit of fuel burned) in the products, and $\overline{h}_{fg}$ is the enthalpy of vaporization (or latent heat) of water at 25 °C, namely 43,961 kJ/kmol H_2O.

The relation between HHV and LHV can be rewritten as:

$$HHV = LHV + \left(m \times h_{fg}\right)_{H_2O} \text{ kJ/unit fuel} \tag{2.45}$$

where m is the mass of H_2O (per unit of fuel burned) in the products, and h_{fg} is the latent heat of water at 25 °C, namely 2442.3 kJ/kg H_2O.

Example 8 Calculate the enthalpy of combustion of gaseous octane (C_8H_{18}) at 25 °C and 1 atm, using the date for enthalpy-of-formation. Consider the water in the products to be in the liquid form.

Solution

The stoichiometric combustion equation for C_8H_{18} is:

$$C_8H_{18} + 12.5O_2 + 12.5 \times 3.76N_2 \rightarrow 8CO_2 + 9H_2O \ (\ell) + 47N_2$$

At 25 °C and 1 atm, the standard reference states are reached by both the reactants and the products. Additionally, since N_2 and O_2 are stable elements, they have zero enthalpy of formation.

After that, C_8H_{18}'s enthalpy of combustion becomes:

$$\overline{h}_c = |Q| = H_R - H_P$$

$$\overline{h}_c = \sum N_R \overline{h}_R^\circ - \sum N_P \overline{h}_{f,P^\circ} = \left[N\overline{h})_{C_8H_{18}} + 0 + 0\right] - \left[N\overline{h})_{CO_2} + N\overline{h}_f^\circ)_{H_2O} + 0\right]$$

Using $\overline{h}_f^\circ$ values from tables, we get

$$\overline{h}_c = 1(-208450) - 8(-393520) - 9(-285830) = 5512180 \text{ kJ/kmol } C_8H_{18}$$

Which is practically matching the recorded value of 5,512,200 kJ in the table. Because the water in the products is to be considered in the liquid phase, this heat of combustion is the HHV of C_8H_{18}. To obtain HHV per kilogram, we divide by the molecular weight, that is:

$$HHV = \frac{5512200}{114} = 48353 \text{ kJ/kg } C_8H_{18}$$

To obtain the LHV per kg, we first obtain N_{H_2O}, which is the kmol of H_2O in the products/kg fuel.

$$N = 9/114 = 0.0789 \, \text{kmol} \, H_2O/\text{kg} \, C_8H_{18}$$

$$LHV = HHV - \left(N \times \overline{h}_{fg}\right)_{H_2O} \text{kJ/kg fuel}$$

$$LHV = 48353 - 0.0789 \times 43961 = 44882 \, \text{kJ/kg} \, C_8H_{18}$$

2.5.3 First-Law Analysis of Reacting Systems

The first law of thermodynamics holds true for systems that react as well as those that don't. It is more practical to restate the first-law in a fashion that communicates the changes in chemical energies directly because chemically interacting systems include changes in their chemical energy.

2.5.3.1 Steady-State Steady-Flow (SSSF) Systems

If the changes in kinetic and potential energies are considered insignificant, the conservation of energy equation for a chemically reacting SSSF system may be expressed as:

$$Q - W = H_P - H_R \tag{2.46}$$

where H_P signifies the total enthalpy of the products (the exit streams) and H_R signifies the total enthalpy of the reactants (the streams at inlet). Then H_P and H_R can be articulated in terms of the specific enthalpies and mole numbers of their components as:

$$H_P = \sum N_P[\overline{h}_f^{\circ} + \left(\overline{h} - \overline{h}^{\circ}\right)]_P \tag{2.47}$$

$$H_R = \sum N_R[\overline{h}_f^{\circ} + \left(\overline{h} - \overline{h}^{\circ}\right)]_R \tag{2.48}$$

where $\overline{h}_f^{\circ}$ is the enthalpy of formation (chemical energy) at the standard reference state of 25 °C and 1 atm, and $\left(\overline{h} - \overline{h}^{\circ}\right)$ is the sensible enthalpy relative to 25 °C and 1 atm as well.

Therefore, the first law of thermodynamics for chemically reacting SSSF systems is:

$$Q - W = \sum N_P[\overline{h}_f^{\circ} + \left(\overline{h} - \overline{h}^{\circ}\right)]_P - \sum N_R[\overline{h}_f^{\circ} + \left(\overline{h} - \overline{h}^{\circ}\right)]_R \tag{2.49}$$

2.5.3.2　Closed Systems

The first law of thermodynamics for a chemically reacting stationary closed system can be expressed as:

$$Q - W = (U_P - U_R)$$

where U_P represents the total internal energy of the products and U_R represents the total internal energy of the reactants.

Utilizing the definition of enthalpy ($u = h - pv$):

$$Q - W = \sum N_P[\bar{h}_f^\circ + \left(\bar{h} - \bar{h}^\circ\right) - p\bar{v}]_P - \sum N_R[\bar{h}_f^\circ + \left(\bar{h} - \bar{h}^\circ\right) - p\bar{v}]_R \quad (2.50)$$

The $(p\bar{v})$ terms are negligibly small for solids and liquids and can be replaced by $(\bar{R}T)$ for ideal gases.

Example 9 A steady-state combustion chamber receives liquid ethanol (C_2H_5OH) at 25 °C and 1 atm. The ethanol burns completely, and dry air enters at 327 °C and 1 atm. The equivalency ratio is 0.8 and the fuel flow rate is 22.68 kg/h. Combustion products release at 1027 °C and 1 atm. Determine ignoring the effects of kinetic and potential energy.

The mass-based air–fuel ratio (a) and the heat transfer rate (b), expressed in kW.

Solution

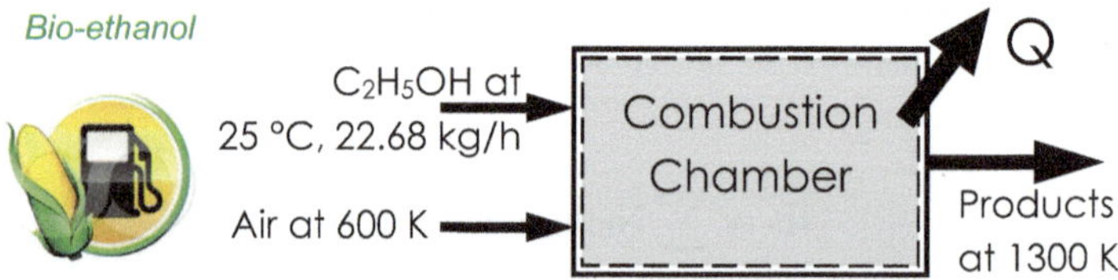

The equivalence ratio $\varphi = 0.8$　　→　　$\lambda = (1/\varphi) = 1.25$
Assume 1 kmol of ethanol, the combustion equation is:

$$C_2H_5OH + 1.25(3)(O_2 + 3.76N_2) \rightarrow 2CO_2 + 3H_2O + 0.75O_2 + 14.1N_2$$

$$C_2H_5OH + 3.75O_2 + 14.1N_2 \rightarrow 2CO_2 + 3H_2O + 0.75O_2 + 14.1N_2$$

$$\frac{A}{F} = \frac{(3.75 \times 32) + (14.1 \times 28)}{(2 \times 12) + (6 \times 1) + (1 \times 16)} = 11.19 \frac{kg\,air}{kg\,ethanol}$$

Enthalpies of Reactants

	N	T (K)	$\overline{h}_f^\circ$	$(\overline{h}_T - \overline{h}_{298})$	$N\left[\overline{h}_f^\circ + (\overline{h}_T - \overline{h}_{298})\right]$
C_2H_5OH	1	298	$-277{,}690$	0	$-277{,}690$
O_2	3.75	600	0	9247	34,676
N_2	14.1	600	0	8891	125,363
					$\Sigma = -117{,}651$

Enthalpies of Products

	N	T (K)	$\overline{h}_f^\circ$	$(\overline{h}_T - \overline{h}_{298})$	$N\left[\overline{h}_f^\circ + (\overline{h}_T - \overline{h}_{298})\right]$
CO_2	2	1300	$-393{,}520$	50,158	$-686{,}724$
H_2O	3	1300	$-241{,}820$	38,903	$-608{,}751$
O_2	0.75	1300	0	33,351	25,013
N_2	14.1	1300	0	31,501	444,164
					$\Sigma = -826{,}298$

Applying the 1st law of thermodynamics with $W = 0$ and neglecting the effects of kinetic and potential energies,

$$Q = \overline{H}_P - \overline{H}_R = (-826298) - (-117651) = -708647 \text{ kJ/kmol fuel}$$

$$|q| = \frac{|Q|}{M_{fuel}} = \frac{708647}{46} = 15405 \text{ kJ/kg fuel}$$

$$\dot{Q} = \dot{m}_{fuel} \times q = \frac{22.68}{3600} \times 15405 = 97 \text{ kJ/s} = 97 \text{ kW}$$

Example 10 Liquid octane (C_8H_{18}) is burnt inside an internal combustion engine along with 125% theoretical air. At 25 °C, the fuel and air enter, and at 900 K, the combustion products exit the engine's exhaust ports. 98% of the carbon is thought to burn into CO_2, with the remaining carbon burning into CO. This engine's heat transfer is exactly equivalent to the work it performs. Calculate: (a) The engine's power output if its burn rate is 0.005 kg/s.

(b) The analysis and composition of the combustion products, as well as their dew point temperature.

Solution

The reaction equation can be expressed as:

$$C_8H_{18} + (1.25 \times 12.5)O_2 + (1.25 \times 12.5 \times 3.75)N_2 \rightarrow (0.98 \times 8)CO_2$$
$$+ (0.02 \times 8)CO + 9H_2O + [(1.25 \times 12.5) - (0.98 \times 8) - 0.5(0.02 \times 8) - (0.5 \times 9)]O_2$$
$$+ 58.75N_2$$
$$C_8H_{18} + 15.625O_2 + 58.75N_2 \rightarrow 7.84CO_2 + 0.16CO + 9H_2O + 3.205O_2 + 58.75N_2$$

$$H_R = \overline{h}^{\circ}_{f_{C_8H_{18}}} + 0 + 0 = -249950 \, \text{kJ/kmol} \, C_8H_{18}$$

$$H_P = 7.87(-393520 + 28041) + 0.16\,(-110530 + 18397) + 9(-241820 + 21924)$$
$$+ \, 3.205(0 + 19246)$$
$$+ \, 58.75\,(0 + 18221) \; = -3726993 \, \text{kJ/kmol} \, C_8H_{18}$$

$$H_P - H_R = -3726993 - (-249950) = -3477043 \, \text{kJ/kmol} \, C_8H_{18}$$

Applying the first law of thermodynamics on the engine:
$Q - W = H_P - H_R$.

The heat transfer from this engine is just equal to the work done by the engine, then:

$$W = |Q|$$
$$2W = H_R - H_P = 3477043$$
$$W = 1738521.5 \, \text{kJ/kmol} \, \text{fuel}$$
$$w = \frac{1738521.5}{114} = 15250 \, \text{kJ/kg} \, C_8H_{18}$$

Power output from the engine $=$
$$\dot{W} = w \times \dot{m}_{fuel} = 15250 \times 0.005 = 76.25 \, \text{kW}$$

2.5.4 Adiabatic Flame Temperature

The chemical energy created during a combustion process is either employed internally to raise the temperature of the combustion products or transmitted as heat to the surroundings in the absence of any work interactions or changes in kinetic or potential energies. The temperature of the products is maximal and is known as the adiabatic flame temperature, or AFT, of the reaction when the heat transfer to the surroundings is zero ($Q = 0$); see Fig. 2.38.

Rephrasing the previous paragraph, the temperature of the products grasps a maximum value (adiabatic flame temperature) if there is:

(1) No heat transfer from the combustion space to the surroundings, i.e., adiabatic combustion chamber ($Q = 0$),

Fig. 2.38 Schematic description of the adiabatic flame temperature

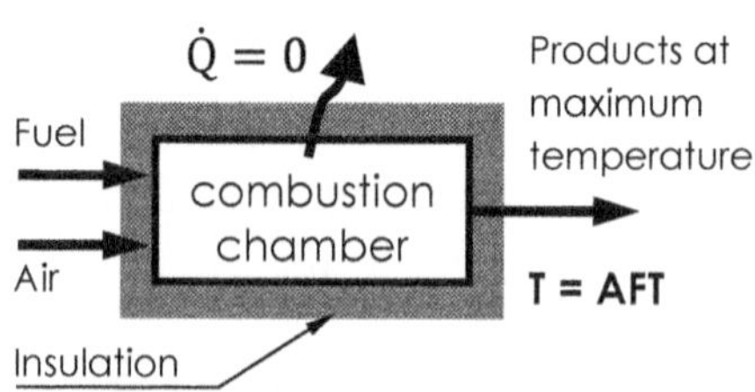

(2)   No work interaction between the combustor and surroundings, and
(3)   No changes in kinetic and potential energies.

For a SSSF combustion process, reconsider the first-law equation:

$$Q - W = H_P - H_R \tag{2.46}$$

Considering $Q = 0$ and $W = 0$, then $H_P = H_R$, or

$$\sum N_P \left[\overline{h}_f^\circ + \left(\overline{h} - \overline{h}^\circ \right) \right]_P = \sum N_R \left[\overline{h}_f^\circ + \left(\overline{h} - \overline{h}^\circ \right) \right]_R \tag{2.51}$$

Because the temperature of the products is unknown before the calculations are made, calculating the enthalpies of products, or HP, is not so simple. Consequently, an iterative method must be used to determine the adiabatic flame temperature.

An alternative, but less accurate, technique is to treat the combustion products as to behave as ideal gases. In this case, the sensible enthalpy of any of the products is

$$h = c_p T, \text{ and } \left(\overline{h} - \overline{h}^\circ \right) = c_p(T - 298) \tag{2.52}$$

Factors Affecting the Adiabatic Flame Temperature

(1)   Using excess air in the combustion process:

Since fuel is the source of chemical energy in the combustion process, then using excess air for the same amount of fuel means that the finite fuel energy will be distributed among a larger quantity of combustion products, which reduces the sensible-energy share of each product. Thus, the products' temperature decreases. This can be verified by inspecting Eq. (2.51). If air enters at 25 °C, then the right-hand side is constant for a given amount of fuel. On the left-hand side, N_P and $\left(\overline{h} - \overline{h}^\circ \right)_P$ are now competing, so greater N_P means lower sensible enthalpies and, hence, lower AFT. The excess air can also be viewed as a coolant here, introduced to reduce the maximum temperature in a combustion chamber.

(2)   Incompleteness of combustion:

If some carbon burns only partially to CO instead of CO_2, then the chemical energy of burning carbon has not been fully liberated. Note that the enthalpy of formation of CO is lower in magnitude than that of CO_2. The natural consequence is the reduction in the temperature of combustion products.

(3)   Dissociation of part of some combustion products at high temperature:

Since AFT is the maximum temperature of combustion products, some species, like CO_2, become unstable and dissociate according to the following reversible reaction:

$$CO_2 \rightleftarrows CO + O \tag{2.53}$$

The dissociation reaction (forward direction) is endothermic, i.e., consumes energy, which reduces the overall temperature of combustion products. The equilibrium of CO_2 dissociation is beyond the scope of the present simplified analysis.

Example 11 Gaseous octane (C_8H_{18}) is completely burned with 200% of theoretical air in a steady-flow combustor. Octane and air enter at 25 °C, 1 atm. If the combustor was well-insulated, determine the temperature of the combustion products.

Solution

The combustion equation is:

$$C_8H_{18} + 25O_2 + 94N_2 \rightarrow 8CO_2 + 9H_2O + 12.5O_2 + 94N_2$$

Enthalpies of Reactants

	N	T (K)	$\overline{h}_f^\circ$	$(\overline{h}_T - \overline{h}_{298})$	$N[\overline{h}_f^\circ + (\overline{h}_T - \overline{h}_{298})]$
C_8H_{18}	1	298	$-$ 208,450	0	$-$ 208,450
O_2	25	298	0	0	0
N_2	94	298	0	0	0
					$\overline{H}_R = -208{,}450$

The temperature of the products is obtained by trial and error.

Assume T = 2000 K,

Enthalpies of Products

	N	T (K)	$\overline{h}_f^\circ$	$(\overline{h}_T - \overline{h}_{298})$	$N[\overline{h}_f^\circ + (\overline{h}_T - \overline{h}_{298})]$
CO_2	8	2000	$-$ 393,520	91,450	$-$ 2,416,560
H_2O	9	2000	$-$ 241,820	72,689	$-$ 1,522,179
O_2	12.5	2000	0	59,199	739,988
N_2	94	2000	0	56,141	5,277,254
					$\overline{H}_P = -20{,}78{,}503$

$$\overline{H}_P - \overline{H}_R = 2286953$$

Assume T = 1600 K,

Enthalpies of Products:

	N	T (K)	$\bar{h}_f^\circ$	$\left(\bar{h}_T - \bar{h}_{298}\right)$	$N\left[\bar{h}_f^\circ + \left(\bar{h}_T - \bar{h}_{298}\right)\right]$
CO_2	8	1600	$-393{,}520$	67,580	$-2{,}607{,}520$
H_2O	9	1600	$-241{,}820$	52,844	$-1{,}700{,}784$
O_2	12.5	1600	0	44,279	553,488
N_2	94	1600	0	41,903	3,938,882
					$\bar{H}_P = -184{,}066$

$$\bar{H}_P - \bar{H}_R = 392516$$

Assume T = 1000 K,

Enthalpies of Products

	N	T (K)	$\bar{h}_f^\circ$	$\left(\bar{h}_T - \bar{h}_{298}\right)$	$N\left[\bar{h}_f^\circ + \left(\bar{h}_T - \bar{h}_{298}\right)\right]$
CO_2	8	1000	$-393{,}520$	33,405	$-2{,}880{,}920$
H_2O	9	1000	$-241{,}820$	25,978	$-1{,}942{,}578$
O_2	12.5	1000	0	22,707	283,838
N_2	94	1000	0	21,460	2,017,240
					$\bar{H}_P = -2{,}522{,}420$

$$\bar{H}_P - \bar{H}_R = -2313970$$

If the products are treated as ideal gases with nearly constant specific heats, the relation between enthalpy and temperature becomes a straight line.

Let $\bar{H}_P - \bar{H}_R = R$ and draw a graph between this "R" value and T; the exact T is obtained at $R = 0$.

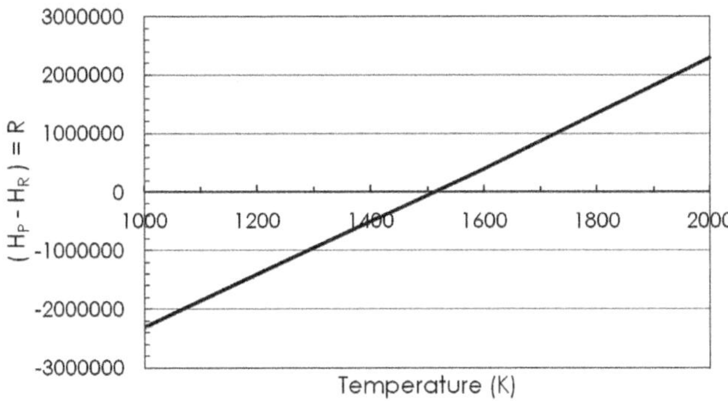

From the graph, the temperature of the products, which is the sought AFT, equals $\cong 1510\ \text{K} \cong 1237\ °\text{C}$.

Example 12

(a)   At 1 atm and 25 °C, methane gas (CH4) gradually enters a gas turbine unit's combustor where it burns alongside air that also enters at the same time. Assuming that the kinetic and potential energy remain constant, ascertain the adiabatic flame temperature in each of the subsequent scenarios:
(a)   Full combustion using all of the theoretical air;
(b)   Full combustion using all of the theoretical air;
(c)   Using 90% theoretical air; incomplete combustion (some CO in the products).

Solution

From the first law of thermodynamics:

$$\sum N_R \left[\bar{h}_f^o + \left(\bar{h} - \bar{h}^o \right) \right]_R = \sum N_P \left[\bar{h}_f^o + \left(\bar{h} - \bar{h}^o \right) \right]_P$$

Case (a); Complete Combustion with 100% Theoretical Air:

$$CH_4(g) + 2\,O_2 + 7.52N_2 \rightarrow CO_2 + 2H_2O + 7.52N_2$$

$$\sum N_R \left[\bar{h}_f^o + \left(\bar{h} - \bar{h}^o \right) \right]_R = 1(-74850) + 0 + 2 \times 0 + 7.52 \times 0 = -74850$$

$$\sum N_P \left[\bar{h}_f^o + \left(\bar{h} - \bar{h}^o \right) \right]_P = 1\left[-393520 + \left(\bar{h}_{CO_2} - \bar{h}_{298} \right) \right] + 2\left[-241820 + \left(\bar{h}_{H_2O} - \bar{h}_{298} \right) \right]$$

$$+7.52\left[0 + \left(\bar{h}_{N_2} - \bar{h}_{298} \right) \right] = -74850$$

Using iteration technique (trial and error), AFT $\approx$ 2330 K.

Case (B); Complete Combustion with 300% Theoretical Air

$$AFT \approx 1140\,K$$

Case (C); Incomplete Combustion

$$CH_4 + 0.9 \times 2\,O_2 + 0.9 \times 7.52\,N_2 \rightarrow a\,CO_2 + b\,CO + 2\,H_2O + 6.768\,N_2$$

Balances of oxygen and carbon yield: a = 0.6 and b = 0.4
Consequently, AFT $\approx$ 2210 K.

Final Remarks on the Adiabatic Flame Temperature (AFT)

(1)   Excess air has a great effect on the value of the adiabatic flame temperature.
(2)   Incompleteness of combustion does affect the adiabatic flame temperature, but to a far less extent than the excess air.
(3)   Excess air is a strong tool to control and monitor the temperature of the combustion products.

2.6 Problems

1. After 171,000 miles, the engine of the family station wagon, "Old Betsy," eventually gave out when Becky was driving it. It is reasonable to conclude that during its lifetime, it traveled 40 mph on average at 1700 RPM. A five-liter V8 engine running on a four-stroke cycle powers the vehicle. Compute:

   (a) The engine's total number of rotations.
   (b) How many spark plug firings have happened throughout the engine as a whole?
   (c) The total number of intake strokes in a single cylinder?

1. 4-stroke, 4-cylinder spark ignition engine with 84 mm bore, 71.5 mm stroke, compression ratio $= 9.1{:}1$, rated power 85 Hp at 5600 rpm, mechanical efficiency 85%. Find the indicated mean effective pressure.

2. A gasoline engine, 4-stroke, 4-cylinder with 65 mm bore, 63.5 stroke rated power 37 Hp at 5000 rpm, $r = 9{:}1$. If the volumetric efficiency is 80%, find the air consumption in kg/min, if the atmospheric pressure and temperature is 1.013 bars, 15 °C respectively.

3. A spark ignition engine, 4-stroke, 4-cylinder running at 4000 rpm, with brake thermal efficiency of 28% and brake horsepower 107. Air to fuel ratio 16.5:1 and volumetric efficiency 75%. If the fuel lower heating value $= 4300$ kJ/kg, find:

(a) Bore and stroke assuming (L/D=1.1).
(b) Brake mean effective pressure (bmep).

4. A six cylinder, four-stroke compression ignition engine of bore 9 cm and stroke 12.5 cm was found to exert a torque of 225 Nm when tested at a speed of 1800 rpm. Fuel of lower heating value 41,200 kJ/kg was consumed at the rate of 0.18 kg/min, and air at the rate of 3.9 kg/min. If the ambient air pressure is 102 kPa and the temperature is 32 °C. Calculate the volumetric efficiency and brake thermal efficiency of the engine. What will the brake mean effective pressure?

5. A 12-cylinder, four-stroke diesel engine is connected directly to a generator to provide electricity. Efficiency of the generator is 0.95. Engine speed is 300 rpm, bmep is 6 bar, cylinder stroke to diameter ratio is 1.5, and the engine brake thermal efficiency is 0.33. In the event that the engine uses 12,000 liters of fuel per day with a specific gravity of 0.75 and a heating value of 41800 kJ/kg. Find the electrical output of the generator in kW (a).

   (b) The stroke length and cylinder diameter.
   (c) Engine ideal air consumption, the ideal air-to-fuel ratio based on atmospheric conditions 1 bar and 27 °C.

6. A multi-cylinder reciprocating diesel engine, an intercooler (I), a power turbine (TP) geared to the engine drive shaft, a turbocharger (with a compressor C and turbine TT mechanically coupled to each other), and other components make

up a diesel system. At the designated sites, the flow patterns for fuel and gas as well as the gas statuses are displayed. Assume that k = cp/cv = 1.333 and that the gas's specific heat at constant pressure, cp, is 1.2 kJ/kgK throughout the system. Engine speed when running is 1900 rev/min.

The fuel has a lower heating value of 42 MJ/kg of fuel. What is the power (in kilowatts) which the turbocharger turbine (T_T) must produce? What is the gas temperature at the exit to the turbocharger turbine? What is the power turbine power output?

7. At 2000 RPM, a diesel engine with four cylinders and a two-stroke cycle, measuring 10.9 cm in bore and 12.6 cm in stroke, generates 88 kW of brake power. Ratio of compression: r = 18:1. Compute:
(a) Engine displacement (cm^3, lit).
(b) Brake mean effective pressure (kPa).
(c) Torque (Nm).
(d) Clearance volume of one cylinder (cm^3)..

8. A 2.4-L, four-cylinder engine running at 3200 RPM uses a four-stroke cycle. The connecting rod length is 18 cm, the compression ratio is 9.4:1, and the relationship between the bore and stroke is L = 1.06 D. Compute:
(a) One cylinder's clearance volume, measured in cm^3, illuminated.
(b) Inches for the bore and stroke.
(c) In m/s, the average piston speed.

9. A 3.5-L, five-cylinder SI engine running at 2500 RPM uses a four-stroke cycle. At this point, the engine's mechanical efficiency is 62%, and each cylinder produces 1000 J of indicated work every cycle. Calculate:
(a) Indicated mean effective pressure (kPa).
(b) Brake mean effective pressure (kPa).
(c) Friction mean effective pressure (kPa).
(d) Brake power in kW and hp.
(e) Torque (Nm).

10. 48 kW of brake power are produced by a 1500 cm^3, four-stroke cycle, four-cylinder CI engine running at 3000 RPM. The air–fuel ratio (AF) is 21:1, and the volumetric efficiency is 0.92. Compute:
(a) Airflow Rate into Engine (kg/s).
(b) Fuel usage (gm/kWh) particular to brakes.
(c) The exhaust flow mass rate (kg/hr).
(d) Brake output (kW/lit) per unit of displacement.

11. A five-liter V6, SI engine running at 2400 RPM powers a pickup vehicle. The volumetric efficiency ηv = 0.91, the compression ratio rc = 10.2:1, and the relationship between the bore and stroke are as follows: stroke L = 0.92 D. Compute:
(a) Length of stroke (cm).
(b) The piston's average speed (m/s).
(c) One cylinder's clearance volume (cm^3).
(d) Engine air flow rate (kg/s).

12. A tiny two-stroke cycle, single-cylinder SI engine with a volumetric efficiency of $\eta v = 0.85$ runs at 8000 RPM. The engine has a displacement of 6.28 cm^3 and is square (bore = stroke). F/A, or fuel-to-air ratio, is 0.067. Compute the following:
(a) Average piston speed (m/s).
(b) Air intake rate into engine (kg/s).
(c) Fuel flow rate into engine (kg/s).
(d) The amount of fuel used per cycle (kg/cycle).
13. Operating at 800 RPM, a single-cylinder, four-stroke cycle CI engine with a 12.9 cm bore and an 18.0 cm stroke consumes 0.113 kg of gasoline in four minutes and produces 76 Nm of torque. Compute:
(a) Brake specific fuel consumption (gm/kWh).
(b) Brake mean effective pressure (kPa).
(c) Brake power (kW).
(d) Specific power (kW/cm^2).
(e) Output per displacement (kW/lit).
(f) Specific volume (lit/kW).
14. What is the difference between saturated and unsaturated hydrocarbon?
15. What are isomers?
16. What is the general formula for alkanes, and what are its advantages?
17. How is the distillation curve obtained?
18. What is the flash point, and what is its important?
19. What is the vapor lock phenomenon?
20. What will happen if the volume of gasoline fuel which vaporizes up to 70 °C becomes very small?
21. Which fuel needs more additives, diesel or gasoline?
22. Calculate the theoretical air–fuel ratio on a mass and mole basis for the combustion of ethanol, C_2H_5OH.
23. Consider a combustion process of decane, $C_{10}H_{22}$, and air, the dry product mole fractions are 86.9% N_2, 1.163% O_2, 10.975% CO_2, and 0.954% CO. Find the equivalence ratio and the percent theoretical air of the reactants.
24. A certain fuel oil has the composition $C_{10}H_{22}$. If this fuel is burned with 150 theoretical air, what is the composition of the products of combustion?
25. On a dry gravimetric basis, Pennsylvania coal comprises 74.2% C, 5.1% H_2, 6.7% O_2, ash, and trace amounts of N_2 and S. This coal is fed, along with steam and oxygen, into a gasifier. On a mole basis, the composition of the departing product gas is determined to be 39.9% CO, 30.8% H_2, 11.4% CO_2, 16.4% H_2O, and trace amounts of CH_4, N_2, and H_2S. For every kilogram of coal, 100 kmol of product gas are produced. How much steam and oxygen are needed?
26. Based on a dry gravimetric examination, a sample of pine bark has the following final analysis: 5.6% H_2, 53.4% C, 0.1% S, 0.1% N_2, 37.9% O_2, and 2.9% ashes. This bark will be burned in a furnace with 100% theoretical air to serve as fuel. Calculate the mass-based air–fuel ratio.

27. Dry air is used to burn liquid propane. The volume percent composition of the combustion products, measured on a dry basis, is 8.6% CO_2, 0.6% CO, 7.2% O_2, and 83.6% N_2, according to a volumetric study. Calculate how much of the theoretical air was utilized in this combustion process.

28. After burning a fuel, CxHy, in dry air, the composition of the product is determined to be 83.1% N_2, 7.3% O_2, and 9.6% CO_2. Determine the percentage of theoretical air consumed and the fuel composition (x/y).

29. The moisture content of many coals from the western United States is high. Take into consideration the following Wyoming coal sample, for which the final analysis is done by mass:

Component	Moisture	H	C	S	N_2	O_2	Ash
% mass	28.9	3.5	48.6	0.5	0.7	12.0	5.8

   This coal is burned in the steam generator of a large power plant with 150% theoretical air. Determine the air-fuel ratio on a mass basis.

30. Pentane (C_5H_{12}) is burnt at 100 kPa constant pressure using 120% theoretical air. The goods are cooled to 20 °C, the room temperature. For each kilogram of fuel, how much mass of water is condensed? Assuming that the air utilized for combustion has a 90% relative humidity, repeat the response.

31. An integrated gasification combined cycle (IGCC) power plant's coal gasifier generates a gas mixture with the volumetric percent composition listed below:

Product	CH_4	H_2	CO	CO_2	N_2	H_2O	H_2S	NH_3
% vol	0.3	29.6	41.0	10.0	0.8	17.0	1.1	0.2

   After cooling this gas to 40 °C and 3 MPa, water scrubbers are used to eliminate the NH_3 and H_2S. Assuming that there is water in the resultant mixture that is fed to the combustors, find the mixture composition and the theoretical air-fuel ratio in the combustors.

32. After analysis, the volumetric composition of the hot exhaust gas from an internal combustion engine is found to be as follows at the engine exhaust manifold. 13% H_2O, 3% O_2, 10% CO_2, 2% CO, and 72% N_2. To remove the carbon monoxide, this gas is injected into an exhaust gas reactor and combined with a specific volume of air. It has been found that at the exit state, a mole fraction of 10% oxygen in the mixture will guarantee that no CO is left. What proportion of flows must enter the reactor?

33. Methanol, CH_3OH, is burned with 200% theoretical air in an engine, and the products are brought to 100 kPa, 30 °C. How much water is condensed per kilogram of fuel?

34. Pentene, C_5H_{10}, is burned with pure oxygen in an SSSF process. The products at one point are brought to 700 K and used in a heat exchanger, where they are cooled to 25 °C. Find the specific heat transfer in the heat exchanger.

35. A SSSF combustor is filled with butane gas, C4H10, and 200% theoretical air at a temperature of 25 °C. At 1000 K, combustion products are released. Determine how much heat is transferred from the combustor to each kilomol of butane burnt.

36. After burning liquid pentane (C_5H_{12}) in dry air, the following dry basis products are measured: 10.1% CO_2, 0.2% CO, 5.9% O_2, and rest N_2. Determine the fuel's enthalpy of formation and the real equivalency ratio.

37. At 25 °C and 200 kPa, there are initially 2 kmol of carbon and 2 kmol of oxygen in a rigid vessel. At a temperature of 1000 K, combustion takes place, producing 1 kmol of carbon dioxide, 1 kmol of carbon monoxide, and surplus oxygen. Determine the final pressure in the vessel and the heat transfer from the vessel during the process.

38. Liquid hydrazine (N_2H_4) and oxygen gas are fed into a combustion chamber in a ratio of 0.5 kg O_2/kg N_2H_4 during a rocket propellant performance test at 100 kPa and 25 °C. It is expected that 100 kJ/kg N_2H_4 of heat will be transferred from the chamber to its surroundings. Determine the temperature of the products exiting the chamber. Assume that only H_2O, H_2, and N_2 are present. The enthalpy of formation of liquid hydrazine is + 50,417 kJ/kmol.

39. Heptene C_7H_{16} burns in an SSSF burner when air and fuel are introduced as gases at the standard reference condition. The products of the mixture—which contains 125% theoretical air—go via a heat exchanger and are chilled to 600 K. Find the heat transfer from the heat exchanger per kmol of heptane burned.

40. In a gas turbine, ethene (C_2H_4) and propane (C_3H_8) are burned as gasses in a 1:1 mol ratio with 100% theoretical air. Fuel is introduced at 25 °C and 1 MPa, and air is mixed with it after being compressed from the environment at 25 °C and 100 kPa to 1 MPa. Because of the turbine's operation, the output temperature and pressure are 800 K and 100 kPa, respectively. Find the mixture temperature before combustion, and the turbine work, assuming it is adiabatic.

41. One fuel substitute for natural gas or petroleum is ethanol (C_2H_5OH), which is often made by fermenting grain. Imagine a process of combustion where liquid ethanol is burned in an SSSF process with 100% theoretical air. At 25 °C, the reactants enter the combustion chamber, and at 60 °C and 100 kPa, the products emerge. Calculate the heat transfer per kmol of ethanol, using the enthalpy of formation of ethanol liquid to be − 257,347 kJ/kmol.

42. Natural gas, considered to be 90% methane and 10% ethane (by volume), and 100% theoretical air each enter the furnace at 25 °C and 100 kPa. The products, thought to be 100% gaseous, exit the furnace at 40 °C and 100 kPa. The furnace is a new, high-efficiency furnace. What is the heat transfer for this process? Compare this to an older furnace where the products exit at 250 °C, 100 kPa.

43. Methane, CH_4, is burned in a SSSF process with two different oxidizers: **Case A**: Pure oxygen, O_2 and **case B**: A mixture of O_2 + x Argon. The reactants are supplied at the reference state, and the products should be at 1800 K for both cases. Find the required equivalence ratio in case (A) and the amount of Argon, x, for a stoichiometric ratio in case (B). ($C_{p(Ar)} = 0.52$ kJ/kg, $M_{Ar} = 29.948$).

44. At the reference state, an engine is filled with a mixture of liquid ethanol and octane, with a mole ratio of 9:1. It also contains stoichiometric air. Utilizing the enthalpy of combustion, the engine expels 30% of its energy as work, 30% as heat loss, and the remaining energy as exhaust. Find the work and heat transfer per kilogram of fuel mixture and also the exhaust temperature.

45. Butane gas C_4H_{10} at 25 °C is mixed with 150% theoretical air at 600 K and is burned in an adiabatic SSSF combustor. What is the temperature of the products exiting the combustor?

46. In a rocket, hydrogen is burned with air, both reactants supplied as gases at the reference state. The combustion is adiabatic, and the mixture is stoichiometric (100% theoretical air). Find the dew-point temperature of the products as well as the adiabatic flame temperature.

47. A stoichiometric mixture of benzene, C_6H_6, and air is mixed to form reactants flowing at 25 °C, 100 kPa. Find the adiabatic flame temperature.

48. A gas turbine is sprayed with liquid n-butane at 25 °C and primary air flowing at 1.0 MPa, 400 K in a stoichiometric ratio. The products are at the adiabatic flame temperature after full combustion, which is too high. Therefore, secondary air is injected at 1.0 MPa, 400 K, resulting in a mixture that is at 1400 K. Show that AFT > 1400 K and find the ratio of secondary to primary air flow.

49. Consider the gas mixture fed to the combustors in the integrated gasification combined cycle (IGCC) power plant having the following volumetric percent composition:

Species	CH_4	H_2	CO	CO_2	N_2	H_2O	H_2S	NH_3
% Vol	3	29.6	41.0	10.0	0.8	17.0	1.1	0.2

   If the adiabatic flame temperature should be limited to 1500 K, what percent theoretical air should be used in the combustors?

50. Acetylene gas at 25 °C, 100 kPa is fed to the head of a cutting torch. Calculate the adiabatic flame temperature if the acetylene is burned with:

   (a) 100% theoretical air at 25 °C.
   (b) 100% theoretical oxygen at 25 °C.

51. Ethene, C_2H_4, burns with 150% theoretical air in a SSSF constant-pressure process with reactants entering at 25 °C, 100 kPa. Find the adiabatic flame temperature.

52. A study is to be made using liquid ammonia as the fuel in a gas-turbine engine. Consider the compression and combustion processes of this engine.

   (a) Air enters the compressor at 100 kPa, 25 °C and is compressed to 1600 kPa where the isentropic compressor efficiency 87%. Determine the exit temperature and the work input per kmol.
   (b) Two kmoles of liquid ammonia at 25 °C and x% theoretical air from the compressor enter the combustion chamber. What is x if AFT is to be fixed at 1600 K?

53. A food processing factory's wet biomass waste is fed into a catalytic reactor where it undergoes an SSSF process to transform it into a low-energy fuel gas that can run the boilers in the processing plant. On a volumetric level, the fuel gas is composed of 50% methane, 45% carbon dioxide, and 5% hydrogen. Determine the lower heating value of this fuel gas mixture per unit volume.
54. Determine the lower heating value of the gas generated from coal, which has the following volumetric percent composition:

Products	CH4	H_2	CO	CO_2	N_2	H_2O	H_2S	NH_3
% Volume	3	29.6	41.0	10.0	0.8	17.0	1.1	0.2

   Do not include the components removed by the water scrubbers ($C_{p,CH4} = 2.254$ kJ/kg K).
55. Determine the higher heating value of the sample Wyoming coal, which has the following ultimate analysis on an as-received basis, by mass:

Component	Moisture	H	C	S	N_2	O_2	Ash
% mass	28.9	3.5	48.9	0.5	0.7	12.0	5.8

   (note $\bar{h}_f^0)_{SO2} = -296842$ kJ/kmol).
56. At 250 °C, blast furnace gas is accessible in a steel mill to be burned to produce steam. This gas's composition is as follows, measured in volume:

Component	CH_4	H_2	CO	CO_2	N_2	H_2O
% volume	0.1	2.4	23.3	14.4	56.4	3.4

   Find the lower heating value (kJ/m^3) of this gas at 250 °C and ambient pressure.

Acknowledgements The authors appreciate the support received for the preparation of this book from the Deanship of Research Oversight and Coordination (DROC) at King Fahd University of Petroleum & Minerals (KFUPM). The support provided by the Interdisciplinary Research Center for Hydrogen Technologies and Carbon Management (IRC-HTCM) on project number INHT2409 is highly appreciated. Also, the support received through the KFUPM Consortium for Hydrogen Future through the projects numbered H2FC2309 and H2FC2315 is highly appreciated. The leading author of this book chapter wishes to extend the acknowledgement to the spirit of Professor Mohammed El-Kassaby, Alexandria University, for his help while being alive before he rested in peace.

References

1. C.R. Ferguson, A.T. Kirkpatrick, *Internal Combustion Engines*, 3rd edn. (Wiley, 2016)
2. W.W. Pulkrabek, *Engineering Fundamentals of the Internal Combustion Engine*, 2nd edn. (Pearson, 2014)

3. J.B. Heywood, *Internal Combustion Engine Fundamentals* (McGraw Hill, New York, 1988)
4. Y.A. Cengel, M.A. Boles, *Thermodynamics: An Engineering Approach*, 8th edn. (McGraw Hill, 2014)

Chapter 3
Fuel/Oxidizer-Flexible Lean Premixed Combustion

3.1 Why Lean Premixed Combustion

As a result of greenhouse gas emissions, one of the biggest issues facing the globe today is global warming (GHG). An essential one of the main causes of the increase in GHG emissions, mostly the carbon dioxide (CO_2), is the burning of fossil fuels. Energy-producing facilities utilize around 80% of fossil fuels for their operations [1]. In order to govern and minimize the greenhouse gas productions, the United Nations Framework Convention on Climate Change (UNFCCC) 21st Conference of the Parties, or COP21, in 2015 adopted the Paris Agreement [2]. The primary goal of this agreement was to cooperate in addressing global warming concerns by limiting the rise in average global temperature to far under 2 °C above pre-industrial points. In order to achieve this goal of minimal CO_2 emissions, industry and the electricity sector must switch to carbon-free fuels. Figure 3.1 from a research by The Global Carbon Project (GCP) illustrates the ongoing rise in CO_2 emissions. From 2018 to 2019, there was a 1.2% increase in total CO_2 emissions in a single year. The amount of carbon dioxide emitted from energy-related applications in 2019 reached a record-breaking 36.8 gigatons [3].

The "Global Energy Statistical Yearbook 2020," published by Enerdata, contains a comprehensive discussion of energy, fossil fuel, and emission statistics for various nations worldwide [4]. In 2017, gas turbines (GTs) generated almost 95% of the electricity in Saudi Arabia through the use of fossil fuels. Saudi Arabia is the world's biggest user of GTs since it ranks among the top 10 nations in the world for both CO2 emissions (534.3 Mt in 2019) and electricity production (350 TWh in 2019).

The annual carbon dioxide emissions in Saudi Arabia, resulting from the combustion of fossil fuels, are depicted in Fig. 3.2. CO_2 emissions rose by 4.2% annually between 2000 and 2019, with a 0.9% increase in just one year, from 2018 to 2019 [4]. These all emphasize how crucial it is to put the effective carbon capture and storage/sequestration into practice for the applications of gas turbines, which is the focus of the current review.

© The Author(s), under exclusive license to Springer Nature Singapore Pte Ltd. 2024

M. A. Nemitallah et al., *Hydrogen for Clean Energy Production: Combustion Fundamentals and Applications*, https://doi.org/10.1007/978-981-97-7925-3_3

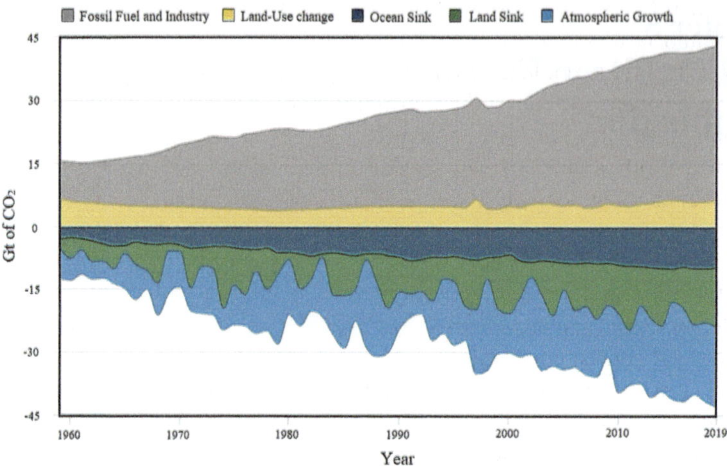

Fig. 3.1 Global carbon budget from 1959 to 2019 [3]

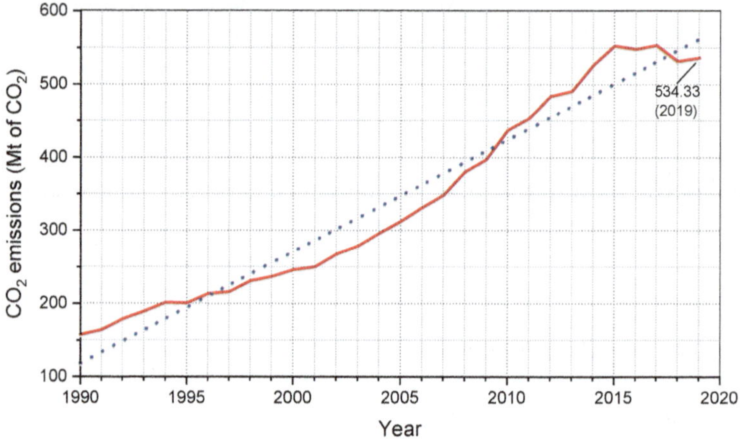

Fig. 3.2 Annual CO_2 emission from fossil fuel combustion in KSA from 1990 to 2019 [4]

There are a number of strategies to control the amount of carbon released into the atmosphere, including increasing the efficiency of industrial plants, consuming less fuel, and/or lowering the fuel's carbon content [5]. Carbon capture and storage/ sequestration, or CCS, technologies can lead to a large decrease in CO_2 emissions. The three basic processes of the various CCS technologies are the safe capture, transportation, and storage of CO_2. To prepare for the following stage, which is transportation, the CO_2 must first be captured, purified, and compressed [6, 7]. Transporting carbon dioxide may be done by ships, road freighters, or pipelines. Finally, the transferred CO_2 may be securely deposited in deep saline aquifers, washed-out oil fields for increased oil recovery procedures, or geographical rock located far below the Earth's surface. Three techniques—pre-, post-, and oxy-fuel combustion—can be used to capture CO_2 [8–11]. By turning fuel addicted to syngas before to entering the combustion region, the pre-combustion process captures some CO_2. Through the use of membrane separation, adsorption, absorption, or cryogenic techniques, the approach of the post-combustion extracts CO_2 from the exhaust/flue gases following combustion [12–15]. However, from the standpoint of carbon capture, using air as the oxidizer (air for combustion) is not advantageous because the main component of GT exhaust is unreacted nitrogen from the air, which lowers the concentration of CO_2 in it and complicates the CO_2 separation process, in addition to raising the linked expenses of carbon capture. The oxyfuel-combustion is the procedure of burning fuel by means of CO_2-diluted oxygen (as opposed to air). This method of oxidizer-flexible combustion method primarily produces H_2O vapor and CO_2 and [16–20], which may easily be recovered at the lowest cost once the latter is condensed out. The primary subject of this review will be oxy-combustion.

In an effort to lower NOx ($NO + NO_2$) emissions, GT manufacturers have transitioned during the past 50 years from conventional non-premixed combustion of air to lean premixed (LPM) technologies of combustion of air. For the purpose of producing clean and efficient energy, several businesses have created innovative, adaptable GT combustors. These include burners with perforated plates, enhanced vortex (EV), dry low NOx (DLN), and most recently, micromixer (MM) burners. Numerous reviews that compile the advancements in combustion technologies have been conducted in recent years. Table 3.1 condenses the summaries of the assessment works pertaining to oxyfuel combustion and highlights the key findings and areas of interest.

This research work goals to evaluate and relate the most recent advancements in gas turbine combustion technologies in order to identify the candidate or candidates that aptitude higher performance considering clean oxy-combustion environments by carbon capture. It is based on the dialogue mentioned above as well as information found in easily accessible literature. This inquiry aims to summarize the operability difficulties of oxy-combustion by exposing thoughts and submissions of LPM oxy-combustion and air-combustion, as well as the enactment and operability of several existing burner technologies with fuel/oxidizer flexibility. Comprehensive reviews and discussions are given to several fuel-flexible combustion methods for clean power generation, such as fuel mixing, ammonia combustion, syngas combustion, H_2 combustion, and H+-enriched combustion.

3.2 Lean Premixed (LPM) Air-Combustion

When it comes to combustion applications like GTs, NOx emissions are a big worry. NOx is formed during combustion by two main mechanisms: the thermal (Zeldovich) mechanism and the quick (Fennimore) process. As can be seen in Fig. 3.3, the Zeldovich mechanism's high temperature dependence on NOx generation accounts

Table 3.1 Outline of recent analysis schoolwork on oxyfuel-combustion in the gas turbines

Researcher(s) and year	Studied areas	Key findings
Nemitallah et al. (2018) [17]	Feasibility of different methods of combustion in gas turbines	• Enhanced-vortex (EV) burners can bring lowest NOx release with broader flammability limit • Higher cost and energy requirement for air splitting (oxygen-separation) in oxyfuel combustion is being main drawbacks of oxy-combustion
Portillo et al. (2019) [18]	Membrane tools for oxygen parting in oxy-combustion system	• Compared to conventional gas turbine combustion, low-emission oxyfuel-combustion power systems can save between 0.5 and 9% and 10.5 and 17.5% more energy and money • The use of oxygen-selective membrane technology can lower the increased cost of producing oxygen with cryogenic air separation units (ASU)
Nemitallah et al. (2019) [19]	• Gas turbine combustion techniques • Burner enterprises for the gas turbines	• H_2-enrichment in the combustion region increases efficiency of combustion and improves fuel economy; • Premixed oxy combustion may offer improved governing over exhaust and flame stabilization as compared to lean premixed (LPM) combustion • The micromixer (MM) burner exhibits greater stability close to blowout limits, whereas the advanced enhanced-vortex (AEV) burner has the widest flame stability overall.
Khallaghi et al. (2020) [20]	• Oxygen production, exhaust gas recirculation, and the impact of operating pressure in GT oxy-fuel combustion systems—oxy-combustion economics using gaseous fuels	• Compared to a traditional gas turbine cycle, the oxy-combustion cycle (Allam cycle) can yield a net efficiency gain of about 50% • A smaller combustion chamber is necessary when the exhaust gas recirculation ratio is lower, indicating that membrane separation may be a feasible substitute for ASUs in the production of O_2 • Using staged combustion can reduce the need for exhaust gas recirculation; conversely, cycle performance can be enhanced by higher operating pressure, albeit at the expense of increased pressure drop

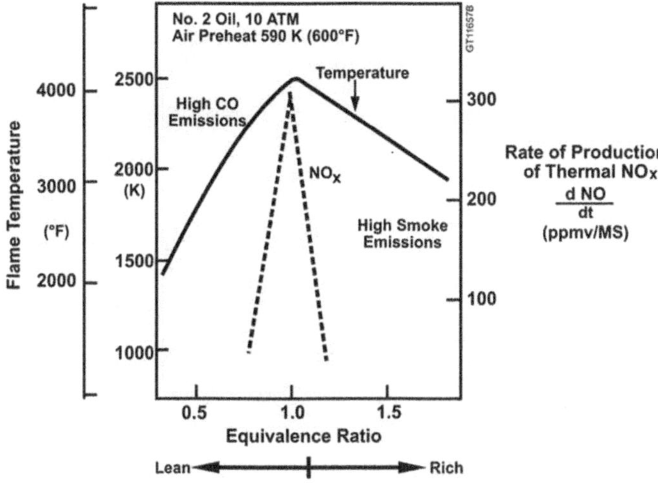

Fig. 3.3 Thermal NOx production rate as function of flame temperature and equivalent ratio [25]

for the term "thermal NOx." In the quick mechanism, when there is an abundance of fuel, the CH radical reacts with N_2 to create NOx [21, 22]. The method of water-injection was first used to regulate the production of NOx in the early 1970s. In order to lower the temperature of the flame and diminish thermal NOx discharges, this water was injected into the combustor. Later, the dry emission control approach, namely LPM combustion, was established as a consequence of the influence of water inoculation on arrangement performance, greater CO emission, and higher steam/ water necessities [23, 24].

3.2.1 Combustion and Emission Characteristics of LPM Systems

Figure 3.4 illustrates how the LPM combustors in GTs divide the inward compressed air expressively inversely from their non-premixed precursors. The later requires a smaller share of the compressor air due to its stoichiometric-rich primary flame zone; the bigger share is therefore used as subordinate air for lining cooling, combustion accomplishment, and downstream watering down. There are stoichiometric areas in non-premixed flames where the local temperature can rise to a point where too much NOx can develop [26, 27]. As an illustration of how lean (and hence colder) the air is marked in Fig. 3.5, LPM combustors, in contrast, feed a larger percentage of the air of the compressor hooked on the flame, leaving less air for lining cooling. The production of thermal NOx is decreased by the LPM flame's lower temperature [28, 29]. Lower flame temperatures, however, may potentially result in insufficient combustion, leaving CO and unburned hydrocarbons in the exhaust stream. When

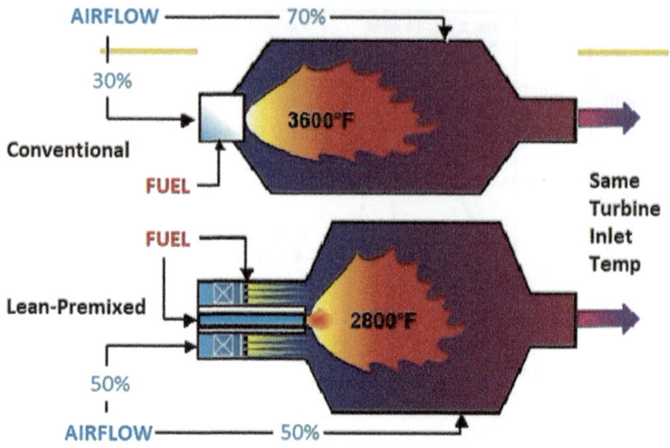

Fig. 3.4 Non-premixed versus lean premixed combustion systems [31]

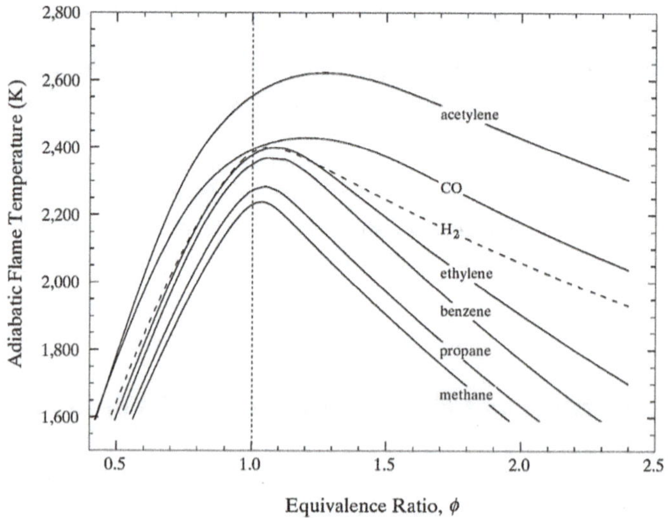

Fig. 3.5 Presentation of sdiabatic flame temperature versus equivalence ratio for some air–fuel mixtures at standard inlet temperature and pressure [32]

building an LPM GT combustor and during its operation, the CO/NOx trade-off depicted in Fig. 3.3 above should be kept in mind [26, 30]. Table 3.2 provides a comparison of LPM and non-premixed flames.

Table 3.2 Differences between lean premixed and non-premixed flames

Type of flame	Attributes	Advantages	Disadvantages
Lean premixed	Fuel and oxidizer are premixed in advance of inflowing the combustor Keeping the equivalence ratio lean	• Significantly lower NOx emissions	• Stability limits are narrow • Turndown ratio*is lower
Non-premixed	Oxidizer and fuel are supplied separately into the combustion zone	• Broader stability limits • Higher turndown ratio[*]	• Excessive NOx emissions

[*] Turndown ratio refers to the operation range of the burner, which is the ratio between maximum and minimum loads of the burner

3.2.2 Flame Instabilities in LPM Combustion

LPM flames are susceptible to instabilities of combustion [33, 34], both static (flashback and LBO) and dynamic (acoustic) instabilities, despite having reduced NOx emissions. At some resonant frequencies, there may be excessive pressure fluctuations as a result of the constructive feedback among the heat-release pressure field and the acoustic field inside the combustor. These combustion instabilities have the potential to negatively impact engine performance and ultimately result in hardware failure.

At temperatures close to the LBO limit, LPM GTs function. Therefore, to stabilize the primary LPM one and stop its blow-off, a fuel-rich pilot flame is provided during load transitions and at low loads [26]. Inadequate anchoring has an impact on LPM flames and can also cause flame blow-off. This is denoted to as static instability or lean flammability [35, 36]. Flashback into the premix ducts happens when the speed of the incoming combustible mixture is less than the flame's burning speed [37, 38]. This is an additional instance of static instability. Thus, one possible cause of flashback is a reduction in air velocity brought on by compressor surge. Another type of flashback is premature auto-ignition. To prevent premature auto-ignition within the premixing region itself, the flow residence time within the premixing region must be shorter than the ignition suspension time of the produced combustible mixture [39].

Combustion instabilities are generally regarded as the biggest problem with LPM combustors [40, 41], and many previous research have looked into them. In a model LPM GT combustor, Yoon et al. [42] looked into combustion instabilities at various inlet mixture velocities. They ascribed the structure and frequency of the fluid dynamical vortex to the beginning of the unstable flame at lower inlet velocities. Bollinger and Williams [43] investigated the effect of the Reynolds number in the range of 3000–35,000 on the turbulent flame speed and demonstrated that the turbulent flame speed in the tube of the burner is a function of the Reynolds number. Higher combustor pressures were investigated by Liu et al. [44], who also calculated the velocity of the turbulent burning at various constant Reynolds numbers (between 6700 and 14,200).

They noticed that when the pressure rises, the turbulent burning velocity drops in a negative exponential way.

3.3 Existing Burner Technologies in LPM Gas Turbines

There are several ways to cut back on NOx emissions. The moist low NOx methodology was main one of the initial approaches [45]. It is possible to reduce the flame temperature and regulate NOx emissions by adding steam or water to the combustor. Because of the higher flow rate, the turbine power output can also be improved simultaneously. Demineralizing the water was not practical or appropriate for numerous combustion applications, therefore it had to be done before using the extra water or steam in the combustion zone. The Original Equipment Manufacturers (OEMs) created the dry LPM combustion notion—which eliminates the need for steam or water—to address this problem. This section examines the many burner technologies now in use that were created with the dry LPM combustion principle in mind.

3.3.1 Dry Low NOx (DLN) Burners

The OEMs created the Dry Low NOx (DLN) LPM GT combustion systems, also known as Dry Low Emissions (DLE), to reduce flame temperature and restrict NOx materialization through lean operation. A two-stage premixed DLN combustor is depicted in Fig. 3.6. Although NG is the main fuel, liquid fuel can also be used in this combustor to get good performance. These commercially available systems can reduce NOx emission to 15–25 ppmvd (parts in each million in a volume basis, dry) when using natural gas as fuel [26]. By premixing the oxidizer and fuel preceding to it entering the combustor, DLN systems prevent the formation of high-temperature stoichiometric regions and the subsequent generation of thermal NOx. This process is based on the principle of LPM combustion. Even at maximum load, this results in significantly lower flame temperatures [46]. However, there is a greater chance of flame breakout and controlling CO emissions might be challenging. Primary (premixed) and secondary (non-premixed) fuel circuits are features of DLN systems. To create a pilot flame, just around 3% of the fuel is inserted in the non-premixed mode. The residual 97% of the fuel is supplied into the swirler through the primary circuit after being premixed upstream with air [47]. As seen in Fig. 3.7, the first cohort of DLN systems (DLN-1) operated in four distinct modes: (i) primary manner; (ii) lean-lean manner; (iii) secondary manner; and (iv) premix manner [26]. Table 3.3 provides a list of these four modes' descriptions.

The venturi and protruding centerbody of the DLN-1 system have been detached from the combustor in the second cohort of DLN schemes (DLN-2). This removed the requirement for these two parts to have cooling air. According to Fig. 3.8, the dual-mode, single-staged combustion scheme with five fuel nozzles is referred to as

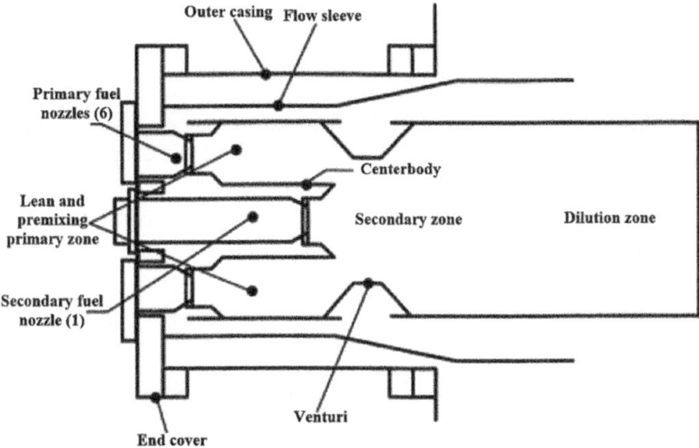

Fig. 3.6 Schematic of DLN burner [48]

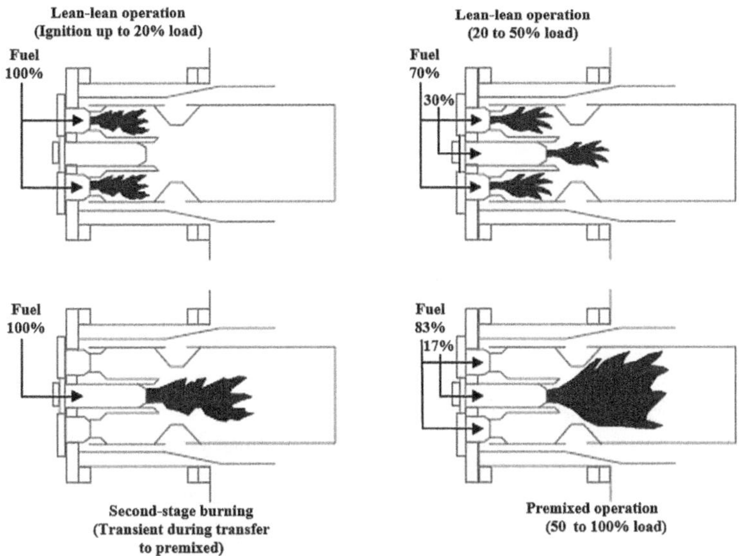

Fig. 3.7 Different operating modes of DLN-1 combustor [48]

DLN-2. DLN-2 can run on liquid fuels in addition to gaseous fuels. When employing gaseous fuel, the combustor operates in the premixed mode for loads between 50 and 100% of the total load after operating in the non-premixed mode for loads under 50%. In contrast, the combustor runs in the non-premixed mode at all loads while using liquid fuel. After that, DLN-2.6 was created, adding a sixth fuel nozzle to the five fuel nozzles of the DLN-2 system in the middle. In contrast to DLN-2, the DLN-2.6 system features extra loading and unloading algorithms and lacks a non-premixed

Table 3.3 Structures and applications of different modes of DLN-1 combustor

Mode	Features	Applications
Primary mode	• Flame remains in the primary stage only • All fuel is supplied through the main nozzles only	• Ignition up to 20% load • Operation and acceleration are restricted to a predefined reference combustion temperature
Lean-Lean mode	• Flames produced in both the primary and secondary stages • Fuel is injected through both primary and secondary nozzles	• To function in intermediate loads (20–50% load)
Secondary mode	• Flame in the secondary stage • All fuel is provided through the secondary nozzle only	• To generate a primary premixing region • For quenching the primary-region flame former to resupplying the fuel • For transition from Lean-Lean mode to Premix mode
Premix mode	• Flame is contained only in the subordinate stage • Fuel is inserted through both main and subordinate nozzles	• Ensures minimum emissions in the premixed mode • To function at middle to full loads (50–100% load)

mode [48]. General Electric has also produced a different DLN system called the H-system, or DLN-2.5. There are three ways to operate this system: full premix mode, piloted premix mode, and non-premixed mode [49].

The fuel and air splits in these systems need to be accurately managed in order to achieve stable combustion under all operating situations. The combustion stability can be significantly impacted by a variety of factors, including fuel calorific/

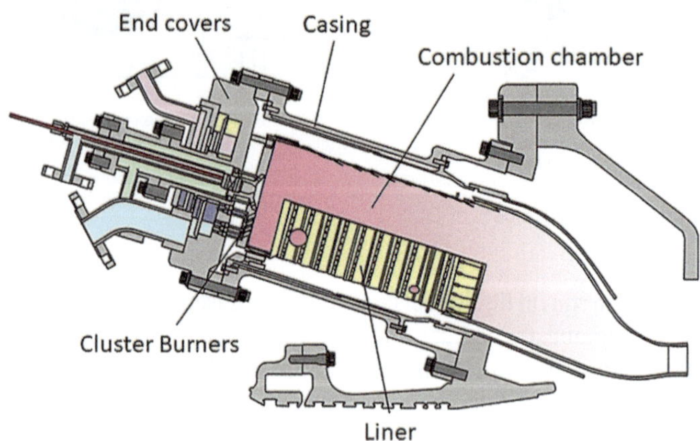

Fig. 3.8 DLN-2 combustion system [50]

heating values, fuel flexibility, ambient temperature, operating circumstances, turbulent chemical interaction, and flow rates [51]. Although the stability range of non-premixed flames is larger, the generation of NOx is notably elevated with higher splits of non-premixed fuel [52].

The DLN systems have operational problems with lean blow-off (LBO), flashback, and combustion instabilities [47]. Power outages and engine stoppage are possible outcomes of these problems. In the conditions of lean premixed of the DLN burners, protections should be reserved to avoid flame extinction [53]. When the fuel–air combination gets thinner, LBO may occur if certain safety measures are not taken. In order to get around this problem, a pilot flame is created, and some fuel is pumped into it in an unmixed state, with the remainder being premixed and burned inside the combustor. Flashback into the premix ducts happens when the velocity of the combustible mixture is less than the local flame speed [54]. Compressor surge-induced reduction in air velocity may potentially be the cause of flashback. Installing fast-responding fuel control valves in the combustor will lessen the impact of flashback. Another type of flashback is premature auto-ignition. To prevent premature auto-ignition within the premixing region itself, the flow residence time within the premixing region must be shorter than the ignition delay time of the produced combustible mixture [39]. As a result, the issue is to properly premix fuel with air in a brief amount of time. Premature auto-ignition can occur for a variety of reasons, including functioning at higher equivalency ratios (higher flame temperatures), adding more highly reactive fuel ingredients (like H_2), raising the temperature of air at the inlet of the combustor, and/or incorrectly estimating the residence time in the premixing zone. In DLN systems, this problem may harm the premixer modules. This auto-ignition problem sometimes necessitates frequent engine shut-offs.

3.3.2 Enhanced Vortex (EV/SEV/AEV) Burners

The limited fuel flexibility and flame temperature range of the DLN system operability resulted in the development of the subsequent cohort of premixing comustors, or Enhanced-Vortex or EnVironmental (EV) burners, as seen in Fig. 3.9. Instead of employing swirlers or center bodies, this burner type stabilizes the flame through vortex breaking up an intensive flow to create a strong inner recirculation zone [55]. The swirl flow design in the combustor is plainly seen in the cross-sectional view of the EV burner. A purely tangential flow is injected from two inlets, and as the flow becomes axial at the burner exit, the divergent burner cross-section causes vortex breakup. Therefore, by increasing the flow before the breakdown point and creating a strong inner circulation zone that stabilizes the created flame, flashback is prevented. Furthermore, the flame is correctly attached due to the aerodynamic fixing of the vortex breakup, removing the possibility of blowout. These are all the main characteristics of the flow configuration of the EV burner.

In order to further enhance the efficiency of the current EV combustion systems, sequential electric vehicle (SEV) burners have been created. As seen in Fig. 3.10 [59],

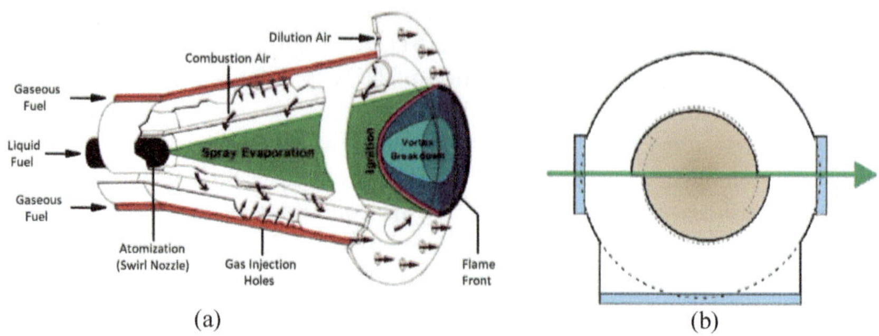

Fig. 3.9 Schematic of EV Burner: **a** components [56] and **b** cross-sectional view [57]

the basic idea behind SEV schemes is to burn the fuel in two phases, first in the annular combustion chamber with EV burners and subsequently in the second chamber with SEV burners [58]. Fuel is burned in the first step in order to warm the compressed air. The first stage's products are then expanded in a high-pressure (HP) turbine, where a two-fold pressure decrease occurs. The remaining fuel is then burned in the second combustor with the help of more cooling air, and the combustion products expand in a four-stage low-pressure turbine. The GT has a high power density since certain versions combine the DLN and SEV technologies into a single unit [60]. The compressor in SEV systems has the ability to produce compressed air at a pressure ratio that is nearly double that of traditional GT compressors. Furthermore, even with the high temperature after the HP turbine, the SEV burners may function without flashback thanks to the greater flame stability of the EV-burner technology, as previously mentioned [61].

Figure 3.11 shows the Advanced EV (AEV) burners, which are intended for use in combustion systems that burn liquid fuels. A nozzle for pouring liquid fuel inside the burner cone is located on a central lance. Unlike EV burners, which have two tangential input slots, AEV burners have four, allowing for a strong tangential flow that shields the cone walls from fuel droplets. To promote greater fuel and air mixing

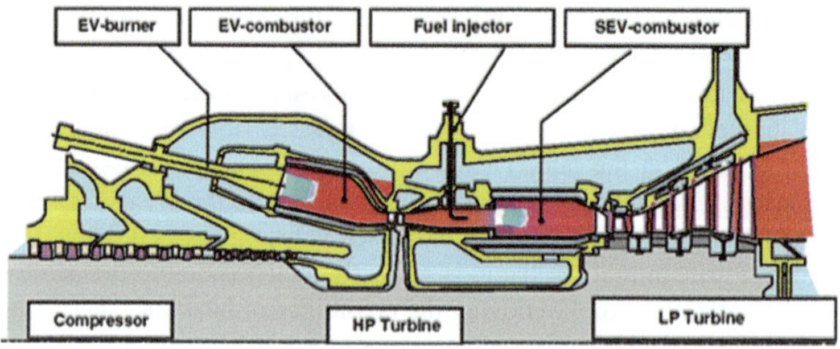

Fig. 3.10 Gas-turbine system utilizing SEV burner [59]

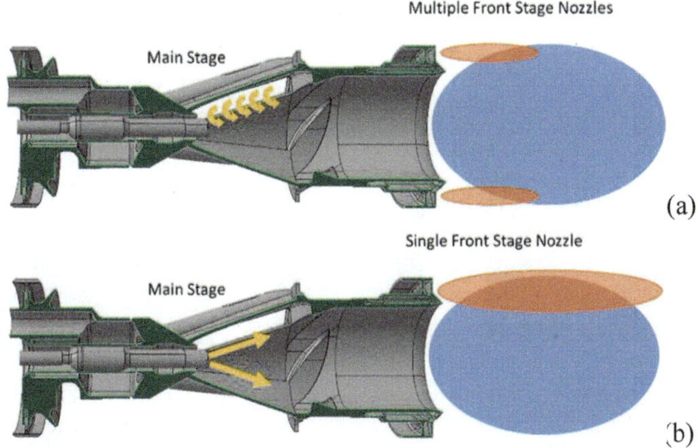

Fig. 3.11 Advanced EV (AEV) burner cross-section with: **a** gaseous fuel and **b** liquid fuel [62]

in the gas phase and a longer evaporation duration, a straight mixing tube is fixed to the cone. The centerline velocity close to the cone exit doubles the bulk velocity value as the tangential inlet flow changes to axial inside the cone. The purpose of maintaining this high axial velocity throughout the mixing tube is to avoid flame flashback. Furthermore, to lessen the chance of flame flashback at the mixing tube wall, extra air is utilized to raise the local axial velocity [45].

3.3.3 Perforated-Plate Burners (PPB)

Both residential and commercial settings frequently utilize perforated-plate burners, or PPBs. A PPB is made out of a platter with a variety of perforations that split the inward flammable mixture into several jets, as the name suggests. Figure 3.12 [63] illustrates how each jet burns in a tiny conical flame, also referred to as a "flamelet." Heat transmission to the burner faceplate and the width and space of the burner jets and holes determine the stability of the combustor. At the cone's departure, the centerline velocity doubles the bulk velocity as the tangential inlet flow changes to axial inside the cone. To avoid flame flashback, this high axial velocity is maintained throughout the mixing tube. Moreover, the local axial velocity is raised by using extra air.

Flame stability is dependent on hole space but not hole diameter, according to Rodrigues et al.'s [64] observations by means of a matrix of PPBs through both CH_4 and propane flames. Smaller hole spacing and steady flames have been noted. On a PPB holding laminar flames, Veetil et al.'s [65] numerical investigation of syngas-air combustion was conducted. For increased hole spacing, they have recorded stronger recirculation and higher flame lift-off. Jithin et al. [66] found that the flame stand-off

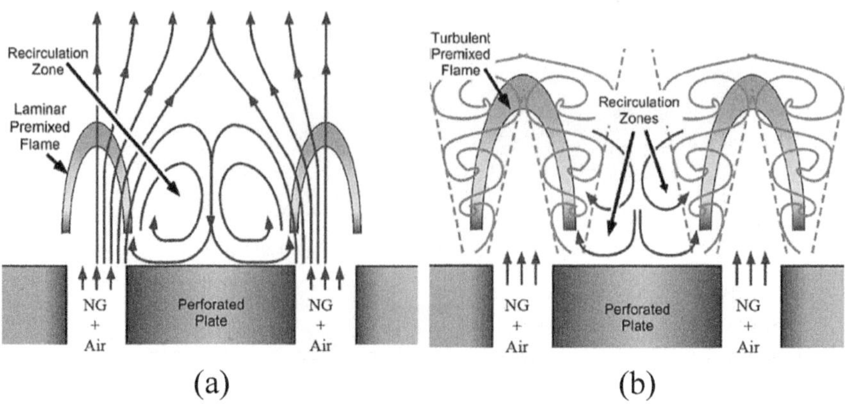

Fig. 3.12 Flow fields in perforated-plate burner of: **a** laminar flames and **b** turbulent flames [63]

remoteness rises with increased hole space when considering laminar air combustion on a PPB. The heat flux distribution from a multi-hole burner carrying methane-air (CH_4-air) flames has been studied experimentally and numerically by Hindasageri et al. [67], taking into account various hole layouts at low Reynolds numbers (Re). They have come to the conclusion that the distribution of heat flux is not significantly affected by the hole arrangement. However, larger hole pitch burners have produced higher heat flux at constant mass flowrates. However, the LBO stability criteria limits the maximum pitch. The thermal properties of PPBs in relation to the separation between the burner and an opposite impingement plate have been studied by Kuntikana et al. [68]. For their burners, they have utilized inline and staggered hole layouts. An rise in the combustible mixture's Reynolds number (Re) has resulted in higher thermal efficiency. The heat flux to the opposite plate distributes more uniformly at greater Re. Because of the highest flame temperature in stoichiometric circumstances, maximal heat transmission has been accomplished. Under fuel-rich operation, the heat flux has not altered, but the thermal efficiency has. Additionally, due of the ambient air entrainment, the study's thermal efficiency decreased as the distance between the burner and the opposite plate increased. According to Lee and Hwang [69], increasing the main hole width of a porous burner having baffle plate burne has been observed to upsurge the temperature and thermal efficiency. According to Moghaddam et al. [70], a multi-hole burner's combustor thermal efficiency increases when the diameter of the burner's primary hole is increased. The motives for blowout and stability of CH_4-air LPM flames fastened on a conductive PPB have been quantitatively investigated by Kedia et al. [71]. They have stated that heat transfer to the plate and flame stretching both have an impact on blowout. Timmermans et al. [72] used hot-wire anemometry and particle image velocimetry to analyze thermo-acoustic instability on a PPB. Their results demonstrate that low-frequency instabilities happen close to the wake in the areas where laminar and turbulent flow transition.

The influence of CO_2 dilution on flame stability under oxy-fuel combustion circumstances in a porous plate burner throughout ranges of equivalency ratio and oxidizer oxygen percent has been studied experimentally and numerically by Habib et al. [5]. According to their findings, as CO_2 dilution increases, the combustion temperature drops and flame fluctuations rise. As the flame moves closer to the combustor centerline, flame stability has improved at high equivalency ratios and flame length has grown at lower equivalency ratios. Researchers Rashwan et al. [73] looked into how reactant premixing level affected flammability limits. It is discovered that the zone of stable operation gets narrower at larger premixing ratios and widens at lower premixing degrees. Figure 3.13 illustrates that the limits of flammability of fuel–air flames are less than those of oxy-fuel flames. In a related study of Rashwan et al. [74], they discovered that the combustor may only be run at oxygen fractions over 29% (by vol.) for the case of partially premixed oxy-fuel flames stabilized across a PPB, and that the fraction of oxygen needs to be more than 36% to reach stable combustion. The limit for LBO for oxy-CH_4 of non-premixed combustion across a PPB has been studied by Habib et al. [75]. They discovered that, at a given firing rate, the LBO limit rises as the equivalency ratio does. A thorough analysis of the PPB's applicability to gas turbines has been conducted by Rodrigues et al. [76] When using gaseous fuel, they have seen this style of burner perform exceptionally well in terms of emissions and combustion efficiency.

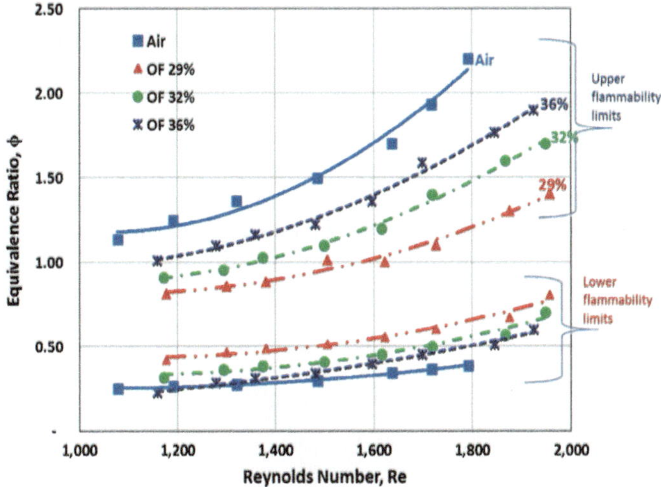

Fig. 3.13 Flammability limits of air–fuel versus oxy-fuel flames stabilized over a perforated-plate burner [73]

3.3.4 *Micromixer (MM) Burners*

A type of burner known as a micromixer (MM) burner mixes the fuel and oxidizer on a tiny scale. As shown schematically in Fig. 3.14, a gap, or plenum, is formed in a conventional MM burner by placing many millimeter-scale straight tubes between the front and back faceplates. Gaseous fuel is contained in the plenum between the straight tubes, which are used to provide the oxidizer. Fuel is injected into the oxidizer flow within each tube through tiny perforations, creating a cross flow configuration that resembles a jet. Each tube serves as a fuel-oxidizer premixer and provides a practically premixed fully grown jet at the front faceplate thanks to the placement of these perforations. Additionally, the distance needed for premixing might be affected by the location of the fuel holes. A hole diameter between one and ten diameters is typically used to describe the mixing length [77]. In the plenum, flow conditioners and baffles are used to guarantee a consistent fuel distribution through the micromixer holes. Combustion and emissions characteristics are mostly determined by the number of premixed jets (i.e., straight tube number) and the distance between these jets on the faceplate. To avoid flame flashback into the tubes, the equivalency ratio and oxidizer-flow pressure drop (Δp) within the MM need to be carefully chosen. MM burners are capable of handling fuel dilution and scaling and staging with ease [17]. Fuel adaptability is one of these burners' most important intrinsic qualities. MM burners are better suited to operate in a wider range of operational circumstances due to all these features.

An earlier version of MM burners is the lean direct injection (LDI) method. To achieve quick mixing in LDI, fuel is delivered into the oxidizer through a number of points. Thus, MM burners use the LDI method, although with a little premixing zone. With careful design, MM burners can be made to operate similarly to LDI. To reduce the dangers of flashback in H2 combustion, LDI has been employed [78, 79]. Three configurations were used: coflow mixing [81], jet-like cross-flow mixing [82], and swirl-based mixing in small cups with radial and axial inflow [80].

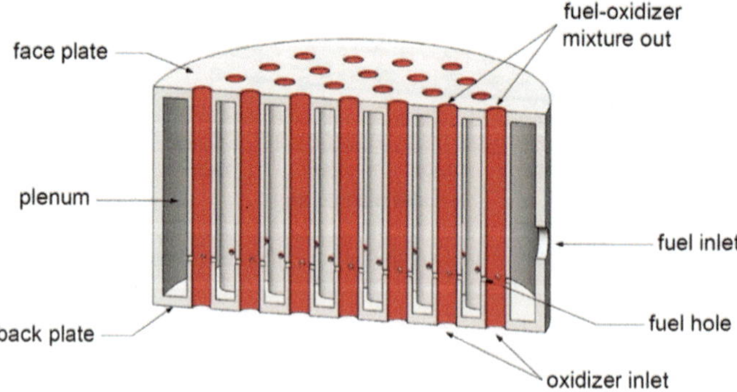

Fig. 3.14 Schematic diagram of a micromixer burner

A durability test of an MM burner was conducted by York et al. [77] under firing conditions similar to those of a full load F-class GT from General Electric. The test lasted for over 100 h. Air–fuel combustion has been applied to various NG, nitrogen, and H2 fuel mixtures. In the investigation, they used an H2/nitrogen fuel blend that included more than 90% (by vol.) H2 to produce ultralow NOx emission (single-digit ppm). The impact of momentum-flux ratio on flame anchoring and NOx emissions in an MM burner has been studied by Funke et al. [83].

They have shown a close association between the generation of NOx and the flamelet anchoring mechanism using air-H2 combustion. The MM of an integrated gasification combined cycle (IGCC) using H2-rich syngas-air combustion at 0.6 MPa medium pressure was investigated by Dodo et al. [84]. They have utilized fuels with three different carbon capture rates (0, 30, and 50%) that contained CH4, H2, and nitrogen. All three rates have demonstrated stable combustion, with single-digit NOx emissions recorded in each instance.

Figure 3.15 shows the GTCP36-300 tiny auxiliary power unit (APU), which uses the MM principle. It has 1600 little injectors and requires about 1600 kW of H2 burning to produce 335 kW of shaft power (21% efficiency) through the production of electrical and pneumatic power. An annular reverse-flow combustion chamber contains an integrated MM burner in the experiment [85, 86]. According to a different study, this MM combustor can operate on pure H2 as well as H2-rich syngas with minimal NOx emissions in both scenarios [87]. The GTCP36-300 APU was tested by Funke et al. [88] using an injector diameter of 0.84 mm. They conducted experiments using H2 fuel and compared the outcomes with the findings of Refs. [85, 86]'s 0.3 mm baseline case. The usage of 0.84 mm nozzles has been observed to result in an approximately 390% increase in power output in each nozzle. Figure 3.16 illustrates that, in all operational circumstances, NOx emissions are less than 4.0 ppm.

In order to examine the impacts of burner design and combustion modeling on flame shape, flow field, and temperature distribution, Ayed et al. [89] looked at a

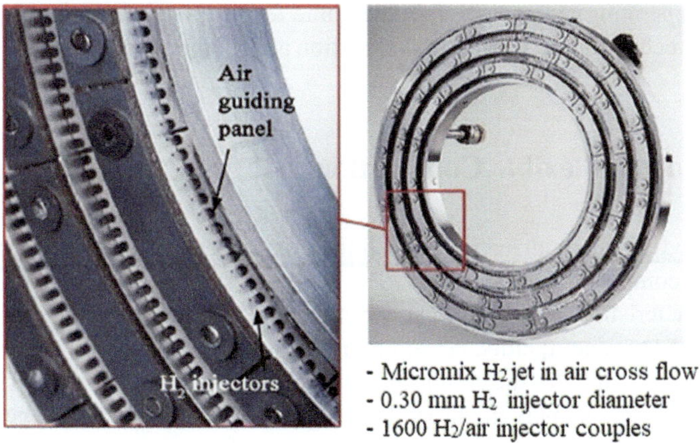

Fig. 3.15 Micromixer combustor for the GTCP36-300 APU [86]

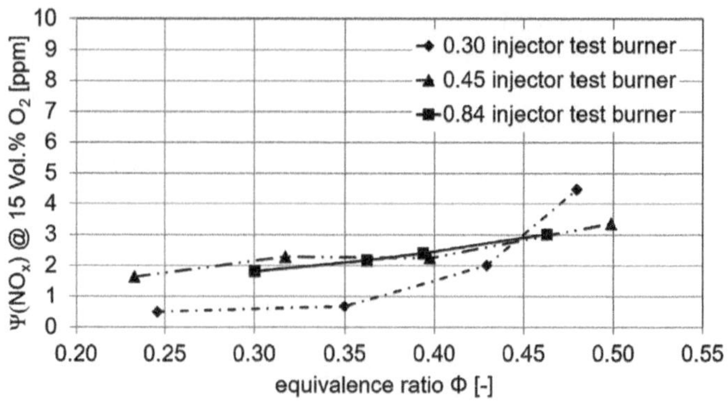

Fig. 3.16 NOx emission at different loads using injectors of different diameters in the GTCP36-300 APU [88]

low-energy–density burner configuration where the burner fuel injector diameter was 0.3 mm. They found that controlling the cooling air flow surrounding the flame could potentially lower NOx emissions for a given geometric parameter [90]. Their research has been used to design a burner with a high energy density and a 1.0 mm injector diameter. The heat rate increased as a result by more than 11 times for each injector. A comparable burner was investigated mathematically and experimentally by Funke et al. [91], who found reduced NOx emissions at the greater energy density. The development of MM burners in terms of flame structure, NOx generation, and geometric factors is an ongoing process. Using the heat from the combustion products, Badra et al. [92] have presented a combustor design for non-premixed flames that is based on the MM principle. One of the earliest investigations into oxy-fuel combustion with an MM-like burner was carried out by Aliyu et al. [93]. They have reported that, as shown in Fig. 3.17, the adiabatic flame temperature (Tad) is the primary parameter governing flame stability and structure. It has been discovered that when operating at constant Tad, the flame form stays unchanged. The various GT burner technologies covered in this section are compiled in Table 3.4.

3.4 Oxidizer-Flexible Combustion

The air used as an oxidizer in the majority of modern combustion systems has a constant composition of roughly 79% N_2 and 21% O_2 by volume. For better flame characteristics, increased heat-transfer efficiency, higher processing speeds, and improved product quality, certain systems use O_2-enriched air. O_2-enrichment also makes it possible for the combustion system to have more flexibility or a higher degree of freedom, namely the flexibility to adjust and regulate the oxygen fraction above the set 21% of air. As we'll talk about next, oxy-fuel combustion is an additional type of oxidizer-flexible combustion in which the oxidizer is CO_2-diluted oxygen.

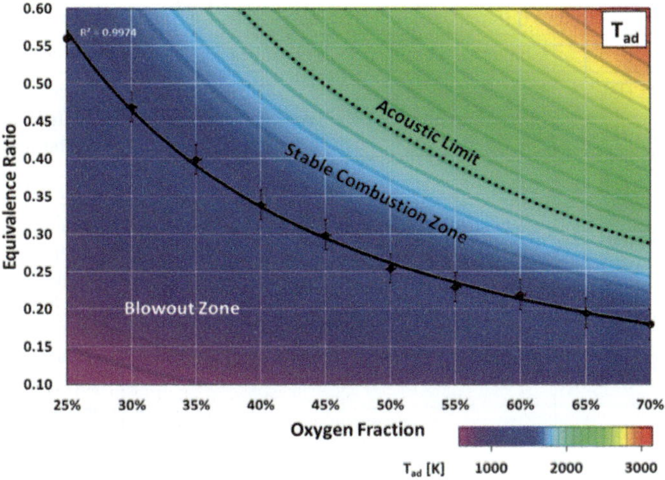

Fig. 3.17 Stability map of a MM-like oxy-fuel combustor operating at fixed jet velocity, plotted against the contours of constant adiabatic flame temperature [93]

3.4.1 Oxy-fuel Combustion Technology

In contrast to the carbon capture methods that include pre- and post-combustion, oxy-fuel combustion uses oxygen as an oxidant instead of air to produce an exhaust stream that is mainly composed of CO_2 and H_2O. This allows for more affordable carbon collection through straightforward H_2O condensation [47]. Figure 3.18's diagrams show basic oxy-combustion systems. However, using pure oxygen results in extremely high flame temperatures [94, 95]. Even the most robust materials ever discovered are challenged by firing temperatures beyond ~ 1300 °C (2370 °F) [26]. As a result, pure oxygen should not be used in GTs immediately [96]; instead, the oxy-fuel flame needs to be diluted. To preserve the natural simplicity of carbon capture, however, no additional substances than CO_2 and H_2O may be added to the exhaust stream. As shown in Fig. 3.18, this diluting effect of recirculating part of the effluent CO_2 [97] is achieved by diluting the oxy-fuel flame. However, adding CO_2 as a diluent instead of N_2 causes large concentrations of CO_2 in the flame. Since CO_2 is not as inert as N_2, it significantly affects the reaction mechanism and kinetics, as will be covered later. Higher CO_2 concentrations in the O_2/CO_2 oxidizer mixture cause an increase in CO emission [98], as seen in Fig. 3.19. However, if low-nitrogen fuels are utilized [101–103], oxy-fuel combustion is nearly NOx-free in the absence of air-based nitrogen [99, 100]. Therefore, oxy-fuel combustion with carbon capture becomes a true zero-emission technology if steps are made to limit CO emissions [104, 105].

Table 3.4 Comparison of different burner technologies for gas-turbine combustion applications

Burner type	Characteristics	Advantages	Disadvantages
Dry low NOx (DLN)	First implementation of LPM combustion	• Low CO and NOx emissions; advanced DLN burners can achieve NOx emissions < 10 ppm as well	• To create an air–fuel mixture upstream of the burner, an extended premix tube is needed, meaning it is less compact • Requires the use of a swirler or centerbody for flame stability; sensitive to flashback and premature auto-ignition • Instabilities in flames at low loads • At various phases, a non-premixed pilot flame is required (high NOx)
Enhanced vortex (EV, SEV, AEV)	Flame stabilization by vortex breakdown	• Reduced NOx emissions • Flame stabilization can be achieved without a centerbody or swirler • Reduced flashback as a result of increased core flow • Good performance is anticipated under oxy-fuel combustion conditions • Dual-fuel (gaseous/ liquid) capabilities (AEV only) • Non-premixed pilot flame is not required	• Variations in turbulent burning rate • Unsteady variation in the anchor point and flame surface

(continued)

Table 3.4 (continued)

Burner type	Characteristics	Advantages	Disadvantages
Perforated plate (PPB)	Fuel/oxidizer mixture supplied through small holes into the combustor	• Design simplicity • Used in both industrial and household applications • Flame symmetry	• Vulnerable to different combustion instabilities (e.g., flame quenching, blow-off, and flashback)
Micromixer (MM)	Fuel and oxidizer are premixed inside numerous small tubes then supplied to the combustion zone	• Capable of withstanding changes in fuel composition, i.e., superior fuel flexibility • Scalability and staging flexibility • Expected to function well in oxy-fuel combustion settings	• Carefully choose the operating conditions (equivalency ratio and Δp throughout the burner) to avoid flashback • Continuous development for usage in GTs; dynamic combustion instabilities close to stoichiometric condition

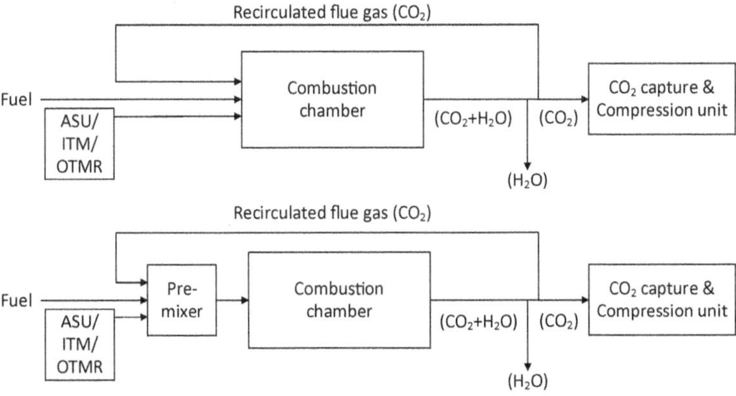

Fig. 3.18 Schematics of non-premixed (top) and premixed (bottom) oxy-combustion systems

3.4.2 Oxy-combustion in Model Combustion Systems

Oxyfuel combustion can lower fuel consumption rates in addition to lowering emissions in conventional systems like boilers and GTs. Studies on the burning of oxy-CH_4 in industrial water-tube boilers have demonstrated that the rates of fuel and oxygen consumption are less than those of burning air fuel [9, 98]. The stability of a non-premixed oxy-CH_4 flame inside a model GT combustor was investigated by Habib et al. [106]. They discovered that less than a minimum oxygen percentage

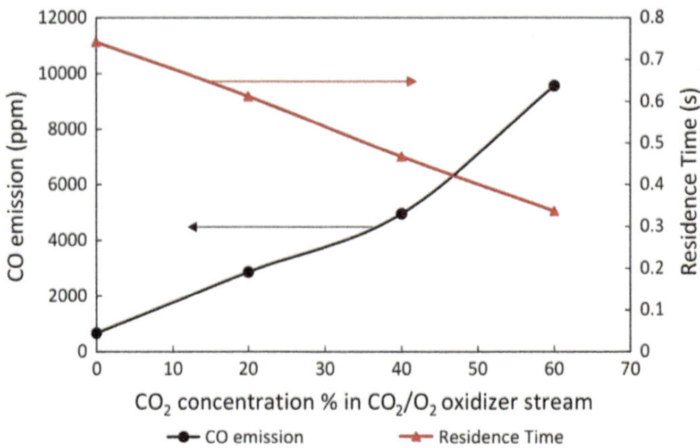

Fig. 3.19 Effect of CO_2 concentration in O_2/CO_2 oxidizer on CO emission and residence time [98]

of 21% is required to maintain a steady flame inside the combustor. It has been determined that the adiabatic flame temperature is the main factor controlling the stability and structure of premixed oxy-CH_4 flames [107]. This is found to be true for oxy-CH_4 ($CH_4/O_2/CO_2$) and oxygen-enriched air-CH_4 ($CH_4/O_2/N_2$) flames in a premixed swirl-stabilized model GT combustor, regardless of the oxygen percentage and equivalency ratio [108]. The stability of premixed oxy-CH_4 flames under fixed swirl, laminar flow velocity, bulk flow velocity, and adiabatic flame temperature was tested by Jourdaine et al. [109] and compared with that of air-CH_4 flames. Again, similar flame forms have been noted for both kinds of flames. Nonetheless, it has been discovered that the operability range of oxy-flames is lesser than that of air-flames.

Kutne et al. [110] conducted an experimental comparison of the stabilities of partially premixed oxy-CH_4 and air-CH_4 swirl flames in a model GT combustor. They discovered that, in contrast to the equivalency ratio, the oxygen fraction significantly affects flame shape and stability. Wider stability was observed in oxygen-fuel flames at higher oxygen percentages. A related study [111] that examined the effects of adiabatic flame temperature and equivalency ratio (Ø) revealed that the oxy-CH_4 flame remains in the outer shear layer (OSL) whereas the air-CH_4 flame vanishes at Ø = 0.6.

Figure 3.20's dotted sections demonstrate that while the oxy-flame continues in both the inner and outer shear layers, the air-flame only stabilizes in the inner shear layer. Because CO_2 and N_2 have distinct transport qualities, it has been observed that the extinction strain rate (resistance to extinction) at Ø = 0.6 is larger in the oxy-flame than in the air-flame. While the air flame extinguished in the OSL, the higher extinction strain rate assures the stability of the oxy-flame in the OSL. Up to Ø = 0.64, the behavior of oxy-flames stays the same; after that, an increase in Ø causes the extinction strain rate in the air-flame to grow and in the oxy-flame to drop [111].

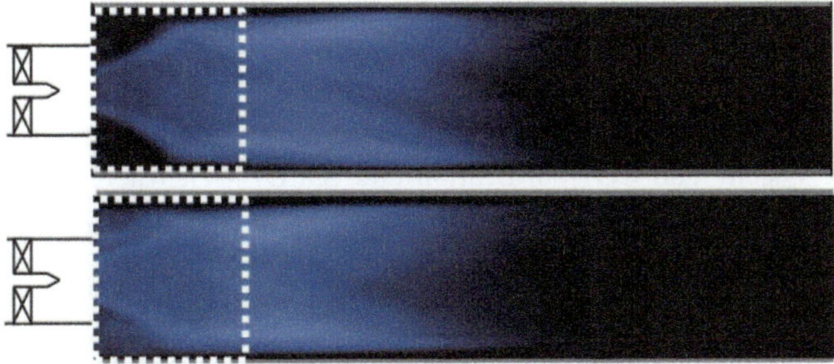

Fig. 3.20 Comparison of mean flow configuration in a swirl-stabilized combustor at $\emptyset = 0.6$ for air-CH$_4$ flame (top) and oxy-CH$_4$ flame (bottom) [111]

3.4.3 Comparison of Air-Combustion and Oxy-combustion Concepts

As the oxidant in air-combustion systems, air has a set chemical makeup (21% O2 and 79% N$_2$ by volume). N$_2$ primarily functions as a diluent, whereas O2 is the real oxidizer. Because it is unable to alter the N$_2$ to O$_2$ ratio and air is a cheap and plentiful resource, the only way to regulate flame temperature is by lean operation, which involves injecting surplus air. When the combustion system is functioning safely beyond of its LBO limit, the related surplus oxygen aids in the prevention of CO burnout. Thus, in air-combustion systems, the equivalency ratio serves as the primary control knob. However, in oxy-combustion systems—which use oxygen that is free of nitrogen—the situation is very different. In specialized air-separation units (ASUs), oxygen is collected from the air; however, these machines require power consumption, which results in a loss of efficiency and energy in power-generation applications [112]. Thus, extracted oxygen is a valuable resource that needs to be handled carefully, unlike air. As a result, oxy-combustion systems ought to be run in close proximity to stoichiometry; they are not permitted to run lean. One of the causes of the high CO emissions shown in Fig. 3.8 is that fuel must burn in the restricted amount of oxygen that is available. Flue gas recirculation (FGR) or exhaust gas recirculation (EGR) is the process of recirculating a significant amount of the effluent CO$_2$ back into the combustor to dilute the flame and control the temperature of the oxy-fuel flame [113]. Thus, the primary control knob in oxy-combustion systems is the oxygen percentage in the O$_2$/CO$_2$ oxidizer. Take note that while FGR is optional in air-combustion systems, it is required in oxy-combustion systems. In conclusion, there are significant differences between air- and oxy-combustion in terms of combustor designs and operability space/conditions [19]. Even with pre- or post-combustion carbon capture, air-combustion systems emit more NOx and CO than oxy-combustion systems do, on account of the inefficiency of these processes compared to carbon capture in oxy-combustion systems [114]. Since most oxy-combustion produces nox, CO is the single main output. Moreover, the lack of nitrogen in the exhaust, in addition to FGR, lowers the exhaust flow rate and, as a result, the equipment size used for exhaust treatment. FGR and H_2O condensation can also recover a significant amount of heat from exhaust gasses [115]. However, in terms of capital costs, oxy-combustion systems need extra units, including the ASU, FGR system, and CO_2 purification unit (CPU) [47], where effluent CO_2 is compressed and purified before being used or stored. These components are not normally present in traditional air-combustion systems.

From the standpoint of combustion, there are additional obstacles in the way of the effective implementation of oxy-combustion technology. In comparison to air-combustion, the higher CO2 content (~ 70% by mass) in the combustor lowers the chemical kinetics rates, which in turn lowers the laminar burning velocity [9, 116] and the combustion efficiency [19, 117]. When comparing the CO_2-rich oxy-fuel premixed reactant mixture to an equivalent air-combustion system operating at the same equivalency ratio, a stable flame is achieved with greater oxygen (above 21%) [118, 119]. A comparative numerical evaluation of a porous combustor indicates that oxy-combustion with an oxygen percentage of 30% by volume can produce an adiabatic flame temperature that is equivalent to air combustion under the same operating conditions [120]. Furthermore, because CO_2 is denser than N_2, dilution affects the jet velocity, gas density, and pressure drop [73, 121, 122]. Furthermore, compared to N_2, CO_2 has a higher heat capacity. This affects the flame stability by reducing the flame temperature and burning velocity [97, 123]. Gas transport features can also have an impact on the latter [124]. Soot production in the oxy-fuel combustion system has been found to be lower than in the air-combustion system [125, 126]. This reduction in soot formation can be achieved by increasing the amount of CO_2 in the reactor.

Using propane as fuel in a swirl-stabilized combustor, it has been discovered that the critical velocity ratio—the ratio of velocities between the fuel and the oxidizer—is six times higher in oxy-combustion than in air-combustion [124]. The oxy-propane flame immediately transitions from the "attached flame" to the "no flame" regime when the operating circumstances approach flame blowout because of this greater critical velocity ratio. In contrast, the flame in air combustion changes from "attached flame" to "lifted flame" and then enters the "no flame" regime. Irrespective of the equivalency ratio, these distinct behaviors take place [127].

3.5   Existing Oxy-combustion Power-Generation Systems

In order to meet the pressing demand for inexpensive, safe, dependable, and secure sources of clean energy on a worldwide scale, oxy-combustion with carbon capture offers a crucial avenue [128]. For instance, adding post-combustion carbon capture to an already-existing air–fuel application may cost an estimated 260 million euros in total up front, with an additional 43 million euros in operating and maintenance expenses per year. On the other hand, converting the same application from air combustion to oxy combustion with carbon capture needs a total capital investment of about 217 million euros, which is 16% less, and about 37 million euros in yearly operational costs, which is 14% less [129]. Compared to air–fuel systems with post-combustion carbon capture, which have a carbon-capture efficiency of 90%, oxy-fuel operation allows for 100% carbon capture [114]. However, the technique of oxy-combustion is still in its infancy. There are currently no large-scale oxy-fuel power plants; instead, oxy-fuel facilities are pilot-scale or intended solely for demonstration. However, as will be discussed later, the technology has been effectively tested on a range of fossil fuels. The fact that there will probably be a significant global reliance on fossil fuels for many years to come [130] indicates how promising it is to use oxy-combustion with carbon capture in large-scale zero-emission power plants (ZEPPs).

3.5.1 *Oxy-combustion of Gaseous Fuel*

A combined cycle of an NG-fired GT plant emits less than half the CO2 emissions per unit electricity of a similar coal-fired plant. More carbon footprint reduction is made possible by the development of carbon capture technologies. The Graz oxy-combustion cycle and the semi-closed oxy-combustion combined cycle (SCOC-CC) are the two contemporary gaseous-fuel oxy-combustion designs based on combined cycle [131, 132]. The Graz cycle uses steam from the steam cycle to cool the turbine and combustion chamber. Much of the exhaust gas that is cycled back into the combustion chamber contains the steam and CO_2 produced during combustion [133]. The SCOC-CC differs from the conventional combined cycle, as shown in Fig. 3.21, in that the gas cycle operates the combustion chamber nearly at stoichiometric conditions and employs pure O_2 as an oxidant in place of air to reduce the excess temperatures. Its five main elements are the following: the flue gas condenser; CO_2 compression; topping cycle; bottoming cycle air separation unit; and this SCOC-CC. The steam turbine, steam condenser, and heat recovery steam generator (HRSG) are components of the bottoming cycle; the gas turbine, combustion chamber, and compressor are components of the topping cycle. The gas turbine's waste gas is directed to the HRSG where it cools and produces steam for the steam turbine. In contrast to the Graz cycle, the SCOC-CC separates the exhaust gas's steam after the HRSG. After cooling, the flue gas enters the flue gas condenser, where the CO_2 and water vapor are separated. For transit, storage, and capture, the residual CO_2 is transferred to a CO_2 compression unit; approximately 93% of the CO_2 is returned to the compressor [133, 134].

NET Power has created a 50 MWth facility powered by high-pressure natural gas [135]. Its sole design goal is to capture all CO_2 produced when running on natural gas. The objective is to quickly create a demonstration facility that is less in size than the planned large-scale (500 MWth) power plant [135]. The demonstration facility's objective is to evaluate various control methods, 100% carbon capture capability, and system cycle operability limits. Recycled CO_2 and oxygen from an ASU are combined at the Allam 200-bar facility [135] and compressed to 200 bar. They have

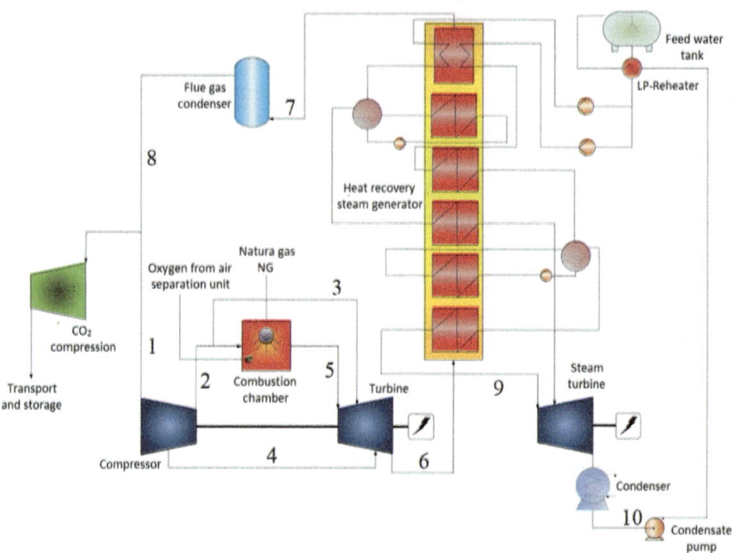

Fig. 3.21 Schematic diagram of a semi-closed oxy-combustion combined cycle (SCOC-CC [134]

shown that it is possible to run a non-premixed burner at supercritical pressure with a 15–30% oxygen percentage while still collecting the leftover flue gas. Similarly, while running on NG and other synthetic gases, a 20 MWth prototype power plant has been deployed to study performance and operability limits under various scenarios [136]. Two NG-fired combustion chambers are used by the burner; one serves as a gas generator and the other as a reheat combustor. The creation of the first generation of ZEPPs is the result of this technology's successful demonstration.

3.6   Fuel-Flexible Combustion in Gas Turbines

The capacity of a combustion application to provide satisfactory performance when run with varied fuels or fuel compositions is referred to as "fuel flexibility." Fuel-flexible systems include the dual-fuel GT. If an NG-fired GT is able to withstand NG composition changes, it is also regarded as fuel-flexible. Many variables can affect how a GT operates, including the kind and composition of fuel, the makeup of the oxidizer, the loading situation, the type of flame, the flame-stabilization mechanism, the design of the combustor, and the permissible amount of emissions. For GT applications, a great deal of research is being done to develop fuel-flexible combustion systems [137–140]. Manufacturers are investing in the development of flexible fuel-powered vehicles (GTs) that are adaptable to a variety of fuel types, loading scenarios, and flame modes. Both numerical and experimental research is done on the adaptation of LPM combustion to various fuels, the related combustion instabilities, and the ensuing emissions.

The fuel type should be chosen appropriately to sustain the ideal flame temperature. Variations in operating circumstances can result in varying flame temperatures across different fuels. The adiabatic flame temperature (Tad) values for a few chosen fuels under stoichiometric operation are listed in Table 3.5 [141, 142]. While designing the combustors, GT manufacturers pay close attention to Tad because it not only correlates with dynamic flame instabilities but also determines the combustion efficiency and the cooling needs of the combustor liner [19]. Tad also describes the combustor stability map. Experimental evidence has established that virtually identical flame macrostructures are seen in stable flames of the same Tad, even when the oxygen percentage and equivalency ratio are not maintained at constant levels [107, 108, 124].

Because different fuel compositions have distinct thermo-physical and chemical properties, changing the fuel composition can have an impact on both laminar and turbulent burning velocities [143]. The stability of premixed flames is significantly influenced by the laminar burning velocity [124, 144]. As stated in Table 3.2 above, non-premixed flames exhibit superior fuel flexibility since their operability range is broader and more flexible than premixed flames [26]. The dynamic and static (flashback and blow-off) combustion instabilities are also influenced by the composition of the fuel.

One crucial criterion in the characterization of fuel-flexible GTs is the Wobbe Index (WI). It is defined as the ratio of the fuel's square root of specific gravity to its lower heating value (LHV) per unit mass and is applicable to gaseous fuels. The latter can be acquired by calculating the fuel to air density ratio [145].

$$WI = \frac{LHV}{\sqrt{\frac{\rho_{fuel}}{\rho_{air}}}} \tag{3.1}$$

Since WI does not account for the temperature of fuel at combustor inlet, the Modified Wobbe Index (MWI) is introduced as expressed in Eq. (3.2) [146].

Table 3.5 Adiabatic flame temperature of some fuels [141, 142]

Fuel	Chemical formula	T_{ad} (K)
Methane	CH_4	2223
Propane	C_3H_8	2261
Hydrogen	H_2	2370
Carbon Monoxide	CO	2381
Syngas (50%CO-50% H_2 to 95%CO-5%H_2)	$(CO + H_2)$	2408–2415

$$MWI = \frac{LHV}{\sqrt{T_{fuel} \times \frac{\rho_{fuel}}{\rho_{air}}}} \tag{3.2}$$

The MWI is displayed for several industrial fuels in Fig. 3.22. The MWI map shows standard NG (mostly CH_4) in the center. At almost constant LHV, the MWI rises with increasing greater hydrocarbon (C2+) concentration. Because propane and butane make up the majority of liquefied petroleum gas (LPG), its MWI is greater. Conversely, with almost constant MWI, increasing the H_2 content (i.e., H_2 enrichment) raises the LHV. When CO and/or inert components (such as H_2O, CO_2, and N_2), such as syngas and coke-oven gas (COG), increase, both MWI and LHV drop.

The MWI ranges for the various industrial fuels are taken from Fig. 3.22 and are displayed in Table 3.6. Because the makeup of NG varies from place to place, the group as a whole lacks a distinct MWI value. Table 3.7 lists the WI and MWI for various gaseous pure-substance fuels. A large body of research views the WI or MWI as a means of validating and defending fuel-flexible operation [147]. Various low- or high-quality fuels can be used with GTs, depending on the situation and the fuel's availability. In addition to being a helpful tool for sizing the fuel circuit and injectors during the combustor design phase, the MWI offers a numerical evaluation of a GT's fuel-flexibility [148, 149]. For instance, the fuel system in a wide-Wobbe GT that runs on COG or NG must take into consideration the fact that far higher COG flow rates are required to deliver the same thermal input energy as NG. Therefore, for coke-oven-gas operation, larger fuel passageways and injection orifices are required.

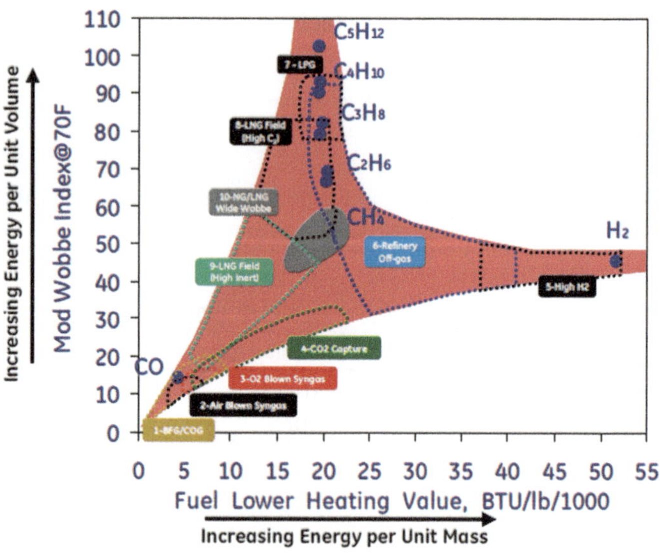

Fig. 3.22 Modified Wobble Index (MWI) of different industrial fuels [19]

It may be necessary to use separate systems for NG and COG if the orifices are too big to inject NG at a speed that allows for sufficient penetration and mixing into the airflow.

Literature on fuel flexibility of GT combustors is replete with studies. A variety of liquid and gaseous fuels, together with propane, H_2-enriched CH_4, and diluted CH_4, in addition to CH_4 as well as kerosene and ethanol as liquid fuels, have been evaluated in a distributed-combustion scheme with swirl flow [150–152]. The findings, obtained without altering the fuel injector or utilizing a spray mechanism, have nearly identical stability and emission features. Premixed oxy-CH_4 and premixed oxy-propane flames in the same combustor were evaluated in a recent investigation of a swirl-stabilized model of gas turbine combustor [153]. Irrespective of the fraction of oxygen and equivalency ratio values, both types of fuel have shown nearly indistinguishable flame morphologies at the same Tad, but Tad has a direct impact on flame speed [153]. Subsections on H_2, syngas, fuel-blended and ammonia combustion are included below.

Table 3.6 Variety of Modified Wobbe Index (MWI) for chosen available industrial fuels [145]

Fuel type	Series of MWI
Natural gas (~CH_4)	48–53
Syngas	24–29
High-H_2 fuels	40–48
LPG	72–87

Table 3.7 Wobbe Index (WI) and Modified Wobbe Index (MWI) values for different fuels [141, 143]

Fuel (Gaseous)	Wobbe Index (WI)	Modified Wobbe Index (MWI)
Butylene-1	88.5	82.5
Carbon monoxide	12.8	12.8
Ethane	68.2	62.5
Ethylene	63.8	60.0
Hydrogen	48.2	40.7
Iso-butane	92.0	84.7
Methane	53.3	47.9
n-butane	92.3	85.1
Propane	81.1	74.5
Propylene	77.0	71.9

3.6.1 Hydrogen and Hydrogen-Enriched Combustion

As was already indicated, the combustion properties and flame stability are impacted when the oxy-fuel flame is thinned by recirculated CO_2. Furthermore, the reaction kinetics are negatively impacted by raising the CO_2 content in the combustor [154]. The consequence was a drop in efficiency of combustion and laminar burning speed [117, 118]. However, enriching hydrocarbon fuels with H_2 can help to reduce some of these operational issues associated with oxy-fuel combustion by increasing the stability limits and improving the reaction kinetics [155]. In recent years, there has been an increased interest in using H_2 as a primary fuel, in addition to enriching frequently utilized fuels. The combustion of LDI H_2 micromix is a proven method. Clean exhaust free of CO_2, CO, and unburned hydrocarbons is produced by burning H_2 [77]. Pure H_2 combustion has even been shown to provide lower NOx emissions. [156]. In order to improve the GT's turndown and expand its operating limits, H_2 may likewise take the place of hydrocarbon fuel at small share loads [157]. Research reveals that the development rates of CO, NOx, and soot are reduced when H_2 is added to hydrocarbon fuels [158, 159]. A decrease in the rates of CH_2O and CH_3CHO generation [162], as well as the surface development rate of soot [160, 161], is said to be the reason why adding H_2 lowers aldehyde emissions. H_2-enrichment expands the flammability limits even further [94, 163], and H_2 addition in GTs can delay LBO [150]. The average and instantaneous characteristics of turbulent premixed flames may be significantly prejudiced by H_2-enrichment. This effect can be quite strong even with very little H_2 [164].

The response rate [166] and unstretched laminar burning velocity [165] can both be raised by H_2-enrichment. This is explained by the advanced molecular diffusivity of H_2 and the increased production of O, H, and OH radicals in H_2-rich flames [157]. When 20% H_2 is added to CH_4, there is a 20% rise in the mole fraction of OH [167]. The burning velocity of fuel-rich mixes remains unchanged when H_2 is added, while it increases in lean mixtures [168]. Additionally, the proportion of laminar to turbulent burning speeds (SL/ST), or the increase in ST larger than that of SL, improves, which is known as the combustion intensity [162, 169]. The addition of H_2 lessens the ignition delay time [170]. Cheng et al.'s proposed mixing law relates this improvement to the mole fraction of H_2 in fuel [171]. Because H_2 is more combustible than other gases, the reaction-zone temperature rises with increasing hydrogen fraction (HF) [107]. The warmer reaction region expands rapidly, recirculating comparatively cooler products hooked on the reaction region, which causes the overall adiabatic flame temperature to fall with growing HF [155].

An optimal swirl number of 0.51 was reported by Cheng et al. [172] after studying premixed swirl-stabilized H_2-air and H_2-N_2-air in an IGCC. At equivalency ratios of 0.17 and constant for varying bulk velocities, the LBO limit was reached. It has been discovered that the Damkohler number (Da) is useful for predicting the LBO limit. The LBO has been seen for various H_2/CO/CH_4 fuel mix compositions with a Damkohler number of 0.4 [173]. Experimental and numerical studies on the steadiness of premixed H_2-enriched oxyfuel-CH_4 flames in a model of GT combustor

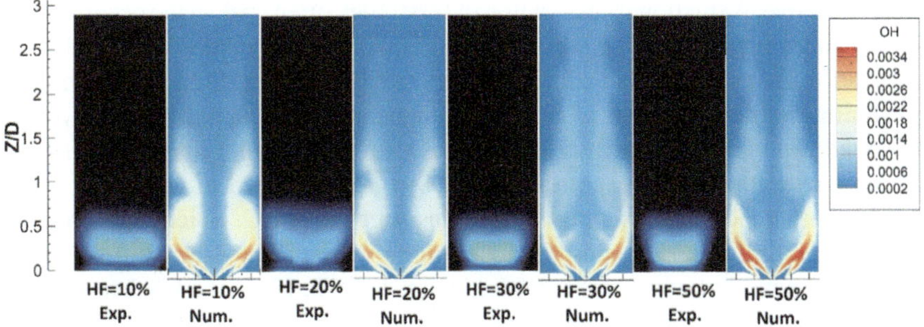

Fig. 3.23 Comparisons of experimentally captured and numerically calculated flame shapes at different HF and fixed oxygen fraction of 30% [174]

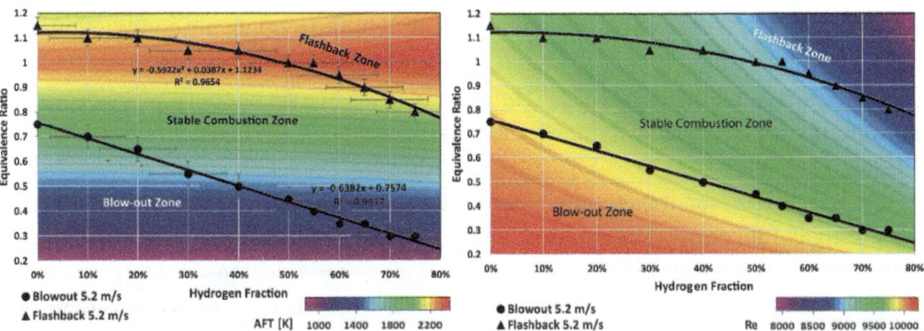

Fig. 3.24 Stability map of a model gas-turbine combustor over ranges of hydrogen fraction and equivalence ratio at fixed throat velocity of 5.2 m/s and fixed oxygen fraction of 30%, plotted on the contours of constant adiabatic flame temperature (left) and Reynolds number (right) [94]

have been carried out by the authors of the current work [94, 174]. As illustrated in Fig. 3.23, the computed flame shape derived from numerical simulations has been compared with experimentally obtained flame images. It has been discovered that H_2-enrichment reduces the likelihood of the flame quenching and raises the critical strain rate. Furthermore, as HF rises, the the Damkohler number rises and Karlovitz number falls. According to the study, the adiabatic flame temperature has less of an influence on the combustor's static stability limitations than the Reynolds number, as seen in Fig. 3.24.

3.6.2 Syngas Combustion

Syngas, which stands for synthetic gas, is mainly made up of CO and H_2 with varying quantities of CO_2, H_2O, and N_2. Syngas burns with a substantially different behavior than natural gas (mostly CH_4). Syngas has a larger H_2 content than NG, which makes it more reactive. A fractional gasification of fossil fuel, often coal, in the occurrence of oxygen and steam at a specific pressure and temperature can yield syngas. Following gasification, many harmful contaminants such as sulfur oxides (SOx), NOx, and H_2S are eliminated from the syngas. In order to produce CO_2-rich syngas—which is then converted to H_2-rich syngas by removing CO_2—fuel reforming is another step in the pre-combustion carbon-capture choice. Syngas is utilized in combined cycles with high efficiency and low emissions, for instance the IGCC, which profit from the high temperature of combustion that goes along with it. Water gas shift technique may likewise be applied to syngas to yield pure H_2 [175, 176]. The Fischer–Tropsch conversion is another method for turning syngas into liquid fuels [177, 178].

Syngas's combustion behavior is greatly influenced by its changing composition. Fuel-flexible GT combustors that run on syngas are therefore required. Syngas has a greater flame temperature, hence LPM burning should be used to regulate the temperature and thereby lower NOx emissions [179]. Higher adiabatic flame temperature is also produced by oxy-syngas combustion, much like syngas-air combustion does [180]. Lower CO content has been revealed to upsurge the laminar burning speed of syngas flames (see Fig. 3.25 [181]). The accurateness of a set of combustion processes and kinetics for describing syngas combustion has been studied by Alzahrani et al. [182].

The larger laminar flame speeds have been noted at advanced preheat temperatures, syngas fuel equivalency ratios, and H_2 content. It has been discovered, meanwhile, that as operating pressure increases, flame speed decreases. Delattin et al.'s research [158] examined the impact of pressure, equivalency ratio, H_2 content, and preheat temperature on NO emission during syngas combustion. In Fig. 3.26, various syngas combustion methods are taken into consideration together with changes in adiabatic flame temperature and NO emission with equivalency ratio. Because of

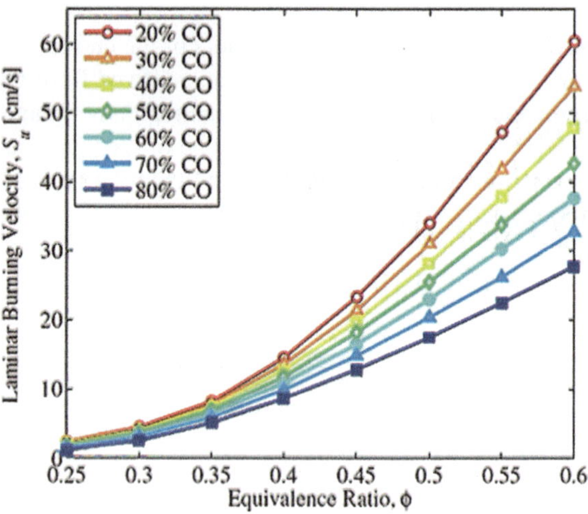

Fig. 3.25 Effects of CO content and equivalence ratio on the laminar burning velocity of syngas flames [181]

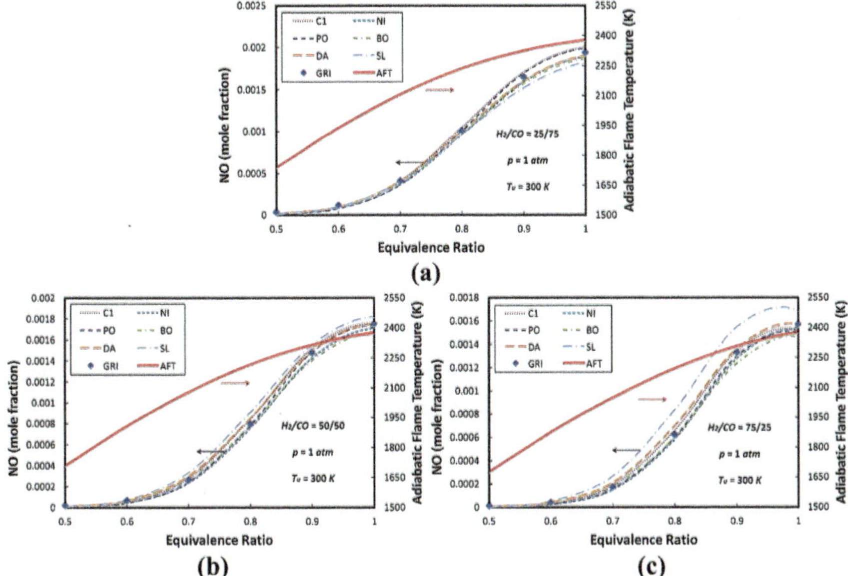

Fig. 3.26 Predictions of NO emission and adiabatic flame temperature at different equivalence ratios for H₂/CO syngas compositions of **a** 25/75, **b** 50/50, and **c** 75/25 at operating conditions of 300 K and 1 atm [182]

the rising fuel concentration in the mixture, which boosts the temperature during combustion, the mole fraction of NO always upsurges with equivalency ratio.

A substitute to coal-fired plants for power production such as the IGCC is the Allam-ZC cycle, which was suggested in a current research on syngas-fueled oxy-fuel power cycles. At turbine inlet parameters of 25 MPa per 1000 °C, condensation temperature of around 25 °C, and 0.245 kWh per kg O_2 specific power consumption of the ASU, the net efficiency of this Allam cycle has been reported to be 47.3%. When 100% of the carbon is captured, NOx emissions are found to be nearly negligible [183].

3.6.3 Ammonia Combustion

As demonstrated previously, using H_2 in combustion has several benefits and meets the primary emission standards for the production of electricity. H_2 transportation over vast distances and storage present certain difficulties, nevertheless. In contrast, ammonia (NH_3) is easier to store and transport because, conferring to the National Institute of Standards and Technology (NIST) database [184], it has a condensation pressure and boiling temperature that are similar to those of propane. Moreover, since 17.8% of the mass of NH_3 is H_2, it can be utilized as fuel. Nitrogen can be created by separating it from air and combining it with renewable H_2. NH_3 is generated in large quantities (180 Mt per year) [185] and is having a long history of usage as a raw material, in the fertilizer business, and as a refrigerant, thus its treatment, storage, and safety protocols are engrained. Even with these benefits, a few problems need to be resolved before ammonia can be fully utilized as a substitute fuel for gas turbines. Among the many negative aspects of ammonia combustion include increased NOx emissions, health and safety hazards, explosion hazards, and ammonia's corrosive properties [186, 187].

For gas turbine applications, current research has examined the combustion features of clean ammonia [188, 189] and mixtures of ammonia with further fuels, for example natural gas [188, 190–196] and hydrogen [191, 197–203], in varying fuel conformations. An experimental investigation on micro-GTs to generate energy from NH_3-based fuels, such as CH_4/NH_3 mixtures, kerosene/ NH_3 mixtures, and clean NH_3, was conducted by Kurata et al. [188]. Although the NOx emission in their study work is higher than that of a standard natural gas-fired gas turbine, it is nevertheless deemed acceptable in terms of emission rules. It is important to note that they have been using a certain fuel injection sequence, as seen in Fig. 3.27, to flinch the gas turbine each time. This sequence requires burning kerosene to drive the turbine before ammonia can be inoculated and burnt in the combustor. Rich-quick-lean combustion has been initiated to be more appropriate for lower emissions in additional work using a swirl-stabilized micro-GT combustor fueled by an $NH_3/$ CH_4 blend, resilient 2 ppmv CO, 49 ppmv NOx, and nearly negligible emissions of HCN, NH_3 and N_2O. With an NH3 proportion of less than 30%, the claimed combustion efficiency is 99.8%. NH_3-CH_4-air flames burn at a higher velocity than NH_3-air flames [195], although the NH3-air flames emit substantially less NOx. An increase in NH_3 mole fraction has been observed in NOx emissions in alternative experiment involving NH_3-CH_4-air combustion in a laminar premixed burner. Nevertheless, the NOx emissions began to decline at a mole fraction of 0.5 [196]. At the Fundamental Research Institute of Electric Power Industry in Japan, the combustion performance of a straight single-burner trial furnace co-fired with pulverized coal and NH3 has also been successfully tested [204].

In addition to natural gas, ammonia-enhanced hydrogen is likewise being researched for usage in gas turbines through combustion. The stability of combustion is significantly influenced by the ratio at which hydrogen and ammonia are combined. After experimenting with various blends, Valera-Medina et al. [201, 203] and Xiao et al. [202] testified unstable combustion at a 50%:50% ratio, though combustion is seen to be stable at a 70–30% NH_3-H_2 proportion by volume. Recirculating the exhaust gas can lower the NOx emission release with H_2-enriched ammonia, although this has little effect on the initial H_2 flame [201]. Moreover, the rise in inlet temperatures is responsible for increased NOx emission. Reduced flame size can also be caused by high inlet temperature [203].

Fig. 3.27 Start-up sequence for 50 kW model micro gas turbine with ammonia combustion [188]

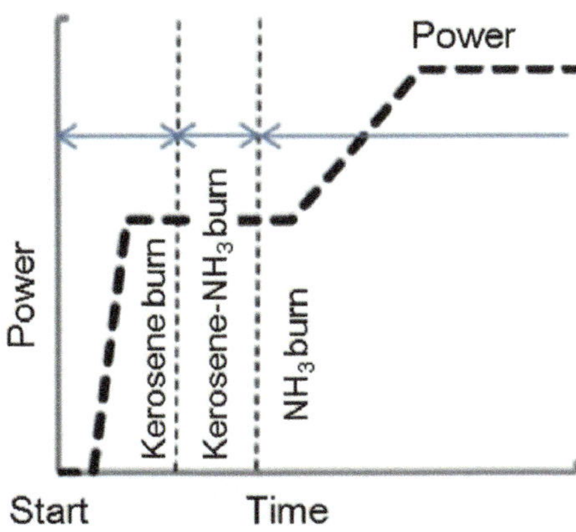

3.6.4   Fuel-Blended Combustion

The fuel blending is the process of combining two or more types of fuel in order to lower fuel costs and increase fuel efficiency. This method works particularly well in petrochemical and industrial facilities when a large supply of process gas byproducts, such hydrogen, is available. In order to improve combustion characteristics, increase efficiency of combustion, and lower CO_2 and pollutant discharged emissions, this progression gas may be utilized with natural gas. Commercially available fuels that are less expensive can also be used through blending. A basic diagram of fuel blending in applications of gas turbine is shown in Fig. 3.28. In this system, available process gas is combined with natural gas in a blending system, and the resulting blended fuel is then injected into the combustion chamber to produce energy. The 7FA gas turbine from General Electric, which uses DLN 3.6 technology, is a prime example of how to benefit from fuel-blended combustion [205].

Fig. 3.28 Schematic of implementing fuel blending in gas-turbine applications [205]

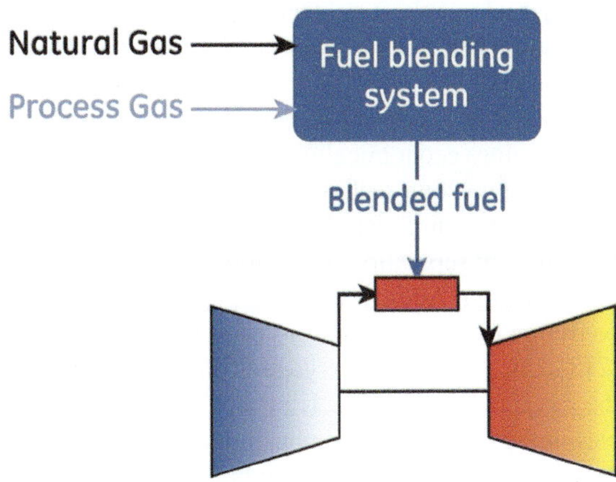

Over the years, an excessive portion of research has been done to study fuel blended combustion. Asfar and Hamed [206] examined the combustion performances of blends of kerosene and alcohol and diesel in a continuous-combustion system; they reported lower exhaust soot concentration and improved thermal efficiency. On the other hand, NOx emissions have somewhat increased. The influences of natural gas propane (NG-propane) blending on the emissions of pollutants in a premixed GT combustor have been studied in an experiment [207]. It has been noted that neither the equivalency ratio change nor adding an amount of 5% propane (by volume) significantly affected the amount of NOx emissions. An experimental investigation has been conducted on the performance of a micro-GT engine having 25 kW using mixtures of biodiesel and bioethanol [208]. In comparison to the reference scenario using only diesel, it is discovered that the thermal efficiency of the engine increases at a proportion of 80% to 20% of biodiesel-bioethanol; however, all emission echelons growth in this case, with the exception of NOx. In a compression-ignition engine, the combustion conduct of diesel fuel mixed with fresh vegetable oils, such as soybean, jatropha, and leftover cooking oils, has been investigated [209]. It has been noted that the exhaust-gas temperature measured while using pure diesel is lower than when using various diesel/oil mixtures. The mixes' CO emissions are realized to be minimal at near 80% load but to rise quickly at advanced loads.

Based on the examination of various GT burners and their ability to burn diverse gaseous fuel mixes above, it can be seen that the MM and swirl-stabilized burners perform well when using an oxidizer/fuel flexible strategy. The primary results of several tests using various oxidizers and fuels in MM burners and swirl-stabilized are listed in Table 3.8.

3.7 Future Perspectives

The near-zero emissions, enhanced flame characteristics, and superior fuel economy of oxy-fuel combustion technology have made it a viable contender for next-generation gas turbine technology. Even though there are currently a number of pilot plants [135, 216, 217] operating on oxy-combustion, more research into this technology is necessary to provide more better oxy-combustion power generation that are more economically and more viable than the currently accessible unadventurous combustion techniques. The numerous aspects of oxyfuel-combustion that this study has discovered should be the focus of future research, including cleaner fuel, oxygen separation that is more economical and power-efficient, and improved burner design. Additionally, existing power plants must be retrofitted in order to be converted from unadventurous power plants to the oxyfuel-combustion power plants [218]. Energy-intensive cryogenic air splitting items are typically utilized to extract the oxygen from air that is needed for oxy-combustion (ASUs). The main obstacles in the way of oxy-combustion application are the high related cost and energy requirement. Ion-transport membranes (ITMs) that are Oxygen-selective are used in membrane-based ASUs to extract oxygen from a heated air stream that cause heating

Table 3.8 Significant studies on fuel-flexible combustion in MM burners and swirl-stabilized

Burner type	Fuel	Combustion technique	Investigation method	Main findings	References
Swirl-stabilized burner	CH_4	Premixed oxy-combustion	Experimental	• Equivalency ratios between 29 and 70% showed stable flames • The flashback limit can be determined more effectively using reaction kinetic rate than using Re • As Re increases, the operation limit expands by shifting the blowout limit to leaner conditions	[210]
	CH_4	Non-premixed oxy-combustion	Experimental	• A steady flame was observed at $\emptyset = 0.65$ • Flame stability is significantly impacted by an oxygen fraction below 25% combustion is found to be impossible at oxygen fractions below 21%	[106]

(continued)

Table 3.8 (continued)

Burner type	Fuel	Combustion technique	Investigation method	Main findings	References
	CH_4	Non-premixed oxy-combustion	Experimental	• At higher °C and fuel flow rates with the same Re and oxygen percentage, the $CH_4/O_2/CO_2$ flame is reported to be extra stable than the $CH_4/O_2/N_2$ flame • The stability of the $CH_4/O_2/N_2$ and $CH_4/O_2/CO_2$ flames was comparable at increasing oxygen percentages	[211]
	H_2-enriched CH_4	Premixed air-combustion	Experimental	• Mixtures with varying fuel compositions exhibited a steady downward trend at lower Ø • Examined flame form in both acoustically connected and uncoupled states	[212]

(continued)

Table 3.8 (continued)

Burner type	Fuel	Combustion technique	Investigation method	Main findings	References
	H_2-enriched CH_4	Non-premixed oxy-combustion	Experimental and numerical	• H_2 enrichment improves flame stability; $\emptyset \geq$ 0.95 prevents H2-enriched oxy-combustion • It is not recommended to use combustion for vane angles higher than 75° • Swirlers having 70° or 65° vane angles showed superior stability than those with a 55° vane angle	[213, 214]
	H_2-enriched CH_4	Premixed oxy-combustion	Experimental and numerical	• At higher HF, a distinctive V-shaped flame develops • Greater CO emissions when HF increases • $\emptyset$ significantly affects Tad's flame form	[94, 174]

(continued)

Table 3.8 (continued)

Burner type	Fuel	Combustion technique	Investigation method	Main findings	References
	C_3H_8	Non-premixed air- and oxy-combustion	Experimental	• A stable flame is detected for oxy-C_3H_8 at swirl number of unity and close to stoichiometry • The critical velocity of oxy-C3H8 combustion is six times greater than that of air-C3H8 combustion • The critical velocity defines the transition from attached flame to lifted flame	[127]
	C_3H_8	Premixed oxy-combustion	Experimental	• The LBO limit conforms to the constant Tad contour • At the same Tad, the flame shapes are the same • A correlation between Tad and flame speed was created for the characterization of oxy-C_3H_8 flames; blowout and flashback happen in oxy-C_3H_8 flames at lower Ø than in oxy-CH_4 flames at immobile oxygen percent	[153]

Table 3.8 (continued)

Burner type	Fuel	Combustion technique	Investigation method	Main findings	References
	NH_3, NH_3–CH_4	Non-premixed air-combustion	Experimental	• To start the gas turbine at first, kerosene was used • NOx emissions are higher than in a typical combustion system but still within legal limits • The combustion stability is increased by higher air temperatures • Efficiency of combustion varies from 89 to 96% in combustion of NH_3-air • Combustion inlet temperature (CIT) preserved within the assortment of 460–580 °C • CIT normalizes the quantity of unburned NH_3 and NOx emission and at mid-range CIT both lean and rich NH_3 region co-exists • In NH_3–CH_4-air combustion NOx emission augmented with the upsurge in NH_3 but after reaching a certain point the NOx emission reduced with the rise in NH_3	[188]

(continued)

Table 3.8 (continued)

Burner type	Fuel	Combustion technique	Investigation method	Main findings	References
	NH_3–CH_4	Premixed and non-premixed air-combustion	Experimental and numerical	• Low emission with zero HCN, NH_3, N_2O, 2 ppmv CO, 49 ppmv NOx has been attained in fuel rich of NH_3–CH_4–air combustion with attaining combustion efficiency of around 99.8% • Lean-rich NH_3–CH_4-air combustion yielded inferior NOx emission as compared to that in NH_3-case	[195]
	NH_3–CH_4	Premixed air-combustion	Experimental and numerical	• CO emission discharges have been extra than the tolerable level at Ø from 1.15 to 1.25 • Fully premixed combustion technique is identified as unsuitable strategy for NH_3-CH_4 fuel inoculation	[215]

(continued)

Table 3.8 (continued)

Burner type	Fuel	Combustion technique	Investigation method	Main findings	References
	NH_3–H_2	Premixed air-combustion	Experimental and numerical	• Unstable combustion is exhibited at NH_3–H_2 relation of 50%:50% (%vol.) • At 50%:50% blend, stable flame is testified with Ø changing from 0.43 to 0.52 for 31 W of the net thermal power • Recirculation of flue gas abridged NOx emission at the equal time preserved initial H_2 flame	[201, 202]
	NH_3–H_2	Premixed air-combustion	Experimental and numerical	• A more stable flame has been seen at 70% and 30% (vol.) of NH_3 to H_2 • As the temperature of the contribution upsurges, the OH and NOx reactivity increases and the flame size decreases • Greater stability has been shown by the flame erection through NH_3 in the innermost region and H_2 in the outside region	[203]

(continued)

Table 3.8 (continued)

Burner type	Fuel	Combustion technique	Investigation method	Main findings	References
Micro-mixer burner	H_2	Premixed air-combustion	Experimental and numerical	• Crossflow mixing of H_2 and air has been done in micromixer • Inferior NOx emission is attained with intrinsic flashback protection	[89–91]
	CH_4	Premixed oxy-combustion	Experimental	• Tad is considered a determining factor for the stability map and flame form. The distribution of temperature inside the combustor has been observed to be determined by Ø. The oxygen fraction and Ø have an impact on the form of the flame	[93]
	NH_3-CH_4	Premixed air-combustion	Experimental and numerical	• NOx growths with the upsurge in NH_3 fraction in the fuel mixture up to $x_{NH_3} = 0.5$ • NOx emission discharges decreases with the decline in Ø	[196]

of the membrane over and above 650 °C and initiates oxygen permeation. Energy-intensive cryogenic air splitting components are typically used. The penetrated flux is determined by partial pressure differential of oxygen on each side of the two sides of the membrane, the thickness of the membrane, the operating temperature, the mass flow rates of the sweep and feed gases, and the membrane material and fabrication technique [219]. These membranes need to be tested in real power generation, even though the possibility of using ITMs on a small scale has been scrutinized in a numeral studies [18].

Energy-intensive cryogenic air separation units are typically used t The layout of the burner has a substantial effect on the parameters of combustion. Thus, the design of the burner and combustion chamber has a momentous impact on the flexibility of the fuel and/or oxygenator, as well as other combustion parameters. The micromixer (MM), one of the burner know-hows examined in the present research, is thought to be a good burner suitable for the premixed-oxy combustion of a wider range of flue flexible capabilities. The majority of this technology's study has been conducted with models built for the lab. In order to achieve a greater degree of technological readiness for usage in commercial gas turbines, it is necessary to evaluate the of MM burner performance in a whole gas turbine.

In addition to the proven CCS techniques, using cleaner fuels is crucial to halting the increase in CO2 emissions worldwide. Researchers are always investigating novel fuels that emit no emissions. Among them, hydrogen is thought to be the most environmentally benign fuel to burn. The hydrogen enrichment of gaseous fuels in GT combustion has been examined in numerous studies. Fuels enhanced with hydrogen, such syngas and further fuels with hydrogen blends, are gradually replacing natural gas [220, 221]. It is feasible to overcome combustion instabilities, increase operability limitations, and lower NOx emissions by blending hydrogen with conventional fuels [155]. However, burning pure hydrogen as fuel has some further advantages because hydrogen is a plentiful gas in nature and doesn't produce any dangerous greenhouse gasses or hazardous chemicals like NOx, SOx, CO, or CO_2. All that is left over after burning hydrogen is heat energy and water. Despite these benefits, the higher cost, complicated storage and transportation requirements, and extremely volatile characteristics of hydrogen combustion make it less accessible at this time [222]. Thus, it is important to continue researching the possibility of combustion of hydrogen inside the gas turbines in the future.

3.8 Concluding Remarks

A thorough analysis of several technologies of combustion and burner design configurations for lower-emission power production in gas turbines is presented in this chapter. The combustion properties of oxy-combustion and lean premixed (LPM) air-combustion are contrasted and addressed in relations of combustion and emission discharges. The advantages and disadvantages of a number of contemporary technologies of LPM gas-turbine burner that include enhanced-vortex, micromixer burners, dry low NOx, perforated-plate as well as their fuel flexibility and capacity to run more efficiently in oxy-combustion situations, are carefully investigated. The main focus is on flexible combustion of fuel and oxidizer and the related obstacles to its application in order to address the operability problems with oxy-combustion. As with air-combustion systems, it has been discovered that the adiabatic flame temperature (Tad) is the primary factor governing combustor operability under oxy-combustion settings. Thus, in compliance with the widely recognized industry standard among air–fuel gas turbine manufacturers, it is recommended that the future oxy-fuel gas turbine combustors be designed and operated based on adiabatic flame temperature, Tad, principally at medium and high loads away from the limit of blowout. Due to its high cost and energy consumption, separation of oxygen from air via air splitting is the primary obstacle in the way of the economical application of oxy-fuel combustion. The cost of oxygen separation can be reduced thanks to the extremely promising advancements in oxygen-separation membrane technology. Fuel flexibility is strongly advised in applications of oxy-fuel combustion in gas turbines in order to report the intrinsic operability problems associated with oxyfuel-combustion. This does, however, present additional difficulties for burner design in handling various fuel mixtures. Improved fuel economy, increased combustion efficiency, reduced pollutants, and improved flame stability have all been shown using hydrogen-enriched combustion. Because syngas has a flame temperature that is higher than that of natural gas, it requires the LPM process to burn.

The process of ammonia combustion is still in its infancy. However, ammonia exhibits a promising promise for reduced emissions as a hydrogen carrier. The fuel blending allows fuel with the right structure and characteristics to be burned at a higher temperature and with less emissions. Because of their exceptional fuel flexibility and flame stability, the micromixer burners and the enhanced-vortex burners are suggested for use with hydrogen enrichment applications in upcoming oxy-combustion gas turbines.

Acknowledgements The authors appreciate the support received for the preparation of this book from the Deanship of Research Oversight and Coordination (DROC) at King Fahd University of Petroleum and Minerals (KFUPM). The support provided by the Interdisciplinary Research Center for Hydrogen Technologies and Carbon Management (IRC-HTCM) on project number INHE2308 is highly appreciated. Also, the support received through the KFUPM Consortium for Hydrogen Future through the projects numbered H2FC2309 and H2FC2315 is highly appreciated.

Nomenclature

AEV	Advanced enhanced-vortex burner
APU	Auxiliary power unit
ASU	Air separation unit
C_p	Specific heat capacity
CCS	Carbon capture and storage
CFD	Computational fluid dynamics
CH_4	Methane
CO	Carbon monoxide
CO_2	Carbon dioxide
COG	Coke-oven gas
DLN	Dry low NOx

EV	Enhanced-vortex burner
FGR	Flue gas recirculation
GT	Gas turbine
Gt	Gigaton
H_2	Hydrogen
H_2O	Water vapor
HF	Hydrogen fraction
IGCC	Integrated gasification combined cycle
ITM	Ion transport membrane
kW	Kilowatt
LBO	Lean blowout limit
LDI	Lean direct injection
LHV	Lower heating value
LPG	Liquefied petroleum gas
LPM	Lean premixed
MM	Micromixer combustor
Mt	Megaton
MW	Megawatt
MWI	Modified Wobbe index
N_2	Nitrogen
NG	Natural gas
NOx	Nitrogen oxides
O_2	Oxygen
OEM	Original equipment manufacturers
OSL	Outer shear layer
PPB	Perforated-plate burner
ppmvd	Parts per million by volume, dry
Re	Reynolds number
SCOC-CC	Semi-closed oxy-fuel combustion combined cycle
SEV	Sequential enhanced-vortex burner
T_{ad}	Adiabatic flame temperature
TRL	Technology readiness level
TWh	Terawatt-hour
WI	Wobbe index
ZEPP	Zero-emission power plants
Δp	Pressure difference

Greek Symbols

$\varnothing$	Equivalence ratio
ρ	Density

Subscripts

ad	Adiabatic
th	Thermal

References

1. D. Saygin, R. Kempener, N. Wagner, M. Ayuso, D. Gielen, The implications for renewable energy innovation of doubling the share of renewables in the global energy mix between 2010 and 2030. Energies **8**(6), 5828–5865 (2015). https://doi.org/10.3390/en8065828
2. United Nations, The Paris Agreement (2015)
3. Z. Hausfather, Analysis: global fossil-fuel emissions up 0.6% in 2019 due to China (2019)
4. Enerdata. Global Energy Statistical Yearbook (2020), https://yearbook.enerdata.net/co2-fuel-combustion/CO2-emissions-data-from-fuel-combustion.html
5. M.A. Habib, F. Tahir, M.A. Nemitallah, W.H. Ahmed, H.M. Badr, Experimental and numerical analysis of oxy-fuel combustion in a porous plate reactor. Int. J. Energy Res. **39**(9), 1229–1240 (2015). https://doi.org/10.1002/er.3323
6. S. L'Orange Seigo, S. Dohle, M. Siegrist, Public perception of carbon capture and storage (CCS): a review. Renew. Sustain. Energy Rev. **2014**(38), 848–863 (2018). https://doi.org/10.1016/j.rser.2014.07.017
7. Carbon Capture and Storage Association, http://www.ccsassociation.org/what-is-ccs/
8. M.A. Rajhi, R. Ben-Mansour, M.A. Habib, M.A. Nemitallah, K. Andersson, Evaluation of gas radiation models in CFD modeling of oxy-combustion. Energy Convers. Manag. **81**, 83–97 (2014). https://doi.org/10.1016/j.enconman.2014.02.019
9. R. Ben-Mansour, M.A. Habib, M.A. Nemitallah, M. Rajhi, K.A. Suara, Characteristics of oxyfuel and air-fuel combustion in an industrial water tube boiler. Heat Transf. Eng. **35**(16–17), 1394–1404 (2014). https://doi.org/10.1080/01457632.2014.888920
10. M. A. Habib, M. A. Nemitallah, M. El-Nakla, Current status of CHF predictions using CFD modeling technique and review of other techniques especially for non-uniform axial and circumferential heating profiles. Ann. Nuclear Energy 188–207 (2014). https://doi.org/10.1016/j.anucene.2014.03.016
11. M.A. Habib, E.M.A. Mokheimer, S.Y. Sanusi, M.A. Nemitallah, Numerical investigations of combustion and emissions of syngas as compared to methane in a 200 MW package boiler. Energy Convers. Manage. **83**, 296–305 (2014). https://doi.org/10.1016/j.enconman.2014.03.056
12. S. van Loo, E.P. van Elk, G.F. Versteeg, The removal of carbon dioxide with activated solutions of methyl-diethanol-amine. J. Pet. Sci. Eng. **55**(1–2), 135–145 (2007). https://doi.org/10.1016/j.petrol.2006.04.017
13. H.T. Hwang, A. Harale, P.K.T. Liu, M. Sahimi, T.T. Tsotsis, A membrane-based reactive separation system for CO_2 removal in a life support system. J. Memb. Sci. **315**(1–2), 116–124 (2008). https://doi.org/10.1016/j.memsci.2008.02.018
14. J. D. Figueroa, T. Fout, S. Plasynski, H. McIlvried, R. D. Srivastava, Advances in CO_2 capture technology-the U.S. Department of Energy's Carbon Sequestration Program. Int. J. Greenh. Gas Control **2**(1), 9–20 (2008). https://doi.org/10.1016/S1750-5836(07)00094-1
15. L.M. Romeo, I. Bolea, J.M. Escosa, Integration of power plant and amine scrubbing to reduce CO_2 capture costs. Appl. Therm. Eng. **28**(8–9), 1039–1046 (2008). https://doi.org/10.1016/j.applthermaleng.2007.06.036
16. C.Y. Liu, G. Chen, N. Sipöcz, M. Assadi, X.S. Bai, Characteristics of oxy-fuel combustion in gas turbines. Appl. Energy **89**(1), 387–394 (2012). https://doi.org/10.1016/j.apenergy.2011.08.004

17. M.A. Nemitallah, S.S. Rashwan, I.B. Mansir, A.A. Abdelhafez, M.A. Habib, Review of novel combustion techniques for clean power production in gas turbines. Energy Fuels **32**(2), 979–1004 (2018). https://doi.org/10.1021/acs.energyfuels.7b03607

18. E. Portillo, B. Alonso-Fariñas, F. Vega, M. Cano, B. Navarrete, Alternatives for oxygen-selective membrane systems and their integration into the oxy-fuel combustion process: a review. Sep. Purif. Technol. **229**(January), 115708 (2019). https://doi.org/10.1016/j.seppur.2019.115708

19. M. A. Nemitallah, A. A. Abdelhafez, A. Ali, I. Mansir, M. A. Habib, Frontiers in combustion techniques and burner designs for emissions control and CO_2 capture: a review. Int. J. Energy Res. 1–33 (2019). https://doi.org/10.1002/er.4730

20. N. Khallaghi, D.P. Hanak, V. Manovic, Techno-economic evaluation of near-Zero CO_2 emission gas-fired power generation technologies: a review. J. Nat. Gas Sci. Eng. **2020**(74), 103095 (2019). https://doi.org/10.1016/j.jngse.2019.103095

21. C.P. Fenimore, Studies of fuel-nitrogen species in rich flame gases. Symp. Combust. **17**(1), 661–670 (1979). https://doi.org/10.1016/S0082-0784(79)80065-X

22. C.T. Bowman, Kinetics of pollutant formation and destruction in combustion. Prog. Energy Combust. Sci. **1**(1), 33–45 (1975). https://doi.org/10.1016/0360-1285(75)90005-2

23. L.B. Davis, R.M. Washam, Development of a dry low NOx combustor. Am. Soc. Mech. Eng. (1989). https://doi.org/10.1115/89-gt-255

24. P. P. Walsh, P. Fletcher, Gas turbine engine components, in *Gas Turbine Performance* (Blackwell Science Ltd), pp. 159–291 (2004). https://doi.org/10.1002/9780470774533.ch5

25. R. Pavri, G. D. Moore, Gas turbines emissions and control. GE Power Syst. 17(GER-4211), 29–43 (2003)

26. W. R. Bender, Lean pre-mixed combustion, in *Gas Turbine Handbook*, pp. 217–227

27. X. Gao, F. Duan, S.C. Lim, M.S. Yip, NOx Formation in hydrogen-methane turbulent diffusion flame under the moderate or intense low-oxygen dilution conditions. Energy **59**, 559–569 (2013). https://doi.org/10.1016/j.energy.2013.07.022

28. J. Warnatz, U. Maas, R. W. Dibble, *Combustion: Physical and Chemical Fundamentals, Modeling and Simulation, Experiments, Pollutant Formation* (2006). https://doi.org/10.1007/978-3-540-45363-5

29. Z. S. Spakovszky, E. M. Greitzer, I. A. Waitz, D. Quattrochi, Thermodynamics and propulsion http://web.mit.edu/16.unified/www/FALL/thermodynamics/notes/node111.html

30. Y. Ohkubo, Low-NOx combustion technology. Spec. Issue Core Technol. Micro Gas Turbine Cogener. Syst. **41**(1), 12–23 (2006)

31. R. Kurz, K. Brun, C. Meher-Homji, J. Moore, F. Gonzalez, Gas turbine performance and maintenance, in *Forty-Second Turbomachinery Symposium* (2013)

32. C.K. Law, *Combustion Physics* (Cambridge University Press, New York, 2006)

33. A. F. Ghoniem, S. Park, A. Wachsman, A. Annaswamy, D. Wee, H. Murat Altay, Mechanism of combustion dynamics in a backward-facing step stabilized premixed flame. Proceedings Combust. Inst. **30**(2), 1783–1790 (2005). https://doi.org/10.1016/j.proci.2004.08.201

34. H. Murat Altay, D.E. Hudgins, R.L. Speth, A.M. Annaswamy, A.F. Ghoniem, Mitigation of thermoacoustic instability utilizing steady air injection near the flame anchoring zone. Combust. Flame **157**(4), 686–700 (2010). https://doi.org/10.1016/j.combustflame.2010.01.012

35. B. Li, B. Shi, X. Zhao, K. Ma, D. Xie, D. Zhao, J. Li, Oxy-fuel combustion of methane in a swirl tubular flame burner under various oxygen contents: operation limits and combustion instability. Exp. Therm. Fluid Sci. **2018**(90), 115–124 (2017). https://doi.org/10.1016/j.expthermflusci.2017.09.001

36. S.S. Rashwan, M.A. Nemitallah, M.A. Habib, Review on premixed combustion technology: stability, emission control, applications, and numerical case study. Energy Fuels **30**(12), 9981–10014 (2016). https://doi.org/10.1021/acs.energyfuels.6b02386

37. A. Benim, K. Syed, *Flashback Mechanisms in Lean Premixed Gas Turbine Combustion*, 1st edn. (Academic Press, 2014).

38. S. Taamallah, N.W. Chakroun, H. Watanabe, S.J. Shanbhogue, A.F. Ghoniem, On the characteristic flow and flame times for scaling oxy and air flame stabilization modes in premixed swirl combustion. Proc. Combust. Inst. **36**(3), 3799–3807 (2017). https://doi.org/10.1016/j.proci.2016.07.022

39. T. Asai, S. Dodo, M. Karishuku, N. Yagi, Y. Akiyama, A. Hayashi, Performance of multiple-injection dry low-nox combustors on hydrogen-rich syngas fuel in an IGCC pilot plant. J. Eng. Gas Turbines Power **137**(9), 1–11 (2015). https://doi.org/10.1115/1.4029614

40. J. Runyon, R. Marsh, P. Bowen, D. Pugh, A. Giles, S. Morris, Lean methane flame stability in a premixed generic swirl burner: isothermal flow and atmospheric combustion characterization. Exp. Therm. Fluid Sci. **2018**(92), 125–140 (2017). https://doi.org/10.1016/j.expthermflusci.2017.11.019

41. K.T. Kim, Combustion instability feedback mechanisms in a lean-premixed swirl-stabilized combustor. Combust. Flame **171**, 137–151 (2016). https://doi.org/10.1016/j.combustflame.2016.06.003

42. J. Yoon, M.K. Kim, J. Hwang, J. Lee, Y. Yoon, Effect of fuel-air mixture velocity on combustion instability of a model gas turbine combustor. Appl. Therm. Eng. **54**(1), 92–101 (2013). https://doi.org/10.1016/j.applthermaleng.2013.01.032

43. L. Bollinger, D. Williams, Effect of Reynolds number in turbulent-flow range on flame speeds of Bunsen Buerner flames. Naca 13 (1949)

44. C.C. Liu, S.S. Shy, M.W. Peng, C.W. Chiu, Y.C. Dong, High-pressure burning velocities measurements for centrally-ignited premixed methane/air flames interacting with intense near-isotropic turbulence at constant Reynolds numbers. Combust. Flame **159**(8), 2608–2619 (2012). https://doi.org/10.1016/j.combustflame.2012.04.006

45. K. Döbbeling, J. Hellat, H. Koch, 25 years of BBC/ABB/Alstom lean premix combustion technologies. J. Eng. Gas Turbines Power **129**(1), 2–12 (2007). https://doi.org/10.1115/1.2181183

46. M. Boyce, *Gas Turbine Engineering Handbook*, 4th edn. (Butterworth-Heinemann, 2011)

47. M. A. Nemitallah, M. A. Habib, H. M. Badr, *Oxyfuel Combustion for Clean Energy Applications; Green Energy and Technology* (Springer International Publishing, Cham, 2019). https://doi.org/10.1007/978-3-030-10588-4

48. L. B. Davis, S. H. Black, Dry low NOX combustion systems for GE heavy-duty gas turbines. GE Power Syst. No. GER-3568G (2000)

49. R. K. Matta, G. D. Mercer, R. S. Tuthill, Power systems for the 21st century—"H" gas turbine combined-cycles. GE Power Syst. No. GER-3935B GE (2000)

50. T. Asai, Y. Akiyama, S. Dodo, Development of a state-of-the-art dry low NOx gas development of a state-of-the-art dry low NOx gas turbine combustor for IGCC with CCS turbine combustor for IGCC with CCS. Intech 3–30 (2017). https://doi.org/10.5772/66742

51. D. Dunn-Rankin, *Lean Combustion: Technology and Control.* (Academic Press, 2008)

52. B. Asgari, E. Amani, A multi-objective CFD optimization of liquid fuel spray injection in dry-low-emission gas-turbine combustors. Appl. Energy **203**, 696–710 (2017). https://doi.org/10.1016/j.apenergy.2017.06.080

53. Q. Tang, H. Liu, M. Li, M. Yao, Optical study of spray-wall impingement impact on early-injection gasoline partially premixed combustion at low engine load. Appl. Energy **185**, 708–719 (2017). https://doi.org/10.1016/j.apenergy.2016.10.108

54. D. Ebi, N.T. Clemens, Experimental investigation of upstream flame propagation during boundary layer flashback of swirl flames. Combust. Flame **168**, 39–52 (2016). https://doi.org/10.1016/j.combustflame.2016.03.027

55. T. Sattelmayer, M.P. Felchlin, J. Haumann, J. Hellat, D. Styner, Second-generation low-emission combustors for abb gas turbines: burner development and tests at atmospheric pressure. J. Eng. Gas Turbines Power **114**(1), 118–125 (1992). https://doi.org/10.1115/1.2906293

56. W. Hubschmid, R. Bombach, A. Inauen, F. Güthe, S. Schenker, N. Tylli, W. Kreutner, Thermoacoustically driven flame motion and heat release variation in a swirl-stabilized gas turbine burner investigated by LIF and chemiluminescence. Exp. Fluids **45**(1), 167–182 (2008). https://doi.org/10.1007/s00348-008-0497-1

57. W. Hubschmid, A. Denisov, F. Biagioli, Acoustic forcing on swirling flow: experiments and simulation. Exp. Fluids **55**(9), 1–12 (2014). https://doi.org/10.1007/s00348-014-1808-3

58. F. Joos, P. Brunner, M. Stalder, S. Tschirren, Field experience with the sequential combustion system of the GT24/GT26. ABB Rev. **5**, 12–20 (1998)

59. G. Müller, M. Valk, G. Hosse, H. Heller, Large gas turbines—the insurance aspects (update). Int. Assoc. Eng. Insur. 13 (IMIA WGP13(00)E) (2000)

60. F. Güthe, J. Hellat, P. Flohr, The reheat concept: the proven pathway to ultra-low emissions and high efficiency and flexibility. Proc. ASME Turbo Expo **2**(March), 641–648 (2007). https://doi.org/10.1115/GT2007-27846

61. F. Joos, P. Brunner, B. Schulte-Werning, K. Syed, A. Eroglu, Development of the sequential combustion system for the GT24/GT26 gas turbine family. ABB Rev. **4**, 4–16 (1998)

62. M. Zajadatz, F. Güthe, E. Freitag, T. Ferreira-Providakis, T. Wind, F. Magni, J. Goldmeer, Extended range of fuel capability for GT13E2 AEV burner with liquid and gaseous fuels. J. Eng. Gas Turbines Power **141**(5) (2018). https://doi.org/10.1115/1.4041144

63. R. Cabra, Turbulent jet flames into a vitiated coflow. Nasa, No. March, CR-2004-212887 (2004)

64. J. M. N. Rodrigues, E. C. Fernandes, Stability analysis and flow characterization of multi-perforated plate premixed burners. 17th Int. Symp. Appl. Laser Tech. Fluid Mech. 7–10 (2014)

65. J. Edacheri Veetil, C.V. Rajith, R.K. Velamati, Numerical simulations of steady perforated-plate stabilized syngas air pre-mixed flames. Int. J. Hydrogen Energy **41**(31), 13747–13757 (2016). https://doi.org/10.1016/j.ijhydene.2016.06.120

66. E.V. Jithin, V.R. Kishore, R.J. Varghese, Three-dimensional simulations of steady perforated-plate stabilized propane-air premixed flames. Energy Fuels **28**(8), 5415–5425 (2014). https://doi.org/10.1021/ef401903y

67. V. Hindasageri, P. Kuntikana, R.P. Vedula, S.V. Prabhu, An experimental and numerical investigation of heat transfer distribution of perforated plate burner flames impinging on a flat plate. Int. J. Therm. Sci. **94**, 156–169 (2015). https://doi.org/10.1016/j.ijthermalsci.2015.02.021

68. P. Kuntikana, S.V. Prabhu, Thermal investigations on methane-air premixed flame jets of multi-port burners. Energy **123**, 218–228 (2017). https://doi.org/10.1016/j.energy.2017.01.122

69. P.H. Lee, S.S. Hwang, Formation of lean premixed surface flame using porous baffle plate and flame holder. J. Therm. Sci. Technol. **8**(1), 178–189 (2013). https://doi.org/10.1299/jtst.8.178

70. M.H. Saberi Moghaddam, M. Saei Moghaddam, M. Khorramdel, Numerical study of geometric parameters effecting temperature and thermal efficiency in a premix multi-hole flat flame burner. Energy **125**, 654–662 (2017). https://doi.org/10.1016/j.energy.2017.02.116

71. K.S. Kedia, A.F. Ghoniem, Mechanisms of stabilization and blowoff of a premixed flame downstream of a heat-conducting perforated plate. Combust. Flame **159**(3), 1055–1069 (2012). https://doi.org/10.1016/j.combustflame.2011.10.014

72. J. Timmermans, M. Vanierschot, E. Van Den Bulck, Flow instabilities in the near wake of a perforated plate. May 9–10 (2012)

73. S.S. Rashwan, A.H. Ibrahim, T.W. Abou-Arab, M.A. Nemitallah, M.A. Habib, Experimental investigation of partially premixed methane-air and methane-oxygen flames stabilized over a perforated-plate burner. Appl. Energy **169**, 126–137 (2016). https://doi.org/10.1016/j.apenergy.2016.02.047

74. S.S. Rashwan, A.H. Ibrahim, T.W. Abou-Arab, M.A. Nemitallah, M.A. Habib, Experimental study of atmospheric partially premixed oxy-combustion flames anchored over a perforated plate burner. Energy **122**, 159–167 (2017). https://doi.org/10.1016/j.energy.2017.01.086

75. M. A. Habib, I. B. Mansir, M. A. Nemitallah, A. E. Khalifa, Experimental study on combustion characteristics and lean blow-out limits of non-premixed oxy-methane flames in a porous-plate reactor. Heat Mass Transf. und Stoffuebertragung 3265–3274 (2019). https://doi.org/10.1007/s00231-019-02639-5

76. N. S. Rodrigues, T. Busari, C. B. William, Y. Chen, A. J. North, J. Enrique Portillo, S. E. Meyer, R. P. Lucht, Development and performance of a perforated plate burner under gas

turbine engine-relevant conditions. 2018 Jt. Propuls. Conf. 1–18 (2018). https://doi.org/10.2514/6.2018-4475

77. W. D. York, W. S. Ziminsky, E. Yilmaz, Development and testing of a low NOX hydrogen combustion system for heavy duty gas turbines, in *Combustion, Fuels and Emissions, Parts A and B; American Society of Mechanical Engineers*, pp. 1395–1405 (2012). https://doi.org/10.1115/GT2012-69913

78. N.T. Weiland, T.G. Sidwell, P.A. Strakey, Testing of a hydrogen diffusion flame array injector at gas turbine conditions. Combust. Sci. Technol. **185**(7), 1132–1150 (2013). https://doi.org/10.1080/00102202.2013.781164

79. C. J. Marek, T. D. Smith, K. Kundu, Low emission hydrogen combustors for gas turbines using lean direct injection, in *ASEE Joint Propulsion Conference and Exhibit.* (American Institute of Aeronautics and Astronautics, Tucson, Arizona, 2005), pp. 1–27

80. H. Lee, S. Hernandez, V. McDonell, E. Steinthorsson, A. Mansour, B. Hollon, Development of flashback resistant low-emission micro-mixing fuel injector for 100% hydrogen and syngas fuels. Proc. ASME Turbo Expo **2**, 411–419 (2009). https://doi.org/10.1115/GT2009-59502

81. T. Asai, S. Dodo, H. Koizumi, H. Takahashi, S. Yoshida, H. Inoue, Effects of multiple-injection-burner configurations on combustion characteristics for dry low-NOx combustion of hydrogen-rich fuels, in *Combustion, Fuels and Emissions, Parts A and B*, vol 2. (ASMEDC, 2011), pp. 311–320. https://doi.org/10.1115/GT2011-45295

82. H. H.-W. Funke, S. Boerner, W. Krebs, E. Wolf, Experimental characterization of low NOx micromix prototype combustors for industrial gas turbine applications, in *Combustion, Fuels and Emissions, Parts A and B; Engineers*, A. S. of M., Ed. (ASMEDC, Vancouver, British Columbia, Canada, 2011), pp. 343–353. https://doi.org/10.1115/GT2011-45305

83. H. H. W. Funke, S. Boerner, J. Keinz, K. Kusterer, D. Kroniger, J. Kitajima, M. Kazari, A. Horikawa, Numerical and experimental characterization of low NOX micromix combustion principle for industrial hydrogen gas turbine applications. Proc. ASME Turbo Expo **2**(PARTS A AND B), 1069–1079 (2012). https://doi.org/10.1115/GT2012-69421

84. S. Dodo, T. Asai, H. Koizumi, H. Takahashi, S. Yoshida, H. Inoue, Combustion characteristics of a multiple-injection combustor for dry low-NOx combustion of hydrogen-rich fuels under medium-low pressure, in *Proceedings of ASME Turbo Expo 2011.* (The American Society of Mechanical Engineers, 2011)

85. S. Börner, H. Funke, P. Hendrick, E. Recker, LES of jets in cross-flow and application to the "micromix" hydrogen combustion, in *XIX International Symposium on Air Breathing Engines* (2009)

86. A.H. Ayed, K. Kusterer, H.H.W. Funke, J. Keinz, D. Bohn, CFD based exploration of the dry-low-NOx hydrogen micromix combustion technology at increased energy densities. Propuls. Power Res. **6**(1), 15–24 (2017). https://doi.org/10.1016/j.jppr.2017.01.005

87. H. H. W. Funke, J. Dickhoff, J. Keinz, A. Haj Ayed, A. Parente, P. Hendrick, Experimental and numerical study of the micromix combustion principle applied for hydrogen and hydrogen rich syngas as fuel with increased energy density for industrial gas turbine applications. Energy Procedia **61**(x), 1736–1739 (2014). https://doi.org/10.1016/j.egypro.2014.12.201

88. H.H.W. Funke, N. Beckmann, S. Abanteriba, An overview on dry low NOx micromix combustor development for hydrogen-rich gas turbine applications. Int. J. Hydrogen Energy **44**(13), 6978–6990 (2019). https://doi.org/10.1016/j.ijhydene.2019.01.161

89. A. Haj Ayed, K. Kusterer, H.H.W. Funke, J. Keinz, C. Striegan, D. Bohn, Experimental and numerical investigations of the dry-low-NOx hydrogen micromix combustion chamber of an industrial gas turbine. Propuls. Power Res. **4**(3), 123–131 (2015). https://doi.org/10.1016/j.jppr.2015.07.005

90. A. Haj Ayed, K. Kusterer, H.H.W. Funke, J. Keinz, C. Striegan, D. Bohn, Improvement study for the dry-low-NOx hydrogen micromix combustion technology. Propuls. Power Res. **4**(3), 132–140 (2015). https://doi.org/10.1016/j.jppr.2015.07.003

91. H. H.-W. Funke, J. Keinz, K. Kusterer, A. Haj Ayed, M. Kazari, J. Kitajima, A. Horikawa, K. Okada, Experimental and numerical study on optimizing the dln micromix hydrogen combustion principle for industrial gas turbine applications, in *Combustion, Fuels and Emissions*, vol

4A. (American Society of Mechanical Engineers, 2015), pp. 1–12. https://doi.org/10.1115/GT2015-42043

92. J. A. Badra, A. R. Masri, Micro-mixer/combustor. US 2014/0260313 (2014)

93. M. Aliyu, A. Abdelhafez, S.A.M. Said, M.A. Habib, M.A. Nemitallah, I.B. Mansir, Characteristics of oxyfuel combustion in lean-premixed multihole burners. Energy Fuels (2019). https://doi.org/10.1021/acs.energyfuels.9b02821

94. B.A. Imteyaz, M.A. Nemitallah, A.A. Abdelhafez, M.A. Habib, Combustion behavior and stability map of hydrogen-enriched oxy-methane premixed flames in a model gas turbine combustor. Int. J. Hydrogen Energy **43**(34), 16652–16666 (2018). https://doi.org/10.1016/j.ijhydene.2018.07.087

95. A.F. Ghoniem, Needs, resources and climate change: clean and efficient conversion technologies. Prog. Energy Combust. Sci. **37**(1), 15–51 (2011). https://doi.org/10.1016/j.pecs.2010.02.006

96. M.A. Habib, M. Nemitallah, R. Ben-Mansour, Recent development in oxy-combustion technology and its applications to gas turbine combustors and ITM reactors. Energy Fuels **27**(1), 2–19 (2013). https://doi.org/10.1021/ef301266j

97. C. Shaddix, A. Molina, 6—ignition, flame stability, and char combustion in oxy-fuel combustion, in *Oxy-Fuel Combustion for Power Generation and Carbon Dioxide (CO2) Capture* (2011), pp. 101–124

98. M.R. Shakeel, Y.S. Sanusi, E.M.A. Mokheimer, Numerical modeling of oxy-methane combustion in a model gas turbine combustor. Appl. Energy **228**(June), 68–81 (2018). https://doi.org/10.1016/j.apenergy.2018.06.071

99. M. Schwartz, J. White, A. Sammels, Solid state oxygen anion and electron mediating membrane and catalytic membrane reactors containing them. US Patent 6:033,632 (2000)

100. G. Scheffknecht, L. Al-Makhadmeh, U. Schnell, J. Maier, Oxy-fuel coal combustion-a review of the current state-of-the-art. Int. J. Greenh. Gas Control **5**(SUPPL. 1), 16–35 (2011). https://doi.org/10.1016/j.ijggc.2011.05.020

101. I. Ali, X. Gou, Q. Zhang, J. Wu, E. Wang, Y. Liu, Experimental study on NOx emission characteristics of oxy-biomass combustion. J. Clean. Prod. **199**(x), 400–410 (2018). https://doi.org/10.1016/j.jclepro.2018.07.022

102. K. Andersson, F. Normann, F. Johnsson, B. Leckner, NO emission during oxy-fuel combustion of lignite **x**, 1835–1845 (2008). https://doi.org/10.1021/ie0711832

103. C. Schluckner, C. Gaber, M. Landfahrer, M. Demuth, C. Hochenauer, Fast and accurate CFD-model for NOx emission prediction during oxy-fuel combustion of natural gas using detailed chemical kinetics. Fuel **2020**(264), 116841 (2019). https://doi.org/10.1016/j.fuel.2019.116841

104. J. Oh, D. Noh, Laminar burning velocity of oxy-methane flames in atmospheric condition. Energy **45**(1), 669–675 (2012). https://doi.org/10.1016/j.energy.2012.07.027

105. M. Shah, N. Degenstein, M. Zanfir, R. Kumar, J. Bugayong, K. Burgers, Near zero emissions oxy-combustion CO_2 purification technology. Energy Procedia **4**, 988–995 (2011). https://doi.org/10.1016/j.egypro.2011.01.146

106. M.A. Habib, M.A. Nemitallah, P. Ahmed, M.H. Sharqawy, H.M. Badr, I. Muhammad, M. Yaqub, Experimental analysis of oxygen-methane combustion inside a gas turbine reactor under various operating conditions. Energy **86**, 105–114 (2015). https://doi.org/10.1016/j.energy.2015.03.120

107. A. Abdelhafez, S.S. Rashwan, M.A. Nemitallah, M.A. Habib, Stability map and shape of premixed $CH_4/O_2/CO_2$ flames in a model gas-turbine combustor. Appl. Energy **215**(January), 63–74 (2018). https://doi.org/10.1016/j.apenergy.2018.01.097

108. A. Abdelhafez, M.A. Nemitallah, S.S. Rashwan, M.A. Habib, Adiabatic flame temperature for controlling the macrostructures and stabilization modes of premixed methane flames in a model gas-turbine combustor. Energy Fuels **32**(7), 7868–7877 (2018). https://doi.org/10.1021/acs.energyfuels.8b01133

109. P. Jourdaine, C. Mirat, J. Caudal, A. Lo, T. Schuller, A comparison between the stabilization of premixed swirling CO_2-diluted methane oxy-flames and methane/air flames. Fuel **201**, 156–164 (2017). https://doi.org/10.1016/j.fuel.2016.11.017

110. P. Kutne, B.K. Kapadia, W. Meier, M. Aigner, Experimental analysis of the combustion behaviour of oxyfuel flames in a gas turbine model combustor. Proc. Combust. Inst. **33**(2), 3383–3390 (2011). https://doi.org/10.1016/j.proci.2010.07.008

111. H. Watanabe, S. J. Shanbhogue, A. F. Ghoniem, Impact of equivalence ratio on the macrostructure of premixed swirling CH_4/Air and CH_4/O_2/CO_2 flames. Proc. ASME Turbo Expo 4B (2015). https://doi.org/10.1115/GT201543224

112. A. Sachajdak, J. Lappalainen, H. Mikkonen, Dynamic simulation in development of contemporary energy systems—oxy combustion case study. Energy **181**, 964–973 (2019). https://doi.org/10.1016/j.energy.2019.05.198

113. D.A. Granados, F. Chejne, J.M. Mejía, Oxy-fuel combustion as an alternative for increasing lime production in rotary kilns. Appl. Energy **158**, 107–117 (2015). https://doi.org/10.1016/j.apenergy.2015.07.075

114. L. Zheng, *Oxy-Fuel Combustion for Power Generation and Carbon Dioxide (CO_2) Capture* (2011)

115. M. Aneke, M. Wang, Process analysis of pressurized oxy-coal power cycle for carbon capture application integrated with liquid air power generation and binary cycle engines. Appl. Energy **154**, 556–566 (2015). https://doi.org/10.1016/j.apenergy.2015.05.030

116. M.A. Habib, R. Ben-Mansour, H.M. Badr, S.F. Ahmed, A.F. Ghoniem, Computational fluid dynamic simulation of oxyfuel combustion in gas-fired water tube boilers. Comput. Fluids **56**, 152–165 (2012). https://doi.org/10.1016/j.compfluid.2011.12.009

117. D. M. Wicksall, A. K. Agrawal, R. W. Schefer, J. O. Keller, The interaction of flame and flow field in a lean premixed swirl-stabilized combustor operated on H2/CH4/Air. Proc. Combust. Inst. **30**(2), 2875–2883 (2005). https://doi.org/10.1016/j.proci.2004.07.021

118. T.C. Williams, C.R. Shaddix, R.W. Schefer, Effect of syngas composition and CO_2-diluted oxygen on performance of a premixed swirl-stabilized combustor. Combust. Sci. Technol. **180**(1), 64–88 (2008). https://doi.org/10.1080/00102200701487061

119. P.H. Joo, M.R.J. Charest, C.P.T. Groth, Ö.L. Gülder, Comparison of structures of laminar methane-oxygen and methane-air diffusion flames from atmospheric to 60atm. Combust. Flame **160**(10), 1990–1998 (2013). https://doi.org/10.1016/j.combustflame.2013.04.030

120. F. Tahir, H. Ali, A. A. B. Baloch, Y. Jamil, Performance analysis of air and oxy-fuel laminar combustion in a porous plate reactor. Energies **12**(9) (2019). https://doi.org/10.3390/en12091706

121. Y. Xie, J. Wang, M. Zhang, J. Gong, W. Jin, Z. Huang, Experimental and numerical study on laminar flame characteristics of methane oxy-fuel mixtures highly diluted with CO2. Energy Fuels **27**(10), 6231–6237 (2013). https://doi.org/10.1021/ef401220h

122. S. Rashwan, A. Ibrahim, T. Abou-Arab, Experimental investigation of oxy-fuel combustion of CNG flames stabilized over a perforated plate burner, in *18th IFRF members Conference Flexible Clean Fuel Conversion to India Freising, Germany* (2015)

123. Y.H. Li, G.B. Chen, Y.C. Chao, Effects of flue gas addition on the premixed oxy-methane flames in atmospheric condition. Energy Procedia **75**, 3054–3059 (2015). https://doi.org/10.1016/j.egypro.2015.07.623

124. S. Taamallah, K. Vogiatzaki, F.M. Alzahrani, E.M.A. Mokheimer, M.A. Habib, A.F. Ghoniem, Fuel flexibility, stability and emissions in premixed hydrogen-rich gas turbine combustion: technology, fundamentals, and numerical simulations. Appl. Energy **154**, 1020–1047 (2015). https://doi.org/10.1016/j.apenergy.2015.04.044

125. K. Andersson, R. Johansson, F. Johnsson, B. Leckner, Radiation intensity of propane-fired oxy-fuel flames: implications for soot formation. Energy Fuels **22**(3), 1535–1541 (2008). https://doi.org/10.1021/ef7004942

126. B. Imteyaz, M.A. Habib, R. Ben-Mansour, The characteristics of oxycombustion of liquid fuel in a typical water-tube boiler. Energy Fuels **31**(6), 6305–6313 (2017). https://doi.org/10.1021/acs.energyfuels.7b00489

127. Z. Abubakar, S.Y. Sanusi, E.M.A. Mokheimer, Stability of propane-air and oxyfuel diffusion flames in a swirl-stabilized combustor. Experimental Study. Energy Procedia **142**, 1552–1557 (2017). https://doi.org/10.1016/j.egypro.2017.12.607

128. U.S. DOE. Carbon Capture, Utilization, and Storage: Climate Change, Economic Competitiveness, and Energy Security (2012)
129. H. Gerbelová, M. van der Spek, W. Schakel, Feasibility assessment of CO2 capture retrofitted to an existing cement plant: post-combustion versus oxy-fuel combustion technology. Energy Procedia **2017**(114), 6141–6149 (2016). https://doi.org/10.1016/j.egypro.2017.03.1751
130. World Oil Outlook (2019)
131. J. Marion, A. Brautsch, F. Kluger, M. Larsson, T. Pourchot, Overview of a manufacturer's efforts to commercialize oxy- combustion for steam power plants, in *2nd Oxyfuel Combustion Conference, IEAGHG International Combustion Research Oxyfuel Network*. (Queensland, Australia, 2011)
132. F. Scala, *Fluidized Bed Technologies for Near-Zero Emission Combustion and Gasification* (2013)
133. E. Thorbergsson, T. Grönstedt, A thermodynamic analysis of two competing mid-sized oxyfuel combustion combined cycles. J. Energy **2016**, 1–14 (2016). https://doi.org/10.1155/2016/2438431
134. M. Sammak, K. Jonshagen, M. Thern, M. Genrup, E. Thorbergsson, T. Gro¨nstedt, A. Dahlquist, Conceptual design of a mid-sized semi-closed oxy-fuel combustion combined cycle, in *Cycle Innovations; Fans and Blowers; Industrial and Cogeneration; Manufacturing Materials and Metallurgy; Marine; Oil and Gas Applications*. (ASMEDC, Vancouver, British Columbia, Canada, 2011), pp. 253–261. https://doi.org/10.1115/GT2011-46299
135. R.J. Allam, M.R. Palmer, G.W. Brown, J. Fetvedt, D. Freed, H. Nomoto, M. Itoh, N. Okita, C. Jones, High efficiency and low cost of electricity generation from fossil fuels while eliminating atmospheric emissions. Including Carbon Dioxide. Energy Procedia **37**, 1135–1149 (2013). https://doi.org/10.1016/j.egypro.2013.05.211
136. R. Anderson, F. Viteri, R. Hollis, A. Keating, J. Shipper, G. Merrill, C. Schillig, S. Shinde, J. Downs, D. Davies, M. Harris, Oxy-fuel gas turbine, gas generator and reheat combustor technology development and demonstration, in *Controls, Diagnostics and Instrumentation; Cycle Innovations; Marine*, vol. 3. (ASMEDC, 2010), pp. 733–743. https://doi.org/10.1115/GT2010-23001
137. S. C. W. Smith, C. M. Newton, R. *Gassman, Fuel Flexible Thermoelectric Micro-Generator with Micro-Turbine*. CA 2576129C (2005)
138. D. Mosbacher, J. Haynes, J. Janssen, J. Brumberg, V. Iyer, *Fuel-Flexible Combustion System and Method of Operation*. US 2007/0234735 A1 (2006)
139. J. T. Herbon, M. P. Boespflug, G. A. Bennett, F. T. Jothiprasad, Passive air-fuel mixing prechamber. US 2012/0144832 A1 (2012)
140. S. Black, P. Roll, M. Grames, M. Goodwin, J. Biccum, R. Legault, B. Soares, D. Dorman, *Dual Fuel Direct Ignition Burners*. US 2019/0162410 A1 (2019)
141. G.A. Richards, M.M. McMillian, R.S. Gemmen, W.A. Rogers, S.R. Cully, Issues for low-emission, fuel-flexible power systems. Prog. Energy Combust. Sci. **27**(2), 141–169 (2001). https://doi.org/10.1016/S0360-1285(00)00019-8
142. D. Singh, T. Nishiie, S. Tanvir, L. Qiao, An experimental and kinetic study of syngas/air combustion at elevated temperatures and the effect of water addition. Fuel **94**, 448–456 (2012). https://doi.org/10.1016/j.fuel.2011.11.058
143. N. Bouvet, F. Halter, C. Chauveau, Y. Yoon, On the effective lewis number formulations for lean hydrogen/hydrocarbon/air mixtures. Int. J. Hydrogen Energy **38**(14), 5949–5960 (2013). https://doi.org/10.1016/j.ijhydene.2013.02.098
144. T. Lieuwen, V. McDonell, E. Petersen, D. Santavicca, Fuel flexibility influences on premixed combustor blowout, flashback, auto ignition, and stability. Proc. ASME Turbo Expo **1**(January), 601–615 (2006). https://doi.org/10.1115/GT2006-90770
145. R. D. *Treloar, Gas Installation Technology*, 1st edn. (Blackwell Publishing, 2005)
146. D. M. Erickson, S. A. Day, R. Doyle, Design considerations for heated gas fuel. GE Power Syst. GER-4189B, 20 (2003)
147. K.H. Casleton, R.W. Breault, G.A. Richards, System issues and tradeoffs associated with syngas production and combustion. Combust. Sci. Technol. **180**(6), 1013–1052 (2008). https://doi.org/10.1080/00102200801962872

148. J. Klimstra, *Interchangeability of Gaseous Fuels—The Importance of the Wobbe-Index*, vol. 2015 (1986). https://doi.org/10.4271/861578

149. K. Do¨bbeling, T. Meeuwissen, M. Zajadatz, P. Flohr, Fuel flexibility of the Alstom GT13E2 medium sized gas turbine, in *Combustion, Fuels and Emissions, Parts A and B*, vol. 3. (ASMEDC, 2008), pp. 719–725. https://doi.org/10.1115/GT2008-50950

150. A.E.E. Khalil, A.K. Gupta, Fuel flexible distributed combustion for efficient and clean gas turbine engines. Appl. Energy **109**, 267–274 (2013). https://doi.org/10.1016/j.apenergy.2013.04.052

151. A.E.E. Khalil, A.K. Gupta, K.M. Bryden, S.C. Lee, Mixture preparation effects on distributed combustion for gas turbine applications. J. Energy Resour. Technol. **134**(3), 1–7 (2012). https://doi.org/10.1115/1.4006481

152. A. E. E. Khalil, A. K. Gupta, mixture preparation effects on distributed combustion, in *53rd AIAA Aerospace Science Meeting* (2015), pp. 1–7. https://doi.org/10.2514/6.2015-2075

153. A. Ali, M.A. Nemitallah, A.A. Abdelhafez, I.G. Alsakhawy, M.M. Kamal, M.A.M. Habib, Static stability and combustion characteristics of oxy-propane flames in a premixed fuel-flexible swirl combustor. Energy Fuels (2019). https://doi.org/10.1021/acs.energyfuels.9b03157

154. C. Yin, J. Yan, Oxy-fuel combustion of pulverized fuels: combustion fundamentals and modeling. Appl. Energy **162**, 742–762 (2016). https://doi.org/10.1016/j.apenergy.2015.10.149

155. H.S. Kim, V.K. Arghode, A.K. Gupta, Flame characteristics of hydrogen-enriched methane-air premixed swirling flames. Int. J. Hydrogen Energy **34**(2), 1063–1073 (2009). https://doi.org/10.1016/j.ijhydene.2008.10.035

156. M. Ditaranto, R. Anantharaman, T. Weydahl, Performance and NOx emissions of refinery fired heaters retrofitted to hydrogen combustion. Energy Procedia **2013**(37), 7214–7220 (2015). https://doi.org/10.1016/j.egypro.2013.06.659

157. R. Sankaran, H.G. Im, Effects of hydrogen addition on the markstein length and flammability limit of stretched methane/air premixed flames. Combust. Sci. Technol. **178**(9), 1585–1611 (2006). https://doi.org/10.1080/00102200500536217

158. F. Delattin, G. Di Lorenzo, S. Rizzo, S. Bram, J. De Ruyck, Combustion of syngas in a pressurized microturbine-like combustor: experimental results. Appl. Energy **87**(4), 1441–1452 (2010). https://doi.org/10.1016/j.apenergy.2009.08.046

159. A.E.E. Khalil, A.K. Gupta, Hydrogen addition effects on high intensity distributed combustion. Appl. Energy **104**, 71–78 (2013). https://doi.org/10.1016/j.apenergy.2012.11.004

160. M. Gu, H. Chu, F. Liu, Effects of simultaneous hydrogen enrichment and carbon dioxide dilution of fuel on soot formation in an axisymmetric coflow laminar ethylene/air diffusion flame. Combust. Flame **166**, 216–228 (2016). https://doi.org/10.1016/j.combustflame.2016.01.023

161. S.H. Park, K.M. Lee, C.H. Hwang, Effects of hydrogen addition on soot formation and oxidation in laminar premixed C_2H_2/air flames. Int. J. Hydrogen Energy **36**(15), 9304–9311 (2011). https://doi.org/10.1016/j.ijhydene.2011.05.031

162. S. Daniele, P. Jansohn, J. Mantzaras, K. Boulouchos, Turbulent flame speed for syngas at gas turbine relevant conditions. Proc. Combust. Inst. **33**(2), 2937–2944 (2011). https://doi.org/10.1016/j.proci.2010.05.057

163. S. Abdelwahid, M. Nemitallah, B. Imteyaz, A. Abdelhafez, M. Habib, Effects of H_2 enrichment and inlet velocity on stability limits and shape of CH_4/H_2–O_2/CO_2 flames in a premixed swirl combustor. Energy Fuels **32**(9), 9916–9925 (2018). https://doi.org/10.1021/acs.energyfuels.8b01958

164. F. Halter, C. Chauveau, I. Gökalp, Characterization of the effects of hydrogen addition in premixed methane/air flames. Int. J. Hydrogen Energy **32**(13), 2585–2592 (2007). https://doi.org/10.1016/j.ijhydene.2006.11.033

165. C. Tang, Z. Huang, C. Jin, J. He, J. Wang, X. Wang, H. Miao, Laminar burning velocities and combustion characteristics of propane-hydrogen-air premixed flames. Int. J. Hydrogen Energy **33**(18), 4906–4914 (2008). https://doi.org/10.1016/j.ijhydene.2008.06.063

166. E. Hu, Z. Huang, J. He, C. Jin, J. Zheng, Experimental and numerical study on laminar burning characteristics of premixed methane-hydrogen-air flames. Int. J. Hydrogen Energy **34**(11), 4876–4888 (2009). https://doi.org/10.1016/j.ijhydene.2009.03.058

167. R.W. Schefer, Hydrogen enrichment for improved lean flame stability. Int. J. Hydrogen Energy **28**(10), 1131–1141 (2003). https://doi.org/10.1016/S0360-3199(02)00199-4

168. M. Nakahara, H. Kido, Study on the turbulent burning velocity of hydrogen mixtures including hydrocarbons. AIAA J. **46**(7), 1569–1575 (2008). https://doi.org/10.2514/1.23560

169. J. Wang, Z. Huang, C. Tang, H. Miao, X. Wang, Numerical study of the effect of hydrogen addition on methane-air mixtures combustion. Int. J. Hydrogen Energy **34**(2), 1084–1096 (2009). https://doi.org/10.1016/j.ijhydene.2008.11.010

170. S. Gersen, N.B. Anikin, A.V. Mokhov, H.B. Levinsky, Ignition properties of methane/hydrogen mixtures in a rapid compression machine. Int. J. Hydrogen Energy **33**(7), 1957–1964 (2008). https://doi.org/10.1016/j.ijhydene.2008.01.017

171. R.K. Cheng, A.K. Oppenheim, Autoignition in methanehydrogen mixtures. Combust. Flame **58**(2), 125–139 (1984). https://doi.org/10.1016/0010-2180(84)90088-9

172. R.K. Cheng, D. Littlejohn, Laboratory study of premixed H_2-air & H_2-N 2-air flames in a low-swirl injector for ultra-low emissions gas turbines. Proc. ASME Turbo Expo **2**(May), 383–393 (2007). https://doi.org/10.1115/GT2007-27512

173. Q. Zhang, D.R. Noble, T. Lieuwen, Characterization of fuel composition effects in H_2/CO/CH_4 mixtures upon lean blowout. J. Eng. Gas Turbines Power **129**(3), 688–694 (2007). https://doi.org/10.1115/1.2718566

174. M.A. Nemitallah, B. Imteyaz, A. Abdelhafez, M.A. Habib, Experimental and computational study on stability characteristics of hydrogen-enriched oxy-methane premixed flames. Appl. Energy **250**(March), 433–443 (2019). https://doi.org/10.1016/j.apenergy.2019.05.087

175. M. Chaos, F.L. Dryer, Syngas combustion kinetics and applications. Combust. Sci. Technol. **180**(6), 1053–1096 (2008). https://doi.org/10.1080/00102200801963011

176. N. F. Othman, M. H. Boosroh, Effect of H_2 and CO contents in syngas during combustion using micro gas turbine. IOP Conf. Ser. Earth Environ. Sci. **32**(1) (2016). https://doi.org/10.1088/1755-1315/32/1/012037

177. M.E. Dry, The Fischer-Tropsch process: 1950–2000. Catal. Today **71**(3–4), 227–241 (2002). https://doi.org/10.1016/S0920-5861(01)00453-9

178. D.J. Wilhelm, D.R. Simbeck, A.D. Karp, R.L. Dickenson, Syngas production for gas-to-liquids applications: technologies, issues and outlook. Fuel Process. Technol. **71**(1–3), 139–148 (2001). https://doi.org/10.1016/S0378-3820(01)00140-0

179. P. Glarborg, Hidden interactions-trace species governing combustion and emissions. Proc. Combust. Inst. **31**(1), 77–98 (2007). https://doi.org/10.1016/j.proci.2006.08.119

180. M. Ilbas, A. Bektas, S. Karyeyen, Effect of oxy-fuel combustion on flame characteristics of low calorific value coal gases in a small burner and combustor. Fuel **2018**(226), 350–364 (2017). https://doi.org/10.1016/j.fuel.2018.04.023

181. R. L. Speth, Fundamental studies in hydrogen-rich combustion: instability mechanisms and dynamic mode selection. Massachusetts Institute of Technology (2010)

182. F.M. Alzahrani, Y.S. Sanusi, K. Vogiatzaki, A.F. Ghoniem, M.A. Habib, E.M.A. Mokheimer, Evaluation of the accuracy of selected syngas chemical mechanisms. J. Energy Resour. Technol. **137**(4), 1–13 (2015). https://doi.org/10.1115/1.4029860

183. Z. Zhu, Y. Chen, J. Wu, S. Zheng, W. Zhao, Performance study on S-CO$_2$ power cycle with oxygen fired fuel of s-water gasification of coal. Energy Convers. Manag. **199**(July), 112058 (2019). https://doi.org/10.1016/j.enconman.2019.112058

184. Thermophysical properties of fluid systems. NIST Chemistry WebBook, SRD 69 http://webbook.nist.gov/chemistry/fluid/

185. C. Philibert, *Producing Industrial Hydrogen from Renewable Energy* (IEA, Paris, 2017)

186. A. Valera-Medina, H. Xiao, M. Owen-Jones, W.I.F. David, P.J. Bowen, Ammonia for power. Prog. Energy Combust. Sci. **69**, 63–102 (2018). https://doi.org/10.1016/j.pecs.2018.07.001

187. H. Kobayashi, A. Hayakawa, K.D.K.A. Somarathne, E.C. Okafor, Science and technology of ammonia combustion. Proc. Combust. Inst. **37**(1), 109–133 (2019). https://doi.org/10.1016/j.proci.2018.09.029

188. O. Kurata, N. Iki, T. Matsunuma, T. Inoue, T. Tsujimura, H. Furutani, H. Kobayashi, A. Hayakawa, Performances and emission characteristics of NH_3–Air and NH_3–CH_4–Air combustion gas-turbine power generations. Proc. Combust. Inst. **36**(3), 3351–3359 (2017). https://doi.org/10.1016/j.proci.2016.07.088

189. T. Cai, D. Zhao, B. Wang, J. Li, Y. Guan, NO emission and thermal performances studies on premixed ammonia-oxygen combustion in a CO_2-free micro-planar combustor. Fuel **280**(x), 118554 (2020). https://doi.org/10.1016/j.fuel.2020.118554

190. H. Ishaq, I. Dincer, A comprehensive study on using new hydrogen-natural gas and ammonia-natural gas blends for better performance. J. Nat. Gas Sci. Eng. **81**(May), 103362 (2020). https://doi.org/10.1016/j.jngse.2020.103362

191. A. Valera-Medina, S. Morris, J. Runyon, D.G. Pugh, R. Marsh, P. Beasley, T. Hughes, Ammonia, methane and hydrogen for gas turbines. Energy Procedia **75**, 118–123 (2015). https://doi.org/10.1016/j.egypro.2015.07.205

192. S. Li, S. Zhang, H. Zhou, Z. Ren, Analysis of air-staged combustion of NH_3/CH_4 mixture with low NOx emission at gas turbine conditions in model combustors. Fuel **2019**(237), 50–59 (2018). https://doi.org/10.1016/j.fuel.2018.09.131

193. T. Honzawa, R. Kai, A. Okada, A. Valera-Medina, P.J. Bowen, R. Kurose, Predictions of NO and CO emissions in ammonia/methane/air combustion by LES using a non-adiabatic flamelet generated manifold. Energy **186**, 115771 (2019). https://doi.org/10.1016/j.energy.2019.07.101

194. H. Xiao, A. Valera-Medina, P.J. Bowen, Study on premixed combustion characteristics of co-firing ammonia/methane fuels. Energy **140**, 125–135 (2017). https://doi.org/10.1016/j.energy.2017.08.077

195. E.C. Okafor, K.D.K.A. Somarathne, R. Ratthanan, A. Hayakawa, T. Kudo, O. Kurata, N. Iki, T. Tsujimura, H. Furutani, H. Kobayashi, Control of NOx and other emissions in micro gas turbine combustors fuelled with mixtures of methane and ammonia. Combust. Flame **211**, 406–416 (2020). https://doi.org/10.1016/j.combustflame.2019.10.012

196. C. Filipe Ramos, R. C. Rocha, P. M. R. Oliveira, M. Costa, X.-S. Bai, Experimental and kinetic modelling investigation on NO, CO and NH_3 emissions from NH_3/CH_4/air premixed flames. Fuel **254**(June), 115693 (2019). https://doi.org/10.1016/j.fuel.2019.115693

197. H. Nozari, A. Karabeyoğlu, Numerical study of combustion characteristics of ammonia as a renewable fuel and establishment of reduced reaction mechanisms. Fuel **159**, 223–233 (2015). https://doi.org/10.1016/j.fuel.2015.06.075

198. K.D.K.A. Somarathne, S. Hatakeyama, A. Hayakawa, H. Kobayashi, Numerical study of a low emission gas turbine like combustor for turbulent ammonia/air premixed swirl flames with a secondary air injection at high pressure. Int. J. Hydrogen Energy **42**(44), 27388–27399 (2017). https://doi.org/10.1016/j.ijhydene.2017.09.089

199. J. Otomo, M. Koshi, T. Mitsumori, H. Iwasaki, K. Yamada, Chemical kinetic modeling of ammonia oxidation with improved reaction mechanism for ammonia/air and ammonia/ hydrogen/air combustion. Int. J. Hydrogen Energy **43**(5), 3004–3014 (2018). https://doi.org/ 10.1016/j.ijhydene.2017.12.066

200. N.A. Hussein, A. Valera-Medina, A.S. Alsaegh, Ammonia-hydrogen combustion in a swirl burner with reduction of NOx emissions. Energy Procedia **2019**(158), 2305–2310 (2018). https://doi.org/10.1016/j.egypro.2019.01.265

201. A. Valera-Medina, D.G. Pugh, P. Marsh, G. Bulat, P. Bowen, Preliminary study on lean premixed combustion of ammonia-hydrogen for swirling gas turbine combustors. Int. J. Hydrogen Energy **42**(38), 24495–24503 (2017). https://doi.org/10.1016/j.ijhydene.2017. 08.028

202. H. Xiao, A. Valera-Medina, P. Bowen, S. Dooley, 3D simulation of ammonia combustion in a lean premixed swirl burner. Energy Procedia **142**, 1294–1299 (2017). https://doi.org/10. 1016/j.egypro.2017.12.504

203. A. Valera-Medina, M. Gutesa, H. Xiao, D. Pugh, A. Giles, B. Goktepe, R. Marsh, P. Bowen, Premixed ammonia/hydrogen swirl combustion under rich fuel conditions for gas turbines operation. Int. J. Hydrogen Energy **44**(16), 8615–8626 (2019). https://doi.org/10.1016/j.ijh ydene.2019.02.041

204. A. Yamamoto, M. Kimoto, Y. Ozawa, S. Hara, Basic co-firing characteristics of ammonia with pulverized coal in a single burner test furnace
205. R. Jones, J. Goldmeer, B. Monetti, Addressing gas turbine fuel flexibility. GE Energy, GER-4601 ((05/11) revB), 20 (2011)
206. K.R. Asfar, H. Hamed, Combustion of fuel blends. Energy Convers. Manag. **39**(10), 1081–1093 (1998). https://doi.org/10.1016/S0196-8904(97)00034-4
207. D. Straub, D. Ferguson, K. Casleton, G. Richards, Effects of propane/natural gas blended fuels on gas turbine pollutant emissions. 5th US Combust. Meet. 4, 1993–2005 (2007)
208. N. Saifuddin, H. Refal, P. Kumaran, Performance and emission characteristics of micro gas turbine engine fuelled with bioethanol-diesel-biodiesel blends. Int. J. Automot. Mech. Eng. **14**(1), 4030–4049 (2017). https://doi.org/10.15282/ijame.14.1.2017.16.0326
209. P.K. Chaurasiya, S.K. Singh, R. Dwivedi, R.V. Choudri, Combustion and emission characteristics of diesel fuel blended with raw Jatropha, soybean and waste cooking oils. Heliyon **5**(5), e01564 (2019). https://doi.org/10.1016/j.heliyon.2019.e01564
210. M. Nemitallah, S. Alkhaldi, A. Abdelhafez, M. Habib, Effect analysis on the macrostructure and static stability limits of oxy-methane flames in a premixed swirl combustor. Energy **159**, 86–96 (2018). https://doi.org/10.1016/j.energy.2018.06.131
211. M.A. Habib, S.S. Rashwan, M.A. Nemitallah, A. Abdelhafez, Stability maps of non-premixed methane flames in different oxidizing environments of a gas turbine model combustor. Appl. Energy **189**, 177–186 (2017). https://doi.org/10.1016/j.apenergy.2016.12.067
212. S.J. Shanbhogue, Y.S. Sanusi, S. Taamallah, M.A. Habib, E.M.A. Mokheimer, A.F. Ghoniem, Flame macrostructures, combustion instability and extinction strain scaling in swirl-stabilized premixed CH_4/H_2 combustion. Combust. Flame **163**, 494–507 (2016). https://doi.org/10.1016/j.combustflame.2015.10.026
213. M. Aliyu, M.A. Nemitallah, S.A. Said, M.A. Habib, Characteristics of H_2-enriched CH_4-O_2 diffusion flames in a swirl-stabilized gas turbine combustor: experimental and numerical study. Int. J. Hydrogen Energy **41**(44), 20418–20432 (2016). https://doi.org/10.1016/j.ijhydene.2016.08.144
214. S.A. Said, M. Aliyu, M.A. Nemitallah, M.A. Habib, I.B. Mansir, Experimental investigation of the stability of a turbulent diffusion flame in a gas turbine combustor. Energy **157**, 904–913 (2018). https://doi.org/10.1016/j.energy.2018.05.177
215. A. Valera-Medina, R. Marsh, J. Runyon, D. Pugh, P. Beasley, T. Hughes, P. Bowen, Ammonia-methane combustion in tangential swirl burners for gas turbine power generation. Appl. Energy **185**, 1362–1371 (2017). https://doi.org/10.1016/j.apenergy.2016.02.073
216. I. Saanum, M. Ditaranto, A. Schönborn, J. Janczewski, Demonstration plant and combustion system for an oxy-fuel gas turbine cycle. Proc. ASME Turbo Expo Turbomach. Tech. Conf. Expo 1–7 (2016)
217. D. Flin, First fire for La Porte carbon capture demo, https://gasturbineworld.com/first-fire-for-la-porte-carbon-capture-demo/
218. R. E. Anderson, S. Macadam, F. Viteri, D. O. Davies, J. P. Downs, A. Paliszewski, Adapting gas turbines to zero emission oxy-fuel power plants, in *Proceedings of ASME Turbo Expo 2008: Power for Land, Sea and Air GT*. (Berlin, Germany, 2008), pp. 1–11
219. U. Balachandran, M. S. Kleefisch, T. P. Kobylinski, S. L. Morissette, S. Pei, *Oxygen Ion-Conducting Dense Ceramic*. US005639.437A (1997)
220. M. Ball, M. Wietschel, The future of hydrogen—opportunities and challenges. Int. J. Hydrogen Energy **34**(2), 615–627 (2009). https://doi.org/10.1016/j.ijhydene.2008.11.014
221. P.P. Edwards, V.L. Kuznetsov, W.I.F. David, N.P. Brandon, Hydrogen and fuel cells: towards a sustainable energy future. Energy Policy **36**(12), 4356–4362 (2008). https://doi.org/10.1016/j.enpol.2008.09.036
222. Rinkesh. What is Hydrogen Energy? https://www.conserve-energy-future.com/advantages_disadvantages_hydrogenenergy.php

Chapter 4
Stratified and Hydrogen Combustion for Higher Turndown and Lower Emissions

4.1 Carbon Capture Technologies

Global warming is one of the major consequences of the rise in energy consumption on the environment. According to the International Energy Agency (IEA), there will be a 30% increase in energy consumption by 2040 [1], which amounts to the addition of 6700 GW of electricity to the current systems. With regard to greenhouse gases (GHG), carbon dioxide (CO_2) is the primary component and a substantial contributor to global warming, accounting for around 77% of GHG emissions [2]. The average sea level is increasing as a result of global warming, while the northern hemisphere's snow caps have drastically decreased (by an estimated 9% over the past five decades). Fossil fuels, which are the main source of CO_2, are the main source of concern because they are the main source of CO_2 [3]. Additionally, it is anticipated that for the next 50 years, fossil fuels will continue to be a substantial source of energy [4]. It is strongly advised that the atmospheric CO_2 concentration levels do not exceed 450 ppm by 2050 [2] in order to avoid the disastrous effects of global warming. Therefore, it is imperative to combat the threat of global warming, particularly by reducing CO_2 emissions, and to develop strategies for the efficient and cleaner use of fossil fuels that can result in a cleaner environment for future generations.

Gas-turbines are being widely used for power production across the globe because of the benefits they offer such as high cycle efficiency, wider operability limits, and low NOx under premixed conditions. NOx can be produced, inside the combustor, through various mechanisms such as fuel NOx, thermal NOx, and prompt NOx [5, 6]. However, NOx emissions are subjective to the combustion temperature and residence time inside the combustion zone. The principal mechanism of NOx formation within the spots of high temperature is the thermal NOx, which is formed through the Zeldovich mechanism. A minor decrease in the temperature of the elevated spots can have a substantial effect on the NOx formation rate. For instance, for a given oxygen fraction (OF) and percentage of nitrogen (N_2), the amount of thermal

M. A. Nemitallah et al., *Hydrogen for Clean Energy Production: Combustion Fundamentals and Applications*, https://doi.org/10.1007/978-981-97-7925-3_4

NOx produced at 2100 K in few milliseconds will be approximately equal to the thermal NOx produced in a few seconds at 1800 K. Therefore, it is highly desirable to control the temperature inside the combustor in order to avoid hotspots that accelerate and favor the formation of thermal NOx inside the combustor [7]. Therefore, intensive research is going on for the development of advanced technologies for low-temperature combustion in gas turbines, including primarily lean premixed combustion techniques, for the sake of reduction of NOx emissions.

To mitigate these noxious pollutants, researchers have come up with different technologies, including Dry-Low NOx (DLN) or Dry-Low Emission (DLE) burners, EV burners, etc. such burners have depicted promising results in achieving low levels of NOx emissions. The dynamics and challenges associated with these technologies are discussed in the next section. CO_2, as well, is one of the main contributors to the global warming, different technologies have been developed to capture CO_2 and minimize the CO_2 emissions to the environment. There are three main technologies for capturing CO_2 i.e., pre-combustion capture, post-combustion capture, and oxy-fuel combustion as shown in Fig. 4.1.

In pre-combustion method, CO_2 is first separated from the fuel and fuel is converted into combustible gas before it enters the combustor. The gas is then burned inside the combustion chamber [8] and CO_2, which is generated from the fuel, is sent for sequestration [9]. The first step is production of syngas ($H_2 + CO$) from the fossil fuel itself which is done mainly through the steam reforming process. Other processes for production of syngas are gasification and partial oxidation for solid and liquid/gaseous fuels respectively [10]. The facilitation of CO_2 removal is done by the high pressure that is inherent part of the water–gas shift process. Higher pressures and higher CO_2 concentration prior the CO_2/H_2 separation result in the less energy consumption in the compression and separation processes than

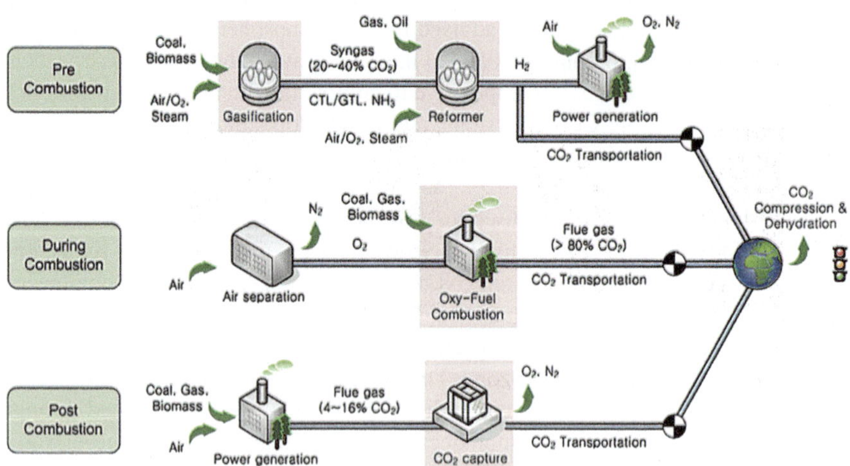

Fig. 4.1 Carbon capture techniques

the post-combustion capture [11]. In post-combustion capture, CO_2 is removed from the exhaust gases of the power plants, which are usually at ambient temperature and pressure. Therefore, post-combustion capture techniques fit easily with the existing power plants without major retrofitting and, consequently, it has an innate advantage of comparatively easier retrofitting as compared to its counterparts [12–14]. The driving potential is comparatively low to capture CO_2 from the exhaust/flue gases due to its lower concentration [15]. However, the considerable volume of CO_2 in the flue gases necessitates the need for heavy equipment resulting in extra capital costs that remains a challenge in order to come up with a commercially viable and cost-effective solution. Other challenge associated with the post-combustion capture include presence of other gases/contaminants in the flue gas, which make the process more complicated and costly when coupled with the existing technologies [16] and the cleaning of flue gases before CO_2 separation [17]. In oxy-fuel combustion, the oxidizer used is pure oxygen rather than air. When burning fuel with pure oxygen, the flue gas usually comprises of CO_2 and H_2O.

In conventional gas-turbine combustion technologies, air is used as an oxidizer which contains N_2, and it takes away considerable amount of heat and lowers down the combustion temperature. Whereas, in oxy-fuel combustion, N_2 is absent which considerably augments the temperature of the flue gas. Among these three technologies, oxy-fuel combustion is a promising technology for CO_2 capture in gas-turbine applications. In oxy-fuel combustion, the oxidizer used is pure oxygen rather than air. When burning fuel with pure oxygen, the flue gas usually comprises of CO_2 and H_2O where CO_2 can be captures by condensing the water. However, oxy-fuel combustion was previously known for its retarded stability and lower efficiency, but with time and advancements in technologies, the efficiencies are being improved. The main advantage of the oxy-fuel combustion compared to the air-combustion is the absence of nitrogen which decreases the NO_x emissions significantly and there are no other major pollutants apart from CO_2. In addition, it is less expensive technology compared to other carbon capture technologies. For example, Gerbelova et al. [18] performed the feasibility assessment of CO_2 capture retrofitted in an existing power plant for post-combustion and oxy-combustion, and they reported that the total capital investment of the post-combustion CO_2 capture system is estimated at 260 million euros and the annual operation and maintenance costs of around 43 million euros. Whereas the oxy-fuel combustion capture requires a capital investment of about 217 million euros and 37 million euros annually for operation and maintenance. Thimsen et al. [19] performed economic comparison of oxy-coal CO_2 capture with pre-combustion and post-combustion technologies and reported that the levelized cost of energy for the oxy-coal cases was approximately 7% lower than levelized cost of energy for the air-fired with post-combustion capture cases employing the same steam cycles. The CO_2 capture for the oxy-coal cases was near 100%, compared with 90% for the post-combustion capture case. Also, Bouillon et al. [20] compared the integrated CO_2 capture from post-combustion and oxy-fuel combustion for coal power plants and reported efficiencies of 34.6% and 36.6% for post-combustion and oxy-fuel combustion, respectively.

Over past few decades the gas turbine manufacturers have upgraded the combustion process to meet the emission criteria [21], from operation diffusion flame to newer novel combustion techniques such as, lean premix combustion (LPM) [22], oxy-fuel combustion [23], and stratified combustion [24, 25] etc. Stratified combustion usually refers to the combustion of fuel and oxidizer with stratification in mixture composition and/or temperature [26, 27]. Stratified or inhomogeneously premixed combustion, in which the flame travels through an inhomogeneous reactant mixture, is another of these attention-seeking modern technologies. The LPM technique also faces its own difficulties, such as flame instabilities and restricted operability under low load situations, despite being able to attain extremely low NOx emissions [28–30]. The concept of premixed stratified-charge combustion has been established in an effort to increase the turndown ratio of a combustor using the LPM approach. In contrast to conventional combustion, stratified combustion uses multiple fuel/oxidizer flows into the combustor to operate. Depending on the equivalence ratio and/or temperature, the flows can either be the same or different. Additionally, some recent studies indicated the emergence of triple inlet stratification combustion [31, 32]. Modern gas turbines rely heavily on staged or stratified combustion to achieve larger flammability limits, higher turndown ratios, better flame anchoring, and lower emissions and fuel consumption. Fuel is staged in the staging technique to produce two successive reaction zones with varying equivalency ratios. The reacting mixture flow that enters the combustor is split into two coaxial streams with different equivalency ratios using the splitting technique. The annular stream is of lower equivalence ratio to keep the overall operating equivalence ratio within the ideal range to maintain the desired low level of emissions, whereas the central stream is of higher equivalence ratio, close to stoichiometry, to keep the flame ignited even under ultra-low loading conditions (i.e., greater turndown ratio). The name of this stratification technique is dual lean premixed combustion (DLPM). By utilizing DLPM technology, it is possible to reduce the total equivalency ratio to values below the benchmark lower flammability limit of LPM combustion while still managing emissions at lower levels. This increases the combustor operability to considerably leaner circumstances.

Fuel flexibility is another effective approach to improve the adversely affected flame stability due the presence of high CO_2 concentrations in oxy-fuel combustion systems. Different fuels have different combustion, emission, and flame stability characteristics. That means, certain fuel compositions, or enrichment with highly reactive clean fuels, can extend flame operability to even lower loading conditions. Syngas, despite being a substitute for natural gas, has different combustion behaviour compared to natural gas. This is because the main constituents of syngas are carbon monoxide and hydrogen, which can produce higher adiabatic flame temperature and wider flame stability. Similarly, enriching hydrocarbon fuels with hydrogen can reduce some of the operational issues associated with the oxy-combustion process, by improving the burning velocity and widening the blowout stability limits [21]. In addition to using hydrogen to boost the kinetics of commonly used fuels, the utilization of hydrogen itself as a primary fuel is getting more attention in recent times. The development of such fuel/oxidizer flexible DLPM burner can lead to wider operability and ultra-low emissions of the combustor for gas turbine applications.

This chapter provides an overview of the combustion methods used in industrial gas turbines to reduce emissions and improve fuel efficiency. In contrast to non-premixed combustion, lean premixed combustion (LPM) technology is a low-temperature combustion approach that controls NOx emissions. One of the most promising LPM-based combustion systems for reducing NOx emissions is the Dry Low NOx (DLN) system. However, DLN combustors struggle to separate CO_2 from the exhaust stream in order to lessen the carbon footprint of gas turbines. This is especially true while running at low load (near blowout). In an effort to solve these challenges, gas turbine manufacturers created enhanced-design burners, such as the Dual Annular Counter Rotating Swirl (DACRS) and Enhanced-Vortex (EV/SEV/AEV) burners, for higher turndown and reduced NOx emissions. The DACRS combustors have nearly twice the volume of traditional burners, giving the fuel plenty of time to burn completely. In EV-burners, the mixing efficiency is increased, leading to enhanced flame stability at low load or start-up conditions. The operations of CO_2 separation and capture from the exhaust stream can be made simpler by modifying the combustors to handle oxy-fuel flames rather than the typical air–fuel flames. This will allow for complete emission control. However, due to the high CO_2 concentrations inside the combustor, oxy-fuel combustion technology has a number of unique difficulties, such as slowed reaction rates, decreased combustion efficiency, and a constrained operating range for stable flames. In order to improve the operability, control the emissions, and turndown ratio of gas turbine combustors, the concept of flame stratification, or heterogenization of the overall equivalency ratio, was originally developed. This idea can be put into practice by either staging the combustion or by dividing the reactants into two coaxial streams with differing equivalency ratios. In order to get a greater turndown ratio, the goal is to have a stream of near-stoichiometric equivalence ratio that serves as a hot pilot and maintains the flame lit even under low load conditions. The stability range of current LPM flames for industrial applications can be expanded with this technology. This chapter presents three staging strategies for stratified combustion (air staging, fuel staging, and combined air–fuel staging), along with each strategy's most recent developments. In this chapter, oxy-fuel combustion technology and stratified combustion technique integration are described as potential future directions that could lead to total control of gas turbine emissions and greater operability turndown ratios. The hydrogen combustion method is a newer approach for reducing emissions from gas turbines. The combustion of hydrogen can lead to improvements in many combustion properties, including soot production, fuel consumption, emissions, and operability of windows. We introduce the most recent advancements and obstacles facing the use of hydrogen gas turbines.

4.2 Lean Premixed Combustion for Gas Turbines

The combustor is the most important part of a gas turbine. It must be designed for sustainable continuous combustion with a flue gas temperature within the operating limits of the turbine materials. The conventional air–fuel combustor designs usually consist of a primary combustion zone for burning the fuel with air, and a secondary zone for diluting the combustion products with more air to control the gas temperature within the operating limit of the turbine. Currently, most of the conventional gas turbines adopt the non-premixed (diffusion) flame type, which is fuel-flexible and stable. However, such kind of flames results in stoichiometric combustion zones and, as a result, high NOx emissions are generated (> 70 ppm for natural gas and > 100 ppm for liquid fuels). Based on that, the idea of premixed combustion was developed to control NOx emissions in modern gas turbine combustors. The Dry Low NOx (DLN) combustor is one of the most promising combustion systems for the control of NOx emissions under lean premixed combustion conditions. The flame temperature is controlled by the excess amount of air premixed with fuel upstream of the flame, resulting in combustion under lean conditions, and, as a result, NOx emissions are reduced. The main disadvantage of the DLN system is its deteriorated flame stability under part load operating conditions. In addition, the flame has higher tendency to flashback. The main challenge facing modern gas turbine manufacturers is thus to design fuel-flexible combustors with stable flames under the different engine loading conditions.

As discussed in the preceding section, a small difference in temperature within the combustion zone can substantially influence the rate of thermal NOx formation, therefore, it is highly recommended to avoid the formation of stoichiometric-rich zones within the flame as observed in the non-premixed combustion mode in order to reduce the NOx emissions. In the non-premixed (diffusion) combustion mode, higher portion of compressed air is fed into the flame in order to decrease the temperature, which rises as a consequence of stoichiometric-rich spots inside the combustion zone and a small amount of air is left for the liner cooling as shown in Fig. 4.2. Whereas, in the lean-premixed combustion mode, most of the air from the compressor is fed in to the flame, making it leaner and only a limited proportion is left for the liner cooling, which avoids the formation of high temperature zones and, consequently, reduces the temperature inside the combustion zone which minimizes the formation of thermal NOx [33]. The lower temperature inside the combustion zone can, in turn, result in incomplete combustion, thereby augmenting the CO emissions as shown in Fig. 4.3. Therefore, this is a trade-off between the different emissions, which needs special attention while designing the lean-premixed combustors.

Apart from the fact that LPM combustion significantly reduce the emissions, they are much susceptible to the static as well as dynamic instabilities [36]. Static instabilities mainly comprise of blowout and flashback. Blowout refers to the state where the flame becomes too weak, lifts-off its base and start propagating in the direction where it ultimately quenches. This usually occurs when the flow velocity is much higher than the burning velocity of the flame. Different correlations are used

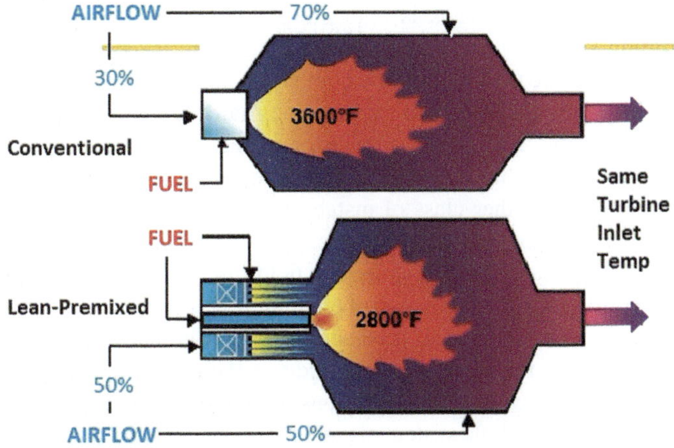

Fig. 4.2 Non-premixed (diffusion) versus lean premixed combustion systems [34]

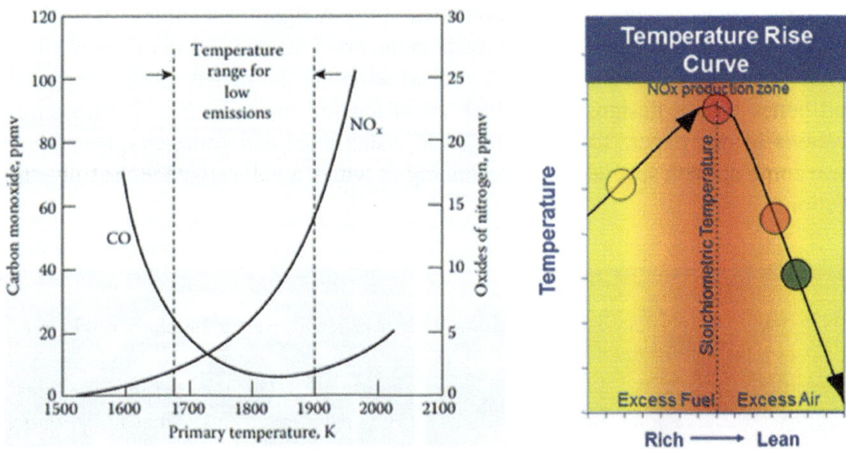

Fig. 4.3 NOx and CO emissions versus primary temperature and temperature versus air to fuel ratio (AFR) [35]

to explain the blowout mechanism and one of them is the Damkohler number (Da) [37–39]. Damkohler number (Da) is the ratio of residence timescale to chemical timescale [40], as expressed in Eq. (4.1), where S_L is the laminar flame speed, α is the thermal diffusivity, d is the characteristic length, and U_{ref} is the flow velocity.

$$Da = \frac{\tau_{res}}{\tau_{chem}} = \frac{S_L^2 d}{\alpha U_{ref}} \tag{4.1}$$

On the other hand, flashback is opposite phenomenon of blowout, where the laminar burning velocity becomes too high compared to the flow velocity that the flame tries to propagate upstream of the combustor in the direction of the incoming flow. Unlike blowout, flashback can cause serious hardware damage when the flame tries to propagate upstream of the combustor mainly because the burners are not designed to handle such high resonant frequencies upstream of the combustor and they ultimately fail. Another class of instability that greatly affects the combustor is the dynamic instability, which is mainly caused by the thermos-acoustic coupling between the heat-release and the pressure fields [41]. The acoustics inside the combustor is nothing but the interaction of pressure waves inside the combustor which can be constructive or destructive. If constructive interference occurs in any one of the pressure waves, it increases the amplitude of the pressure fluctuations. When the combustor operates with such fluctuations, it may result in hardware damage, as shown in Fig. 4.4, or the whole system may collapse [40, 42, 43]. As the risk involved in anything can never be zero, pressure fluctuations are always associated with the combustor even when it is operating in a stable manner. However, a study reported that the stable combustion is obtained when the amplitude of these fluctuations is < 5% of the mean pressure inside the combustor [44]. A significant number of advances on gas turbine combustor had been made in the previous decades to overcome these instabilities in the LPM combustors such as adopting the fuel-flexibility approach, and better burner designs etc., which are discussed throughout in the upcoming sections of this paper. To overcome these instabilities and emissions, researchers have come up with several burner technologies which are discussed in the following sections.

Fig. 4.4 Comparison of a damaged and a new combustor due to combustion acoustics [45]

4.3 Trending Gas Turbine Burner Technologies

4.3.1 Dual Annular Counter Rotating Swirl (DACRS) Burners

The NOx emissions are a primary concern in development of gas-turbine combustors over the decades. Various methods have been developed and adopted to lower down the NOx emissions in the past which have worked up to an extent. One of such promising method is the use of water or steam which lowers down the temperature and reduces the NOx emissions. This technology is mainly prevalent in the stationary sources. But injecting water or steam impose other challenges, despite lowering down the temperature, such as corroding the surfaces because of presence of impurities in water, quenching the CO radicals which makes it counterproductive by decreasing the efficiency [44]. In order to avoid water or steam, LPM combustors are equipped with DLN or DLE technologies [46]. These combustors are capable of decreasing the NOx emissions to the single-digit level [47]. By adhering to emission limits at peak load and controlling the emissions across the range of engine load, certain design requirements must be met to ensure the DLE burners work to their full potential. The system's ability to react quickly to load changes, maintaining minimal combustion acoustics, and ensuring smooth switching from one fuel to another, especially in dual-engine systems, are other critical parameters to ensure combustion stability and a wide operability range under all circumstances [35]. The DLE combustor concept was developed by several manufacturers with the goal of lowering NOx emissions without the use of steam or water injection.

General Electric (GE) adopted the LPM combustion for its LM6000 engine (see Fig. 4.5), which has the highest pressure-ratio in its family. In LM6000, DACRS premixers were adopted which are triple annular design for fuel staging to achieve ultra-lean flame with reduced temperatures [48, 49]. Figure 4.5a shows the LM6000 premixed combustor whereas Fig. 4.5b illustrates the cross-sectional diagram of the DACRS premixer. There is absence of dilution and cooling air. Rather, the turbine nozzle cooling air is utilized to cool the liners convectively from the back side. Also, thermal barrier coatings are applied to maintain acceptable liner metal temperatures. On important thing to note here is that the volume of the DACRS combustors is almost twice the conventional burners. This volume is large enough to provide ample residence time for the reaction of CO to be completed and to make sure that no unburned hydrocarbons are present inside the combustor volume.

4.3.2 Enhanced-Vortex (EV/SEV/AEV) Burners

Enhanced-Vortex (EV) burners are also known as EnVironmental burners [50], which is attributed to their ability to achieve ultralow NOx emissions. Beside other LPM gas-turbine technologies that offer ultralow emissions and outstanding flexibility such as

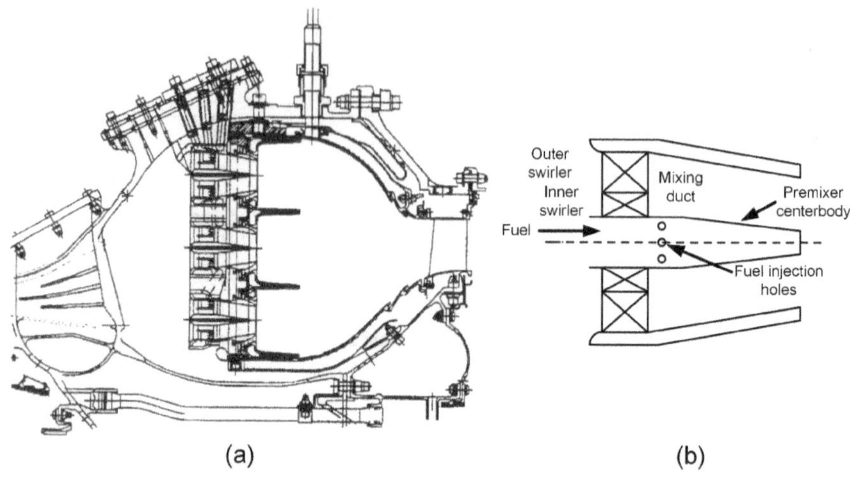

Fig. 4.5 **a** GE LM6000 DLN combustor system, and **b** cross-section of DACRS premixer [48]

micromixers burners, EV-burners are a strong competitor that offer wider flexibility and effective flame stabilization over a wide range of operation that enhances the turndown ratio; thus, they are a potential candidate for the zero-emission gas-turbines. As shown in Fig. 4.6, the geometry of EV-burners comprises of two half-cone shells. One of the shells is offset to a little extent in the radial direction; however, the axes of both half-cones remain parallel to each other. Therefore, two tangential slots of equal width are created between the shells. The stabilization mechanism of EV-burners is based on vortex breakdown which creates a strong inner recirculation zone [51] instead of using a swirler or center body. The gaseous fuel is injected transversely into the combustion air through a row of holes upstream of each air slot, which in fact enhances the effectiveness of mixing the gaseous fuel with air. The mixing effectiveness is an important parameter in ultralow NOx burners. While the combustible mixture moves in the axial direction, the diameter of the burner increases which in turn increases the strength of the swirl. Also, the divergence angle is designed in such a manner to position the vortex breakdown point closer to the burner exit as shown in Fig. 4.7. Moreover, EV-burners can be operated with dual fuels i.e., gaseous fuels and liquid fuels [50], which makes them even a better candidate to be implemented in ultralow NOx applications.

The mixing effectiveness is improved in EV-burners therefore, it results in higher flame stability at low load or startup conditions. To avoid formation of thermal NOx, mixing effectiveness plays an important role by minimizing the probability of formation of hot spots inside the combustion zone and, thus, results in complete combustion. All these characteristics are attributed to the vortex breakdown inside EV-burners which is an abrupt change in the structure of strongly swirling flows. The onset of vortex breakdown depends primarily on swirl number instead of the flow Reynold numbers [53]. However, the crux of the above brief discussion is that the velocity gradients within the vortex get weaker as flow recirculation is distributed

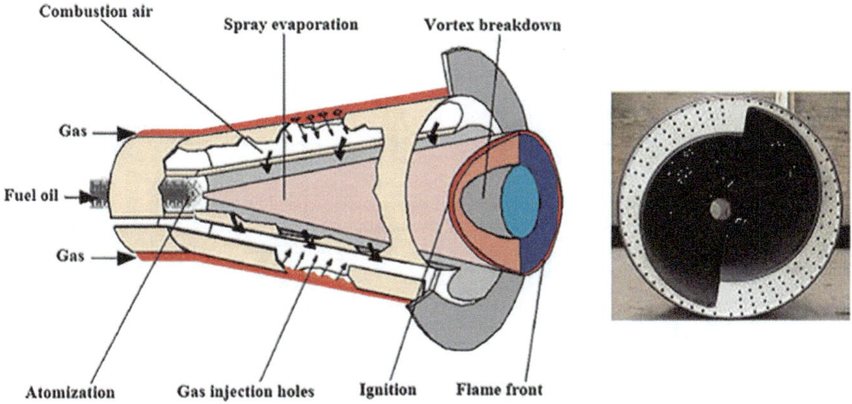

Fig. 4.6 Schematic of the EV-burner [52]

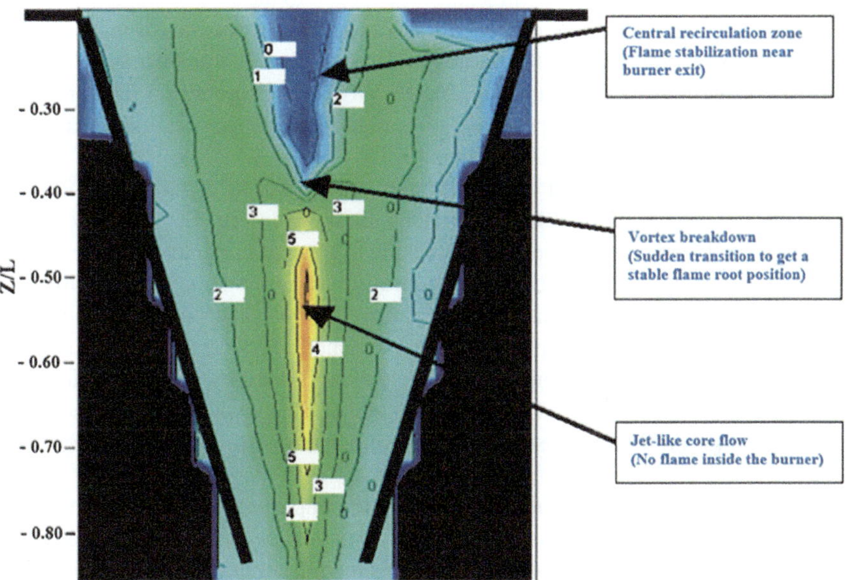

Fig. 4.7 Axial velocity distribution inside an EV-burner [52]

over a large area within the combustor and deceleration of axial flow along the vortex axis [54].

Another variant of the EV-burners is the Sequential EV (SEV) burner that has been developed to increase the efficiency of EV burners. As the name suggests, the combustion process in the SEV burners [55] takes place sequentially in two stages as shown is Fig. 4.8 [56]. First, the combustion process takes place in the combustion chamber that employ EV burners and then, in the second combustion chamber that

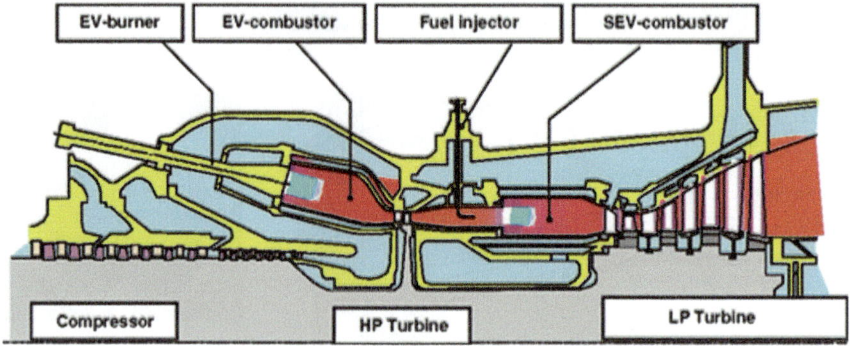

Fig. 4.8 Gas-turbine equipped with Sequential EV (SEV) burner [56]

use SEV burners. In the first stage of combustion, the fuel is burned to preheat the compressed air. The products of first stage are moved to the second stage where they get expanded in the high-pressure turbine and experience a pressure drop nearly by a factor of 2. The remaining fuel is burned in the second stage combustor and expand in a four-stage low-pressure turbine. The SEV burners operate in a flashback free operation despite of high temperatures after high-pressure turbine due to enhanced flame stability offered by the EV-burners [57].

To use the liquid fuels in the EV-burners, another type of burners, known as Advanced EV (AEV) burner, was introduced as shown in Fig. 4.9. A nozzle sprays liquid fuel inside the burner cone along with the four tangential inlet slots to establish a strong tangential flow protecting the cone walls from impingement of fuel droplets. To elongate the evaporation time, which in turn improves the mixing effectiveness between fuel and air in the gas phase, a cone is attached with straight mixing tube. The prevention of flashback is an inherent property of AEV burners which is achieved by keeping high axial velocity across the mixing tube. This is also achieved, however, by using excess air that increases local axial velocity close to the wall of the mixing tube [52].

Numerous studies have been done to look at the improved performance and lower NOx emission characteristics of EV burners. To construct a staged premixed burner that emits even less NOx at lower portion loads than the conventional functioning of an EV-burner, Zajadatz et al. [59] investigated the controlled splitting of gaseous fuel between the main premix circuit and the center pilot lance. The current gas-turbine engines, including the existing fuel distribution system, may readily be retrofitted with this fuel staging approach because it didn't require significant design changes for the EV burner system. To understand the impacts of fuel staging and fuel–air unmixedness on NOx emissions in EV burners with two cones, Cho et al. [60] carried out a numerical analysis. They stated that NOx emissions measured at the burner exit showed a high association with fuel and air unmixedness. To lessen unmixedness and the consequent NOx emissions, the authors suggested fuel splitting between the premix circuit and the central pilot lance. It was also suggested to improve the

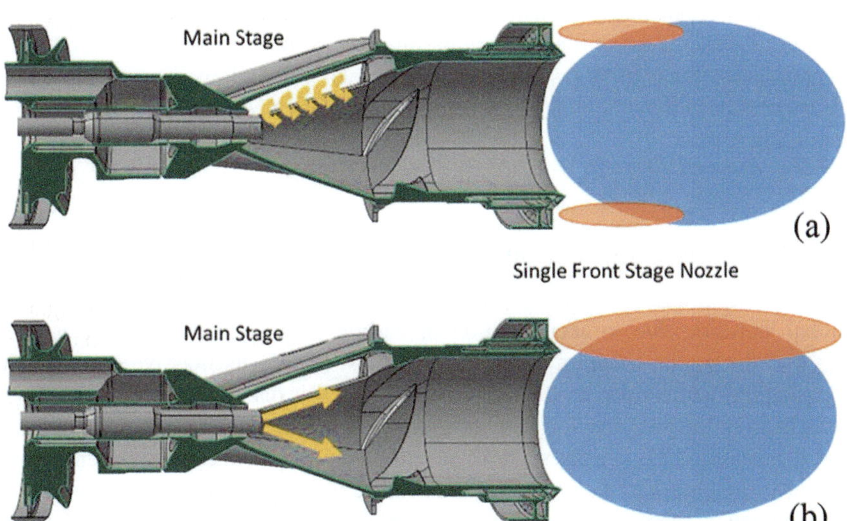

Fig. 4.9 Advanced EV *(AEV)* burner cross-section with: **a** gaseous fuel and **b** liquid fuel [58]

design by extending the time between fuel holes in the premix circuit, which would give fuel and air more time to mix. Using three different burner operating modes—perfectly premixed, in which fuel and air are mixed in the air supply line upstream of the burner, standard, in which fuel is injected through the burner's premix circuit, and staged, in which fuel is divided between the premix circuit and the central pilot lance—Guthe et al. [61] also compared fuel staging. In order to separate the impacts of unmixedness and changing turbulence intensity on combustion instabilities, they used the first mode as a reference scenario that is devoid of any spatial change in equivalency ratio. The staged mode optimized the fuel split for reducing both instabilities and emissions by utilizing the inherent flexibility of EV-burners. They claimed that permitting the flame anchoring point (vortex breakdown) to wander farther inside the burner cone worsens the ability to manage combustion instabilities by raising the pressure drop across the burner. As a result, the staged mode was optimized to regulate the position, variations, and emissions of the flame. The impact of fuel staging on the GT24 Alstom gas-turbine combustor was examined by Guyot et al. [62]. They used two premix stages to inject the fuel, with the first pumping it into the center of the burner air slots and the second injecting it into the center of the burner cone. Throughout the whole operating range, both premix stages were running continuously, eliminating diffusion fires and lowering NOx emissions as a result. During the erratic operating conditions, they reported outstanding combustion stability. The operating window was increased, and it showed low pulsations across the entire working range in addition to low NOx emissions, i.e., less than 5 ppmvd at baseload at 15% O_2. Without using steam or water injection, Steinbach et al. [63]

created a dual fuel EV burner to achieve strict NOx goals. At high pressures of up to 20 bars, they conducted combustion tests and evaluated NOx values under full load conditions. Despite using fuel oil, a liquid fuel, the NOx emissions for fuel oil were 25 ppmvd and for natural gas they were 10 ppmvd. The findings demonstrated that EV burners have the potential to emit significantly less NOx, provided that flawless premixing is used.

4.4 Oxy-fuel Combustion for Gas Turbines

Global warming due to accumulation of carbon dioxide in the atmosphere has captured the world attention in the recent years. Due to its massive atmospheric release rates, which exceed 30 Gt CO_2 per year, CO_2 is one of the greenhouse gases thought to be responsible for 60% of the global warming. Thus, in order to stop global warming, CO_2 capture systems have been developed to capture the enormous amounts of CO_2 emitted from power plants. The oxy-fuel combustion method uses pure oxygen as the oxidizer instead of air, in contrast to pre- and post-combustion processes for carbon capture. In the combustion chamber, using oxygen as an oxidizer results in an extremely high flame temperature [64, 65]. The firing temperature, above 2350 °F, can create a serious challenge in combustor material science. Therefore, this technique of combustion in gas turbines is not appropriate to be used right away [66]. The oxygen concentration at the combustor inlet can be regulated by recirculating a portion of the flue gas [67]. The exhaust of an oxy-combustion system contains mainly water vapor and highly concentrated CO_2. This CO_2 has an important effect on the gas transport properties, heat capacity, and the combustion process as well [67, 68]. Since the air (containing nitrogen) is not used in combustion, the combustion remains nitrogen-free. This eliminates the formation of thermal NOx. Also CO_2 can be easily separated from the exhaust gas through a simple condensation process of water vapor. Near-zero emission from the combustion process can be achieved as some amount of the generated CO_2 can be recirculated back to the combustor and the rest can be sent to CO_2 capture and compression plant [69, 70]. However, there is a chance of NOx formation if there is any presence of N_2 with the oxygen supplied to the combustor or if impure fuel is used [71–73]. The emission of CO increases with the increase in CO_2 in the oxidizer mixture and the lowest emission of CO can be achieved at near stoichiometric operation [74].

Typically, oxy-combustion systems require additional auxiliary units compared to conventional air combustion systems, like air separation unit, flue gas circulation system, and CO_2 purification unit (for CO_2 sequestration). There are also some challenges in adopting oxy-combustion technology. The increase of CO_2 concentration within the flame zone reduces the chemical kinetics rates, the laminar burning velocity, and the combustion efficiency [75, 76]. In order to obtain a stable oxy-flame, the CO_2 rich mixture within the combustor requires more oxygen, above the 21% of air-combustion at a given equivalence ratio [77, 78]. Furthermore, CO_2 is denser than N_2, so mixing with CO_2 affects the jet velocity, gas density, and pressure drop [79–81]. Adding CO_2 of high concentration can increase the O_2/CO_2 mixture's heat capacity. This affects the flame stability making the flame temperature lower and flame speed slower. Moreover, this flame speed can be influenced by gas transport properties as well [82]. The critical velocity ratio (fuel to oxidizer velocity ratio) in oxy-combustion was found to be six times that of normal air combustion [82] and, because of that, the oxy-flames encountered a sudden transition from the attached flame regime to the blowout regime directly without being lifted-off unlike the case of air combustion [82]. This reveals the dependency of the transition between 'attached flame' and 'lifted flame' state on the critical velocity ratio irrespective of the equivalence ratio. Moreover, for unity swirl number at stoichiometric conditions, the oxy-fuel flames showed more stability in the same combustor [83]. It was found that an oxy-combustion system with an oxygen mass fraction of 0.243 in the oxidizer mixture can produce an adiabatic flame temperature that is similar to that of air-combustion in a comparative numerical analysis on a porous/perforated combustor [84].

In contrast to traditional air–fuel combustion, oxy-fuel combustion has unusual kinetics. Oxy-fuel combustion produces flames with greater temperatures, necessitating the recycling of some flue gases to regulate the temperature in the combustion zone [85]. The oxygen fraction (OF) in the oxidizer, a mixture of O_2 and CO_2, should not be less than a defined level for stable combustion, according to a significant amount of research in the field of oxy-fuel combustion [86]. That can be the differences in the thermos-physical properties of N_2 and CO_2 [87, 88]. The CO_2 dilution affects the combustion characteristics in three different ways i.e., it has different radiation effects [89], thermal effects [90], and kinetic effects [91]. CO_2 has been found to adversely affect the flame speed [77] as well as the combustion efficiency [92] and slows down the reaction kinetics within the combustion zone [93, 94]. This is because CO_2 has higher volumetric heat capacity as compared to N_2; therefore, it takes more heat from the flame as compared to N_2, thereby it decreases the flame temperature and, consequently, the flame speed and flame stability are adversely affected. Moreover, higher concentrations of CO_2 in oxy-fuel combustion also result in decreased laminar flame speed and deteriorated stability [76]. Another problem associated with the oxy-fuel combustion is the higher temperature which results in CO_2 dissociation. The dissociation of CO_2 is itself an endothermic process which takes away heat from the flame and adversely affects the reaction kinetics [95]. All the factors mentioned greatly affect the combustion characteristics of oxy-fuel combustion resulting in deteriorated flame stability and narrower window for stable operation as compared to the air–fuel combustion. This is observed in the premixed combustion [23], non-premixed combustion [81], and partially premixed combustion [96]. Therefore, it can be inferred from the above argument that in order to switch the combustor from air–fuel operation to oxy-fuel operation without modifying the geometry of the combustor, the OF of 21% (by vol.) would not be sufficient for the flame to be sustained [78, 79]. Nemitallah and Habib [86] and Ditaranto and Hals [97] reported that the oxy-fuel flames cannot sustain with OF less than 21% and an oxygen fraction of 30% is needed in order to obtain similar characteristics to the air-based combustion [98]. This finding was validated by Abdelhafez et al. [99]

when they reported that the adiabatic flame temperature for methane-air flames at an equivalence ratio of unity is obtained at OF $= 30\%$ in oxy-methane flames. Beside this, other studies have reported different values of minimum OF needed for stable operation of oxy-fuel flames such as 25% [100], 27–35% [101]. Whereas, Song et al. [102] reported that 36% OF is needed in oxy-fuel flames to obtain temperature profiles identical to those of the air flames.

Several studies have been performed and showed that the oxy-fuel combustion experiences retarded stability as compared to the air–fuel flames. Moreover, higher OF is needed to compensate for the heat loss due to the addition of CO_2. Liu et al. [103] investigated the characteristics of oxy-fuel combustion in a gas-turbine combustor and reported that the higher value of OF ultimately results in higher flammability limits and wider stability maps. Shi et al. [104] experimentally investigated the oxy-methane flames under a range of OF. They considered OF ranging from 21 to 86% (by vol.) and reported that higher OF results in earlier flame flashback, whereas lower OF results in earlier blowout. With the addition of O_2, the laminar flame speed increased with a reported range of OF for sustainable flame of 50%-86% as shown in Fig. 4.10.

In order to explore the impact of oxidizer composition on flame stability, Nemitallah and Habib [86] did an experimental and numerical investigation of oxy-methane diffusion flames. They found that when the OF was dropped below 25%, the stability of the flame was negatively impacted. Poor burnout and an unstable raised flame were noted when the OF in the O_2/CO_2 stream was tuned to 21%,

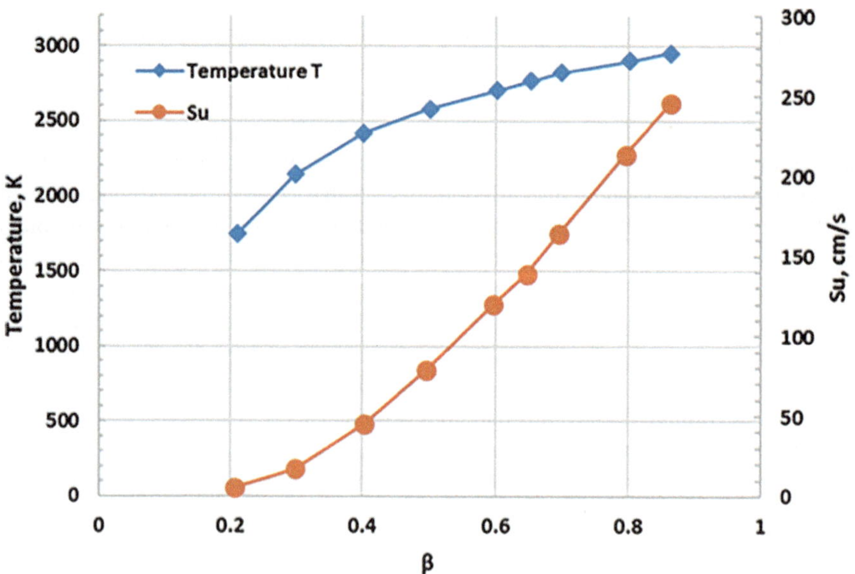

Fig. 4.10 Effect of oxygen fraction (OF) on adiabatic flame temperature (AFT) and laminar flame speed (LFS) [104]

according to Heil et al. [105]. However, full burnout of the flame was seen at 27% OF, and a stable flame was shown at 34% OF. Even under stoichiometric conditions, Abdelhafez et al. [99] demonstrated that no stable oxy-flames could be produced in a swirl-stabilized premixed gas-turbine model combustor with oxygen fractions below 22%. They also noted that the impact of the equivalency ratio (Φ) is substantially less than that of the oxygen fraction (OF), and that stable flames can be produced under much leaner conditions with higher OFs. Hu et al. [106] examined the effects of OF, equivalence ratio, and dilution ratio as well as the laminar flame speed of oxy-methane and methane-air flames under atmospheric circumstances. No stable oxy-flames could be produced, as demonstrated by Abdelhafez et al. The laminar flame speed was observed to be a quadratic function of OF. Additionally, when other variables remained constant, the laminar flame speed of N_2-diluted flames was around five times that of CO2-diluted flames. In their investigation of oxy-propane and air-propane flames at various OFs and equivalency ratios, Ali et al. [107] found that air-propane flames had a leaner blowout. They also stated that both flames blow out at about 0.2 as seen in Figs. 4.11 and 4.12, where the influence of the oxidizer is dominating at higher OFs and minimal at lower OFs.

Since the oxy-fuel combustion still in the development phase, its readiness can be assessed using the technology redness level (TRL) system. There are 9 levels in TRL system starting from 1 to 9, where TRL 1 represents the basic principles level and TRL 9 corresponds to the full-scale commercial development level. Table 4.1 lists the TRL levels for coal-powered oxy-combustion power plants. Majority of the oxy-combustion plants of less than 5 MWe in size have been develop before 2005 and falls into the TRL level of 5 or lower. On the other hand, larger testing facilities can be ranked between TRL level 5 and 6, like Oxy-Coal UK" 15-MW$_{th}$ test rig at Renfrew, Alstom's boiler simulator furnace (BSF), 30-MWth test rig of Clean Environment Development Facility at Ohio, and 40-MWth Doosan Babcock testing facility. These test facilities are partially integrated systems like oxygen supply systems, exhaust gas filtration system, and boiler testing system. Oxy-CFB Pilot plant by CIUDEN, Lacq Pilot Project by Total, and Schwarze Pumpe pilot plant by Vattenfall can be categorized into TRL-6. Large scale pilot project such as Callide oxyfuel facility can be ranked as TRL-7, while plants like White Rose project are classified as TRL-8. Oxy-combustion technologies are currently under trial phase and at TRL-8 since 2016 at Full-scale commercial projects that can be categorized into TRL-9 are yet to be developed [108].

4.5 Premixed Oxy-fuel Combustion for Gas Turbines

Due to the enormous rise in global energy demand, fossil fuels will be used for a longer period of time while renewable energy sources are developed and put into use, necessitating the creation of technology that reduce carbon emissions [109]. Pre-combustion, post-combustion, and oxy-combustion are examples of carbon capture systems [88, 110]. Because they necessitate substantial alterations to the current

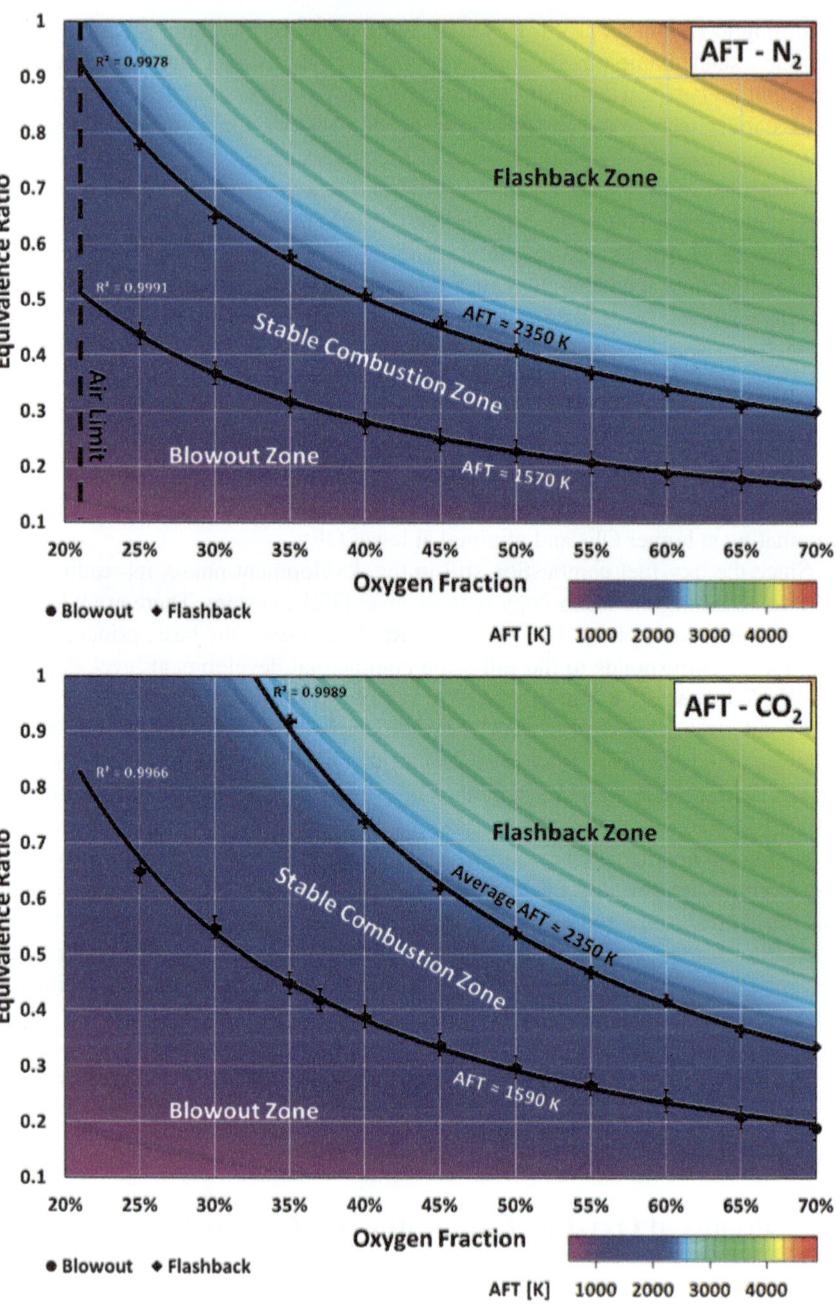

Fig. 4.11 Stability maps of oxy-propane (top) and air-propane (bottom) flames plotted on the contours of adiabatic flame temperature (AFT) [107]

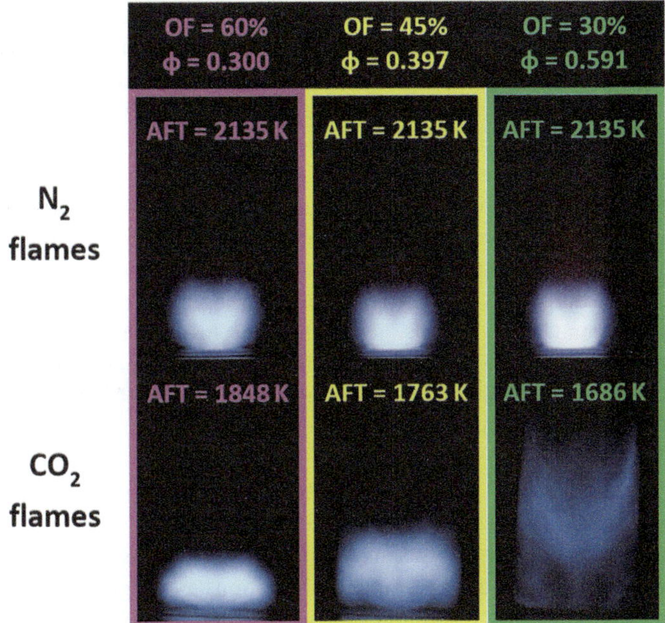

Fig. 4.12 Comparison of flame macrostructure for oxy-propane and air-propane flames [107]

Table 4.1 Technology readiness level (TRL) for oxyfuel combustion in coal-fired power plants with carbon capture [108]

TRL	Name of TRL phase and corresponding R&D	Oxy-combustion projects	Facility development phase
9	Commercial full-scale deployment		Commercial
8	Commercial sub-scale demonstration plant	White Rose Project	Demonstration
7	Pilot plant	Callide Oxy-fuel Project	Industrial Scale Pilot
6	Component prototype demonstration	CIUDEN's TDP Pilot Plants, Lacq Pilot Plant, Schwarze Pumpe Pilot Plant	Industrial Scale Pilot
5	Component prototype development	Various large-scale burner test facilities (i.e., Alstom, B&W, Doosan Babcock)	Industrial Scale Pilot
4	Laboratory component testing		Bench
3	Analytical 'proof of concept'		Bench
2	Application formulation		Bench
1	Basic principles observed		Bench

setups and plants, pre- and post-combustion capture techniques are both expensive [88, 110]. This, however, is not the case with oxy-combustion, which requires less modification and, thus, less capital investment [88]. Fuel is burned in a mixture of pure oxygen and recycled exhaust gas. The recycled effluent CO_2 helps reducing the combustion temperature [111]. Replacing the air-based N_2 by CO_2 will yield three effects: changing the thermal and transport properties of the combustion process, changing the reaction rates due to the chemical interactions of CO_2, and increasing the rate of radiation heat transfer due to the increased CO_2 concentrations [112]. Oxy-combustion enjoys better heat transfer characteristics than air-combustion due to the absence of nitrogen, since nitrogen consumes a portion of the heat released in the combustion zone [113]. The physical differences between CO_2 and N_2 are what cause the disparities between oxy-fuel and air–fuel combustion in terms of heat and mass transport characteristics [101]. Due to higher CO_2 and H_2O concentrations and a different CO_2/H_2O ratio than air–fuel combustion, oxygen combustion has a higher rate of radiation heat transfer [114, 115].

Combustion stability is a significant problem that prevents the development of more lean-premixed (LPM) gas turbines. Lean operations seek to increase power plant efficiency while lowering NOx and CO emissions. Because of weaker combustion properties, the leaner side of premixed combustion has a larger chance of blowout [116]. Premixed flames near blowout demonstrate a transient behavior of intermittent local extinction. When it comes to critical applications such as aircrafts, the stability of the flame is very important, and lean blowout posed a major safety hazard there [116], hence the flammability limits need to be accurately quantified. Combining the concepts of oxy-fuel combustion and premixed combustion can fully control the emissions; however, different flame characteristics along with changes in flame stabilization behavior are expected. In the case of oxy-methane, flame stability is limited within the range of OF from 29 to 70% (by vol.) in the oxidizer mixture, which narrows down the operational map [96]. CO_2 adversely affects the chemical kinetics of the flame. Experimental results indicated that increasing the OF leads to more robust flames that are bright in color [96]. Improved reaction kinetics were also observed upon increasing the OF in oxy-methane flames. A study found that increasing the premixing ratio in air combustion reduced the environmentally harmful emissions of CO and NO [117]. It was reported that increasing the inlet velocity of the premixed reactants entering an oxy-methane combustor leads to a reduced window of operation, with the flashback limit shifting towards the leaner side [117]. It was also noticed that as lean blowout is approached, the maximum heat release area transitions to a lower place. In addition, as the equivalence ratio is decreased, the influence of interaction with the outer recirculation zone is more evident. The flame starts to take an M-shape at lower equivalence ratios, instead of a V-shape at higher equivalence ratios [118].

The main disadvantage of the DLN system is its deteriorated flame stability under part load operating conditions. The main challenge facing modern gas turbine manufacturers is thus to design fuel/oxidizer-flexible combustors of ultra-low emissions with stable flames under the different engine loading conditions. Stratified combustion is considered as one of the most promising techniques towards the overcome of such challenge. Stratified combustion usually refers to the combustion of fuel and oxidizer with stratification in mixture composition and/or temperature.

4.6 Stratified Combustion for Gas Turbines

The combustion systems of the different power production applications are either of the premixed or non-premixed (diffusion) type. Each type has its own merits and demerits in terms of combustion characteristics and flame stability while varying the loading conditions. However, in practice, both the types can co-exist in the same combustor, making the flame partially premixed. This is attributed to the fact that obtaining perfect premixing is tough due to combustion instabilities and limited length available for mixing. As a result, supplying mixture can suffer from inhomogeneity in composition. The fluctuation in mixture composition, i.e., equivalence ratio in gas turbine combustion can lead to performance degradation of the turbine by initiating thermo-acoustic instabilities. In extreme cases, this can even cause physical damage of the turbine. Stratified-charge combustion is a kind of premixed combustion with controlled non-uniform spatial distribution of equivalence ratio within the combustion zone. This mode of combustion is characterized by wider flame stability limits and lower emissions, especially under low-load operating conditions, and it has been adopted in industrial combustion systems [119–121] and in test facilities as well [122–127]. Stratified combustion is analogous to partially premixed combustion but with controlled non-uniformity in the spatial distribution of equivalence ratio. Stratified combustion mode could be categorized into the back-supported and the front-supported regimes. In the back-supported regime, combustion happen when the combustion products are closer to stoichiometry than the combustion reactants, and the reverse is correct for the case of front-supported regime. Despite the limited availability of experimental data on stratified-charge combustion, numerical models were developed depending on simple extensions of the available models for premixed combustion [122–126, 128]; more extensive models were also proposed [129, 130]. However, experimental data are required for the validation and development of advanced numerical models for stratified-charge combustion. Further clarification is also still needed regarding the effects of distribution of local equivalence ratio on the generated combustion dynamics and the structure of the turbulent lean stratified premixed flames.

The characteristics of such flames have been investigated in several studies under laminar and turbulent flow conditions. Wider flammability limits and higher laminar flame speed have been reported for the stratified laminar flame as compared to its non-stratified counterpart in a number of studies [131, 132]. The presence of a nearby stoichiometric combustion zone enhances the stability of the stratified flame under low load conditions and widens the stability limits [132]. Stratified flames showed higher mean curvature and wrinkling when compared to non-stratified ones [133]. Pasquier et al. [134] studied lean stratified propane flames, and their results showed that flame propagation is mainly function of the stoichiometric history of the reactant mixture, and that the lean mixture zones can burn faster if there are near-stoichiometric combustion spots within the combustor. The different generated flames with different propagation speeds within a combustor adopting stratification result in additional flame stretching, leading to reduced flame thickness [135]. Anselmo et al. [136] investigated the influence of stratification on the flame structure in a low-turbulence slot combustor. The results showed that the stratified flame had higher flame surface density as compared to non-stratified premixed flame.

Testing stratified flames under laminar flow conditions is useful for understanding the basic behavior of the flame; however, for the successful implementation of this technology into actual combustion systems, flames should be tested under higher turbulence levels [137–142]. Measurements of the mean temperature profiles showed lower turbulent burning speed in the stratified flames as compared to the non-stratified ones [137, 140]. Despite the significant effect of stratification on flame curvature in methane-air flames, Vena et al. [139] found that this effect is not significant in the case of isooctane-air flames. They reported that the effect of stratification on flame curvature could be limited only to the cases of locally stoichiometric flames. Pires et al. [142] reported an agreement between the results of laminar and turbulent methane-air stratified flames with gradient changes of stoichiometry. They also reported higher flame speeds in the stratified flames, which was attributed to higher diffusion of heat and radicals from the near-stoichiometric zones to the leaner ones. This allowed back-supported flames to ignite even outside the fuel flammability limit. Despite the lower temperature within the rich zones of the flame, the diffusion of hydrogen atoms from the rich zones to the stoichiometric ones resulted in higher flame propagation speed. Marzouk et al. [143] studied stratified propane-air flames under varied spatial distribution of equivalence ratio, and the results confirmed the conclusions of previous studies, namely that richer flames support the stability of leaner flames, making the latter able to ignite even outside the flammability limit. Higher rates of heat release were also reported for the stratified premixed flames, compared to the non-stratified premixed ones. Richardson et al. [144] studied the behavior of strained flames with two opposite streams, one of fresh reactants and the other of recycled exhaust at different equivalence ratios. They reported that the modified pool flux has a more significant role than the heat on flame behavior with charge stratification.

Direct numerical simulations (DNS) have been used in a number of studies to solve for turbulent stratified flames. Using DNS in the full three-dimensional (3D) domain, Helie and Trouve [145] investigated the behavior of a stratified propane-air flame considering a single-step reaction mechanism. The results showed that charge stratification has no effect on the surface density of the generated flame. Furthermore, the overall reaction rate was reduced for the stratified flame, as compared to the non-stratified one, due to deviation from stoichiometric operation and reduction of local reaction rate. Haworth et al. [146] studied a turbulent stratified propane-air flame at a stoichiometric mean equivalence ratio using two dimensional (2D) numerical simulations and full chemistry and compared the results with a homogeneous (non-stratified) flame. The results showed insignificant difference of heat release between the two flames. Jimenez et al. [147] extended the study by Haworth et al. [146] over a range of mean equivalence ratio. The results showed that the mean equivalence ratio has a significant role in determining the rate of heat release. With acceptable results, a number of studies have been performed on modeling of stratified flames based on extending the available models for premixed combustion [129, 130, 135, 148, 149]. However, the issues of limited experimental data along with model limitations need to be resolved first before the development of properly validated models. Further work is required to quantify the microstructure and stability of stratified premixed flames in comparison to homogeneous premixed flames.

Stratified-charge multi-stage flames have shown distinct combustion characteristics as compared to homogeneous single-stage flames [142, 150–162]. Kyritsis et al. [152, 153] experimentally investigated the flame propagation in a laminar premixed stratified reacting stream and compared the results with those of homogeneous flames under the same overall operating equivalence ratio. The comparisons showed faster flame propagation rates and wider flammability limits, far outside the fuel flammability limit, for the stratified flames as compared to the homogeneous ones. The same conclusions were confirmed in a number of other studies [142, 151, 162]. Zhang and Abraham [161] concluded that the stratified-charge flames have different combustion characteristics as compared to normal homogeneous flames under the same overall operating equivalence ratio. They attributed such differences to the different mass and heat transfer characteristics within the flame zone.

Another type of stratified-charge flames is the triple flame, where the equivalence ratio is gradually varied in a direction perpendicular to flame propagation [157–160]. Triple flames can even propagate faster than homogeneous flames of stoichiometric mixtures under specific conditions [59, 60]. For triple flames, the combustion process is significantly affected by the distribution of equivalence ratio within the combustor [150, 154]. Both triple and stratified flames of variant spatial distribution of equivalence ratio showed wider operability limits and better combustion characteristics when compared to homogeneous flames.

Such combustion techniques have already been adopted in some industrial applications. Stratified combustion has been adopted in internal combustion engines (ICEs) to improve flame stability and to avoid misfire and slow burning of the fuel [163–167]. In the gasoline direct injection (GDI) combustion technique, for example, a stratified charge is generated in the vicinity of the spark plug, resulting in more stable combustion under very lean firing conditions with the cylinder. GDI aims at fine-tuning the requirements of higher fuel economy and stricter emission regulations [168]. Kim et al. [169] reported, based on experimental measurements, that stratification can reduce the NOx emissions of a non-stratified flame by up to 83.4%. The interaction between the rich and lean flames, in terms of transfer of heat and active intermediate combustion species, results in accelerated reaction rates and higher flame speed in the lean flame, making it possible to extend the operation below the fuel flammability limit. Such interaction within the combustion zone is the key parameter determining the behavior of non-homogeneous flames. Actually, both the triple and direct injection flames have different combustion characteristics when compared to dual stage lean premixed combustion (DLPM) flames. However, all of them are characterized by interaction within the combustion zone due to the non-uniform spatial distribution of equivalence ratio. DLPM flames with varied spatial distribution of equivalence ratio can widen the stability range of existing LPM flames for industrial applications.

Staging for stratified combustion is divided into three categories: (a) air-staging, (b) fuel staging, and (c) combined air–fuel staging. The next section provides a brief discussion on various types of stratified combustion along with their respective recent advancements.

4.6.1 Air Staging

In practice, the combustion processes in gas turbine combustors takes place in steps. First, the carbon and hydrogen bond from the hydrocarbon fuel converts through a series of quick progressing reaction, forming carbon monoxide (CO) and water vapor (H_2O). Secondly, the produced CO from the reactions, oxidizes with the remaining oxygen in the air through slower reactions. About two-third of the total chemical energy from the combustion process comes from the first step and the remaining comes from the oxidation of CO. In order to effectively handle the heat release from these reactions and reduce the harmful emissions, the combustion in gas turbine combustors is carried out in stages. These stages are usually known as primary zone, secondary zone and dilution zone, as shown in Fig. 4.13. In the primary zone, fuel is supplied through the injection port and compressed air through the swirler. Air and fuel mixture then get ignited using the phenomenon called 'Aerodynamic Spark Plug'. The equivalence ratio in this zone is kept over stoichiometry to maintain the stability of reaction and prevent blowout. Majority of the heat release from the combustion process (two-third of the total energy) occurs in this zone. The secondary zone is responsible for the remaining of the heat release. Oxidation of the carbon monoxide, formed in the primary zone, is done here. Lastly in the dilution stage, another jet of air is supplied to maintain the exhaust temperature for a level suitable to ensure integrity of turbine blade and prevent the detachment of the blades [170]. Air splitting in stages can also reduce the formation of NOx by controlling the combustor temperature [171].

Rich burn, Quick mix, Lean burn, commonly known as RQL combustors are common examples of staged combustion that uses air staging. It was first introduced in 1980 and adopted by NASA in 1990's as mode of low-NOx combustion for aero-propulsion engines. This combustor was later commercialized by Pratt and Whitney which they referred to as TALON or the 'Technology for Advanced Low NOx' for aero-engine applications. Effective operation with varying and complex fuel-oxidizer composition makes RQL combustors suitable for stationary gas turbine applications as well. In the primary zone, the combustion takes place over stoichiometry, for instance at equivalence ratio (Φ) of 1.8. Rich burning condition in the primary zone decreases the formation of NOx due to reduced oxygen supplied and helps in maintain flame stability through producing hydrocarbon radicals and hydrogen of high energy concentration. Partially oxidized effluents containing partially pyrolyzed

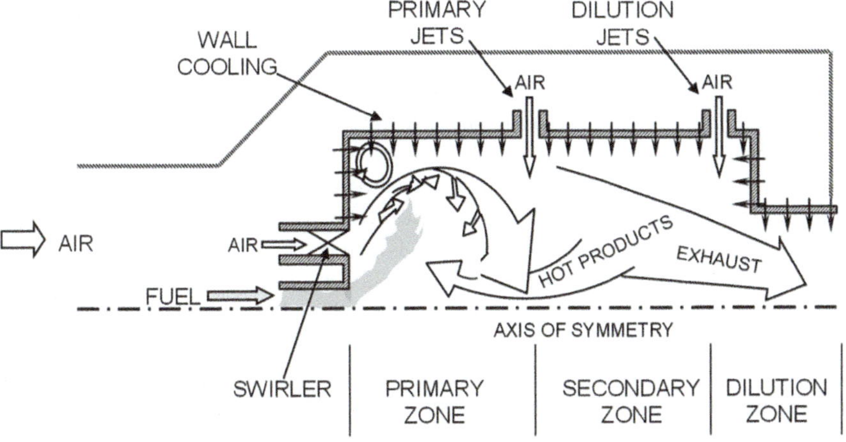

Fig. 4.13 Air staging in gas turbine combustor [170]

hydrocarbon species from the primary zone is then mixed with the jet of air in quick-mix section for a uniform mixture composition. The interaction between the effluent and air in this section leads to the ignition process for the secondary zone. Figure 4.14 shows the laboratory scale combustor with quick-mix section. In the secondary zone, further combustion process takes place in fuel-lean condition (e.g., $\Phi = 0.6$) through the oxidation of CO into CO_2. Leaner operation with relatively lower temperature also helps in eliminating the formation of NOx. Thus, in an ideal condition, the exhaust gas will no elements of pollution like CO and NOx. Finally, the exhaust is again diluted for temperature controlling purpose with an air jet before progressing towards the turbine blade [172].

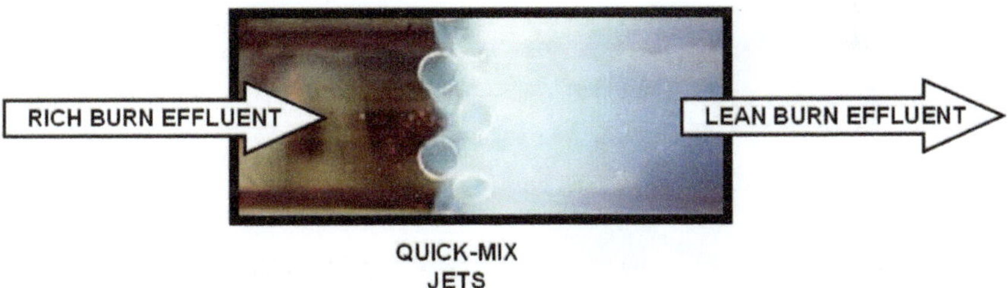

Fig. 4.14 Quick mix section in a laboratory model Rich burn Quick mix Lean burn (RQL) combustor [172]

4.6.2   Fuel Staging

Similar to the air staging, combustion with fuel staging corresponds to the splitting of total fuel supply into different injectors to obtain regions of different equivalence ratios within the combustor [173]. Fuel staging in a gas turbine combustor, can be carried out in both axial and radial direction. Axial fuel staging (AFS) is often known as sequential fuel staging and involves an additional fuel injector in secondary combustion zone [174–176]. AFS increases firing temperature in two zones inside the combustor reducing CO emission and helps in abating the smoke in the combustor exhaust. Inadequate mixing of the air and fuel in the primary zone can cause the formation of finely divided soot particles that leads to smoke production in the exhaust stream. The soot particles can be eliminated with the increase in combustion temperature downstream as an effect of secondary fuel injection in AFS. High firing temperature can also increase the NOx formation during combustion, thus AFS should be carefully implemented to minimize NOx formation inside the combustor [177]. AFS in gas turbine combustor can be found in General Electric (GE)'s 7F-DLN 2.6 + gas turbine combustor (shown in Fig. 4.15), which can deliver 25% less fuel consumption with reduced emission and faster start time. During low load condition, the firing temperature can be controlled through regulating the percentage of fuel splitting or turning off the secondary fuel staging entirely [178, 179].

The reactive mixture flow that enters the combustor is split into two coaxial streams with different equivalency ratios in the splitting procedure before being introduced to the flame zone. The annular stream is leaner than the central one in order to maintain the overall operating equivalence ratio within the ideal range to maintain

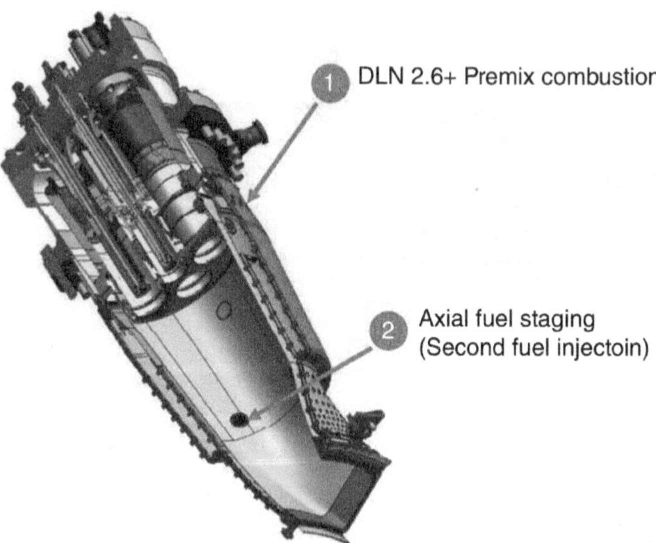

Fig. 4.15 Axial fuel staging in General Electric (GE)'s DLN 2.6+ combustor [178]

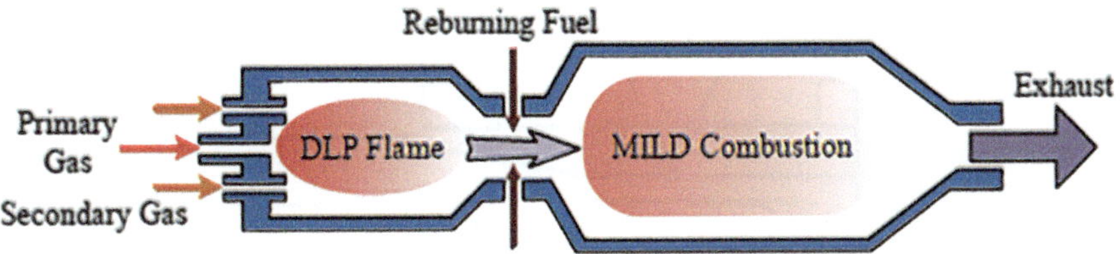

Fig. 4.16 Schematic representation of a combined DLPM-MILD combustor

the environmental performance of the combustor. The central stream is characterized by a higher equivalence ratio to enhance the flame stability under different loading conditions. Dual stage lean premixed combustion is the name of this stratification technology (DLPM). In other words, even if the annular stream is too thin to ignite in a conventional LPM combustor, the stable flame of the richer central stream will sustainably ignite it. This is possible by implementing DLPM technology, which can increase the combustor operability on the leaner side to overall equivalence ratios that are even below the lower flammability limit. As a result, it is possible to reduce part loads and still maintain emission control. If the annular stream is, however, ultra-lean and cannot burn to completion, a stream of reburning fuel can be introduced further downstream in the common exhaust plume of the two streams to complete the combustion of the flue gas under high temperature and low oxygen conditions, which is known as Moderate or Intense Low-oxygen Dilution (MILD) combustion. The system combining both DLPM and MILD combustion is presented in Fig. 4.16. One can consider the DLPM flame as a pre-combustion stage to the MILD flame in the proposed combustion system. Actually, MILD combustion can increase flame stability while emitting less pollution. Prior research, however, has demonstrated that for such a mixed staged-flame MILD combustion system, the NOx emissions at the combustor outlet are mostly controlled by the NOx levels out of the center DLPM flame [180, 181]. Based on this, it is necessary to optimize the functioning of the DLPM flame to reduce NOx emissions and supply a proper hot gas mixture for the downstream MILD flame.

Under no-load/idle condition and transient condition, LPM combustors can suffers from flame blowouts which hampers the power generation from gas turbine. Partially premixed combustion with a pilot flame is an effective way to address the flame blowout during no-load and transient condition [182–184]. This technique is falls into the category of radial fuel staging. One of the common examples of pilot flame-assisted combustion is General Electric's first generation DLN combustor (illustrated in Fig. 4.17) which takes the advantages of the radial fuel splitting for stabilization of the lean premixed mixtures flames. About 97% of the total fuel is injected trough the primary fuel circuit for LPM mixture and the remaining is used for pilot flame through secondary fuel circuit [185, 186]. Combustion in DLN burner takes place in four separate modes: (a) primary mode, (b) lean-lean mode, (c) secondary mode, and (d) premix mode. Figure 4.18 depicts the four modes in pilot flame assisted DLN combustor. Primary mode is used for 0–20% load turbine start-up and the

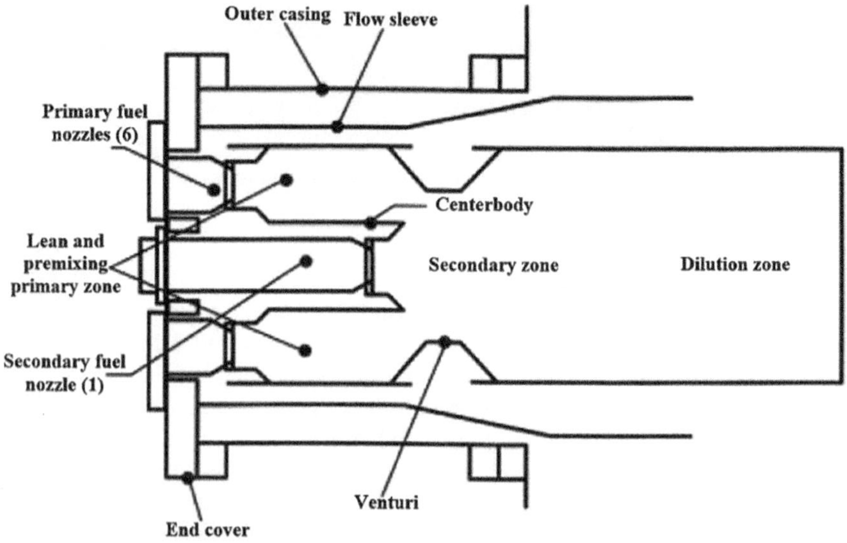

Fig. 4.17 Schematic diagram of the General Electric's first generation DLN combustor [186]

flame in this mode is present only in primary zone. In lean-lean mode, fuel injection occurs through both primary and secondary nozzle and operates between 02–50% load. Secondary mode is used for transition from lean-lean to premix mode. During transition, the flame present only in secondary injector. The premix mode holds the flame in both the injectors and used for load between 50 and 100% load [186].

Emara et al. [182] investigated the performance a pilot flame-assisted conical swirl burner designed by ABB, in which the vortex generated by the conical swirl setup breaks down at the burner mouth and stabilizes the flame through creating recirculation zones as illustrated in Fig. 4.19. The pilot injector is located at the centre of the conical swirl. The investigation showed improvement in flame stabilization and reduction in NOx emission when 10% of the total fuel was supplied through the pilot injector. Furthermore, less instability in the combustor was observed for equivalence ratio less than 0.583. In the experiment, lifted flame was observed with no fuel through the pilot injector and, with the increase in pilot fuel, the flame turned from the lifted mode to the properly anchored mode, as shown in Fig. 4.20. In a numerical study to predict lean blowout limit with a pilot-assisted flame, Zubrilin et al. [183] reported that introducing pilot flame in the swirl-stabilized combustor can broaden the stability of the combustor. They found that even with a small portion of the total fuel injected through the pilot inlet can help in achieving the minimum equivalence ratio of lean blowout where up to four times reduction of the minimum equivalence is possible as compared to supplying fuel through the main inlet only.

Besides pilot flames, the enhanced lean blowout (ELBO) assisted combustion is also regarded as fuel-staged combustion. A small amount (approximately 10%) of total fuel supply is injected into the mixing duct to enhance the blowout limit. The

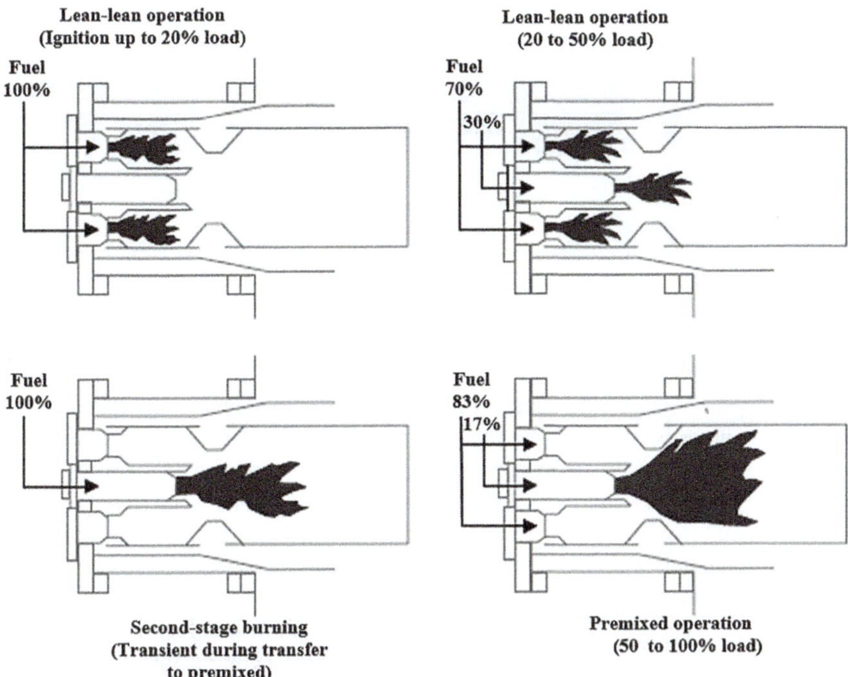

Fig. 4.18 Operating modes in Dry Low NOx (DLN) combustor [186]

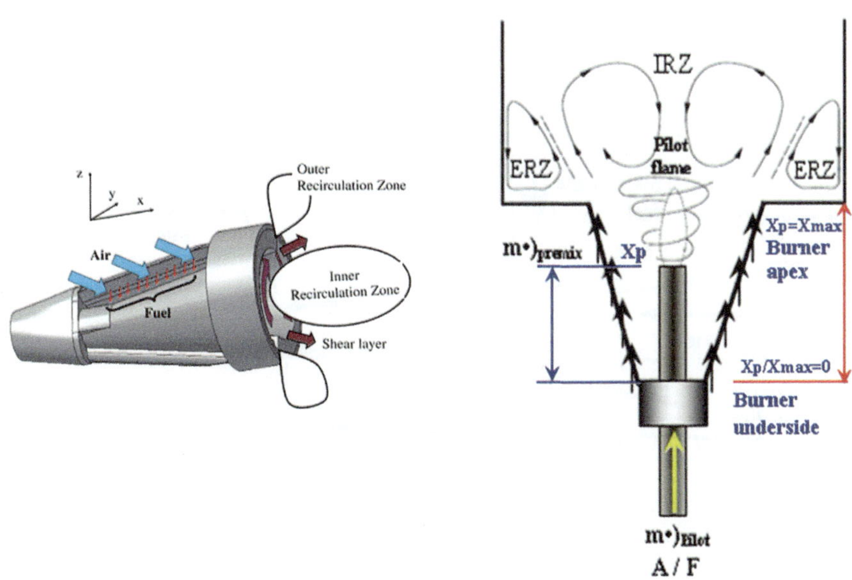

Fig. 4.19 Schematic of conical swirl-stabilized burner (left), and flame recirculation zones (right) pilot-flame assisted ABB swirl-stabilized burner [182]

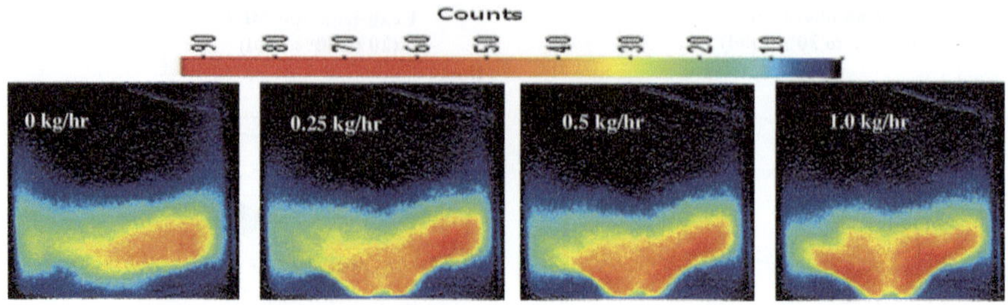

Fig. 4.20 Effect of mass flow rates (0. 0.25, 0.5, and 1.0 kg/hr) of pilot flame in flame anchoring depicted using OH- chemiluminescence [182]

fuel injection holes/ports are known as ELBO holes meanwhile the injected fuel through the holes is called ELBO fuel. Figure 4.21 shows the schematic diagram of ELBO fuel supply with its corresponding parts in the combustor. Injecting ELBO fuel increases the local fuel–air ratio in the mixing section in-between fresh incoming and burnt mixture. Moreover, in some configurations, ELBO setup can incorporate axial fuel staging as well. In addition to flame stability, ELBO with axial fuel staging can helps in managing acoustic issues related to fuel injection [49].

The implementation of ELBO fuel is also beneficial in low-load operation. For a low-load operation in legacy Dry Low Emission (DLE) combustors, initially only a few cups/cans of the combustor were used with circumferential fuel injection, while keeping the remaining cans turned off. This causes the creation of localized cold zone and increase of CO emission, and the system requires more combustor volume to achieve complete combustion. A variable ELBO feature in the premixer is proposed by Durbin et al. [187] to address these issues by creating smaller diffusion

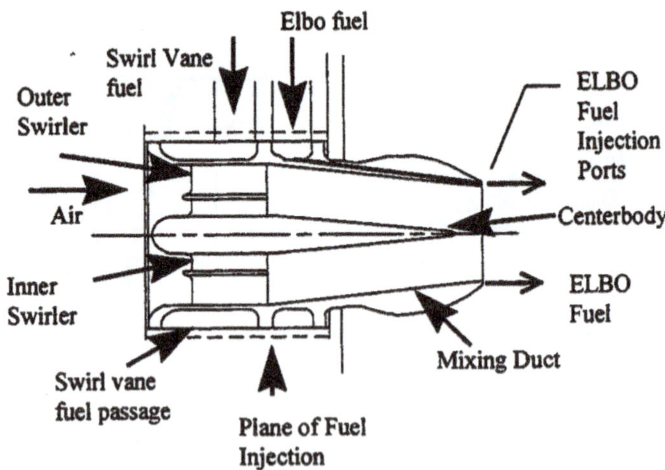

Fig. 4.21 Schematic of enhanced lean blowout (ELBO) fuel injection system for swirl stabilized burner [49]

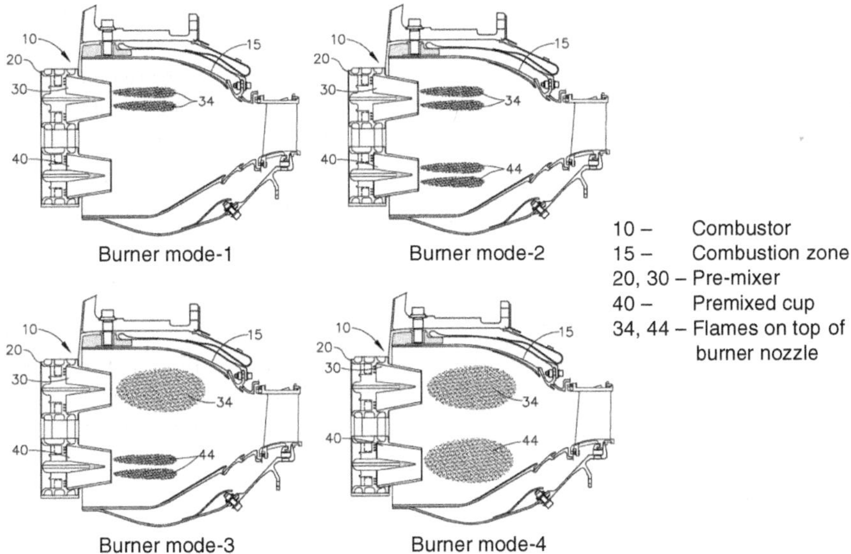

Fig. 4.22 Operating modes in variable enhanced lean blowout (ELBO) combustor [187]

flame to tackle the cold zone and act as source of ignition for the premixed flame. The proposed system is capable of operating in four different modes (shown in Fig. 4.22): (a) burner mode 1: from start up to 15% load, (b) burner mode 2: for load range 15–50%, (c) burner mode 3: for load range 50–75%, and lastly (d) burner mode 4: from 75% load to full power [187].

4.6.3 Combined Air–Fuel Staging

Combined air–fuel staging refers to the co-existence of fuel-oxidizer streams with different equivalence ratios within the same combustor. Combined staging approach is becoming popular in recent years due to its inherit advantage in terms of combustion stability and emission control. The common examples of combined staging are Beihang Axial Swirl Independently-Stratified (BASIS) burner [188, 189], Cambridge Stratified Swirl Burner [25, 190], Cambridge slot burner [191], Technology of Low Emission of Stratified swirl (TeLESS) combustor [192], non-separated stratified swirl burner [193], axially staged combustor [194] etc. A dual swirl stratification burner, developed based on the principle of Lean Premix Pervaporated (LPP) combustion, was investigated by Han et al. [188, 189]. The combustor employed two swirlers; inner swirler contained of 8 straight blades at angle of 40°, convergent-divergent throat, and a center-body. The outer swirler contained 20 blades at angle of 30°, as illustrated in Fig. 4.23. The lip of the outer inlet is raised by 11.7 mm. The separation due to the lip area in-between the inner and outer inlet led to the creation of lip-recirculation zone. This burner configuration is known as Beihang Axial Swirl Independently Stratified or BASIS burner. Three separate flame were observed, stratified flame, V-shaped flame, and lifted flame. The stratified flames were found to be more stable compared to the other two flames. Figure 4.24 shows the flame shapes in the BASIS burner where the S, V, L, SR and Φ_{total} stand for stratified, V-shaped, lifted, stratification ratio, and total equivalence ratio. The stratification ration is usually referred to the ratio of inner equivalence ratio and outer equivalence ratio of the stratified stream. As shown in Fig. 4.24, increasing the SR from 0.25 to 2.5, the flames start the transition from the lifted to the attached flame mode, and again to lifted flame, except for the flame of $\Phi_{total} = 0.6$. For the flames with $\Phi_{total} = 0.6$, no transition from attached flame to lifted flame with the increase in SR was reported. The study also reported that the flame macrostructure was heavily influenced by the equivalence ratio of inner fuel-oxidizer stream.

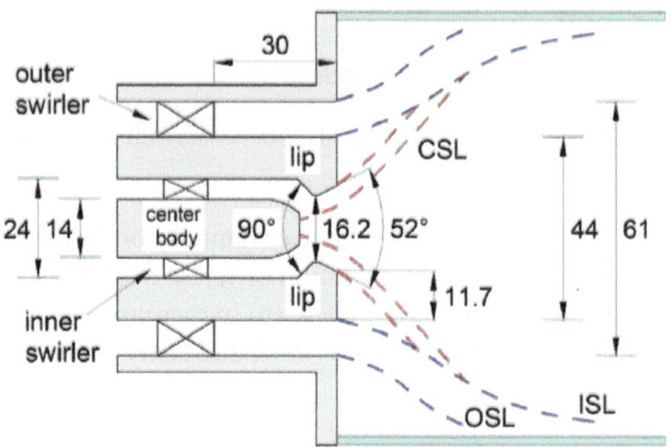

Fig. 4.23 Schematics of BASIS burner [189]

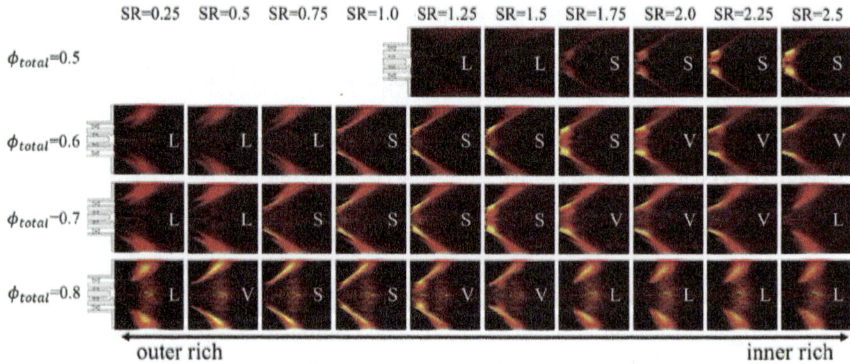

Fig. 4.24 Time-averaged flame shapes in Beihang Axial Swirl Independently Stratified (BASIS) burner [189]

To address the combustion instability and emission, Beihang University developed another staged combustor called Technology of Low Emissions of Stirred Swirl (TeLESS) combustor [192]. The proposed combustor included two swirl-stabilized inlets as shown in Fig. 4.25a, where the inner dome shaped inlet was used for pilot flame and the outer inlet was regarded as main intel. The combustor was reported to be capable of operating at three separate modes: low-power mode, circumferentially staged mode (CSM), and high-power mode. In low-power mode, the annular combustor array operated with only the central deme inlet, which is very sensitive to the equivalence ratio in terms of NOx emission and combustor efficiency. The optimum Φ was reported to be 0.36 in order to obtain the allowable NOx emission and highest possible combustor efficiency in low-power mode. In CSM, every two combustors had one operated with pilot flame and the other with both pilot and main flame. Combustion efficiency and NOx emission in CSM was found decreasing with the reduction in fuel stage ratio, which is the ratio between mass flow rates of fuel in pilot stream and total flow. Suitable fuel stage ratio in this stage was found between 1.5 and 35%. Lastly in high-power mode, both pilot and main flames were used to achieve higher power generation. Combustion efficiency in high-power mode was observed at around 99.9% for single combustor setup. The stages in three modes are provided in Fig. 4.25b. A second variant of the combustor was also developed later which was called TeLESS-II [195] to achieve further improvement over the existing model. Both the variants of TeLESS were aimed towards their application in aviation gas turbine combustion.

Stratified dual-swirl combustors are also advantageous in addressing hermos-acoustic oscillations. Arndt et al. [196, 197] examined the hermos-acoustic behaviour of SFB 606 (Sonderforschungsbereich 606) dual-swirl model gas turbine combustor, where the swirl number for the inner and outer inlet are 0.76 and 1.06, respectively. The dual-swirl SFB 606 burner schematics is provided in Fig. 4.26. The study reported insignificant effect of stratification on the flow field and mean shape of the flame in a global lean-condition ($\Phi = 0.75$). However, the influence of stratification on the acoustics was found evident in the experiment. The amplitudes of different nodes increased with the stratification level. Considerable stratification was observed in the upstream of flame root. Lower OH* CL fluctuations were observed between the inner shear layer and inner recirculation zone, as shown in Fig. 4.27. This fluctuation increased and the flame stratification became clear with the increase in the level off stratification. The root of the flames was observed to be lifted by 10 mm and the main flame zone existed between 30 and 40 mm from the nozzle obtaining the characteristic V-shape. Sweeney et al. [25, 190] developed a premixed-stratified burner which consisted of two annular concentric inlet and a bluff body in the center, as shown in Fig. 4.28. The inner annular inlet of the burner injects the fuel–air mixture axially while the outer inlet employs swirling effect on the second stream. The amount of CO and H_2 was reported to rise with the increase in flame stratification in reference to the homogeneously premixed configuration (equivalence ratio same for both inlets). The level of swirling also effects the flame shape, where the moderately swirled flames possessed relatively compact flames than the other ones.

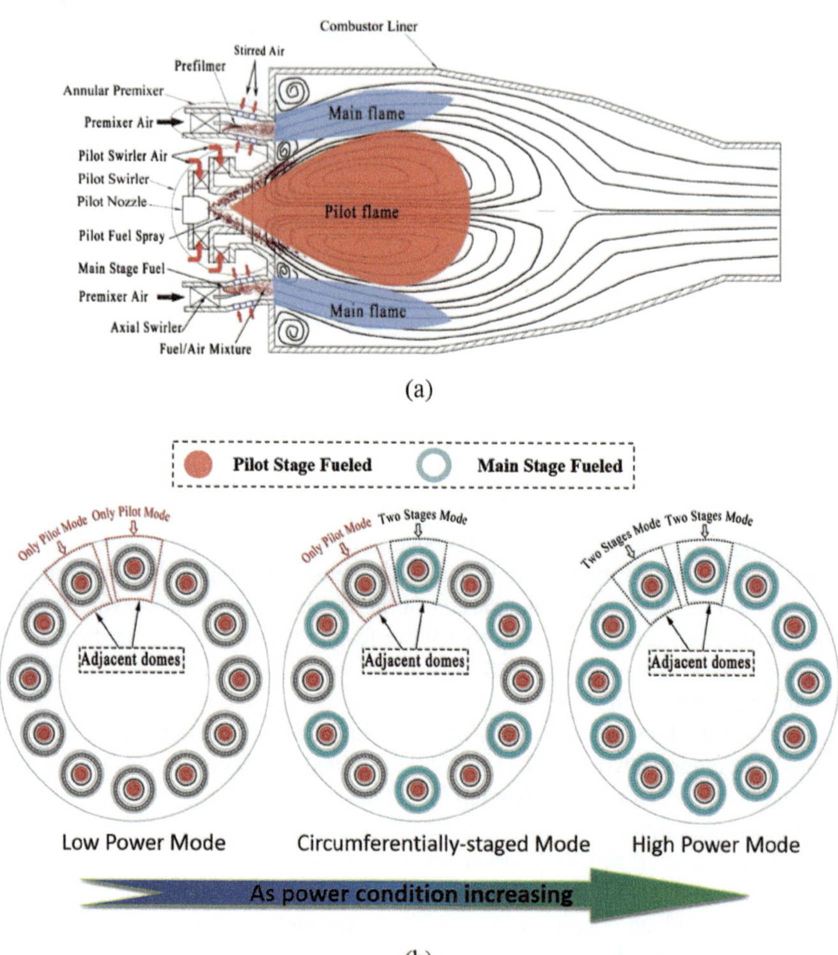

Fig. 4.25 Schematic of Technology of Low Emissions of Stirred Swirl (TeLESS) combustor: **a** single module burner, and **b** different operational modes of TeLESS combustor [192]

Stratification in a Multi-nozzle swirl-stabilized burner was investigated in a study by Samarasinghe et al. [198]. As shown in Fig. 4.29, the setup consisted of five swirl-stabilized inlets with 2-inch diameter and a center-body with 1-inch diameter. Fully premixed fuel-oxidizer mixture was used in the combustor and the stratification effect was obtained by increasing the fuel through the inlet stream of the central inlet. Figure 4.30 depicts the chemiluminescence intensity of the burner in different staging and equivalence ratios, where the intensity is interaction areas of the flames due to heat release rates. Figure 4.30a, b show similar unstable flames with no staging employed. With the introduction of staging level of 1.4 and 4.4, the intensity can be seen increasing as shown in Fig. 4.30(c) dand (d), respectively. The other

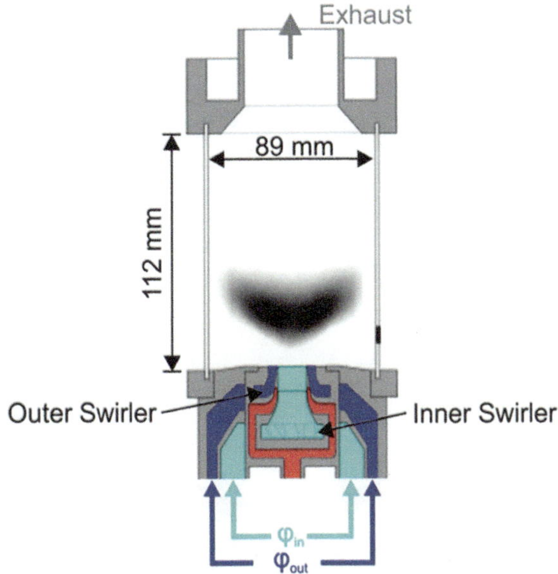

Fig. 4.26 Schematic diagram of duel-swirl SFB 606 burner [197]

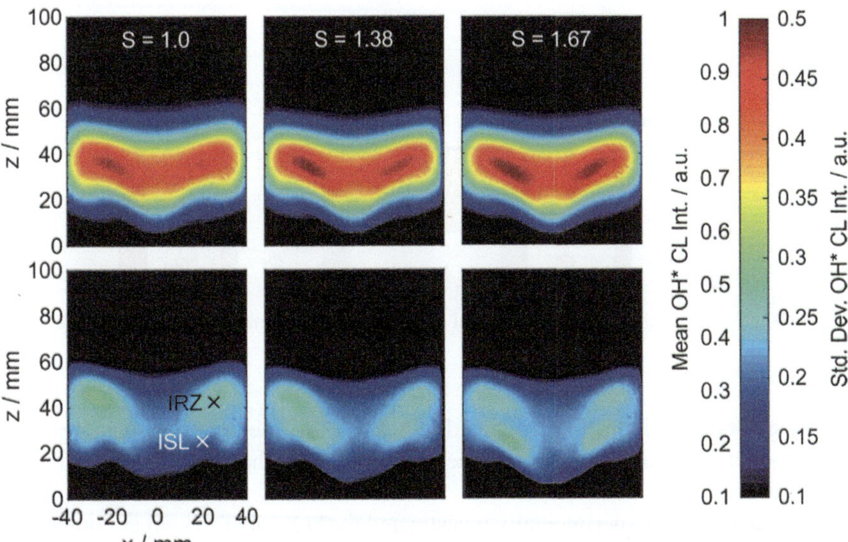

Fig. 4.27 Mean (top row) and standard deviation (bottom row) of the OH* chemiluminescence signal for the various stratification ratios in duel-swirl SFB 606 burner [197]

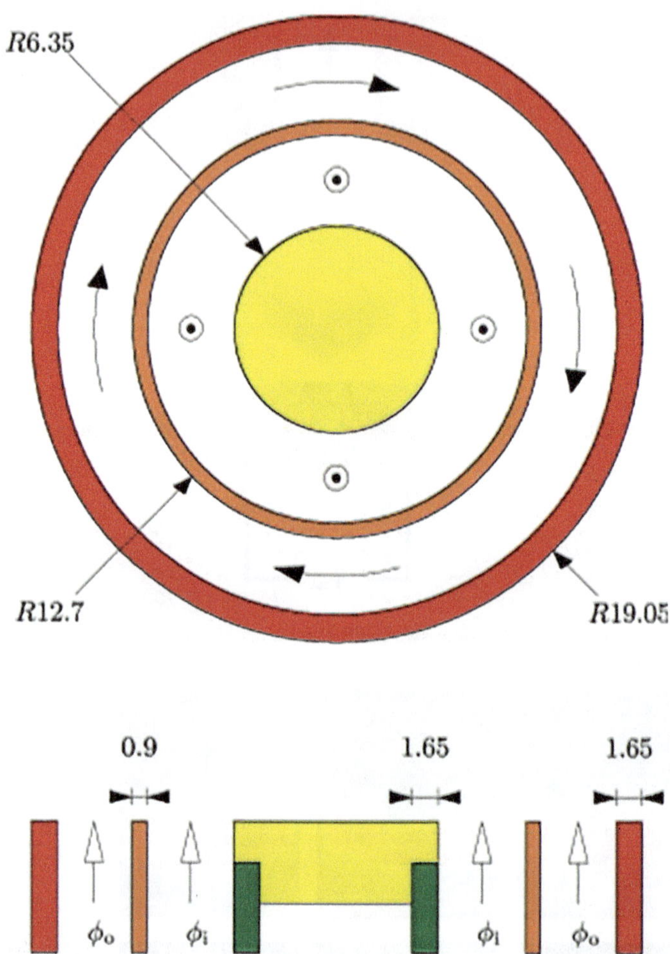

Fig. 4.28 Plan view (top) and cross-sectional view (bottom) of Cambridge Stratified Swirl Burner [25]

flames can be found almost unchanged indicating that the stratification suppresses the self-excited instabilities through the central flame or the flame-flame interaction mechanism.

Zeng et al. [199] experimentally investigated the behaviour of syngas combustion in a dual stage counter-rotating swirler system. The inner swirl jet nozzle and the outer counter rotating swirler were at Swirl number of 0.74 and 0.92, respectively. The composition of $CO–H_2$ was 30–70 mol percentage. The results showed that the higher H_2 content increased the mass flow rate in the fuel and increased the flow are as well in the central area. With the higher H_2 content, the flame becomes imperative, as the flame lifts off in combustion of hydrogen rich fuels. Stratified combustion may contain the potential for better flame anchoring as it uses pilot flame for the anchoring

Fig. 4.29 Multi-nozzle swirl-stabilized combustor setup developed by Samarasinghe et al. [198]

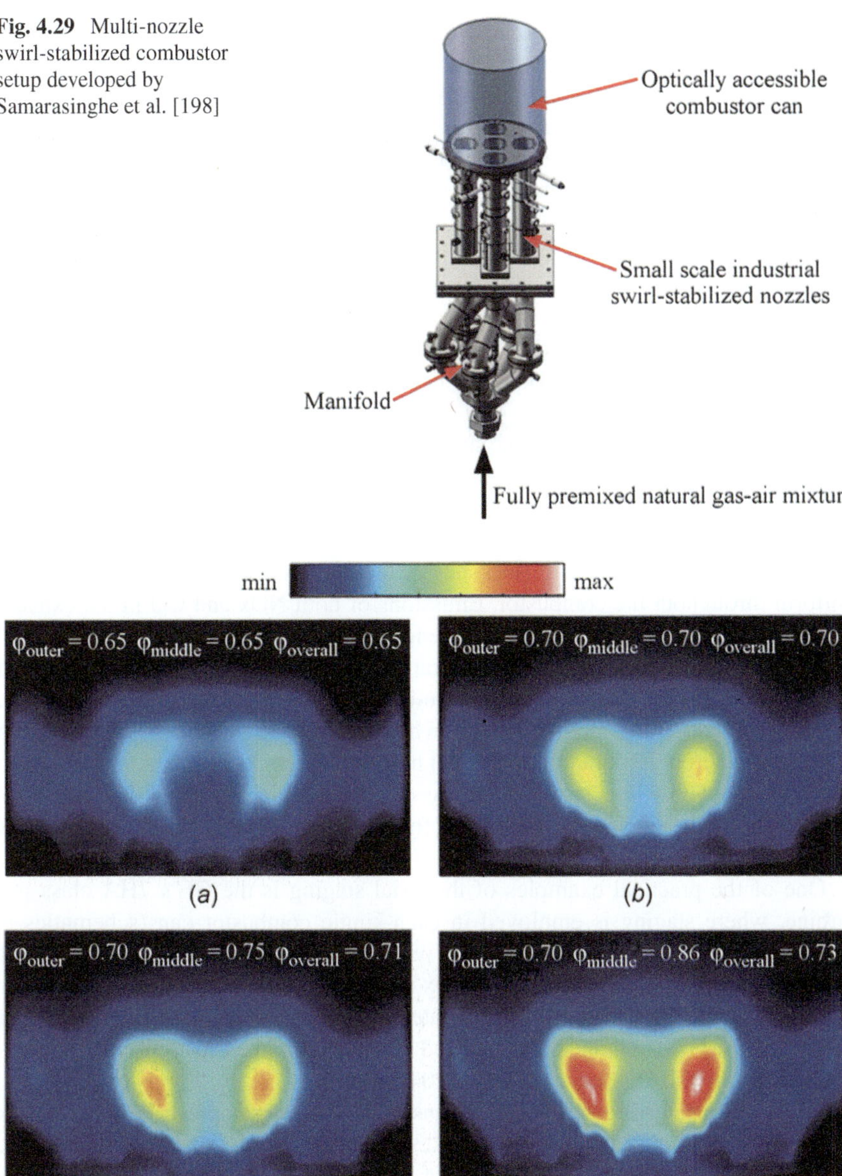

Fig. 4.30 Time-averaged CH* chemiluminescence images of the flame in Multi-nozzle swirl-stabilized combustor setup at percent staging: **a** $\Phi_{overall} = 0.65$, percent staging = 0, **b** $\Phi_{overall} = 0.70$, percent staging = 0, **c** $\Phi_{overall} = 0.71$, percent staging = 1.4, and **d** $\Phi_{overall} = 0.73$, percent staging = 4.4 [198]

purpose. Ihme [200] computationally examined the rapid compression machine in integrated combined cycle for syngas fuels. The results from the study reported that, the stratification of syngas mixture can significantly influence the auto-ignition process. García-Armingol et al. [201] investigated the emission and the stability of the premixed syngas combustion. They found that the utilization of staging strategy was found beneficial in dealing with the combustion issues usually observed in lean premixed syngas combustion. Staging of the syngas fuels showed considerable improvement in flame stability by minimizing the possibility for flame flashback. Therefore, the pressure fluctuations caused by flashback could also be lowered with syngas fuel staging. The study also reported that staging could be useful in broadening the stability range and reducing the CO and NOx products in the emission.

Zhang et al. [202] examined the stratification of Moderate or Intense Low-oxygen Dilution (MILD) burner (see Fig. 4.31), where the combustor consisted on a central nozzle for non-premixed pilot flame surrounded by four direction injection premixed flow nozzles. The central nozzle contained a 35° half-angle divergent cone. Air for pilot flame is injected through a swirler with 4 blades at 40° and the ratio between the main air and the pilot air is 4:1. Fuel-oxidizer stream in the central nozzle was shut-off after the flame stabilized. Results showed that the distribution of OH radicals was uniform throughout the combustor. Emissions of both NOx and CO in the exhaust were reported below 10 ppm for equivalence ratio ranging from 0.63 to 0.8 at 15% O_2 correction. The performance of the stratified MILD burner was further enhanced by introducing additional fuel-oxidizer mixture at downstream of the combustor. Figure 4.32 depicts the modified setup, where the secondary fuel-oxidizer mixture was injected through four circumferential holes in a premixed state. At atmospheric pressure, about 35% decrease in NOx was observed at higher overall Φ. The equivalence ratio of the secondary stream was found responsible for the increase in NOx at fixed Φ of the primary stream [203].

One of the practical examples of the axial staging is the GE's 7HA class gas turbine, where staging is employed in each single combustor can (schematics is shown in Fig. 4.33). The combustor was reported to be fuel-flexible and capable of enhancing turndown and decreasing NOx emission as compared to the combustor without staging effect. Figure 4.34 illustrates the normalized NOx emission with and without staging over a range of normalized exit temperature. The NOx emission was found to be around 25 ppm at 15% O_2 [204].

Kawasaki Heavy Industries' L30A gas turbine combustor can be regarded as combined type staged combustor. The L30A burner consist of two burner units, the first unit (primary zone) consists of a pilot nozzle and a swirl-stabilized main inlet. The pilot nozzle burns the fuel in diffusion flame with air supplied through a swirler. The main burner in first burner unit is a premixed-swirl annular burner surrounding the pilot nozzle. Lastly, the additional burner unit (secondary zone) contains four circumferential swirl-stabilized premixed burners for employing the staging effect. Figure 4.35 shows the schematics of L30A gas turbine combustor with different components. This arrangement is also called low emission combustion chamber (LECC) and capable of reducing NOx and CO emissions and enhancing the operability limit. NOx emission remains blow 25 ppm within the whole range of

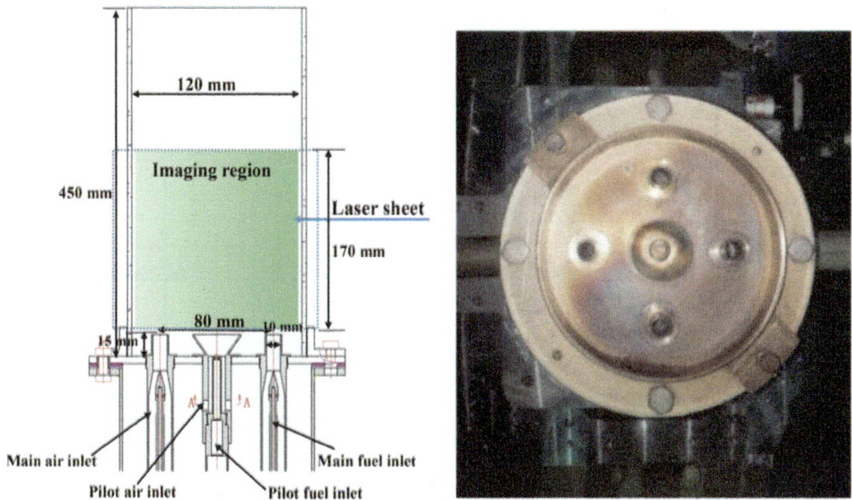

Fig. 4.31 Schematics (left) and image of burner plate (right) of stratified moderate or intense low-oxygen dilution (MILD) combustor developed by Zhang et al. [202]

operation with this combustor, while CO emission can be reduced to as low as 10 ppm. The amplitudes of pressure oscillations can be maintained below 80% of the permissible value at 30% load [205].

4.7 Stratified Oxy-fuel Combustion: A Way Forward

Combination of stratified flames ensures stable combustion and at lower combustor temperature, which prevents the formation of NOx [206–208]. Both the mass flow rate and the equivalence ratio play important role in characteristics of stratified combustion. Increase in the mass flow rate of the pilot flame makes the flame stronger with longer reaction zone [193]. In addition, the pilot flame can be beneficial in anchoring the flame [189]. In a high-pressure configuration, the equivalence ratios can affect heavily on the combustion efficiency and the emission [192]. One study reported V-shaped structures for the pilot flame-only and M-shaped structure for combustion with both pilot and main flame [188]. Majority of the gas turbines now a days employ some form of air/fuel/combined staging for air–fuel combustion.

Oxy-fuel combustion is one of the most promising carbon-capture technologies for controlling CO_2 emissions with minimum modifications of the combustion system. The produced CO_2-rich effluent stream of this process simplifies the capturing process, as compared to the other carbon-capture technologies. However, oxy-fuel combustion comes with its own challenges in terms of flame stabilization, especially at lower oxygen concentrations in the oxidizer mixture. Integrating the stratified-charge dual lean premixed (DLPM) technique with the oxy-combustion

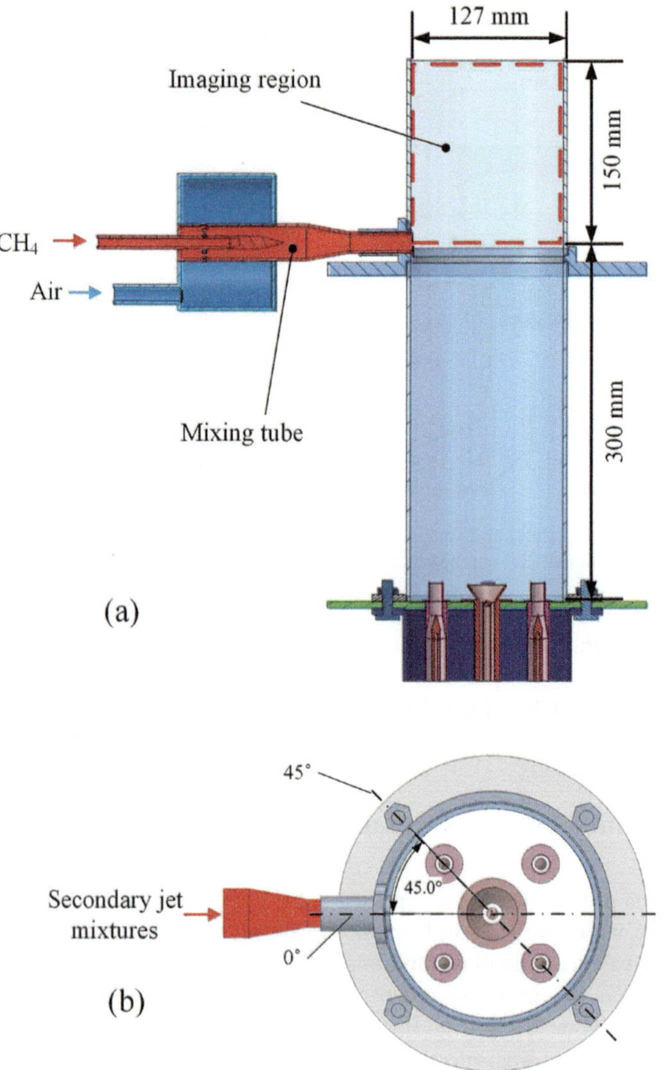

Fig. 4.32 **a** Cross-sectional and **b** planar schematic diagram of stratified moderate or intense low-oxygen dilution (MILD) combustor with axial staging assembly [203]

technology can, however, result in total control of gas turbine emissions while widening the range of flame stability especially under low load operating conditions. Even though the potential of stratified combustion is reflected through numerous on air–fuel combustion, in case of oxy-fuel combustion, a few investigations have been performed. Furthermore, the most of these are staged oxy-fuel combustion investigated for industrial furnaces [209, 210]. Therefore, future research is needed to identify the characteristics of stratified oxy-combustion to enhance the performance

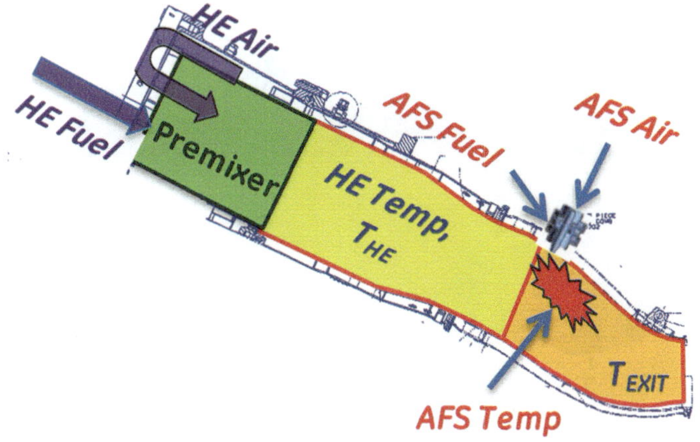

Fig. 4.33 Schematic diagram of the GE's 7HA class gas turbine combustor [204]

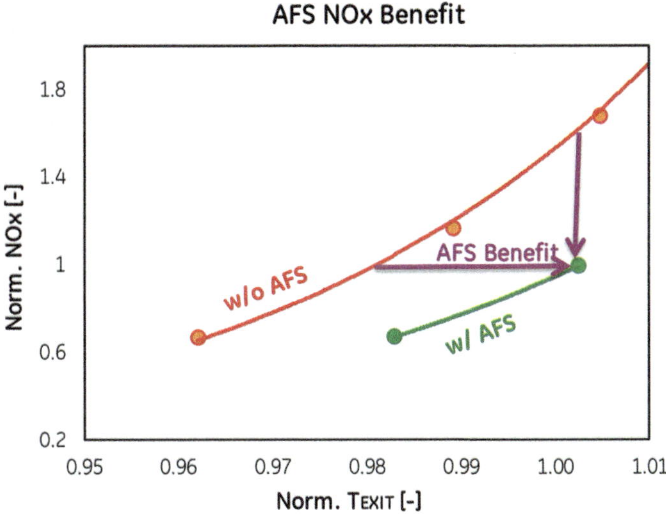

Fig. 4.34 NOx emission at various exit temperature with and without AFS implementation in GE's 7HA class gas turbine combustor [204]

and counter the instabilities of oxy-fuel combustion for proper application of this technology in the industrial field.

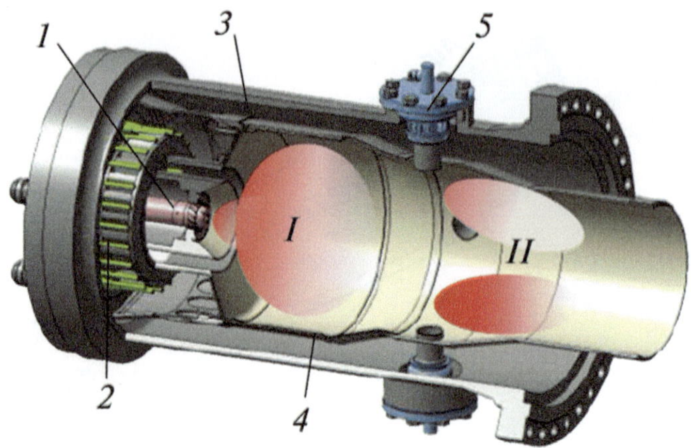

Fig. 4.35 Longitudinal section of L30A low-emission gas turbine combustor, showing: (1) Pilot burner, (2) main burner of burner unit-1, (3) burner unit-1 casing with holes for air, (4) flame tube, (5) burner unit-2, and (I, II) first and second combustion stages [205]

4.8 Advances Toward the Hydrogen Gas Turbine

The increasing demand of the fossil fuels has considerably led to the increase in the pollution therefore, intense research is going on to search alternative energy sources that can help in reducing the environmental pollution and mitigate the increase of global warming. The most important factor in achieving sustainability is to search for a fuel that has zero emission characteristics. Therefore, the most ideal candidate for this purpose is hydrogen. Despite hydrogen is widely used in the industries for many purposes, hydrogen is also considered to be the most potential candidate for the fuels in the power systems such as gas turbines in the near future [211]. Hydrogen has a low molecular weight and it can be produced through various techniques, since it occurs in many different forms in nature [212, 213]. The use of hydrogen, as a fuel, has a long history since it was used as a rocket fuel in US in 1960s. It has also been reported that hydrogen was used as a fuel in automobiles at that time. The interest in hydrogen as a potential source of fuel started in the 1970s when the price of fossil fuels considerably escalated. By 1990s, the hydrogen economy was revisited with considerable advancements by that time [214]. In the recent years, hydrogen has gained much more attention as a fuel since it has considerably been reviewed as a potential source for clean energy production [215–218].

The production of hydrogen all around the world has approximately reached 50 million tonnes. Most of the produced hydrogen is consumed by the producers at their facilities. The majority of hydrogen (around 95%) is produced from different processes utilizing the hydrocarbons whereas the remaining 5% is produced using the process of electrolysis [219, 220]. Figure 4.36 shows the sources of the hydrogen production as well as the consumption areas of the hydrogen. It can be seen from the figure that most of the hydrogen is consumed in methanol production. Hydrogen

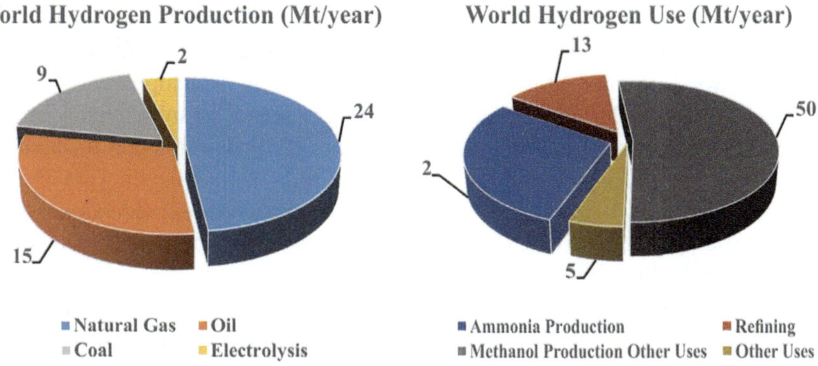

Fig. 4.36 Production sources of hydrogen (left), and areas of consumption of produced hydrogen (right) [222]

has occupied a considerable share in the market, and it has a potential to take its due share as a fuel in the near future. Parra et al. [221] performed tecno-economic analysis of the hydrogen storage. In their study, they considered the production of hydrogen using the electrolyser and the consumption of electrical supply. They also considered the integration of hydrogen in different available technologies and they concluded that the integration of hydrogen in different processes can substantially help in reducing the CO_2 emissions and decrease the potential of global warming.

4.8.1 Characteristics of Hydrogen Combustion

Due to the numerous advantages that hydrogen offers, hydrogen has been found as a viable blending agent, which possesses the required characteristics (higher burning velocity and wider flammability limits) and results in lower NO_x emissions. Hydrogen has a high calorific value (141.8 MJ/kg), which is approximately three times that of CH_4 [223], thereby providing favourable temperatures. Because of the innate advantages that hydrogen offers, it is regarded as the future energy carrier. Enriching the combustion process with hydrogen is, thus, an excellent method for the improvement of various combustion characteristics, such as operability window, emissions, fuel consumption, and soot formation.

Although the combustion of hydrogen is CO_2 free, it can result in generation of NO_x in substantial amount. NO_x is strongly monitored in many developed countries since it is responsible for environmental pollution which can result in smog or acid rains. Studies have reported that by switching from methane to pure hydrogen, substantial increase has been noticed in the NO_x levels even for the low NOx burners [224]. Another important characteristic of the hydrogen combustion is the higher flame temperature. The higher flame temperature is one of the main parameters in the formation of NOx where the nitrogen is oxidized by oxygen in higher temperature atmosphere. Since this formation of NOx is because of the higher temperature, this is known as thermal NOx and it is highly sensitive to the temperature. A minute increase in the combustion temperature has the potential to substantially increase the rate of formation of thermal NOx. Todd et al. [225] reported that for hydrogen fuelled GE 6 FA test combustor, higher values of NOx up to 200 ppm were observed at 15% O_2 whereas Brunetti et al. [226] reported that their observed values were as high as 800 ppm at 15% O_2. These values were reported when the combustor was fuelled with 85–90% hydrogen.

In order to overcome the issue of higher temperature associated with the combustion of hydrogen, it is usually preferred to enrich the primary fuel with hydrogen in order to improve its operability window due to its higher laminar flame speed. Since the presence of CO_2 deteriorates the stability of its retarded thermos-physical properties, the addition of hydrogen significantly improves the combustion characteristics of the combustor. There have been several studies on the enrichment of hydrogen in the primary fuel, usually methane. For example, Schefer [227] experimentally studied the combustion characteristics of a swirl-stabilized premixed flame with H_2 enrichment over wide ranges of operating conditions. He reported that the addition of H_2 (up to 20%) in the CH_4/air mixture significantly extended the lean stability limits of the burner. Ghoniem et al. [228] experimentally studied the effect of H_2 enrichment on a C_3H_8/air flame. They reported that H_2 enrichment increased the NO_x emissions because of the rise in temperature, but on the other hand it also extended the flammability limit, allowing for leaner operation, which consequently improved the combustor turndown ratio and decreased the ultimate NOx emissions. A similar study on swirl-stabilized premixed flames in a 50-kW combustor was done by Tuncer [229], who reported that H_2 enrichment enhanced the lean flammability limit of CH_4 flames and consequently reduced NOx emissions because of the lower adiabatic flame temperature associated with lean operation. The drawback was, however, early flashback because of the higher burning velocity of H_2 due to the increased flame propagation speed. Ferrières et al. [230] experimentally and numerically studied natural-gas/H_2/O_2/N_2 flames at three hydrogen fractions of 0, 20, and 60% and reported reductions in the mole fractions of carbon and hydrocarbon species at 60% H_2 due to enhanced kinetics, but the presence of H_2 favoured the formation of CO.

4.8.2 Recent Developments in Hydrogen Gas Turbines

The interest in hydrogen fuelled gas turbines is emerging in the market because of the due to a spike in the development of the combustors fuelled on hydrogen or high hydrogen content fuels since hydrogen can be a viable option for mitigating the emissions if the low NOx combustors are developed. Several attempts have been made by various researchers in academia as well as in the industry to develop such technologies. For example, Sun et al. [231] performed numerical investigation on the hydrogen/air combustion for a jet-stabilized combustor and reported that the complete combustion can be achieved except for the stoichiometric conditions where the combustion efficiency was less than 65%. Ditaranto et al. [232] made an advancement in the hydrogen fired gas turbine cycle that they incorporated the concept of exhaust gas recirculation. They reported that by incorporating exhaust gas recirculation, lower temperatures were achieved resulting in lower NOx emissions. They simulated different scenarios for exhaust gas recirculation and all options demonstrated improved efficiency. They also reported that dry exhaust gas recirculation is more preferred compared to the wet exhaust gas recirculation despite of the fact that higher recirculation rate is required. By doing so, they reported that 1% gain in the efficiency was achieved when compared to the base case. Kahraman et al. [233] compared the performance of gas turbine combustor fuelled by hydrogen as well as jet-A fuel. They assessed the combustion characteristics in the Rolls-Royce Nene turbojet combustor for different values of excess air ratio. They reported that the combustion efficiency and pressure drop increases with the increase in the excess air ratio. They also added that the combustion of hydrogen is seen to have positive effect on the pressure drop values, the temperature at the exit of combustion chamber, the CO_2 values as well as the unburned hydrocarbons. However, they also added that the lower excess air ratios are beneficial for limiting the NOx emissions and vice versa. Koç et al. [234] analysed the simple and recuperative gas turbine cycles using natural gas and hydrogen as fuels. They highlighted the threshold values of pressure for which the efficiencies were higher for each cycle. They added that although the costs associated with hydrogen were higher, but their performance was improved compared to the natural gas cycles with zero CO_2 emissions. Hussain et al. [29] performed experimental investigation on the micromixer gas-turbine to analyse the effect of hydrogen enrichment in case of oxy-fuel combustion. They reported that oxygen fraction values as low as 13% can be achieved with hydrogen fraction up to 65%. Their results also showed that the burner could handle the flames with hydrogen fraction less than 90%, which is almost near to the case of pure hydrogen combustion for future hydrogen gas turbines.

4.9 Concluding Remarks

In this chapter, a comprehensive review is presented on the recent advances of gas turbine combustors for lower emissions and higher turndown ratio. State of the art burner designs along with recent combustion techniques are presented in this chapter for clean and efficient combustion in gas turbines. Lean premixed combustion (LPM) technique can significantly reduce the NOx emissions, due to the reduced combustion temperature, down to the single digit level when implemented in Dry Low NOx (DLN) burners. However, this technique can be associated with combustion acoustics and difficulties to sustain the environmental performance at low load conditions. Recent burner designs, like the Dual Annular Counter Rotating Swirl (DACRS) and Enhanced-Vortex (EV/SEV/AEV) burners, can enhance the mixing effectiveness resulting in wider operability limits of the combustor. In order to sustain both flame stability and environmental performance at part load and startup conditions, the concept of flame stratification, i.e., heterogenization of the overall equivalence ratio through staging or splitting the reactants stream, was introduced. This technique resulted in significant improvement in the turndown ratio of the LPM combustors at acceptable levels of CO_2 and NOx emissions. However, in order to completely control the emissions of NOx while capturing easily the resulting CO2, the oxy-fuel combustion technique was introduced. Despite the controlled emissions, oxy-fuel combustion technique has its own challenges in terms of reduced combustor operability limits, flame instabilities, and reduced rates of reactions due to the elevated concentrations of CO_2. Combining the techniques of stratified and oxy-fuel combustion can aide in overcoming the combustion instabilities and acquire the benefits of both staging and NOx free combustion of gas turbine combustors. However, future research is needed to identify the characteristics of stratified oxy-combustion combustors for proper application of this technology in the industrial field. Due to its better combustion characteristics (higher burning velocity and wider flammability limits) with no CO_2 emissions, hydrogen is regarded as the future energy carrier, especially for gas turbines. Alternatively, due to excessive high temperature of generated in hydrogen combustion, the NOx emission levels are significantly higher. Utilizing pure hydrogen under oxy-combustion conditions can be an effective way for clean and effective combustion in gas turbines. However, combustion with pure hydrogen also introduces some additional challenges including flame instabilities, excessive combustion temperature, and safety related issues. At the current stage, hydrogen-enriched low carbon fuels (e.g., natural gas) are recommended for gas turbines incorporating stratified oxy-combustion. More research and development are required in order to overcome the challenges related to hydrogen gas turbine, which should be the future of gas turbines.

Acknowledgements The authors appreciate the support received for the preparation of this book from the Deanship of Research Oversight and Coordination (DROC) at King Fahd University of Petroleum & Minerals (KFUPM). The support provided by the Interdisciplinary Research Center for Hydrogen Technologies and Carbon Management (IRC-HTCM) on project number INHE2308 is highly appreciated. Also, the support received through the KFUPM Consortium for Hydrogen Future through the projects numbered H2FC2309 and H2FC2315 is highly appreciated.

Nomenclatures

2D	Two-dimensional
3D	Three-dimensional
AEV	Advanced environmental vortex burner
AFR	Air fuel ratio
AFS	Axial fuel staging
AFT	Adiabatic flame temperature

BASIS	Beihang axial swirl independently stratified
CO	Carbon monoxide
CO_2	Carbon dioxide
D	Characteristic length
Da	Damkohler number
DACRS	Dual annular counter-rotating
DLN	Dry low NOx
DLE	Dry low emission
DLPM	Dual lean premix
DNS	Direct numerical simulations
ELBO	Enhanced lean blowout
ESP	Electrostatic precipitator
EV	Environmental vortex or enhanced-vortex burner
FGD	Flue gas desulfurization
GDI	Gasoline direct injection
GHG	Greenhouse gases
GT	Gas turbine
GW	Gigawatt
H_2	Hydrogen
H_2O	Water vapor
IEA	International energy agency
ICE	Internal combustion engine
K	Kelvin
LECC	Low emission combustion chamber
LPM	Lean premixed
MEA	Mono ethanolamine
MILD	Moderate or intense low-oxygen dilution
N_2	Nitrogen
NOx	Nitrogen oxides
O_2	Oxygen
OF	Oxygen fraction
ppm	Parts per million
ppmvd	Parts per million by volume, dry
RQL	Rich burn, Quick mix, Lean burn
SEV	Sequential environmental vortex burner
S_L or LFS	Laminar flame speed
TALON	Technology for Advanced Low NOx
TeLESS	Technology of Low Emission of Stratified swirl
U_{ref}	Flow velocity
α	Thermal diffusivity
Φ	Equivalence ratio

References

1. IEA, *World Energy Outlook 2017* (n.d.)
2. IPCC, *Climate Change—The Physical Science Basis* (2007). https://doi.org/10.1017/CBO 9781107415324.004
3. R. Priddle, *World Energy Outlook 1998* (1998). https://doi.org/10.1111/jam.12213
4. J.N. Armor, Addressing the CO_2 dilemma. Catal. Lett. **114**, 115–121 (2007). https://doi.org/10.1007/s10562-007-9063-3
5. A.N. Hayhurst, A.D. Lawrence, Emissions of nitrous oxide from combustion sources. Prog. Energy Combust. Sci. **18**, 529–552 (1992)
6. N. Gascoin, Q. Yang, K. Chetehouna, Thermal effects of CO_2 on the NOx formation behavior in the CH_4 diffusion combustion system. Appl. Therm. Eng. **110**, 144–149 (2017). https://doi.org/10.1016/j.applthermaleng.2016.08.133
7. Y. Kang, S. Wei, P. Zhang, X. Lu, Q. Wang, X. Gou, X. Huang, S. Peng, D. Yang, X. Ji, Detailed multi-dimensional study on NOx formation and destruction mechanisms in dimethyl ether/air diffusion flame under the moderate or intense low-oxygen dilution (MILD) condition. Energy **119**, 1195–1211 (2017). https://doi.org/10.1016/j.energy.2016.11.070
8. R. Pardemann, B. Meyer, Pre-combustion carbon capture. Handb. Clean Energy Syst. 1–28 (2015). https://doi.org/10.1002/9781118991978.hces061
9. P. Babu, H.W.N. Ong, P. Linga, A systematic kinetic study to evaluate the effect of tetrahydrofuran on the clathrate process for pre-combustion capture of carbon dioxide. Energy **94**, 431–442 (2016). https://doi.org/10.1016/j.energy.2015.11.009
10. D. Jansen, M. Gazzani, G. Manzolini, E. Van Dijk, M. Carbo, Pre-combustion CO_2 capture. Int. J. Greenhouse Gas Control **40**, 167–187 (2015). https://doi.org/10.1016/j.ijggc.2015.05.028
11. T. Lockwood, A comparative review of next-generation carbon capture technologies for coal-fired power plant. Energy Procedia **114**, 2658–2670 (2017). https://doi.org/10.1016/j.egypro.2017.03.1850
12. G.P. Hammond, J. Spargo, The prospects for coal-fired power plants with carbon capture and storage: a UK perspective. Energy Convers. Manag. **86**, 476–489 (2014). https://doi.org/10.1016/j.enconman.2014.05.030
13. K. Yoro, P. Sekoai, The potential of CO_2 capture and storage technology in South Africa's coal-fired thermal power plants. Environments **3**, 24 (2016). https://doi.org/10.3390/environments3030024
14. D.Y.C. Leung, G. Caramanna, M.M. Maroto-Valer, An overview of current status of carbon dioxide capture and storage technologies. Renew. Sustain. Energy Rev. **39**, 426–443 (2014). https://doi.org/10.1016/j.rser.2014.07.093
15. M.K. Mondal, H.K. Balsora, P. Varshney, Progress and trends in CO_2 capture/separation technologies: a review. Energy **46**, 431–441 (2012). https://doi.org/10.1016/j.energy.2012.08.006
16. T.C. Merkel, H. Lin, X. Wei, R. Baker, Power plant post-combustion carbon dioxide capture: an opportunity for membranes. J. Memb. Sci. **359**, 126–139 (2010). https://doi.org/10.1016/j.memsci.2009.10.041
17. N.S. Sifat, Y. Haseli, A critical review of CO_2 capture technologies and prospects for clean power generation. Energies **12** (2019). https://doi.org/10.3390/en12214143
18. H. Gerbelová, M. van der Spek, W. Schakel, Feasibility assessment of CO_2 capture retrofitted to an existing cement plant: post-combustion vs. oxy-fuel combustion technology. Energy Procedia **114**, 6141–6149 (2017). https://doi.org/10.1016/j.egypro.2017.03.1751
19. D. Thimsen, J. Wheeldon, D. Dillon, *Economic Comparison of Oxy-coal Carbon Dioxide (CO_2) Capture and Storage (CCS) with Pre- and Post-combustion CCS* (Woodhead Publishing Limited, 2011). https://doi.org/10.1533/9780857090980.1.17
20. P.A. Bouillon, S. Hennes, C. Mahieux, ECO_2: post-combustion or oxyfuel-a comparison between coal power plants with integrated CO_2 capture. Energy Procedia **1**, 4015–4022 (2009). https://doi.org/10.1016/j.egypro.2009.02.207

21. M.A. Haque, M.A. Nemitallah, A.A. Abdelhafez, I.B. Mansir, M.A.M. Habib, Review of fuel/oxidizer-flexible combustion in gas turbines. Energy Fuels **34**, 10459–10485 (2020). https://doi.org/10.1021/acs.energyfuels.0c02097

22. W.R. Bender, Lean pre-mixed combustion, in: *Gas Turbine Handbook*, pp. 217–227. https://www.netl.doe.gov/sites/default/files/gas-turbine-handbook/3-2-1-2.pdf

23. M.A. Nemitallah, S.S. Rashwan, I.B. Mansir, A.A. Abdelhafez, M.A. Habib, Review of novel combustion techniques for clean power production in gas turbines. Energy Fuels **32**, 979–1004 (2018). https://doi.org/10.1021/acs.energyfuels.7b03607

24. B. Lawler, D. Splitter, J. Szybist, B. Kaul, Thermally stratified compression ignition: a new advanced low temperature combustion mode with load flexibility. Appl. Energy **189**, 122–132 (2017). https://doi.org/10.1016/j.apenergy.2016.11.034

25. M.S. Sweeney, S. Hochgreb, M.J. Dunn, R.S. Barlow, The structure of turbulent stratified and premixed methane/air flames I: non-swirling flows. Combust. Flame **159**, 2896–2911 (2012). https://doi.org/10.1016/j.combustflame.2012.06.001

26. D. Winkler, W. Geng, G. Engelbrecht, P. Stuber, K. Knapp, T. Griffin, Staged combustion concept for gas turbines. J. Glob. Power Propuls. Soc. **1**, CVLCX0 (2017). https://doi.org/10.22261/cvlcx0

27. A.N. Lipatnikov, Stratified turbulent flames: recent advances in understanding the influence of mixture inhomogeneities on premixed combustion and modeling challenges. Prog. Energy Combust. Sci. **62**, 87–132 (2017). https://doi.org/10.1016/j.pecs.2017.05.001

28. A.H. Ibrahim, S.Y. Kamal, T.W. Abou-Arab, M.A. Nemitallah, M.A. Habib, H. Kayed, Operability of fuel/oxidizer-flexible combustor holding hydrogen-enriched partially premixed oxy-flames stabilized over a perforated plate burner. Energy Fuels **34**, 8653–8665 (2020). https://doi.org/10.1021/acs.energyfuels.0c01407

29. M. Hussain, A. Abdelhafez, M.A. Nemitallah, A.A. Araoye, R. Ben-Mansour, M.A. Habib, A highly diluted oxy-fuel micromixer combustor with hydrogen enrichment for enhancing turndown in gas turbines. Appl. Energy **279**, 1–11 (2020). https://doi.org/10.1016/j.apenergy.2020.115818

30. B. Imteyaz, M. Habib, M. Nemitallah, A. Abdelhafez, R. Ben-Mansour, Operability of a premixed combustor holding hydrogen-enriched oxy-methane flames: an experimental and numerical study. Int. J. Energy Res. 1–15 (2020). https://doi.org/10.1002/er.5998

31. A.M. ELKady, S.T. Evulet, *Fuel-Flexible Tripple-Counter-Rotatig Swirler and Method of Use*, US20080163627A1 (2008)

32. A.M. ELKady, S.T. Evulet, *Triplpe Annular Counter Rotating Swirler*, US20080115501A1 (2008)

33. M.A. Nemitallah, A.A. Abdelhafez, M.A. Habib, *Approaches for Clean Combustion in Gas Turbines* (Springer International Publishing, Cham, 2020). https://doi.org/10.1007/978-3-030-44077-0

34. R. Kurz, K. Brun, C. Meher-Homji, J. Moore, F. Gonzalez, Gas turbine performance and maintenance, in *Forty-Second Turbomachine Symposium* (2013)

35. A.H. Lefebvre, D.R. Ballal, *Gas Turbine Combustion: Alternative Fuels and Emissions* (2010). https://doi.org/10.1002/1521-3773(20010316)40:6<9823::AID-ANIE9823>3.3.CO;2-C

36. F.H. Verhoek, Thermodynamics and rocket propulsion. J. Chem. Educ. **46**, 140 (1969). https://doi.org/10.1021/ed046p140

37. C.R. Shaddix, T.C. Williams, R.W. Schefer, *Effect of Syngas Composition and CO_2-Diluted Oxygen on Performance of a Premixed Swirl-Stabilized Combustor* (2007), pp. 889–897. https://doi.org/10.1115/GT2007-28286

38. L. Figura, J.G. Lee, B.D. Quay, D.A. Santavicca, *The Effects of Fuel Composition on Flame Structure and Combustion Dynamics in a Lean Premixed Combustor* (2007), pp. 181–187. https://doi.org/10.1115/GT2007-27298

39. P. Strakey, T. Sidwell, J. Ontko, Investigation of the effects of hydrogen addition on lean extinction in a swirl stabilized combustor. Proc. Combust. Inst. **31**, 3173–3180 (2007). https://doi.org/10.1016/j.proci.2006.07.077

40. Q. Wang, L. Hu, S.H. Yoon, S. Lu, M. Delichatsios, S.H. Chung, Blow-out limits of nonpremixed turbulent jet flames in a cross flow at atmospheric and sub-atmospheric pressures. Combust. Flame **162**, 3562–3568 (2015). https://doi.org/10.1016/j.combustflame.2015.06.012

41. S. Ducruix, T. Schuller, D. Durox, S. Candel, Combustion dynamics and instabilities: elementary coupling and driving mechanisms. J. Propuls. Power. **19**, 722–734 (2003). https://doi.org/10.2514/2.6182

42. S. Taamallah, Z.A. Labry, S.J. Shanbhogue, A.F. Ghoniem, Thermo-acoustic instabilities in lean premixed swirl-stabilized combustion and their link to acoustically coupled and decoupled flame macrostructures. Proc. Combust. Inst. **35**, 3273–3282 (2015). https://doi.org/10.1016/j.proci.2014.07.002

43. L. Crocco, Research on combustion instability in liquid propellant rockets. Symp. Combust. **12**, 85–99 (1969). https://doi.org/10.1016/S0082-0784(69)80394-2

44. Y. Huang, V. Yang, Dynamics and stability of lean-premixed swirl-stabilized combustion. Prog. Energy Combust. Sci. **35**, 293–364 (2009). https://doi.org/10.1016/j.pecs.2009.01.002

45. S.S. Rashwan, M.A. Nemitallah, M.A. Habib, Review on premixed combustion technology: stability, emission control, applications, and numerical case study. Energy Fuels **30**, 9981–10014 (2016). https://doi.org/10.1021/acs.energyfuels.6b02386

46. M.P. Boyce, *Combustors*, 4th edn. (Elsevier, 2012)

47. M. M. Schorr, C. Joel, *Gas Turbine NOx Emissions Approaching Zero-Is it Worth the Price?* (GE Power Generation GER, 1999), pp. 1–10

48. N.D. Joshi, M.J. Epstein, S. Durlak, S. Marakovits, P.E. Sabla, Development of a fuel air premixer for aeroderivative dry low emissions combustors. Proc. ASME Turbo Expo. **3** (1994). https://doi.org/10.1115/94-GT-253

49. N.D. Joshi, H.C. Mongia, G. Leonard, J.W. Stegmaier, E.C. Vickers, Dry low emissions combustor development, in: *Volume 3: Coal, Biomass and Alternative Fuels; Combustion and Fuels; Oil and Gas Applications; Cycle Innovations* (American Society of Mechanical Engineers, 1998). https://doi.org/10.1115/98-GT-310.

50. M. Zajadatz, D. Pennell, S. Bernero, B. Paikert, R. Zoli, K. Döbbeling, *Development and Implementation of the AEV Burner for the Alstom GT13E2* (2012), pp. 351–360. https://doi.org/10.1115/GT2012-68471

51. T. Sattelmayer, M.P. Felchlin, J. Haumann, J. Hellat, D. Styner, Second-generation low-emission combustors for ABB gas turbines: burner development and tests at atmospheric pressure. J. Eng. Gas Turbines Power **114**, 118–125 (1992). https://doi.org/10.1115/1.2906293

52. K. Döbbeling, J. Hellat, H. Koch, 25 years of BBC/ABB/Alstom lean premix combustion technologies. J. Eng. Gas Turbines Power **129**, 2–12 (2007). https://doi.org/10.1115/1.2181183

53. S. Leibovich, Vortex stability and breakdown—survey and extension. AIAA J. **22**, 1192–1206 (1984). https://doi.org/10.2514/3.8761

54. M. Escudier, Vortex breakdown: observations and explanations. Prog. Aerosp. Sci. **25**, 189–229 (1988). https://doi.org/10.1016/0376-0421(88)90007-3

55. F. Joos, P. Brunner, M. Stalder, S. Tschirren, Field experience with the sequential combustion system of the GT24/GT26. ABB Rev. 12–20 (1998)

56. G. Müller, M. Valk, G. Hosse, H. Heller, Large gas turbines—the insurance aspects (Update). Int. Assoc. Eng. Insur. **13** (2000). http://www.imia.com/wp-content/uploads/2013/05/WGP-1300-Large-Gas-Turbines.pdf

57. F. Joos, P. Brunner, B. Schulte-Werning, K. Syed, A. Eroglu, *Development of the Sequential Combustion System for the ABB GT24/GT26 Gas Turbine Family* (1996). https://doi.org/10.1115/96-GT-315

58. M. Zajadatz, F. Güthe, E. Freitag, T. Ferreira-Providakis, T. Wind, F. Magni, J. Goldmeer, Extended range of fuel capability for GT13E2 AEV burner with liquid and gaseous fuels. J. Eng. Gas Turbines Power. **141** (2018). https://doi.org/10.1115/1.4041144

59. M. Zajadatz, R. Lachner, S. Bernero, C. Motz, P. Flohr, Development and design of alstom's staged fuel gas injection EV burner for NOx reduction, in: *Volume 2: Turbo Expo 2007* (ASMEDC, 2007), pp. 559–567. https://doi.org/10.1115/GT2007-27730

60. C.H. Cho, G.M. Baek, C.H. Sohn, J.H. Cho, H.S. Kim, A numerical approach to reduction of NOx emission from swirl premix burner in a gas turbine combustor. Appl. Therm. Eng. **59**, 454–463 (2013). https://doi.org/10.1016/j.applthermaleng.2013.06.004

61. F. Guethe, R. Lachner, B. Schuermans, F. Biagioli, W. Geng, A. Inauen, S. Schenker, R. Bombach, W. Hubschmid, Flame imaging on the ALSTOM EV-burner: thermo acoustic pulsations and CFD-validation, in *44th AIAA Aerospace Science Meeting Exhibit* (American Institute of Aeronautics and Astronautics, 2006). https://doi.org/10.2514/6.2006-437

62. D. Guyot, T. Meeuwissen, D. Rebhan, Staged premix EV combustion in Alstom's GT24 gas turbine engine, in: *Volume 2: Combustion and Fuels Emission Parts A and B* (American Society of Mechanical Engineers, 2012), pp. 1537–1545. https://doi.org/10.1115/GT2012-70102

63. C. Steinbach, T. Ruck, J. Lloyd, P. Jansohn, K. Döbbeling, T. Sattelmayer, T. Strand, ABB's advanced EV burner—a dual fuel dry low NOx burner for stationary gas turbines, in *Proceedings ASME Turbo Expo* (American Society of Mechanical Engineers, 1998). https://doi.org/10.1115/98-GT-519

64. B.A. Imteyaz, M.A. Nemitallah, A.A. Abdelhafez, M.A. Habib, Combustion behavior and stability map of hydrogen-enriched oxy-methane premixed flames in a model gas turbine combustor. Int. J. Hydrogen Energy **43**, 16652–16666 (2018). https://doi.org/10.1016/j.ijhydene.2018.07.087

65. A.F. Ghoniem, Needs, resources and climate change: clean and efficient conversion technologies. Prog. Energy Combust. Sci. **37**, 15–51 (2011). https://doi.org/10.1016/j.pecs.2010.02.006

66. M.A. Habib, M. Nemitallah, R. Ben-Mansour, Recent development in oxy-combustion technology and its applications to gas turbine combustors and ITM reactors. Energy Fuels **27**, 2–19 (2013). https://doi.org/10.1021/ef301266j

67. C. Shaddix, A. Molina, 6-Ignition, flame stability, and char combustion in oxy-fuel combustion, in *Oxy-Fuel Combustion for Power Generation and Carbon Dioxide Capture* (2011), pp. 101–124. https://www.sciencedirect.com/science/article/pii/B97818456967195 00065?via%3Dihub

68. Y.H. Li, G.B. Chen, Y.C. Chao, Effects of flue gas addition on the premixed oxy-methane flames in atmospheric condition. Energy Procedia **75**, 3054–3059 (2015). https://doi.org/10.1016/j.egypro.2015.07.623

69. J. Oh, D. Noh, Laminar burning velocity of oxy-methane flames in atmospheric condition. Energy **45**, 669–675 (2012). https://doi.org/10.1016/j.energy.2012.07.027

70. M. Shah, N. Degenstein, M. Zanfir, R. Kumar, J. Bugayong, K. Burgers, Near zero emissions oxy-combustion CO_2 purification technology. Energy Procedia **4**, 988–995 (2011). https://doi.org/10.1016/j.egypro.2011.01.146

71. I. Ali, X. Gou, Q. Zhang, J. Wu, E. Wang, Y. Liu, Experimental study on NOx emission characteristics of oxy-biomass combustion. J. Clean. Prod. **199**, 400–410 (2018). https://doi.org/10.1016/j.jclepro.2018.07.022

72. K. Andersson, F. Normann, F. Johnsson, B. Leckner, *NO Emission During Oxy-Fuel Combustion of Lignite* (2008), pp. 1835–1845. https://doi.org/10.1021/ie0711832

73. C. Schluckner, C. Gaber, M. Landfahrer, M. Demuth, C. Hochenauer, Fast and accurate CFD-model for NOx emission prediction during oxy-fuel combustion of natural gas using detailed chemical kinetics. Fuel **264**, 116841 (2020). https://doi.org/10.1016/j.fuel.2019.116841

74. M.R. Shakeel, Y.S. Sanusi, E.M.A. Mokheimer, Numerical modeling of oxy-methane combustion in a model gas turbine combustor. Appl. Energy **228**, 68–81 (2018). https://doi.org/10.1016/j.apenergy.2018.06.071

75. M.A. Nemitallah, A.A. Abdelhafez, A. Ali, I. Mansir, M.A. Habib, Frontiers in combustion techniques and burner designs for emissions control and CO_2 capture: a review. Int. J. Energy Res. **43**, 7790–7822 (2019). https://doi.org/10.1002/er.4730

76. D.M. Wicksall, A.K. Agrawal, R.W. Schefer, J.O. Keller, The interaction of flame and flow field in a lean premixed swirl-stabilized combustor operated on H_2/CH_4/air. Proc. Combust. Inst. **30**(II), 2875–2883 (2005). https://doi.org/10.1016/j.proci.2004.07.021

77. T.C. Williams, C.R. Shaddix, R.W. Schefer, Effect of syngas composition and CO_2-diluted oxygen on performance of a premixed swirl-stabilized combustor. Combust. Sci. Technol. **180**, 64–88 (2008). https://doi.org/10.1080/00102200701487061

78. P.H. Joo, M.R.J. Charest, C.P.T. Groth, Ö.L. Gülder, Comparison of structures of laminar methane-oxygen and methane-air diffusion flames from atmospheric to 60atm. Combust. Flame **160**, 1990–1998 (2013). https://doi.org/10.1016/j.combustflame.2013.04.030

79. Y. Xie, J. Wang, M. Zhang, J. Gong, W. Jin, Z. Huang, Experimental and numerical study on laminar flame characteristics of methane oxy-fuel mixtures highly diluted with CO_2. Energy Fuels **27**, 6231–6237 (2013). https://doi.org/10.1021/ef401220h

80. S. Rashwan, A. Ibrahim, T. Abou-Arab, Experimental investigation of oxy-fuel combustion of CNG flames stabilized over a perforated plate burner, in *18th IFRF Members Conference—Flexible and Clean Fuel Conversion to Industry* (Freising, Germany, 2015), p. 25

81. I.A. Ramadan, A.H. Ibrahim, T.W. Abou-Arab, S.S. Rashwan, M.A. Nemitallah, M.A. Habib, Effects of oxidizer flexibility and bluff-body blockage ratio on flammability limits of diffusion flames. Appl. Energy **178**, 19–28 (2016). https://doi.org/10.1016/j.apenergy.2016.06.009

82. S. Taamallah, K. Vogiatzaki, F.M. Alzahrani, E.M.A. Mokheimer, M.A. Habib, A.F. Ghoniem, Fuel flexibility, stability and emissions in premixed hydrogen-rich gas turbine combustion: technology, fundamentals, and numerical simulations. Appl. Energy **154**, 1020–1047 (2015). https://doi.org/10.1016/j.apenergy.2015.04.044

83. Z. Abubakar, S.Y. Sanusi, E.M.A. Mokheimer, Stability of Propane-air and oxyfuel diffusion flames in a swirl-stabilized combustor; an experimental study. Energy Procedia **142**, 1552–1557 (2017). https://doi.org/10.1016/j.egypro.2017.12.607

84. F. Tahir, H. Ali, A.A.B. Baloch, Y. Jamil, Performance analysis of air and oxy-fuel laminar combustion in a porous plate reactor. Energies **12** (2019). https://doi.org/10.3390/en12091706

85. M.A. Habib, H.M. Badr, S.F. Ahmed, R. Ben-Mansour, K. Mezghani, S. Imashuku, G.J. la O', Y. Shao-Horn, N.D. Mancini, A. Mitsos, P. Kirchen, A.F. Ghoneim, A review of recent developments in carbon capture utilizing oxy-fuel combustion in conventional and ion transport membrane systems. Int. J. Energy Res. **35**, 741–764 (2011). https://doi.org/10.1002/er.1798

86. M.A. Nemitallah, M.A. Habib, Experimental and numerical investigations of an atmospheric diffusion oxy-combustion flame in a gas turbine model combustor. Appl. Energy **111**, 401–415 (2013). https://doi.org/10.1016/j.apenergy.2013.05.027

87. M.A. Habib, S.A. Salaudeen, M.A. Nemitallah, R. Ben-Mansour, E.M.A. Mokheimer, Numerical investigation of syngas oxy-combustion inside a LSCF-6428 oxygen transport membrane reactor. Energy **96**, 654–665 (2016). https://doi.org/10.1016/j.energy.2015.12.043

88. M.A. Nemitallah, A study of methane oxy-combustion characteristics inside a modified design button-cell membrane reactor utilizing a modified oxygen permeation model for reacting flows. J. Nat. Gas Sci. Eng. **28**, 61–73 (2016). https://doi.org/10.1016/j.jngse.2015.11.041

89. K. Lee, H. Kim, P. Park, S. Yang, Y. Ko, CO_2 radiation heat loss effects on NOx emissions and combustion instabilities in lean premixed flames. Fuel **106**, 682–689 (2013). https://doi.org/10.1016/j.fuel.2012.12.048

90. Y. Lafay, B. Taupin, G. Martins, G. Cabot, B. Renou, A. Boukhalfa, Experimental study of biogas combustion using a gas turbine configuration. Exp. Fluids **43**, 395–410 (2007). https://doi.org/10.1007/s00348-007-0302-6

91. K.K. Gupta, A. Rehman, R.M. Sarviya, Bio-fuels for the gas turbine: a review. Renew. Sustain. Energy Rev. **14**, 2946–2955 (2010). https://doi.org/10.1016/j.rser.2010.07.025

92. J. Zhang, J. Mi, P. Li, F. Wang, B.B. Dally, Moderate or intense low-oxygen dilution combustion of methane diluted by CO_2 and N_2. Energy Fuels **29**, 4576–4585 (2015). https://doi.org/10.1021/acs.energyfuels.5b00511

93. A. Amato, B. Hudak, P. D'Carlo, D. Noble, D. Scarborough, J. Seitzman, T. Lieuwen, Methane oxycombustion for low CO2 cycles: blowoff measurements and analysis. J. Eng. Gas Turbines Power **133**, 1–9 (2011). https://doi.org/10.1115/1.4002296

94. J. Oh, D. Noh, E. Lee, The effect of CO addition on the flame behavior of a non-premixed oxy-methane jet in a lab-scale furnace. Appl. Energy **112**, 350–357 (2013). https://doi.org/10.1016/j.apenergy.2013.06.033

95. W. Jerzak, M. Kuźnia, Experimental study of impact of swirl number as well as oxygen and carbon dioxide content in natural gas combustion air on flame flashback and blow-off. J. Nat. Gas Sci. Eng. **29**, 46–54 (2016). https://doi.org/10.1016/j.jngse.2015.12.054

96. S.S. Rashwan, A.H. Ibrahim, T.W. Abou-Arab, M.A. Nemitallah, M.A. Habib, Experimental investigation of partially premixed methane–air and methane–oxygen flames stabilized over a perforated-plate burner. Appl. Energy **169**, 126–137 (2016). https://doi.org/10.1016/j.apenergy.2016.02.047

97. M. Ditaranto, J. Hals, Combustion instabilities in sudden expansion oxy-fuel flames. Combust. Flame **146**, 493–512 (2006). https://doi.org/10.1016/j.combustflame.2006.04.015

98. P. Kutne, B.K. Kapadia, W. Meier, M. Aigner, Experimental analysis of the combustion behaviour of oxyfuel flames in a gas turbine model combustor. Proc. Combust. Inst. **33**, 3383–3390 (2011). https://doi.org/10.1016/j.proci.2010.07.008

99. A. Abdelhafez, M.A. Nemitallah, S.S. Rashwan, M.A. Habib, Adiabatic flame temperature for controlling the macrostructures and stabilization modes of premixed methane flames in a model gas-turbine combustor. Energy Fuels **32**, 7868–7877 (2018). https://doi.org/10.1021/acs.energyfuels.8b01133

100. M.A. Habib, M.A. Nemitallah, P. Ahmed, M.H. Sharqawy, H.M. Badr, I. Muhammad, M. Yaqub, Experimental analysis of oxygen-methane combustion inside a gas turbine reactor under various operating conditions. Energy **86**, 105–114 (2015). https://doi.org/10.1016/j.energy.2015.03.120

101. C. Yin, J. Yan, Oxy-fuel combustion of pulverized fuels: combustion fundamentals and modeling. Appl. Energy **162**, 742–762 (2016). https://doi.org/10.1016/j.apenergy.2015.10.149

102. Y. Song, C. Zou, Y. He, C. Zheng, The chemical mechanism of the effect of CO_2 on the temperature in methane oxy-fuel combustion. Int. J. Heat Mass Transf. **86**, 622–628 (2015). https://doi.org/10.1016/j.ijheatmasstransfer.2015.03.008

103. C.Y. Liu, G. Chen, N. Sipöcz, M. Assadi, X.S. Bai, Characteristics of oxy-fuel combustion in gas turbines. Appl. Energy **89**, 387–394 (2012). https://doi.org/10.1016/j.apenergy.2011.08.004

104. S.I.B. Shia, Z. Zhua, N. Wanga, P. Lub, An experimental study on oxy-fuel combustion of methane under various oxygen mole fractions, in *8th International Symposium on Coal Combustion* (2015).

105. P. Heil, D. Toporov, H. Stadler, S. Tschunko, M. Förster, R. Kneer, Development of an oxycoal swirl burner operating at low O_2 concentrations. Fuel **88**, 1269–1274 (2009). https://doi.org/10.1016/j.fuel.2008.12.025

106. X. Hu, Q. Yu, J. Liu, N. Sun, Investigation of laminar flame speeds of $CH_4/O_2/CO_2$ mixtures at ordinary pressure and kinetic simulation. Energy **70**, 626–634 (2014). https://doi.org/10.1016/j.energy.2014.04.029

107. A. Ali, M.A. Nemitallah, A. Abdelhafez, M. Hussain, M.M. Kamal, M.A. Habib, Comparative analysis of the stability and structure of premixed $C_3H_8/O_2/CO_2$ and $C_3H_8/O_2/N_2$ flames for clean flexible energy production. Energy **214**, 1–10 (2021). https://doi.org/10.1016/j.energy.2020.118887

108. R. Stanger, T. Wall, R. Spörl, M. Paneru, S. Grathwohl, M. Weidmann, G. Scheffknecht, D. McDonald, K. Myöhänen, J. Ritvanen, S. Rahiala, T. Hyppänen, J. Mletzko, A. Kather, S. Santos, Oxyfuel combustion for CO_2 capture in power plants. Int. J. Greenh. Gas Control. **40**, 55–125 (2015). https://doi.org/10.1016/j.ijggc.2015.06.010

109. T.F. Wall, Combustion processes for carbon capture. Proc. Combust. Inst. **31**(I), 31–47 (2007). https://doi.org/10.1016/j.proci.2006.08.123

110. M. Aneke, M. Wang, Process analysis of pressurized oxy-coal power cycle for carbon capture application integrated with liquid air power generation and binary cycle engines. Appl. Energy **154**, 556–566 (2015). https://doi.org/10.1016/j.apenergy.2015.05.030

111. M. Aliyu, M.A. Nemitallah, S.A. Said, M.A. Habib, Characteristics of H_2-enriched CH_4-O_2 diffusion flames in a swirl-stabilized gas turbine combustor: Experimental and numerical study. Int. J. Hydrogen Energy **41**, 20418–20432 (2016). https://doi.org/10.1016/j.ijhydene.2016.08.144

112. F. Liu, H. Guo, G.J. Smallwood, The chemical effect of CO_2 replacement of N_2 in air on the burning velocity of CH_4 and H_2 premixed flames. Combust. Flame **133**, 495–497 (2003). https://doi.org/10.1016/S0010-2180(03)00019-1

113. D.A. Granados, F. Chejne, J.M. Mejía, C.A. Gómez, A. Berrío, W.J. Jurado, Effect of flue gas recirculation during oxy-fuel combustion in a rotary cement kiln. Energy **64**, 615–625 (2014). https://doi.org/10.1016/j.energy.2013.09.045

114. L. Wang, N.E. Endrud, S.R. Turns, M.D. D'Agostini, A.G. Slavejkov, A study of the influence of oxygen index on soot, radiation, and emission characteristics of turbulent jet flames. Combust. Sci. Technol. **174**, 45–72 (2002). https://doi.org/10.1080/00102200290021245

115. T. Wall, Y. Liu, C. Spero, L. Elliott, S. Khare, R. Rathnam, F. Zeenathal, B. Moghtaderi, B. Buhre, C. Sheng, R. Gupta, T. Yamada, K. Makino, J. Yu, An overview on oxyfuel coal combustion-State of the art research and technology development. Chem. Eng. Res. Des. **87**, 1003–1016 (2009). https://doi.org/10.1016/j.cherd.2009.02.005

116. T. Muruganandam, S. Nair, Y. Neumeier, T. Lieuwen, J. Seitzman, Optical and acoustic sensing of lean blowout precursors, in: *38th AIAA/ASME/SAE/ASEE Joint Propulsion Conference and Exhibit* (American Institute of Aeronautics and Astronautics, 2002). https://doi.org/10.2514/6.2002-3732

117. M. Nemitallah, S. Alkhaldi, A. Abdelhafez, M. Habib, Effect analysis on the macrostructure and static stability limits of oxy-methane flames in a premixed swirl combustor. Energy **159**, 86–96 (2018). https://doi.org/10.1016/j.energy.2018.06.131

118. J. Runyon, R. Marsh, P. Bowen, D. Pugh, A. Giles, S. Morris, Lean methane flame stability in a premixed generic swirl burner: isothermal flow and atmospheric combustion characterization. Exp. Therm. Fluid Sci. **92**, 125–140 (2018). https://doi.org/10.1016/j.expthermflusci.2017.11.019

119. A. Mansour, *Gas Turbine Fuel Injection Technology* (2005), pp. 141–149. https://doi.org/10.1115/GT2005-68173

120. A.C. Alkidas, Combustion advancements in gasoline engines. Energy Convers. Manag. **48**, 2751–2761 (2007). https://doi.org/10.1016/j.enconman.2007.07.027

121. M.C. Drake, D.C. Haworth, Advanced gasoline engine development using optical diagnostics and numerical modeling. Proc. Combust. Inst. **31**(I), 99–124 (2007). https://doi.org/10.1016/j.proci.2006.08.120

122. C. Straub, A. Kronenburg, O.T. Stein, R.S. Barlow, D. Geyer, Modeling stratified flames with and without shear using multiple mapping conditioning. Proc. Combust. Inst. **37**, 2317–2324 (2019). https://doi.org/10.1016/j.proci.2018.07.033

123. S. Schneider, D. Geyer, G. Magnotti, M.J. Dunn, R.S. Barlow, A. Dreizler, Structure of a stratified CH_4 flame with H_2 addition. Proc. Combust. Inst. **37**, 2307–2315 (2019). https://doi.org/10.1016/j.proci.2018.06.205

124. F.C.C. Galeazzo, B. Savard, H. Wang, E.R. Hawkes, J.H. Chen, G.C. Krieger Filho, Performance assessment of flamelet models in flame-resolved les of a high Karlovitz methane/air stratified premixed jet flame. Proc. Combust. Inst. **37**, 2545–2553 (2019). https://doi.org/10.1016/j.proci.2018.09.025

125. C. Duwig, C. Fureby, Large eddy simulation of unsteady lean stratified premixed combustion. Combust. Flame **151**, 85–103 (2007). https://doi.org/10.1016/j.combustflame.2007.04.004

126. A.X. Sengissen, J.F. Van Kampen, R.A. Huls, G.G.M. Stoffels, J.B.W. Kok, T.J. Poinsot, LES and experimental studies of cold and reacting flow in a swirled partially premixed burner with and without fuel modulation. Combust. Flame **150**, 40–53 (2007). https://doi.org/10.1016/j.combustflame.2007.02.009

127. W. Meier, P. Weigand, X.R. Duan, R. Giezendanner-Thoben, Detailed characterization of the dynamics of thermoacoustic pulsations in a lean premixed swirl flame. Combust. Flame **150**, 2–26 (2007). https://doi.org/10.1016/j.combustflame.2007.04.002

128. Z. Tan, R.D. Reitz, An ignition and combustion model based on the level-set method for spark ignition engine multidimensional modeling. Combust. Flame **145**, 1–15 (2006). https://doi.org/10.1016/j.combustflame.2005.12.007

129. P.D. Nguyen, L. Vervisch, V. Subramanian, P. Domingo, Multidimensional flamelet-generated manifolds for partially premixed combustion. Combust. Flame **157**, 43–61 (2010). https://doi.org/10.1016/j.combustflame.2009.07.008

130. E. Knudsen, H. Pitsch, A general flamelet transformation useful for distinguishing between premixed and non-premixed modes of combustion. Combust. Flame **156**, 678–696 (2009). https://doi.org/10.1016/j.combustflame.2008.10.021

131. T. Kang, D.C. Kyritsis, Methane flame propagation in compositionally stratified gases. Combust. Sci. Technol. **177**, 2191–2210 (2005). https://doi.org/10.1080/00102200500240836

132. C. Galizzi, D. Escudié, Experimental analysis of an oblique laminar flame front propagating in a stratified flow. Combust. Flame **145**, 621–634 (2006). https://doi.org/10.1016/j.combustflame.2005.12.001

133. B. Rendu, E. Samson, A. Boukhalfa, An experimental study of freely propagating turbulent propane/air flames in stratified inhomogeneous mixtures. Combust. Sci. Technol. **176**, 1867–1890 (2004). https://doi.org/10.1080/00102200490504490

134. N. Pasquier, B. Lecordier, M. Trinité, A. Cessou, An experimental investigation of flame propagation through a turbulent stratified mixture. Proc. Combust. Inst. **31**(I), 1567–1574 (2007). https://doi.org/10.1016/j.proci.2006.07.118

135. V. Robin, A. Mura, M. Champion, O. Degardin, B. Renou, M. Boukhalfa, Experimental and numerical analysis of stratified turbulent V-shaped flames. Combust. Flame **153**, 288–315 (2008). https://doi.org/10.1016/j.combustflame.2007.10.008

136. P. Anselmo-Filho, S. Hochgreb, R.S. Barlow, R.S. Cant, Experimental measurements of geometric properties of turbulent stratified flames. Proc. Combust. Inst. **32**(II), 1763–1770 (2009). https://doi.org/10.1016/j.proci.2008.05.085

137. B. Böhm, J.H. Frank, A. Dreizler, Temperature and mixing field measurements in stratified lean premixed turbulent flames. Proc. Combust. Inst. **33**, 1583–1590 (2011). https://doi.org/10.1016/j.proci.2010.06.139

138. M.S. Sweeney, S. Hochgreb, M.J. Dunn, R.S. Barlow, A comparative analysis of flame surface density metrics in premixed and stratified flames. Proc. Combust. Inst. **33**, 1419–1427 (2011). https://doi.org/10.1016/j.proci.2010.05.069

139. P.C. Vena, B. Deschamps, G.J. Smallwood, M.R. Johnson, Equivalence ratio gradient effects on flame front topology in a stratified iso-octane/air turbulent V-flame. Proc. Combust. Inst. **33**, 1551–1558 (2011). https://doi.org/10.1016/j.proci.2010.06.041

140. F. Seffrin, F. Fuest, D. Geyer, A. Dreizler, Flow field studies of a new series of turbulent premixed stratified flames. Combust. Flame **157**, 384–396 (2010). https://doi.org/10.1016/j.combustflame.2009.09.001

141. C. Galizzi, D. Escudié, Experimental analysis of an oblique turbulent flame front propagating in a stratified flow. Combust. Flame **157**, 2277–2285 (2010). https://doi.org/10.1016/j.combustflame.2010.07.008

142. A. Pires Da Cruz, A.M. Dean, J.M. Grenda, A numerical study of the laminar flame speed of stratified methane/air flames. Proc. Combust. Inst. **28**, 1925–1932 (2000). https://doi.org/10.1016/S0082-0784(00)80597-4

143. Y.M. Marzouk, A.F. Ghoniem, H.N. Najm, Dynamic response of strained premixed flames to equivalence ratio gradients. Proc. Combust. Inst. **28**, 1859–1866 (2000). https://doi.org/10.1016/S0082-0784(00)80589-5

144. E.S. Richardson, V.E. Granet, A. Eyssartier, J.H. Chen, Effects of equivalence ratio variation on lean, stratified methane-air laminar counterflow flames. Combust. Theory Model. **14**, 775–792 (2010). https://doi.org/10.1080/13647830.2010.490881

145. J. Hélie, A. Trouvé, Turbulent flame propagation in partially premixed combustion. Symp. Combust. **27**, 891–898 (1998). https://doi.org/10.1016/S0082-0784(98)80486-4

146. D.C. Haworth, R.J. Blint, B. Cuenot, T.J. Poinsot, Numerical simulation of turbulent propane-air combustion with nonhomogeneous reactants. Combust. Flame **121**, 395–417 (2000). https://doi.org/10.1016/S0010-2180(99)00148-0

147. C. Jiménez, B. Cuenot, T. Poinsot, D. Haworth, Numerical simulation and modeling for lean stratified propane-air flames. Combust. Flame **128**, 1–21 (2002). https://doi.org/10.1016/S0010-2180(01)00328-5

148. K. Bray, P. Domingo, L. Vervisch, Role of the progress variable in models for partially premixed turbulent combustion. Combust. Flame **141**, 431–437 (2005). https://doi.org/10.1016/j.combustflame.2005.01.017

149. V. Robin, A. Mura, M. Champion, P. Plion, A multi-dirac presumed pdf model for turbulent reactive flows with variable equivalence ratio. Combust. Sci. Technol. **178**, 1843–1870 (2006). https://doi.org/10.1080/00102200600790763

150. N. Il Kim, J. Il Seo, K.C. Oh, H.D. Shin, Lift-off characteristics of triple flame with concentration gradient. Proc. Combust. Inst. **30**, 367–374 (2005). https://doi.org/10.1016/j.proci.2004.07.001

151. S. Balusamy, A. Cessou, B. Lecordier, Laminar propagation of lean premixed flames ignited in stratified mixture. Combust. Flame **161**, 427–437 (2014). https://doi.org/10.1016/j.combustflame.2013.08.023

152. T. Kang, D.C. Kyritsis, Phenomenology of methane flame propagation into compositionally stratified, gradually richer mixtures. Proc. Combust. Inst. **32**(I), 979–985 (2009). https://doi.org/10.1016/j.proci.2008.06.007

153. T. Kang, D.C. Kyritsis, Departure from quasi-homogeneity during laminar flame propagation in lean, compositionally stratified methane-air mixtures. Proc. Combust. Inst. **31**(I), 1075–1083 (2007). https://doi.org/10.1016/j.proci.2006.07.051

154. N. Il Kim, J. Il Seo, Y.T. Guahk, H.D. Shin, The propagation of tribrachial flames in a confined channel. Combust. Flame **146**, 168–179 (2006). https://doi.org/10.1016/j.combustflame.2006.04.001

155. S.W. Grib, M.W. Renfro, Propagation speeds for interacting triple flames. Combust. Flame **187**, 230–238 (2018). https://doi.org/10.1016/j.combustflame.2017.09.018

156. G.R. Ruetsch, L. Vervisch, A. Liñán, Effects of heat release on triple flames. Phys. Fluids **7**, 1447–1454 (1995). https://doi.org/10.1063/1.868531

157. X. Han, S.K. Aggarwal, K. Brezinsky, Effect of unsaturated bond on NOx and PAH formation in n-heptane and 1-heptene triple flames. Energy Fuels **27**, 537–548 (2013). https://doi.org/10.1021/ef301671q

158. S.H. Chung, Stabilization, propagation and instability of tribrachial triple flames. Proc. Combust. Inst. **31**(I), 877–892 (2007). https://doi.org/10.1016/j.proci.2006.08.117

159. R. Owston, J. Abraham, Structure of hydrogen triple flames and premixed flames compared. Combust. Flame **157**, 1552–1565 (2010). https://doi.org/10.1016/j.combustflame.2010.01.014

160. T. Plessing, P. Terhoeven, N. Peters, M.S. Mansour, An experimental and numerical study of a laminar triple flame. Combust. Flame **115**, 335–353 (1998). https://doi.org/10.1016/S0010-2180(98)00013-3

161. J. Zhang, J. Abraham, A numerical study of laminar flames propagating in stratified mixtures. Combust. Flame **163**, 461–471 (2016). https://doi.org/10.1016/j.combustflame.2015.10.020

162. L. Bartolucci, S. Cordiner, V. Mulone, V. Rocco, Natural gas partially stratified lean combustion: analysis of the enhancing mechanisms into a constant volume combustion chamber. Fuel **211**, 737–753 (2018). https://doi.org/10.1016/j.fuel.2017.09.100

163. N. Sharma, A.K. Agarwal, Effect of the fuel injection pressure on particulate emissions from a gasohol (E15 and M15)-fueled gasoline direct injection engine. Energy Fuels **31**, 4155–4164 (2017). https://doi.org/10.1021/acs.energyfuels.6b02877

164. J. Lee, S. Chu, J. Kang, K. Min, H. Jung, H. Kim, Y. Chi, The classification of gasoline/diesel dual-fuel combustion based on the heat release rate shapes and its application in a light-duty single-cylinder engine. Int. J. Engine Res. **20**, 69–79 (2019). https://doi.org/10.1177/1468087418817676

165. H. Kahila, A. Wehrfritz, O. Kaario, V. Vuorinen, Large-eddy simulation of dual-fuel ignition: Diesel spray injection into a lean methane-air mixture. Combust. Flame **199**, 131–151 (2019). https://doi.org/10.1016/j.combustflame.2018.10.014

166. A. García, J. Monsalve-Serrano, R. Sari, N. Dimitrakopoulos, M. Tunér, P. Tunestål, Performance and emissions of a series hybrid vehicle powered by a gasoline partially premixed combustion engine. Appl. Therm. Eng. **150**, 564–575 (2019). https://doi.org/10.1016/j.applth ermaleng.2019.01.035

167. H. Huang, D. Lv, J. Zhu, Z. Zhu, Y. Chen, Y. Pan, M. Pan, Development of a new reduced diesel/natural gas mechanism for dual-fuel engine combustion and emission prediction. Fuel **236**, 30–42 (2019). https://doi.org/10.1016/j.fuel.2018.08.161

168. S. Park, S. Woo, H. Oh, K. Lee, Effects of various lubricants and fuels on pre-ignition in a turbocharged direct-injection spark-ignition engine. Energy Fuels **31**, 12701–12711 (2017). https://doi.org/10.1021/acs.energyfuels.7b01052

169. T.Y. Kim, C. Park, S. Oh, G. Cho, The effects of stratified lean combustion and exhaust gas recirculation on combustion and emission characteristics of an LPG direct injection engine. Energy **115**, 386–396 (2016). https://doi.org/10.1016/j.energy.2016.09.025

170. S. Samuelsen, Conventional type combustion, in *Gas Turbine Handbook*, pp. 209–217. https://netl.doe.gov/sites/default/files/gas-turbine-handbook/3-2-1-1.pdf

171. U. Borchert, J.A. Szymczyk, Fluidic analyses of a model gas turbine combustion chamber with and without combustion under actual operating conditions, in: XX International Conference on Research and Practice Didact. Model and Mechanical Engineering (Stralsund, 2011)

172. S. Samuelsen, Rich burn, Quick-mix, lean burn (RQL) combustor, in *Gas Turbine Handbook*, pp. 227–233

173. P. Kiameh, Power Generation Handbook: Selection, Applications, Operation, and Maintenance (McGraw-Hill Handbooks, 2003)

174. D.W. Cihlar, C.P. Willis, M.A. Singley, J.G. Reed, Axial Fuel Staging System for a Gas Turbine Combustor (European Patent-EP 3214374 A1), EP 3214374, 2019.

175. M.B. Wilson, J. Harper, Y.H. Chong, C.E. Jones, Axial Fuel Staging System for Gas Turbine Combustors, US20190178498 (2019)

176. W. Shao, Z. Wang, X. Liu, Z. Zhang, Y. Xiao, Y. Zhao, Numerical and experimental parametric study of emission characteristics in an axial fuel staging system. Energy Fuels **33**, 12723–12735 (2019). https://doi.org/10.1021/acs.energyfuels.9b02293

177. M. Boyce, Gas turbine performance test, in *Gas Turbine Engineering Handbook*, 4th edn. (Elsevier, 2012), pp. 769–802

178. S. Gülen, The hall of fame, in *Gas Turbines Electric Power Generation* (Cambridge University Press, 2019), pp. 626–650. https://doi.org/10.1017/9781108241625.022

179. Axial staging in combustor upgrade helps to deliver emission compliance at 25% load. Turbomach. Int. (2018). https://www.turbomachinerymag.com/ge-power-introduces-new-dln2-6flex-upgrade-solution-for-7f-gas-turbines/

180. S. Hayashi, H. Yamada, M. Makida, Extending low-NOx operating range of a lean premixed-prevaporized gas turbine combustor by reaction of secondary mixtures injected into primary stage burned gas. Proc. Combust. Inst. **30**(II), 2903–2911 (2005). https://doi.org/10.1016/j.proci.2004.08.112

181. S. Hayashi, H. Yamada, NOx emissions in combustion of lean premixed mixtures injected into hot burned gas. Proc. Combust. Inst. **28**, 2443–2449 (2000). https://doi.org/10.1016/S0082-0784(00)80658-X

182. A. Emara, A. Lacarelle, C.O. Paschereit, Pilot flame impact on flow fields and combustion performances in a swirl inducing burner, in *45th AIAA/ASME/SAE/ASEE Joint Propulsion Conference and Exhibit* (2009), pp. 1–13. https://doi.org/10.2514/6.2009-5015

183. I.A. Zubrilin, N.I. Gurakov, S.G. Matveev, Lean blowout limit prediction in a combustor with the pilot flame. Energy Procedia **141**, 273–281 (2017). https://doi.org/10.1016/j.egypro.2017.11.105

184. X. Fu, F. Yang, Z. Guo, Combustion instability of pilot flame in a pilot bluff body stabilized combustor. Chin. J. Aeronaut. **28**, 1606–1615 (2015). https://doi.org/10.1016/j.cja.2015.08.018

185. M.A. Nemitallah, M.A. Habib, H.M. Badr, Oxyfuel *Combustion for Clean Energy Applications* (Springer International Publishing, Cham, 2019). https://doi.org/10.1007/978-3-030-10588-4

186. L.B. Davis, S.H. Black, Dry low NOX combustion systems for GE heavy-duty gas turbines, in *GE Power Systems* (2000)
187. M.D. Durbin, M.A. Mueller, L.K. Blakemen, D.A. Lind, System and Method for Flame Stabilization, US 9719685 B2 (2017)
188. X. Han, D. Laera, D. Yang, C. Zhang, J. Wang, X. Hui, Y. Lin, A.S. Morgans, C.-J. Sung, Flame interactions in a stratified swirl burner: flame stabilization, combustion instabilities and beating oscillations. Combust. Flame **212**, 500–509 (2020). https://doi.org/10.1016/j.combus tflame.2019.11.020
189. X. Han, D. Laera, A.S. Morgans, C.J. Sung, X. Hui, Y.Z. Lin, Flame macrostructures and thermoacoustic instabilities in stratified swirling flames. Proc. Combust. Inst. **37**, 5377–5384 (2019). https://doi.org/10.1016/j.proci.2018.06.147
190. M.S. Sweeney, S. Hochgreb, M.J. Dunn, R.S. Barlow, The structure of turbulent stratified and premixed methane/air flames II: swirling flows. Combust. Flame **159**, 2912–2929 (2012). https://doi.org/10.1016/j.combustflame.2012.05.014
191. R.S. Barlow, G.H. Wang, P. Anselmo-Filho, M.S. Sweeney, S. Hochgreb, Application of Raman/Rayleigh/LIF diagnostics in turbulent stratified flames. Proc. Combust. Inst. 32(I), 945–953 (2009). https://doi.org/10.1016/j.proci.2008.06.070
192. L. Li, Y. Lin, Z. Fu, C. Zhang, Emission characteristics of a model combustor for aero gas turbine application. Exp. Therm. Fluid Sci. **72**, 235–248 (2016). https://doi.org/10.1016/j.exp thermflusci.2015.11.012
193. C.T. Chong, S.S. Lam, S. Hochgreb, Effect of mixture flow stratification on premixed flame structure and emissions under counter-rotating swirl burner configuration. Appl. Therm. Eng. **105**, 905–912 (2016). https://doi.org/10.1016/j.applthermaleng.2016.03.164
194. N.S. Rodrigues, O. Busari, W.C.B. Senior, C.T. McDonald, Y.T. Chen, A.J. North, W.R. Laster, S.E. Meyer, R.P. Lucht, NOx reduction in an axially staged gas turbine model combustor through increase in the combustor exit Mach number. Combust. Flame **212**, 282–294 (2020). https://doi.org/10.1016/j.combustflame.2019.10.039
195. B. Wang, C. Zhang, X. Hui, Y. Lin, J. Li, Influence of sleeve angle on the LBO performance of TeLESS-II combustor, in: *52nd AIAA/SAE/ASEE Joint Propulsion Conference* (American Institute of Aeronautics and Astronautics, Reston, Virginia, 2016), pp. 1–8. https://doi.org/ 10.2514/6.2016-4693
196. C.M. Arndt, M. Severin, C. Dem, M. Stöhr, A.M. Steinberg, W. Meier, Experimental analysis of thermo-acoustic instabilities in a generic gas turbine combustor by phase-correlated PIV, chemiluminescence, and laser Raman scattering measurements. Exp. Fluids **56**, 1–23 (2015). https://doi.org/10.1007/s00348-015-1929-3
197. C.M. Arndt, C. Dem, W. Meier, Influence of fuel staging on thermo-acoustic oscillations in a premixed stratified dual-swirl gas turbine model combustor, flow. Turbul. Combust. **106**, 613–629 (2021). https://doi.org/10.1007/s10494-020-00158-6
198. J. Samarasinghe, W. Culler, B.D. Quay, D.A. Santavicca, J. O'Connor, The effect of fuel staging on the structure and instability characteristics of swirl-stabilized flames in a lean premixed multinozzle can combustor. J. Eng. Gas Turbines Power **139**, 1–10 (2017). https:// doi.org/10.1115/1.4037461
199. Q. Zeng, D. Zheng, Y. Yuan, Counter-rotating dual-stage swirling combustion characteristics of hydrogen and carbon monoxide at constant fuel flow rate. Int. J. Hydrogen Energy **45**, 4979–4990 (2020). https://doi.org/10.1016/j.ijhydene.2019.12.068
200. M. Ihme, On the role of turbulence and compositional fluctuations in rapid compression machines: autoignition of syngas mixtures. Combust. Flame **159**, 1592–1604 (2012). https:// doi.org/10.1016/j.combustflame.2011.11.022
201. T. García-Armingol, Á. Sobrino, E. Luciano, J. Ballester, Impact of fuel staging on stability and pollutant emissions of premixed syngas flames. Fuel **185**, 122–132 (2016). https://doi. org/10.1016/j.fuel.2016.07.086
202. H. Zhang, Z. Zhang, Y. Xiong, Y. Liu, Y. Xiao, Experimental and numerical investigations of MILD combustion in a model combustor applied for gas turbine, in *Volume 4B: Combustion Fuels, and Emissions* (American Society of Mechanical Engineers, 2018), pp. 1–10. https:// doi.org/10.1115/GT2018-76253

203. X. Zheng, Y. Xiong, Y. Liu, F. Lei, Y. Xiao, Emission characteristics and visualization of an axial-fuel-staged MILD combustor. Combust. Sci. Technol. 1–22 (2020). https://doi.org/10.1080/00102202.2020.1751619

204. H. Karim, J. Natarajan, V. Narra, J. Cai, S. Rao, J. Kegley, J. Citeno, Staged combustion system for improved emissions operability& flexibility for 7HA class heavy duty gas turbine engine. Proc. ASME Turbo Expo. **4A–2017**, 1–10 (2017). https://doi.org/10.1115/GT2017 63998

205. L.A. Bulysova, A.L. Berne, V.D. Vasil'ev, M.N. Gutnik, M.M. Gutnik, Study of sequential two-stage combustion in a low-emission gas turbine combustion chamber. Therm. Eng. **65**, 806–817 (2018). https://doi.org/10.1134/S0040601518110010

206. N. Kousheshi, M. Yari, A. Paykani, A.S. Mehr, G.F. de la Fuente, Effect of syngas composition on the combustion and emissions characteristics of a syngas/diesel RCCI engine. Energies **13** (2020). https://doi.org/10.3390/en13010212

207. S.A. Burtsev, I.A. Eletskiy, D.S. Kochurov, Gas stratification application in closed-cycle gas turbines, in *AIP Conference Proceedings* (2019), p. 070007. https://doi.org/10.1063/1.513 3218

208. P. Koutmos, G. Paterakis, E. Dogkas, C. Karagiannaki, The impact of variable inlet mixture stratification on flame topology and emissions performance of a premixer/swirl burner configuration. J. Combust. **2012** (2012). https://doi.org/10.1155/2012/374089

209. A. Gopan, B.M. Kumfer, J. Phillips, D. Thimsen, R. Smith, R.L. Axelbaum, Process design and performance analysis of a staged, pressurized oxy-combustion (SPOC) power plant for carbon capture. Appl. Energy **125**, 179–188 (2014). https://doi.org/10.1016/j.apenergy.2014.03.032

210. L. Duan, L. Li, D. Liu, C. Zhao, Fundamental study on fuel-staged oxy-fuel fluidized bed combustion. Combust. Flame **206**, 227–238 (2019). https://doi.org/10.1016/j.combustflame.2019.05.008

211. A. Görgülü, H. Yağlı, Y. Koç, A. Koç, Comprehensive analysis of the effect of water injection on performance and emission parameters of the hydrogen fuelled recuperative and non-recuperative gas turbine system. Int. J. Hydrogen Energy **45**, 34254–34267 (2020). https://doi.org/10.1016/j.ijhydene.2020.09.038

212. D. Batet, F.T. Zohra, S.B. Kristensen, S.J. Andreasen, L. Diekhöner, Continuous durability study of a high temperature polymer electrolyte membrane fuel cell stack. Appl. Energy **277**, 115588 (2020). https://doi.org/10.1016/j.apenergy.2020.115588

213. C. Jeppesen, S.S. Araya, S.L. Sahlin, S.J. Andreasen, S.K. Kær, An EIS alternative for impedance measurement of a high temperature PEM fuel cell stack based on current pulse injection. Int. J. Hydrogen Energy **42**, 15851–15860 (2017). https://doi.org/10.1016/j.ijhydene.2017.05.066

214. J. Rifkin, *The Hydrogen Economy: The Creation of the Worldwide Energy Web and the Redistribution of Power on Earth* (Penguin, 2003)

215. H.T. Arat, M.G. Sürer, S. Gökpinar, K. Aydin, Conceptual design analysis for a lightweight aircraft with a fuel cell hybrid propulsion system. Energy Sources, Part A Recover. Util. Environ. Eff. 1–15 (2020). https://doi.org/10.1080/15567036.2020.1773966

216. B. Tanç, H.T. Arat, Ç. Conker, E. Baltacioğlu, K. Aydin, Energy distribution analyses of an additional traction battery on hydrogen fuel cell hybrid electric vehicle. Int. J. Hydrogen Energy **45**, 26344–26356 (2020). https://doi.org/10.1016/j.ijhydene.2019.09.241

217. H.T. Arat, M.K. Baltacioglu, B. Tanç, M.G. Sürer, I. Dincer, A perspective on hydrogen energy research, development and innovation activities in Turkey. Int. J. Energy Res. **44**, 588–593 (2020). https://doi.org/10.1002/er.5031

218. B. Tanç, H.T. Arat, E. Baltacıoğlu, K. Aydın, Overview of the next quarter century vision of hydrogen fuel cell electric vehicles. Int. J. Hydrogen Energy **44**, 10120–10128 (2019). https://doi.org/10.1016/j.ijhydene.2018.10.112

219. R. Kothari, D. Buddhi, R.L. Sawhney, Comparison of environmental and economic aspects of various hydrogen production methods. Renew. Sustain. Energy Rev. **12**, 553–563 (2008). https://doi.org/10.1016/j.rser.2006.07.012

220. J.D. Holladay, J. Hu, D.L. King, Y. Wang, An overview of hydrogen production technologies. Catal. Today **139**, 244–260 (2009). https://doi.org/10.1016/j.cattod.2008.08.039
221. D. Parra, L. Valverde, F.J. Pino, M.K. Patel, A review on the role, cost and value of hydrogen energy systems for deep decarbonisation. Renew. Sustain. Energy Rev. **101**, 279–294 (2019). https://doi.org/10.1016/j.rser.2018.11.010
222. *A Green, Safe, Efficient and Inexpensive Hydrogen Production System that Operates at Room Temperature in Air* (n.d.)
223. T.A. Adams, L. Hoseinzade, P.B. Madabhushi, I.J. Okeke, Comparison of CO_2 capture approaches for fossil-based power generation: review and meta-study. Processes **5** (2017). https://doi.org/10.3390/pr5030044
224. M. Dutka, M. Ditaranto, T. Løvås, NOx emissions and turbulent flow field in a partially premixed bluff body burner with CH_4 and H_2 fuels. Int. J. Hydrogen Energy **41**, 12397–12410 (2016). https://doi.org/10.1016/j.ijhydene.2016.05.154
225. D.M. Todd, Demonstrated Applicability of Hydrogen Fuel for Gas Turbines (2000)
226. I. Brunetti, N. Rossi, S. Sigali, G. Sonato, S. Cocchi, ENEL's Fusina zero emission combined cycle: experiencing hydrogen combustion. Powergen Eur. (2010)
227. R.W. Schefer, Hydrogen enrichment for improved lean flame stability. Int. J. Hydrogen Energy **28**, 1131–1141 (2003). https://doi.org/10.1016/S0360-3199(02)00199-4
228. A.F. Ghoniem, A. Annaswamy, S. Park, Z.C. Sobhani, Stability and emissions control using air injection and H_2 addition in premixed combustion. Proc. Combust. Inst. **30**(II), 1765–1773 (2005). https://doi.org/10.1016/j.proci.2004.08.175
229. O. Tuncer, The effect of hydrogen enrichment of methane fuel on flame stability and emissions, in *Proceedings 2013 International Conference on Renewable Energy Research and Applications (ICRERA 2013)* (2013), pp. 103–108. https://doi.org/10.1109/ICRERA.2013.6749734
230. S. de Ferrières, A. El Bakali, B. Lefort, M. Montero, J.F. Pauwels, Experimental and numerical investigation of low-pressure laminar premixed synthetic natural gas/O_2/N_2 and natural gas/H_2/O_2/N_2 flames. Combust. Flame **154**, 601–623 (2008). https://doi.org/10.1016/j.combustflame.2008.04.018
231. H. Sun, P. Yan, Y. Xu, Numerical simulation on hydrogen combustion and flow characteristics of a jet-stabilized combustor. Int. J. Hydrogen Energy **45**, 12604–12615 (2020). https://doi.org/10.1016/j.ijhydene.2020.02.151
232. M. Ditaranto, T. Heggset, D. Berstad, Concept of hydrogen fired gas turbine cycle with exhaust gas recirculation: assessment of process performance. Energy **192**, 116646 (2020). https://doi.org/10.1016/j.energy.2019.116646
233. N. Kahraman, S. Tangöz, S.O. Akansu, Numerical analysis of a gas turbine combustor fueled by hydrogen in comparison with jet-a fuel. Fuel **217**, 66–77 (2018). https://doi.org/10.1016/j.fuel.2017.12.071
234. Y. Koç, H. Yağlı, A. Görgülü, A. Koç, Analysing the performance, fuel cost and emission parameters of the 50 MW simple and recuperative gas turbine cycles using natural gas and hydrogen as fuel. Int. J. Hydrogen Energy **45**, 22138–22147 (2020). https://doi.org/10.1016/j.ijhydene.2020.05.267

Chapter 5
Application of Lean Premixed Combustion for Emission Control in Different Combustors

5.1 Characteristics of Lean Premixed Oxy-combustion in Swirl Burner

Thermal power facilities are thought to benefit from minimizing their CO_2 emissions by burning oxygen-rich fuels. As this method is combined with carbon capture and sequestration (CCS) technology, it is expected that the capturing process will be made easier. The high CO_2 concentration in the exhaust stream is blamed for this [1, 2]. The primary challenge in this situation is to install oxy-fuel burners in place of the existing air–fuel combustors with the fewest changes and performance deficiencies possible. In oxy-combustion thermal power plants, O_2 parting units were frequently utilized to separate O_2 from air. Though, these parting items use an inordinate amount of the production power in the oxy-combustion power plants for the separation of oxygen [3]. As a result, operating oxy-fuel combustion systems close to the equivalence ratio of one is preferred (i.e. near stoichiometry). These burning arrangements are also made to allow for flue gas recirculation with the purpose of regulating the burning temperature [4, 5]. Along with the advantage of easy CO_2 detention under oxy-combustion situations, merging premixed burning and oxy-combustion may lead to total governing of NOx releases. However, using CO_2 as a dilution gas in oxy-combustion as a substitute of N_2 in regular air burning creates extra hurdles for flame stabilization. This is explained by the fact that CO_2 and N_2 have different physical characteristics. In order to accommodate these changes, the burner design must be modified. Additionally, CO_2 helps to chemical reactions in high-temperature processes by radiating in the infrared region and becoming active. This could reduce the amount of reaction kinetics present in the flame and impose additional limitations on flame steadiness [6, 7]. A detailed analysis based on experimental inquiry and numerical modeling is offered on the burning and stability characteristics of premixed fuels after the general characteristics of lean premixed combustion are presented in this part. The impact of the equivalence ratio was investigated in this study from the blowout boundary to the flashback limit. O_2/CO_2 oxidizer's oxygen content was maintained at 60% vol. Additionally, the approaching combustible mixture's bulk throat speed was preserved at 5.2 m/s. The LES computer model was authenticated by assessments with experimental data for radial and axial temperature profiles, anticipated OH* concentration maps, and graphic flame appearance.

5.1.1 General Features of Lean Premixed Combustion

There is diminutive data in the works [7–14] about investigations of the combustion properties of entirely premixed CH_4 flames burned in O_2/CO_2 oxy-mixtures. In our most recent papers, [15, 16], experimental studies of premixed oxy-fuel burning are supported with modified features. In these investigations, the flame stability bounds, and certain premixed flame appearances are examined. These consist of the adiabatic flame temperatures and laminar flame speed of a model premixed gas turbine combustor. The results of the trials (performed at a fixed input velocity) showed that the flames burst out and flashback at a certain adiabatic flame temperature (AFT). Marsh et al. [8] used measurements of the stability boundaries, the location of the flame stabilization, and discharges at a burner capacity of 37.5 KW to assess the methane burning in N_2 and CO_2 dilution gases. The experiment covered several oxygen fractions that ranged from 0.2 to 0.7. They claimed that the location of the flame and the amount of heat released are significantly more affected by CO_2 dilution than by N_2. This was attributed to the differences between CO_2 and N_2's transport characteristics. High concentrations of CO are released along with the dissipating gases as a result of the dilution of CO_2. Because of the thermal dissociation of CO_2, the flame temperature is lower. However, increasing N_2 dilution causes an increase in NOx emissions. Comparative experimental research involving premixed CO_2 diluted and N_2 diluted flames was published by Jourdain et al. [9]. Investigating their stability processes is the study's goal. The study showed that CO_2 and N_2 diluted flames display identical geometries for a given swirl number while keeping the AFT, the equivalence ratio, and the constant velocity of the entering jet. That revision has demonstrated that flames that are CO_2-diluted can be stabilized using a similar flame alignment to flames with N_2-dilution deprived of modifying the hardware. Experimental research on premixed oxy-flames and air-flames in a swirl-stabilized combustor was conducted by Watanabe et al. [10]. They maintained the same equivalence ratio, Reynolds number, and AFT in order to compare the flames of air and oxygen. They believed that at high equivalence ratios, the flame stabilizes inside the shear layers of the flame. The Processing Vortex Core is a fundamental aspect of whirling flow (PVC). It is recognized as a source of illogical frequency that causes annoying sounds [13]. When the zone of recirculation becomes unstable, the PVC occur and begin to spin around the axis of symmetry.

Additionally, Weigand et al. recently conducted an numerical and experimental analysis of swirl flames in a model of a gas turbine combustor [17, 18]. They report their findings and analysis in terms of two key areas. The first includes the flow field of combustion, temperature, the flame's macrostructure, and species concentration; the second includes the chemical interactions of turbulence. Using planar laser induced fluorescence (PLIF), they transmitted the flame macrostructure and displayed the OH and CH radicals. They also emphasized that the OH radical molar concentration inside the combustor zone is the most accurate technique to report flame macrostructure. Lean premixed air-propane flame stability in a rearward facing step combustor was assessed by Nemitallah et al. [19] using the large eddy simulation (LES) turbulent approach. According to the authors, at higher equivalence ratios, the flame apparently stabilizes itself quite strongly on the side of the combustor back step. At lower equivalent ratios, the flame becomes lifted, and the distance of lift-off grows as the equivalence ratio drops. The main objective of this work is to examine the flow-field, stability, and combustion characteristics of premixed $CH_4/O_2/CO_2$ flames in a swirl-stabilized combustion chamber. It was specifically intended to demonstrate in what way the operability boundaries, visual flame look (flame morphology), and burning properties fluctuate as a function of the stationary operational equivalence ratio at 60% O.F. Such analysis was carried out using a computational code built on the LES model in the complete 3-D field of the combustor. Additionally, experimental data were made to describe the flame's temperature patterns and flame structure. The created computer model was effectively validated using measurements taken from the identical combustor.

5.1.2 Features of Swirl Burner

The goal is to examine how the equivalence ratio affects the morphology and burning properties of premixed swirl-stabilized $CH_4/O_2/CO_2$ flames at atmospheric pressure and without preheating. This is accomplished with a fixed oxygen fraction (O.F.) of 60% and a mutual 5.2 m/s inlet bulk velocity. The O.F. is characterized as

$$O.F. = \frac{\dot{V}_{O_2}}{\dot{V}_{O_2} + \dot{V}_{CO_2}} \tag{5.1}$$

$\dot{V}_{CO_2}$ and $\dot{V}_{O_2}$ present the volume flow rates of CO_2 and O_2, (in SLPM) respectively. For a thorough description of in what way the flow rates of CH_4, CO_2, and O_2 were attuned for each data point to preserve the equivalent inlet bulk velocity for every given mixture of oxygen percentage and equivalence ratio, the reader is also denoted to earlier readings by the authors [15, 16].

Diagrams of the combustor and test setup used to analyze premixed swirl-stabilized flames of $CH_4/O_2/CO_2$ are shown in Figs. 5.1 and 5.2. There are earlier studies [15, 20, 21] that provide thorough explanations of the combustor's design, size, and supportive equipment. But it's important to note that the combustor was created to validate a nearly flawless premixed reactant combination ($CH_4 + O_2 + CO_2$). This prevents differences in local equivalence ratio at the burner throat, that

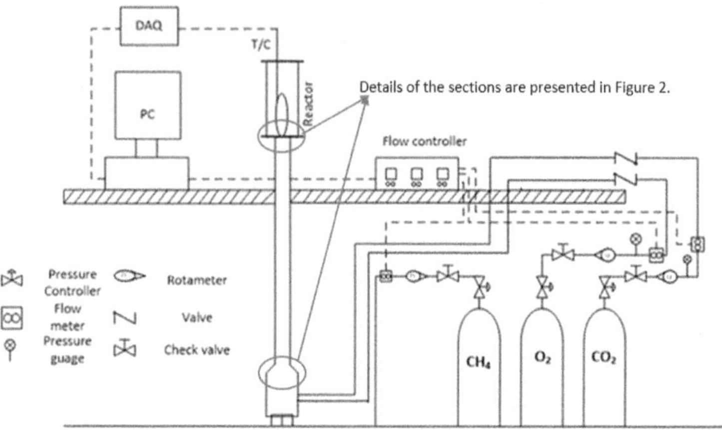

Fig. 5.1 Schematic diagram of the experimental test setup

would greatly affect the macrostructure and stability of the flame. The provided experimental and computational results can also be directly evaluated thanks to flawless
premixing. In particular, all the numerical simulations performed here have taken
into account properly premixed inlet blends deprived of any dissimilarities in the
local equivalence ratio. By taking pictures of the visual flame's presence with a
10-Megapixel camera set at 1/60-s shutter speed, 5.6 f-stop, and 1600 ISO, the
visual flame's shape was empirically assessed. Using an R-type (Pt-PtRh13%) thermocouple with a junction diameter of 1 mm, local hot-gas-path temperatures were
also recorded at designated locations inside the combustor. The model developed
by Brohez et al. [22] was used to correct the readings for radiation losses to the
surroundings. The power density of the combustor was between 3.0 and 5.5 MW/
m^3/bar, which is comparable to the bulk of industrial gas turbines that typically
operate between 3.5 and 20 MW/m3/bar [23]. This is very significant to highlight.

The mass flow rates of CH_4, CO_2, and O_2 are altered in proportion to one another
in order to maintain a consistent bulk flow velocity at the burner's throat. This means
that for each grouping of and OF, all three flow rates are related. Even when both
data points share the same or OF, every single movement between two data points
necessitates the change of flow rates. The oxygen fraction was examined one value
at a time from 15 to 70% in steps of 5%, in order to plot flame stability maps. The
range of the equivalence ratio was 0.1–1.0. For each OF, the blowout and flashback
limits of flame are determined using the method below. Fuel lines have emergency
shut-off valves to prevent flashbacks and for safety. These valves immediately halt
the flow of fuel. You can refer to previously published work [15] for more details on
the apparatus.

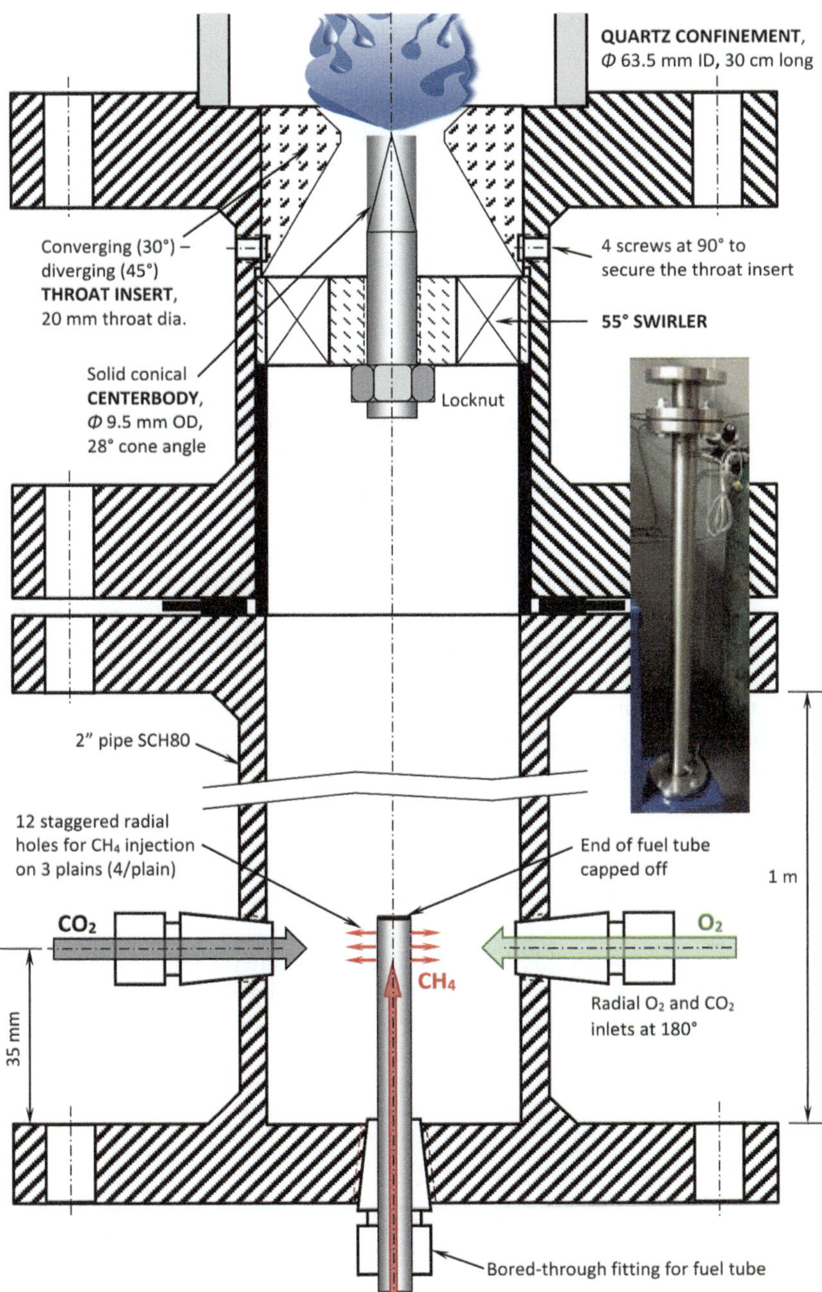

Fig. 5.2 Schematic diagram of the premixing combustor

5.1.3   Modeling of Premixed Oxy-fuel Flames

The computational geometry, grid independence analysis, models, and governing equations that were utilized to mimic the swirling $CH_4/O_2/CO_2$ flames under study are presented in this part. When simulating flames stabilized over a 55° swirler, the premixed burning model was utilized, which is a blend of the viscous model of large eddy simulation (LES) and the presumptive probability density function (PDF). Using Gambit, a full 3-D numerical simulation of the model gas-turbine combustor was created, and Ansys-Fluent 19.2 was utilized to run the simulations. Figure 5.3 shows the premixed swirl-stabilized gas-turbine model combustor together with all of its dimensions and swirler position. The combustion chamber measures 30 cm in length.

For the 3-D geometry, quadrant meshing was chosen, using 28k tetrahedral cells to reduce computing time. To ensure that the outcomes are unaffected by the mesh size, a mesh independence investigation was conducted. Three alternative mesh sizes, namely 28k, 33k, and 36k, are compared in Fig. 5.4. In order to cut computational time and expense, 28k cells were employed in this work even though adding more cells results in no appreciable modifications, as seen in the Figure.

The conversation equation can be difficult to solve, especially for reactive turbulent streams that have a large range of length and time scales. Due to turbulent combustion's multi-scale and multi-physics properties, modeling methodologies are needed to simplify the pathetic description of the complex physical event. There are three distinct methods that can be distinguished: Large-Eddy Simulation (LES), Direct Numerical Simulation (DNS), and Reynolds-Averaged Navier–Stokes (RANS); Denis and Vervisch [24] analyzed the specifics of each methodology. In the case of LES, the equations are filtered rather than averaging the influence of turbulence: The effect of the tiniest turbulent eddies is approximated using sub-filter-scale models, whereas the bigger turbulent eddies are resolved and computed explicitly because they are more geometry-dependent and have a greater impact on the flow. Thus, in contrast to RANS, LES turbulent fluctuations are partially resolved, which lowers the modeling uncertainty but raises the computing expense. When energy is transferred from big to small eddies by turbulence, it dissipates along the following

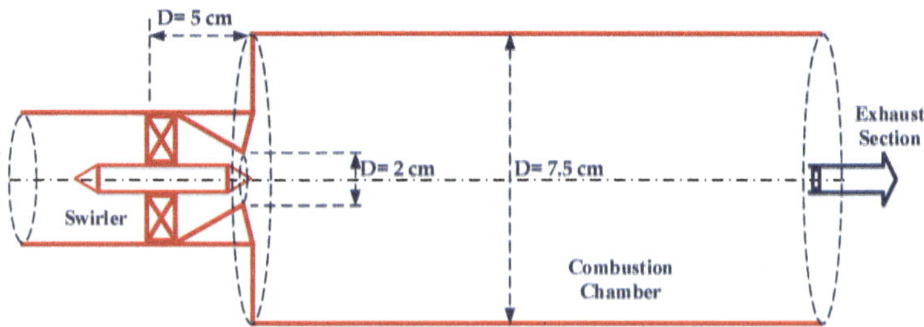

Fig. 5.3 Sizes of the combustor geometry

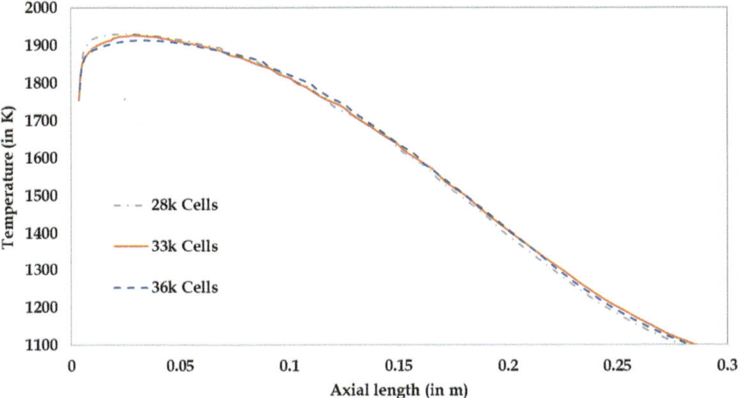

Fig. 5.4 Grid-independence study for different values of the total number of cells

spectrum [25]:

$$E(k) = C_k \varepsilon^{2/3} k^{-5/3} \tag{5.2}$$

The Kolmogorov constant, C_k, is used. In LES, the different flow quantities, Φ, are filtered either in the physical space or the spectral one (by suppressing any components higher than a certain cut-off length) (weighted averaging in a given volume). The following is how filtered quantities are stated:

$$\overline{\Phi}(x) = \int \Phi(x')g(x - x')dx' \tag{5.3}$$

g is a filtering function. The Favre, or filtered density-weighted quantity, Φ, in variable-density flows, such as those in the current investigation, can be represented as follows:

$$\overline{\rho}\overline{\Phi}(x) = \overline{\rho\Phi} \tag{5.4}$$

Frequently, the spatial filter is the grid. By applying the Favre filtering process to each term in the conservation equations, such as the continuity equation, energy equation, momentum equation, and species transport equation, the governing equations for LES are derived. The following equations are displayed below:

Continuity Equation [19, 26]:

$$\frac{\partial \overline{\rho}}{\partial t} + \frac{\partial \left(\overline{\rho}\tilde{u}_j\right)}{\partial x_j} = 0 \tag{5.5}$$

where $\tilde{u}_j$ is the cleaned velocity vector.

Conservation equations of momentum [19, 26]:

$$\frac{\partial \overline{\rho} \tilde{u}_i}{\partial t} + \frac{\partial \left(\overline{\rho} \tilde{u}_j \tilde{u}_i\right)}{\partial x_j} = -\frac{\partial \overline{P}}{\partial x_i} + \frac{\partial \overline{\sigma}_{ji}}{\partial x_j} + \frac{\partial \tau_{ij}}{\partial x_j} \tag{5.6}$$

where $\overline{\sigma}_{ji}$ and τ_{ij} refer to the strained viscous-stress tensor and the equivalent sub-grid scale (SGS) term, respectively.

Species Transport Equation [19, 26]:

$$\frac{\partial \overline{\rho} \tilde{Y}_k}{\partial t} + \frac{\partial \left(\overline{\rho} \tilde{Y}_k \tilde{u}_j\right)}{\partial x_j} = \frac{\partial}{\partial x_j}\left(\overline{\rho} D_k \frac{\partial \tilde{Y}_k}{\partial x_j}\right) - \frac{\partial}{\partial x_j}\left[\overline{\rho}\left(Y_k u_j - \tilde{Y}_k \tilde{u}_j\right)\right] + \overline{\dot{\omega}_k} \tag{5.7}$$

where Y_k and ω_k stand for the specie mass fraction and reaction rate, respectively, and D is the molecular diffusivity.

Conservation of Energy [19, 26]:

$$\frac{\partial \overline{\rho} \tilde{E}}{\partial t} + \frac{\partial \left(\overline{\rho} \tilde{u}_j \tilde{E} + \tilde{u}_j \overline{P}\right)}{\partial x_j} = -\frac{\partial \overline{q}_j}{\partial x_j} + \frac{\partial \overline{\sigma_{jk} u_k}}{\partial x_j} + \frac{\gamma R}{\gamma - 1} \frac{\partial \tau_j T}{\partial x_j} + \frac{\partial \tau_{jk} u_k}{x_j} \tag{5.8}$$

where q is the filtered heat flux and E is the filtered total specific energy. Emanuel et al. [27], Herná et al. [28] provide more information on modeling the unsolved terms. The finite volume and under-relaxation method offered by Ansys Fluent was used to discretize each of the aforementioned transport equations.

Chemical reactions are limited to narrow reacting layers at small scales during turbulent combustion, making them impossible to resolve on standard LES grids. Therefore, the majority of turbulence-chemistry interactions must be included. Predicting reaction rates in premixed turbulent combustion is exceedingly challenging because the reaction rates are highly nonlinear functions of temperature and species mass fractions. The sub-grid model of turbulent combustion is thus an essential part of LES and is required to account for the impact of turbulence-chemistry interactions at the under-resolved scales on the reaction rate. Researchers have used a variety of strategies to predict the filtering reaction rates that correlate to a particular flame regime.

The probability density function (PDF) model was used in this investigation. This presumptive-PDF paradigm avoids the requirement to construct a transport equation by averaging the thermochemical properties and the reaction rate over the presumptive probability density function. The reaction chemistry can also be considered using the flamelet generated manifold (FGM) approach. By decreasing the chemical subspace using a set of reference laminar flamelet computations and storing the chemical flame structure inside a lookup table, this method focuses on adding comprehensive chemistry effects within the combustion model. Therefore, this method can lower the computing expense of performing reacting flows with complex chemical kinetics. The presumed-PDF methodology combined with the FGM technique is the main emphasis of this work, and a description of the method may be found in Reference [26].

Table 5.1 The configuration parameters and operating conditions

Oxy-methane combustion ($CH_4/O_2/CO_2$) with O.F. $= 60\%$	
Equivalence ratio, φ	0.350–0.700
Operating pressure	1.0 bar
Combustor power (based on equivalence ratio)	2.25–4.25 kW
Temperature of inlet mixture	300 K
Bulk velocity of inlet mixture	5.20 m/s
Viscosity ratio	10%
Turbulence intensity	5%

For simulations that are intrinsically unstable and dominated by large-scale turbulent structures, such as combustion instabilities, LES is a very appealing option. For responding flows, LES makes sense for a number of reasons. One explanation for this is that the big turbulent eddies are anisotropic, persistent, and geometry-dependent. On the other side, the large eddies can also be used to determine the small eddies because they are isotropic, simpler to model, and more predictable. Additionally, the LES has fewer constants to be adjusted for free jets, swirl, jets in cross flow, buoyancy, etc., in comparison to those of the RANS model, which makes LES a more accurate turbulence model. As grid resolution is raised, the LES modeling error decreases in contrast to RANS. Additionally, the rate at which big eddies combine reactants and products generally regulates combustion, and such large eddies are resolved directly in LES. The LES model successfully handles flows near walls, in contrast to other turbulence models. The assumed probability density function (PDF) and the flamelet-generated manifold were used in the current work to run large-eddy simulations (FGM). Examined is the impact of the equivalence ratio (φ). All simulations were conducted using the same oxygen fraction (O.F.) of 60% and inlet bulk velocity of 5.2 m/s, similar to the experiments. Table 5.1 is a list of the operational factors taken into account.

A mixed thermal boundary condition was used for the combustor wall to account for both convection and radiation heat transfer. A coupled scheme was employed for the pressure–velocity coupling, and a first-order upwind scheme was chosen for the spatial discretization of momentum, turbulent kinetic energy, etc. The criterion of convergence was 10^{-3} for the continuity, species, and momentum equations and 10^{-6} for the energy equation.

5.1.4   Combustion Characteristics of Premixed Oxy-flames

The predicted distribution of the hydroxyl radical (OH*) and the associated experimental visual image of flame shape can be compared for validation. According to Weigand et al. [17, 18], the OH* and CH* radicals are the best representations of the visual flame appearance. Thus, Fig. 5.5 contrasts the experimental flame image

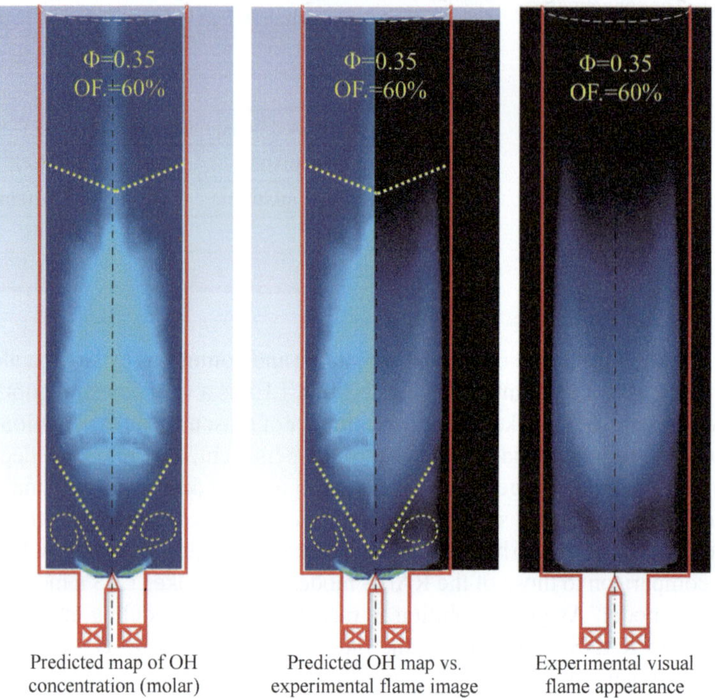

Predicted map of OH   Predicted OH map vs.   Experimental visual
concentration (molar)   experimental flame image   flame appearance

Fig. 5.5 Comparison of the calculated map of OH concentration (molar) and the conforming experimental visible flame shape at O.F. = 60% and $\varphi = 0.350$

at O.F. = 60% and the projected molar concentration map of OH. A tiny preliminary region within the diverging part that is downstream of the burner throat and a greater conical downstream region that is anchored at the centerbody tip are observed experimentally and numerically in excellent agreement. The corner recirculation zone (CRZ), which separates the two zones, circulates combustion products without any reactions and with no OH concentration. This flame form has thus been referred to as "double conical" in earlier study [16, 29, 30].

Following the successful validation of the numerical results, the projected flow fields are analyzed next to quantify the influence of the equivalence ratio at a fixed O.F. of 60%. Figure 5.6 compares the distributions of centerline temperature at different equivalence ratios, such as 0.2, 0.3, 0.4, 0.5, and 0.6, which correspond to AFT of 1389, 1806, 2173, 2501, and 2796 K, respectively. The centerline temperature normally rises because of increasing combustor power, flame heat release, and AFT, as can be seen from the graph. The centerline temperature is initially as low as 500 K and only climbs to roughly 1100 K near the combustor outlet, according to the temperature distribution at = 0.2, which leads to an exciting discovery. This demonstrates the blowout limit, where the flame loses stability inside the combustion chamber. In the conical reaction zone that anchors at the centerbody

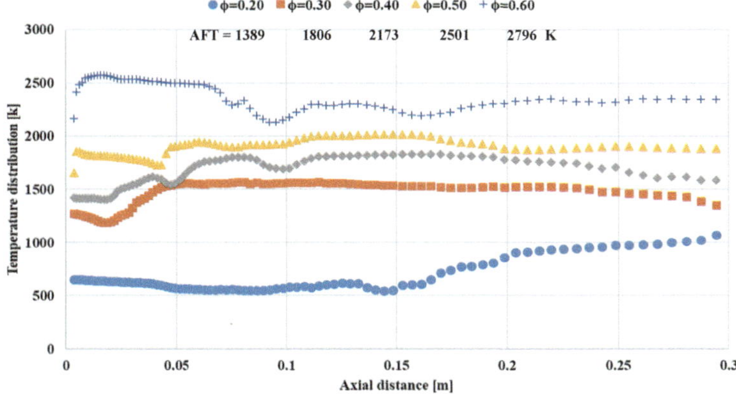

Fig. 5.6 Distribution of centerline temperature at different equivalence ratios

tip in the $\varphi = 0.3$ and 0.4 examples, which mimic the $\varphi = 0.35$ flame shown in Fig. 5.5, the centerline temperature starts out cool and subsequently climbs to reach AFT. Thus, the following part will offer a complete analysis of the flame structure. The early stages of the combustor's temperature changes show that the whirling flow field is three-dimensional. With $= 0.60$, the centerline temperature only reaches an early peak and approaches its AFT value in the flame. This suggests that the flame is quite small and cannot support a significant IRZ.

Figure 5.7 displays the centerline distribution of axial velocity normalized by the value of bulk inflow velocity (5.2 m/s). At all equivalence ratios (apart from 0.6), swirl creates an IRZ in the first 10 cm of the combustor, as evident by the negative axial velocity that suggests reverse flow or recirculation. The fact that the $\varphi = 0.6$ scenario exhibits a different response demonstrates that this case, which can be extrapolated from Fig. 5.6, symbolizes the flashback limit, wherein the flame is at its most condensed shape and cannot support a large IRZ. Be aware that the centerline temperature increases when the axial velocity increases beyond $+ 0.9$ of the bulk inlet value, peaking at $Z = 1$ cm (Fig. 5.6). As a result, responding flow can be found close to the centerline, and in this flow, the bulk inlet velocity predominates in the absence of an IRZ. This suggests that at the burner throat, a delicate balance between the flame speed and bulk velocity takes place before the onset of flashback. The present authors discussed this balance in one of their earlier studies [22], where it was found that the flashback limit of this combustor occurred consistently at 5.2 m/s inside the analyzed φ-OF area since the bulk inflow velocity was set at that speed. The following examination of flame morphology will give more details on the limiting flashback condition. Figure 5.7 also reveals an unusual result: the $\varphi = 0.2$ scenario of the blowout limit has a noticeably significant swirl-induced IRZ, but it is unable to maintain this excessively lean condition.

Figure 5.8 completes Fig. 5.6 by comparing the radial profiles of temperature for various equivalence ratios (i.e., AFTs) and axial locations ($Z = 2.5$, 5, and 10 cm) inside the combustor. Radial temperatures often increase as the combustor power,

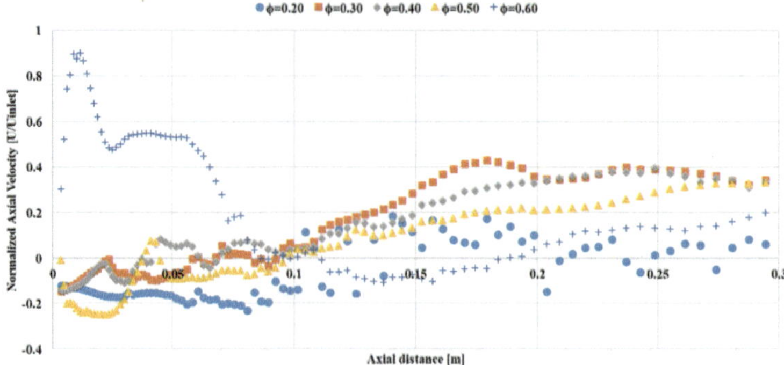

Fig. 5.7 Variation of the axial component of velocity (along the axis) normalized by the bulk inlet velocity (5.2 m/s) at different values of equivalence ratio

flame heat release, and AFT increase. The cold profiles at $\varphi = 0.2$ serve as additional evidence that there is no stable flame at this level. The results of Fig. 5.6 and the presence of the downstream conical region that anchors at the centerbody tip are supported by the hotter radial profiles in the $\varphi = 0.3$ and 0.4 situations (see Fig. 5.5). The analysis of Fig. 5.7, which shows that this flame is the most compact, is supported by the fact that the temperature difference between centerline and wall regions is greatest at $\varphi = 0.6$, particularly at Z = 2.5 and 5 cm.

5.1.5 *Macrostructure of Premixed Oxy-fuel Flames*

This part analyzes the structure of the flame by displaying experimental images of flame appearance and numerical 2-D maps of various parameters, along with comparisons to measure the impact of the equivalence ratio. Three various flame shapes/ morphologies, namely (I) double conical, (II) corner-stabilized, and (III) swirl-stabilized, were characterized in earlier investigations on premixed swirl-stabilized flames [29–31]. (V-shape). These shapes are listed in Table 5.2, along with brief summaries of the related stabilization procedures.

In this work, these flame shapes were created experimentally by changing the equivalence ratio from 0.3 to 0.5 at a constant O.F. = 60%, as shown in Fig. 5.9. The flame adopts the double conical shape (I), which is typical of low-temperature flames nearing the blowout limit, at $\varphi = 0.3$ (AFT = 1806 K). The term "double conical" refers to the fact that a reaction-free (i.e., cold) CRZ divides the flame into two reaction zones; the first region stabilizes in the diverging portion of the burner throat, whereas the second region assumes a conical shape and anchors near the tip of the centerbody. As a result, the flame becomes stable within the outer and inner shear layers. By raising the equivalence ratio, the reaction-free region gradually shrinks ($\varphi = 0.35$, AFT = 1995 K), and the reaction regions combine to form a cup-shaped

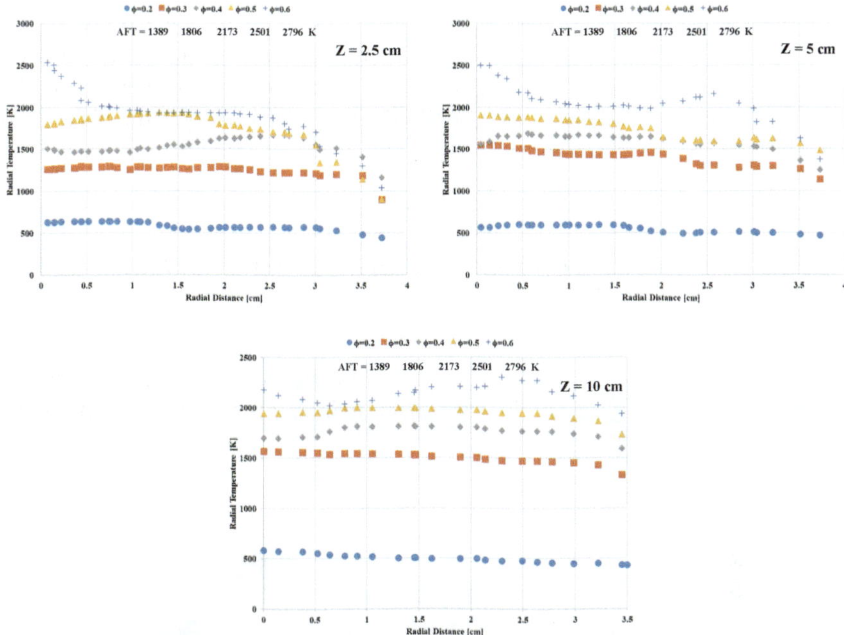

Fig. 5.8 Radial profiles of temperature at different axial locations (Z = 2.5, 5, and 10 cm) inside the combustor and different values of equivalence ratio

Table 5.2 Summary of the appearance description, flame morphology, and stabilization method

	Flame shape	Description of flame morphology and stabilization method
I	Double conical	Flame stabilized within the inner and outer shear layers [29, 30]
II	Corner-stabilized	Flame stabilized along outer shear layer near combustor walls [12, 32]
III	Swirl-stabilized (V-shape)	Typical V-shape of swirl-stabilized flames [33, 34]

flame that stabilizes firmly inside the CRZ ($\varphi = 0.4$, AFT = 2,173 K), hence the term corner stabilized (II). It is evident that this flame is stabilized within the CRZ due to the bright color of CRZ of the $\varphi = 0.40$ flame in the isometric view in Fig. 5.9. The flame changes from the corner-stabilized structure to the usual V-shape (III) of swirl-stabilized flames ($\varphi = 0.50$, AFT = 2501 K) as the equivalence ratio is raised further ($\varphi = 0.45$, AFT = 2342 K). This V-shape, which has significant turbulence intensity and persists up to the flashback limit, is maintained by a powerful IRZ that gets smaller and smaller as flashback approaches ($\varphi = 0.60$, AFT = 2796 K). In the isometrical view of the $\varphi = 0.50$ flame, note how the intense light-blue hue is constrained to the burner throat as the flashback limit is approached.

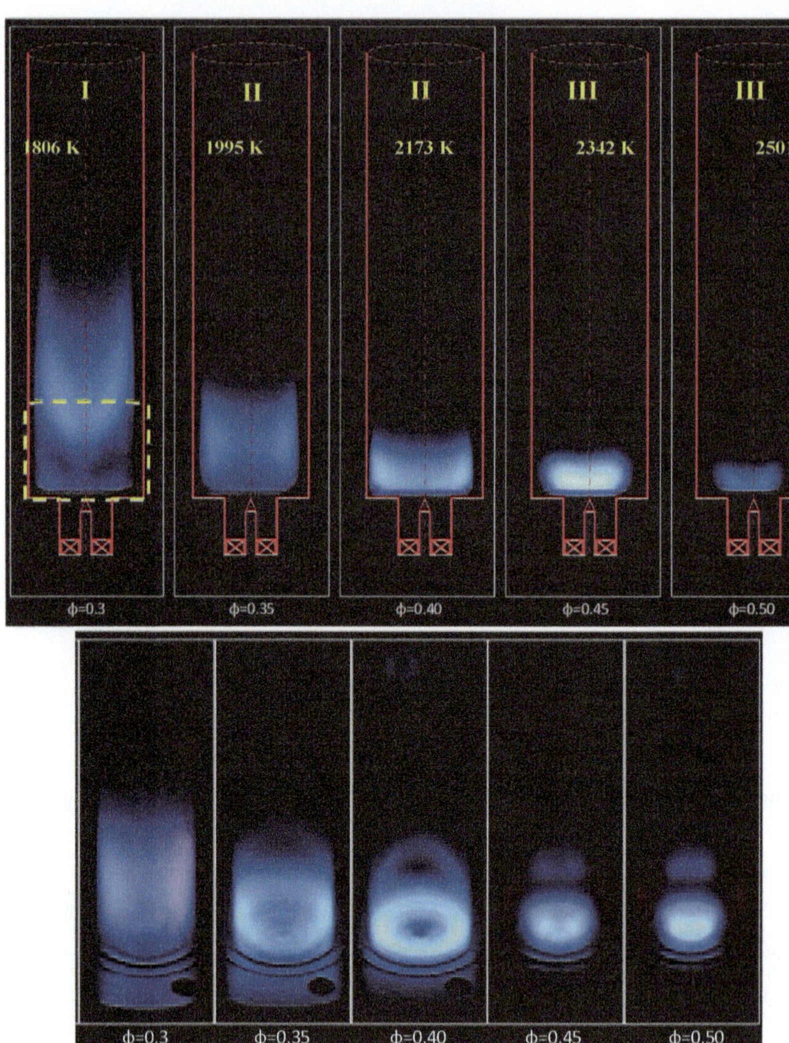

Fig. 5.9 Visible-light images of oxy-methane flames at O.F. = 60% and different values of equivalence ratio. Top: front view, bottom: isometric view

In addition to the blowout and flashback limits (at $\varphi = 0.2$ and 0.6, respectively), Fig. 5.10 displays the anticipated velocity streamlines and vectors overlaid on maps of axial velocity for flames of shape I ($\varphi = 0.3$), shape II ($\varphi = 0.4$), and shape III ($\varphi = 0.5$). The related anticipated maps of the OH* mole fraction are shown in Fig. 5.11. The negative values of axial velocity ("blue" regions) in Fig. 5.10 indicate two distinct IRZs at $\varphi = 0.3$ (double conical flame): one within a reaction region at the diverging section of burner throat, and another within the reaction region that assumes a conical shape with the tip touching the first IRZ. The "green" CRZ divides the two IRZs. These reaction regions are highlighted in Fig. 5.11 by the OH* mole fraction. While the second reaction region takes on a conical shape with a tip that extends upstream towards the first region, the first reaction region is clearly positioned within the diverging part of the burner throat. The two reaction regions are separated by an OH-free CRZ, supporting the conclusions of Fig. 5.10. The second IRZ shrinks (see Fig. 5.10) at $\varphi = 0.4$ (corner-stabilized flame), and the two response regions combine to form a single region that approaches the burner throat and anchors within the "yellow" dominant CRZ. This CRZ has a substantial OH* content, which denotes reactivity, as shown by Fig. 5.11 as well. The transition from corner-stabilization (form II) to swirl-stabilization is explained by the V-shaped flame that goes even closer to the throat when is increased to 0.5 (see Fig. 5.10). This flame has a stronger IRZ and a weaker CRZ (shape III). Now that the CRZ has no OH* radicals, as expected for swirling flames, it can be shown that this flame is IRZ-stabilized (Fig. 5.11). The results of Figs. 5.7 and 5.8 are directly supported by the observation that the IRZ is strongest but most compact with nearly little CRZ at the flashback limit ($\varphi = 0.6$) (see Fig. 5.10). Due to the high turbulence intensity, the concentration of OH* is now noticeably high throughout the whole IRZ (Fig. 5.11). Figure 5.11 demonstrates that the concentration of OH* peaks close to the combustor exit in the $\varphi = 0.2$ case, indicating that reactivity starts very late, thus providing strong evidence that this lean case represents the blowout limit. Despite the significant size of IRZ, it is insufficient to stabilize such excessively lean flame.

Figure 5.13 overlaps the velocity streamlines on the maps of temperature to relate the various flow features to the regions of heat release, whereas Fig. 5.12 shows the anticipated temperature maps for the same five cases ($\varphi = 0.2$–0.6). Two distinct "red" hot regions are visible at $\varphi = 0.3$ (double conical flame), which correspond

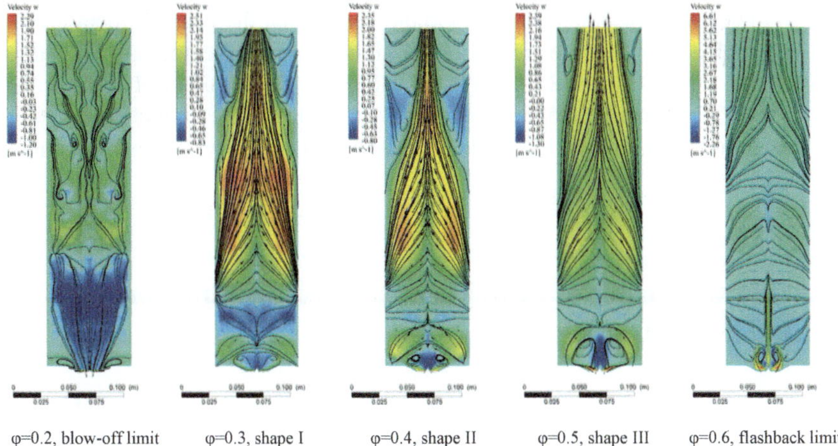

Fig. 5.10 Predicted velocity streamlines and vectors overlapping the maps of axial velocity component at different values of equivalence ratio

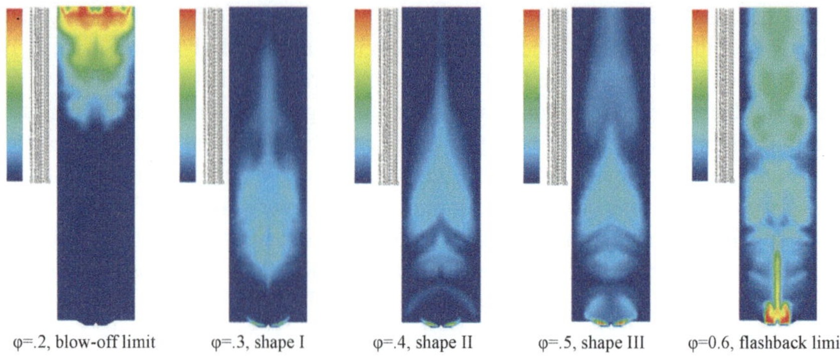

Fig. 5.11 Calculated maps of the mole fraction of OH* radical at different values of equivalence ratio

to the two reaction regions indicated above and are a feature of this flame form (I). As previously mentioned, a reaction-free (cooler) CRZ divides the reaction zones. Despite an increase in combustor power as φ approaches 0.4, the second zone gets colder because it fuses with the first one to form a flame of shape II, which is stabilized by a hotter CRZ, hence the term corner-stabilized. An additional indication that this flame is anchored by its IRZ, as is typical for swirl flames, is the shift to the V-shape that is caused by raising to 0.5. This V-shape has a noticeably hotter IRZ and a significantly colder CRZ. The IRZ is hotter and more compact near the flashback limit ($\varphi = 0.6$), with concentrated heat release that is facilitated by the high turbulence intensity. On the other hand, the CRZ is considerably colder. When the local flame speed at the burner throat surpasses the bulk inlet velocity of fresh reactants due to the high temperature of the IRZ, flashback occurs. Since no flame can anchor inside the combustor, the blowout limit ($\varphi = 0.2$) is again observed as the temperature increases close to the combustor outlet. It is important to note that the peak temperatures in each anticipated flow field are around 160 K higher than the theoretical AFT values shown at the top of Fig. 5.12. Previous research has demonstrated that the actual AFT is 100–200 K lower than its hypothetical value due to the dissociation of CO_2 [8]. This indicates the numerical code's capability to correctly forecast the flames under study.

5.2   Characteristics of Lean Premixed Oxy-combustion in Micromixer Burner

One of the advantageous technologies of carbon capture is oxygen-fuel combustion, which produces an exhaust stream that is predominantly made up of CO_2 and H_2O and almost completely free of nitrogen dioxide [35]. Therefore, after simple condensation of H_2O, high-purity CO_2 can be economically captured [36]. The elimination of

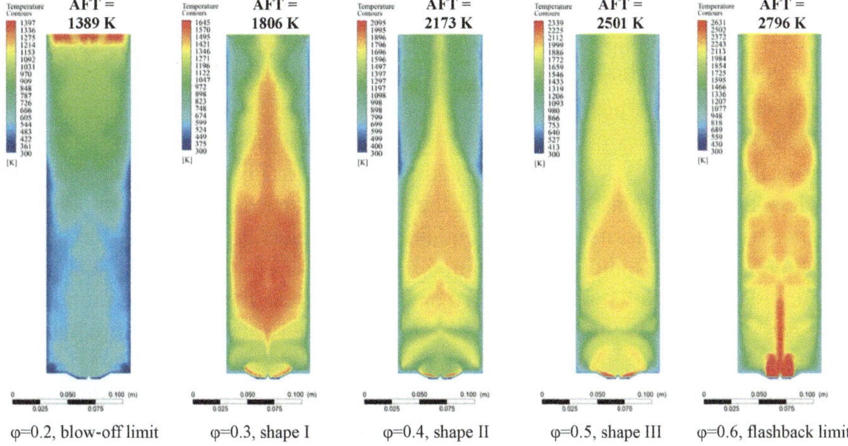

Fig. 5.12 Predicted maps of temperature at different values of equivalence ratio

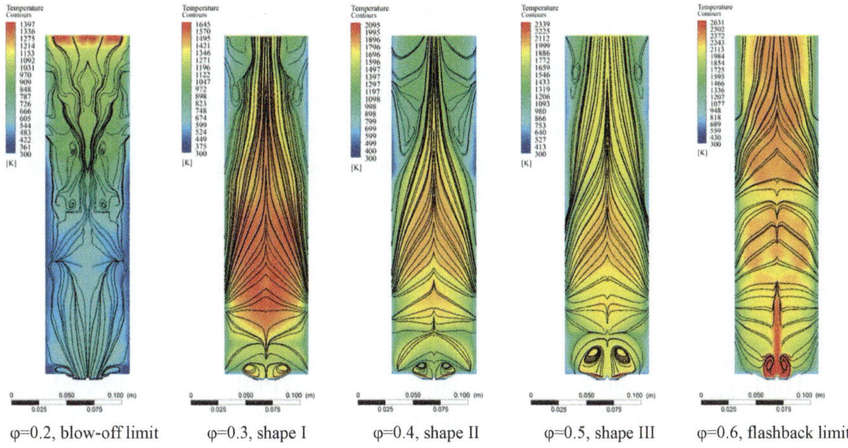

Fig. 5.13 Calculated velocity streamlines overlapping the temperature maps at different values of equivalence ratio

NOx emissions, which are generally formed during air–fuel combustion, is a further noteworthy benefit. Consequently, oxy-fuel applications become almost emission-free. Flue-gas recirculation is used to manage the oxy-flame temperature, which is regulated by the oxygen fraction (OF) in the diluted O_2/CO_2 oxy-mixture [37]. The effluent CO_2 is circulated in part to dilute the inlet O_2 the combustor and lower the flame temperature. Changes in the diluent from N_2 to CO_2 are implied when switching from the conventional air–fuel to the N_2-free oxy-fuel combustion [38]. Contrary to N_2, CO_2 has an adverse effect on the flame chemical kinetics, the flame speed, and the adiabatic flame temperature [39–41]. Oxy-fuel flames also extinguish more readily than their air–fuel counterparts [42]. The minimum OF necessary to endure steady swirl-stabilized oxy-flames was determined to be bigger than the typical 21% value of air [43, 44], often 30% [45–47]. In addition, the OF in oxy-flames has a significant impact on their properties [48].

Because oxy-flames perform worse, burner technologies are required that offer superior flame stability, such as perforated-plate burners (PPBs). These burners have a variety of perforations that separate the bulk inlet reactant mixture into several jets that burn in the form of small neighboring flamelets downwind of the burner. Gas stoves are one example of a common residential or commercial application for air–fuel PPBs, which have been studied extensively in the literature and found to be geometry-flexible. The study of Veetil et al. [49] looked at the relationship between the placement and spacing of holes in a PPB and the flame structure and stability. According to them, smaller hole-to-hole spacing increases stability by reducing the flame's standoff distance from the burner surface and the curve of its base. Inline, star, and staggered configurations of flames stabilized over PPBs were examined by Hindasageri et al. [50] at various pitch-to-diameter ratios and Reynolds numbers. Heat transfer from the flames to the plate was studied when they impinged on it in the opposite direction from the burner. The heat flow to the plate is said to increase as flame-to-flame spacing increases. However, with wider flame spacings, this advantage is limited by inferior flame stability near lean blow-off. Kedia and Ghoniem [51] investigated the blow-off mechanism through 2-D simulations using chemical kinetics. According to Kedia and Ghoniem [51], both the flame stretch and heat transmission from the flame to the burner plate have an impact on the blowout of a laminar methane/air flame stabilized over a PPB. Jithin et al. [52] performed a numerical analysis which revealed that the distance of flame standoff rises with the inlet velocity. The standoff distance of premixed C_3H_8-air flames reduces with an increase in the equivalence ratio (making it less lean), and the heat flux from the flame to the burner faceplate subsequently rises. The flame rests on the plate in the extreme situation of no heat flux (adiabatic plate). Premixed LPG/air flames were tested on ceramic as well as copper (highly conductive) PPBs, and the effects of the plate material, thickness, and hole diameter were studied in the work of Gamal et al. [53]. The results demonstrate that heat transmission through the burner plate is the hidden key control parameter underlying all of these effects. Larger hole sizes were found to be advantageous for flame stability due to the comparable lower velocity of reactant jets at a specific reactant mixture flow rate.

Oxy-flames stabilized on PPBs have been studied under restricted conditions. An examination of partially premixed $CH_4/O_2/CO_2$ and CH_4/air flames stabilized over a PPB was conducted experimentally by Rashwan et al. [54]. They stated that, over the examined range of equivalence ratio, stable oxy-flames could not be obtained below 29% OF. In a recent experimental investigation, Aliyu et al. [55] evaluated premixed oxy-methane flames anchored on a multi-hole burner spanning the range of 25–70% OF. Oxygen fractions even below 25% might have been achievable if the equivalence ratio had been maintained high enough. They acquired stable flames with φ as low as 0.6 at OF = 25%. This demonstrates how multi-hole burners outperform other technologies in the aforementioned studies where the threshold of minimum OF was about 30%.

The enhanced micromix burner technology used in today's air–fuel lean-premixed gas turbines is a descendant to the PPBs discussed above. A micromixer (MM) is a premixer that separates the bulk incoming airflow among a collection of small tubes and injects fuel inside each one to premix with air at the microscopic level. Thus, the downstream reaction region that results from MM includes a number of small, neighboring premixed flamelets that resemble PPBs. MM combustors have been developed and used successfully by a variety of air–fuel gas-turbine manufacturers, including General Electric, Honeywell Garrett, and Mitsubishi Hitachi Power Systems [56–58]. They are intrinsically resistant to blowout and flashback. The air-combustion of hydrogen fuel diluted with nitrogen in an MM combustor was assessed by York et al. [59]. They were able to transmit an 1850-K flame temperature while operating steadily and without flashbacks in this type of combustor. The method may be able to lower emissions and flashback threat, according to the experimental and computational examination by Lee et al. [60] who studied axial and radial micromixing of air and fuel. Asai et al. [61] tested three different fuels with varied H_2 content in order to examine the air-combustion of hydrogen-rich fuels in an MM combustor. For all three fuels, exit temperatures as high as 1775 K and stable operation were reported. A MM combustor comprising about 1600 flamelets was created by Funke et al. [62] to be integrated into an operational gas turbine. Each tube had a diameter of roughly 0.3 mm. They claimed that the use of MM technology enabled stable engine running during acceleration, idle, and start-up. In an MM combustor, Funke et al. [63] investigated the impact of the momentum-flux ratio of fuel to air on the stability of H_2/air flames and found that the MM technology enables safe and low-NOx hydrogen in air combustion. A MM air–fuel combustor operating at atmospheric pressure was used by Dodo et al. [64] to model the conditions of an Integrated Gasification Combined Cycle (IGCC). The hydrogen percentage of the fuel simulants utilized for producing synthetic gas ranged from 40 to 60%. The MM burner ran consistently without incurring any blowout or flashback.

An extensive amount of study has been conducted on the use of PPBs for stabilizing air–fuel flames and their gas-turbine descendant, MMs, as the aforementioned literature survey demonstrates. Despite this technology's higher flame stability [55] and oxidizer flexibility [50, 55], there hasn't been much prior study on the use of PPBs in oxy-combustion, and MMs haven't been tried in these circumstances. This inspired the current authors to build on their previous work by Aliyu et al. [55] and analyze the geometrical flexibility of MMs by studying CH_4/O_2/CO_2 flames on a multi-hole burner that resembles a MM, as shown in this section. Here, two additional geometries are compared to the multi-hole burner of Aliyu et al. [55] to investigate the impacts of hole size and spacing. The trials for this study were carried out in a similar oxygen-fraction range to Aliyu et al. [55] and at a fixed flame-base jet velocity of 5.2 m/s. Some flames were imaged to observe how burner geometry impacts the macrostructure of the flame. Temperature readings were also recorded in the hot plumes of a select few flames to examine the effect of burner geometry on the profile factor of the combustor. This ground-breaking research will show the PPBs and MMs' promising potential for the use of oxy-fuel combustion in zero-emission power production. The information from this study is detailed in the ensuing subsections.

5.2.1   Features of Micromixer Burner

In the current study, the combustion and stability characteristics of premixed oxy-methane flames are investigated using a multi-hole burner that resembles gas-turbine micromixers. Instead of injecting fuel into the burner headend itself (as done in micromixers), it was supplied here into a large premixing plenum located upstream of the burner, together with controlled flows of O_2 and CO_2. By removing any jet-to-jet variation in the equivalence ratio, this simplification was created to guarantee practically near-perfect premixing of the oxy-fuel reactants upstream of the burner. As a result, the sophisticated interior features of a micromixer were simplified in a multi-hole burner. The methodology followed the same steps as Aliyu et al.'s [55] approach. In actuality, the test rig and experimental set-up depicted in Figs. 5.14 and 5.15, respectively, are those employed by Aliyu et al. To keep things simple, the premixing plenum received a metered flow of CO_2 at atmospheric temperature and pressure instead of using exhaust CO_2 recirculation. O_2 and CH_4 were also not preheated.

Figure 5.16 marked HE1, depicts the baseline burner geometry of Aliyu et al. [55]. In the current investigation, this serves as the baseline headend. Here, HE1 was evaluated against HE2 and HE3, two more headends. HE1 (3.18 mm) and HE2 (3.97 mm–25% increase) are used to examine the effect of jet diameter at fixed jet spacing of 5.5 mm, whereas HE2 (5.5 mm) and HE3 (7.0 mm–27% increase) are used to study the impact of spacing at fixed jet diameter of 3.97 mm. The entire flow area is as follows since HE1 has 61 holes while HE2 and HE3 each have 37 holes:

$$A_{\text{HE1}} = N_{\text{HE1}} \times \frac{\pi}{4} D_{\text{HE1}}^2 = 61 \times \frac{\pi}{4} \times 3.18^2 = 485\,\text{mm}^2 \tag{5.9a}$$

$$A_{\text{HE2}} = A_{\text{HE3}} = N_{\text{HE2}} \times \frac{\pi}{4} D_{\text{HE2}}^2 = 37 \times \frac{\pi}{4} \times 3.97^2 = 458\,\text{mm}^2 \tag{5.9b}$$

All three headends thus have approximately the same flow area (within 5%). Consequently, they have the same overall flow rate of premixed reactants for any specified mixture composition since the flame-base jet velocity (designated "v" in Fig. 5.16) remains constant at 5.2 m/s for the entirety of the present investigation, comparable to Aliyu et al. [55]. All three headends also have the same combustor power for a specified mixture composition since the power varies linearly with the fuel flow rate.

The hole array on each headend adopts a hexagonal arrangement to keep consistent spacing between every core jet and its six surrounding jets. Hexagonal symmetry thus prevails in all headends. This layout also produces concentric hexagonal "rings"

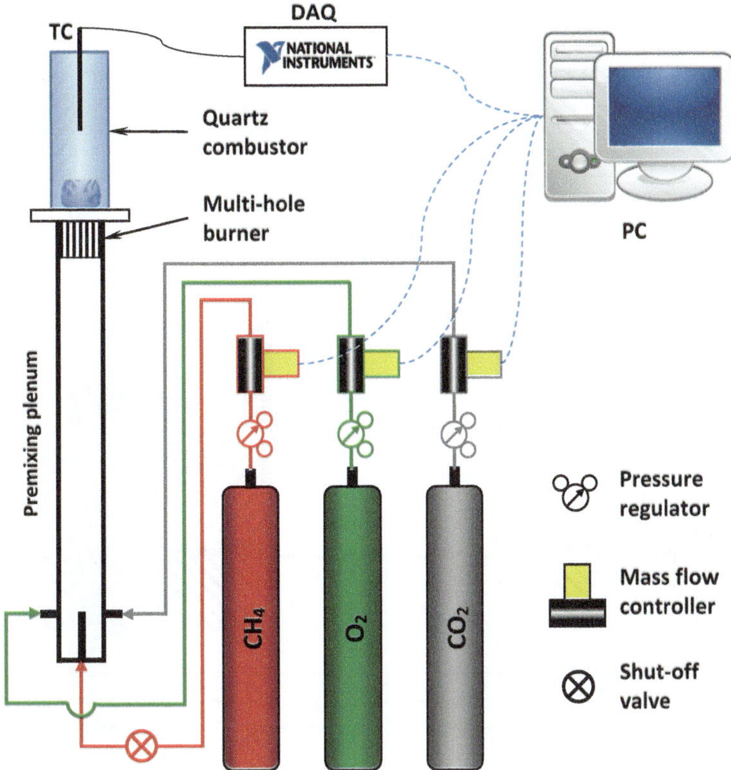

Fig. 5.14 Schematic of experimental setup

around the central jets of each headend (HE1 has four rings, HE2 and HE3 each have three rings), as shown in Fig. 5.16. Although, as previously mentioned, micromixers are naturally resistive to flashback, flame arrestors are embedded in all three headends to inhibit flashback for the operator's safety, and the entry of each hole is purposely reduced in diameter (1.6 mm). However, because of the suitable size of the hole aspect ratio (L/D) downstream of the flame arrestor, the flow can rebound and leave the headend fully developed.

The present investigation was carried out at a constant jet velocity because it has been demonstrated in earlier studies that, when compared to jet Reynolds number, this velocity is the parameter most strongly influencing flame structure [65, 66].

A quartz confinement (see Fig. 5.15) protected the combustor flame from outside air while allowing optical accessibility for photographing the flame structure. Selected flames were photographed using a 3200 ISO sensor, a 1/60 shutter speed, a 5.6 f-stop aperture, and a 50 cm focal length. All gas flowrates (CH_4, O_2, and CO_2) were regulated with full-scale uncertainty of 0.5% using thermal mass flow controllers from Aalborg-Inc. This lead to 0.4% and 0.02 uncertainty in the oxygen

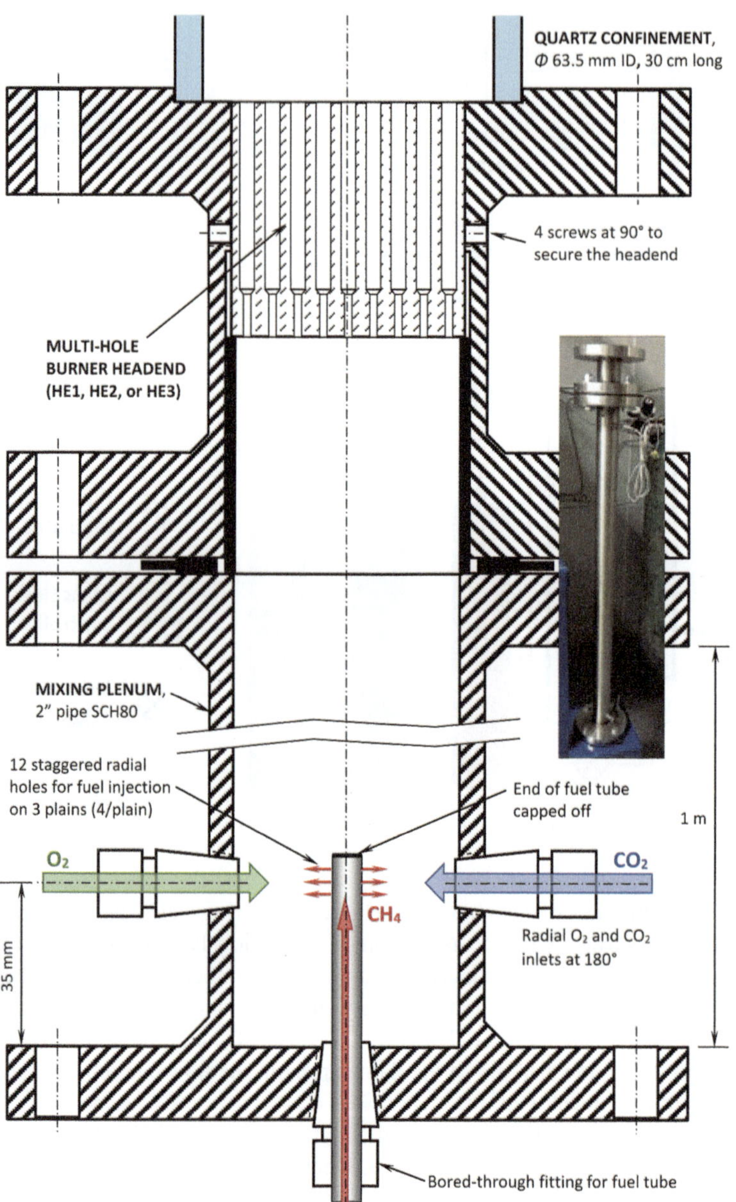

Fig. 5.15 Schematic diagram of the experimental test rig utilizing a multi-hole burner headend

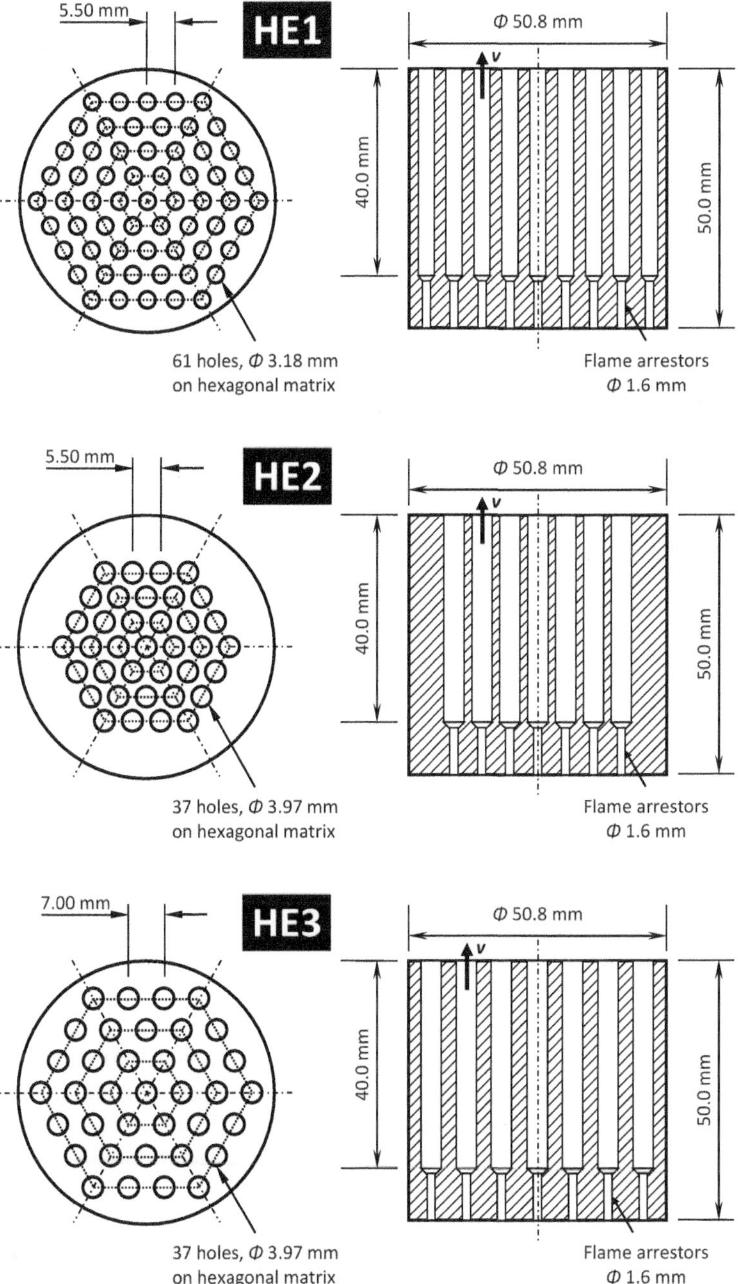

Fig. 5.16 Schematics of the different examined burners to study the effect of geometry. HE1 is the baseline design from Aliyu et al. [55]. HE2 investigates changing the jet diameter for the same spacing as HE1. HE3 examines changing the jet spacing for the same diameter as HE2. Left: top view; right: sectional front view

percentage and equivalence ratio, respectively. It is possible to get more details regarding the experimental facility in the authors' earlier articles [65, 67].

The profile factor of the combustor (i.e., radial profile of temperature near the exit) was estimated by recording the local temperature distribution within the hot plumes of designated flames. The temperature was measured using an R-type thermocouple (Pt/PtRh-13%) with a spherical 1-mm junction. The temperature data was collected using a DAQmx card in conjunction with National Instruments' LabVIEW user interface (NI). The temperature values were corrected using the procedure outlined by Brohez et al. [68] to account for the radiation inaccuracy to the surroundings.

The blowout limits of burners HE2 and HE3 were established over the oxygen-fraction (OF) range of 25–70% (by volume) in the O_2/CO_2 oxidizer at constant 5.2 m/s jet velocity (see Fig. 5.16), similar to the analysis of HE1 in the work of Aliyu et al. [55]. Therefore, the flow rates of CH_4, O_2, and CO_2 were altered simultaneously in order to keep the same jet velocity for any needed inlet combination of equivalence ratio (φ) and OF. The authors point out past studies [65] where readers can get further details on these improvements. Using tabulated data of enthalpy of formation and sensible enthalpy, the oxy-methane combustion reaction was examined to determine the adiabatic flame temperature (AFT) as a function of φ and OF [69]. The following is a form of the reaction equation:

$$CH_4 + \frac{2}{\varphi}\left[O_2 + \left(\frac{1}{OF} - 1\right)CO_2\right] \rightarrow \left[1 + \frac{2}{\varphi}\left(\frac{1}{OF} - 1\right)\right]CO_2$$
$$+ 2\,H_2O + 2\left(\frac{1}{\varphi} - 1\right)O_2 \tag{5.10}$$

Because ultra-lean circumstances (φ ranging from 0.18 to 0.65) were taken into consideration, complete combustion (i.e., minimal CO emission) was assumed for simplicity. Additionally, CO_2 dissociation was ignored because it was thought to only have a tolerable effect on AFT under stoichiometric conditions, according to estimates from earlier studies [70–72].

This study intends to examine the structure and stability of premixed oxy-methane flames and the impact of the geometry of a multi-hole burner (imitating micromixers in gas turbines). Figure 5.16 shows the analysis and comparison of two geometries (HE2 and HE3) to a reference geometry (HE1) from Aliyu et al. [55]. To demonstrate the effects of jet size and spacing through one-to-one comparisons, HE2 and HE3 are added to the methodology, test ranges, and analyses of Aliyu et al. [55]. The stability and operability of the combustion chamber are investigated within the φ-OF space.

Figure 5.17 depicts a sample image of a HE1 flame to help clarify the types of flames multi-hole burners normally produce. The reactant combination ($CH_4 + O_2 + CO_2$) is discharged from 61 pores in the headend in 61 jets. The primary reaction region contains 61 flamelets because each jet produces a little conical flame known as a "flamelet." A secondary region in the shape of a common plume is present downstream of the flamelets. The isometric view was unable to distinguish between the several flamelets, as shown in Fig. 5.17. The camera had to be positioned so that

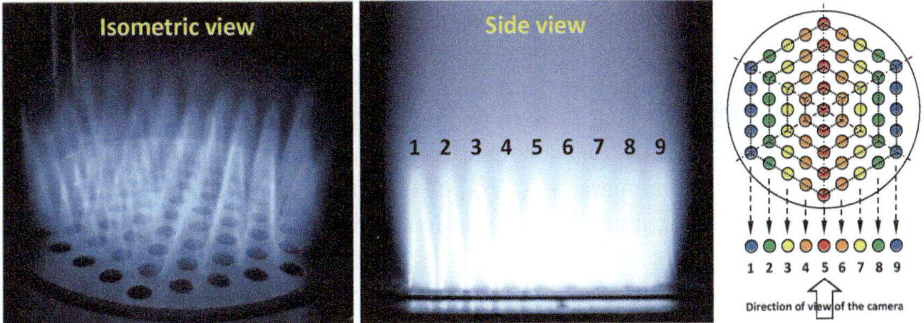

Fig. 5.17 Sample image of a HE1 flame and the direction of view of the camera

it was perpendicular to one of the three axes that make up the headend's hexagonal symmetry. As a result, the camera saw each flamelet in a line of sight as a single flamelet. The 61 flamelets in the schematic top view of Fig. 5.17 are colored to demonstrate how they were condensed to nine distinct lines of sight and showed up as nine flamelets only in the side view of the flame. In addition, the concentric hexagonal rings that encircled the core jet in Fig. 5.16 were visible, since the camera was aligned with one of the three axes of symmetry of the burner headend. Flamelet 5 signifies the central jet in Fig. 5.17, whereas flamelets 4 and 6 represent the first ring, 3 and 7 denote the second ring, 2 and 8 characterize the third ring, and 1 and 9 represent the fourth (peripheral) ring. So, in the sample side-view image of Fig. 5.17, all four rings are lighted and fixed firmly.

5.2.2 *Profile Factor (PF) of the Combustor*

Considering that the three headends' flames occupy varying proportions of the combustor's cross-sectional area (46% for HE1, 28% for HE2, and 43% for HE3), the current analysis begins by casting some insight on the impact of wall heat transmission. The profile factor (PF) of the combustor, which is determined using the following equation [73] based on the radial profile of temperature near the exit section, is taken into consideration in this investigation.

$$PF = \frac{\text{Peak profile temperature} - \text{Mean profile temperature}}{\text{Mean profile temperature} - \text{Reactant temperature at combustor inlet (300 K)}}$$

(5.11)

The cross-sectional area of the combustor is discretized into smaller concentric annular sub-areas to facilitate area-weighted calculation of mean profile temperature. It is a customary procedure in gas turbines to examine the PF because it offers data on how uniform the gas temperature is at turbine inlet and indicates data on the impact of heat losses at the combustor walls. A uniform profile (smaller PF) is beneficial to minimize exposing the turbine to localized hot spots, especially near the blade roots and tips, as they are a cooling issue and places of stress concentration.

The four stable flames that were used for temperature measurements are listed in Table 5.3. While Flames C and D share a value of φ 0.5, Flames A and B share an OF of 60%. The values of φ and OF were carefully selected to provide flames A and D a mutual adiabatic flame temperature (AFT) of 1600 K and flames B and C a common AFT of 1900 K. The recorded peak profile temperatures and the corresponding computed PF values are also listed in Table 5.3 for each headend. Experimental temperature measurements are also valuable for confirming future computational studies of the existing flames.

The radial profiles of temperature near the combustor output are shown in Fig. 5.18 for all three headends under the various operating conditions shown in Table 5.3. All headends create similar bow-shaped profiles and PF values. Both Fig. 5.18 and Table 5.3 show that HE2 demonstrates improved profile uniformity (lower PF) despite only having flames occupying 28% of the cross-sectional area of the combustor. The most asymmetric profiles and largest PF were predicted from HE2, which also exhibited visible hot streaks close to the centerline. Contrarily, HE1, which has flames that fill 46% of the cross-sectional area, produces the greatest PF values. HE2 is affected the least by wall heat transfer in this case because its perimeter flamelets are the furthest away from the walls. Complementary computational examination is necessary to analyze the minor changes in wall heat transfer across the three headends.

Figure 5.18 additionally allows for the following key observations, in addition to the PF analysis:

- Despite the considerable disparities in OF and φ between the left and right plots, each headend displays almost identical profiles on the left and right graphs for a given AFT, i.e., the values of temperature are nearly the same. Because a common AFT results in essentially equal flame morphologies despite the varying φ and OF, it can be identified as the main factor governing the morphology of premixed CH_4 flames at constant inlet velocity. This discovery has been made for premixed oxy-methane and oxygen-enriched air-methane flames as well as for HE1 in the reference work of Aliyu et al. [55] and swirl-stabilized HE1 flames

Table 5.3 Operating conditions of the four flames chosen for temperature measurements

Flame	φ	OF	AFT [K]	Calculated PF			Peak profile temperature [K]		
				HE1	HE2	HE3	HE1	HE2	HE3
A	0.250	60%	~ 1600	0.222	0.111	0.207	1208	1338	1391
B	0.325		~ 1900	0.213	0.122	0.131	1266	1490	1530
C	0.5	42%		0.190	0.116	0.136	1345	1498	1576
D		33%	~ 1600	0.195	0.126	0.172	1206	1373	1399
Percentage of combustor cross-sectional area occupied by flame							46%	28%	43%

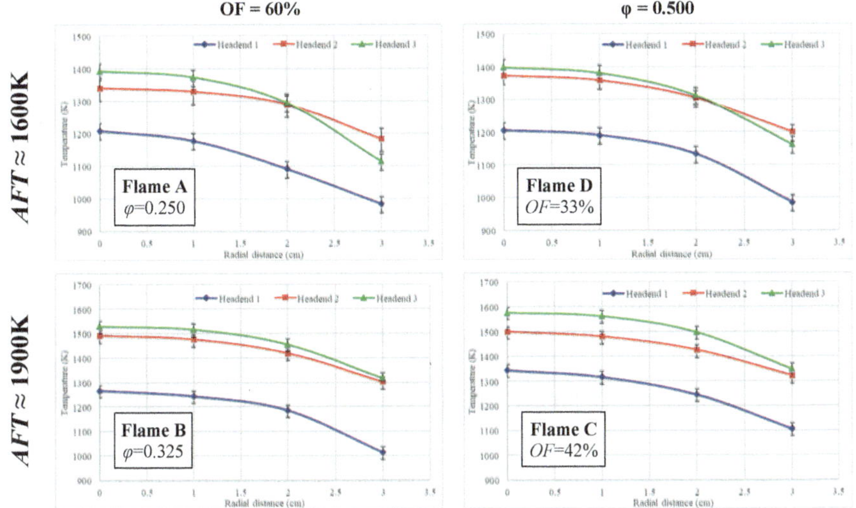

Fig. 5.18 Radial profiles of temperature near the combustor exit for all three headends at different operating conditions

[74]. The uniqueness of the current study is in demonstrating that this conclusion also holds true for the HE2 and HE3 geometries, i.e., that AFT is the key factor dictating the morphology of premixed CH_4 flames for a range of burner geometries/technologies and flame-stabilization techniques.

- When compared to HE2 and HE3, HE1 consistently displays colder temperature profiles. Due to its lower jet diameter, HE1 has the shortest flamelets for a specific set of operating parameters (3.18 mm). Compact flames cause a greater axial difference between the value of flame-tip temperature and the values close to the combustor exit. As a result, the recorded profile temperatures are lower because more heat is lost by the hot gases in the flame plume before they reach the thermocouple. In contrast, the longer flamelets of HE2 and HE3 demonstrate hotter temperature profiles due of the larger jet diameter (3.97 mm). Higher recorded profile temperatures are the result of plume gases losing less heat due to longer flames, which reduce the axial distance between the flame tip and the position of temperature measurements.

5.2.3 Burner Blowout Mechanism

After shedding some insight on how the different headends are impacted by heat transfer to the wall, the analysis moves on to looking at how multi-hole and micromixer burners blow out their fuel. This mechanism is shown in Fig. 5.19 by gradually decreasing AFT (via OF) to approach flame blowout at fixed $\varphi = 0.5$ for all three headends. The analysis begins with an AFT of 1900 K, which is high enough to create powerful, compact flames with all flamelets lighted and steady in each headend. The perimeter flamelets are put out by the blowout mechanism because they lose heat to the walls. The peripheral flamelets quench when the local flame speed falls below the jet velocity. The stable inner flamelets can ignite the peripheral jets further down in the combustor by lighting up the toroidal recirculation region that the periphery jets produce, when the local bulk flow velocity is small enough to match the low flame speed of the quenched jets,. Further reductions in AFT result in the quenching of more flamelet rings and a commensurate increase of the recirculation region up until the combustor is maintained by only a few pilot inner flamelets before the ultimate full blowout.

Because HE1 and HE2 have the same overall combustor power, the larger jet diameter of HE2 is correlated with fewer jets (37 in HE2 versus 61 in HE1). Figure 5.19 makes the advantage of a bigger HE2 jet diameter immediately clear; stability is

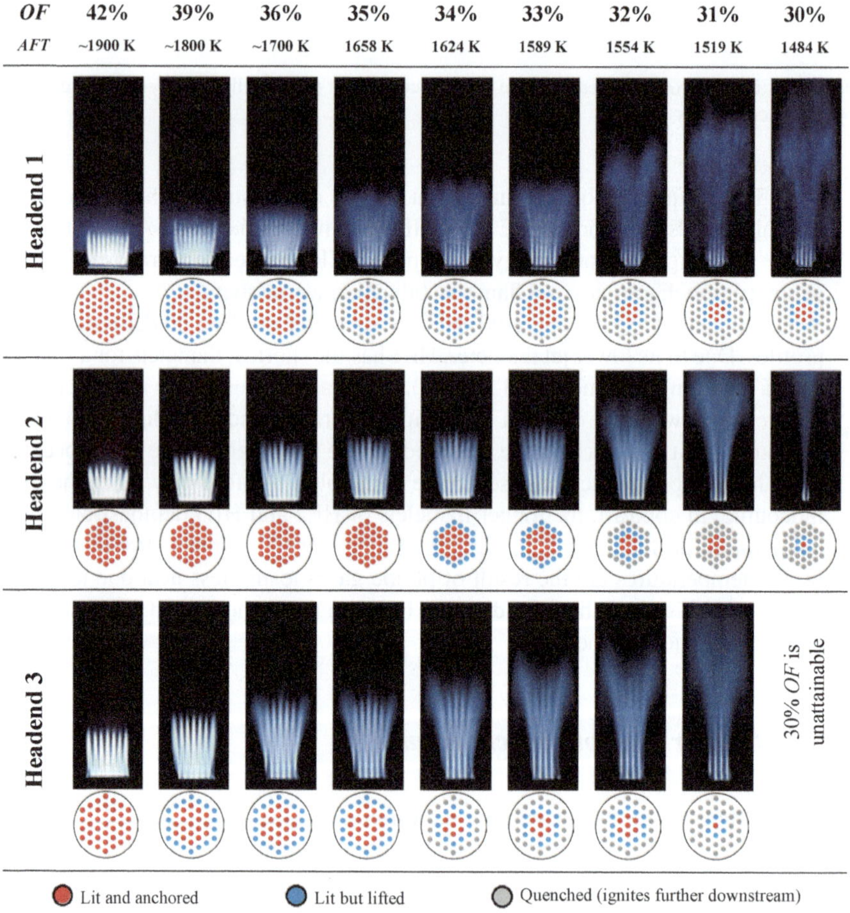

Fig. 5.19 Effect of decreasing the *OF* to approach flame blowout at constant $\varphi = 0.5$

benefited because the bigger perimeter flamelets are less vulnerable to heat loss to the walls. As a result, down to AFT = 1658 K, HE2 preserves the stability of its perimeter (third) ring. In contrast, the tiny flamelets in the fourth and outermost ring of HE1 begin to rise at 1800 K and rise to 1700 K. At 1624 K, the outermost flamelets in HE2 begin to lift, whereas in HE1, the outermost ring is quenched by the same conditions. The perimeter ring of HE2 (18 flamelets) requires an AFT as small as 1554 K, or around 49% of the combustor's output, to be quenched. As a result, HE2 becomes unstable and begins to lose a full additional ring of flamelets for every subsequent 1% reduction in OF. At 1484 K, HE2 has a lifted initial ring and is solely piloted by the middle flamelet (2.7% of the combined power), which causes blowout shortly after. In contrast, HE1 keeps its first ring and core flamelet (11.5% of the combined combustor power) stable down to 1484 K. This analysis makes the conflicting impacts of flamelet number and size quite apparent; first, the advantage of larger flamelets (HE2) presents itself, but later the benefit of more flamelets (HE1) stands out. In essence, compared to 1/37 = 2.7% in HE2, one extinguished HE1 flamelet only equates to 1/61 = 1.6% of the combined power. So, unlike HE1, HE2 cannot tolerate losing flamelets as quickly.

A greater jet spacing in HE3 (7.0 mm) versus HE2 (5.5 mm) is harmful to flame stability (versus 1554 K in HE2) because the outermost (third) flamelet ring of HE3 starts to lift as early as 1800 K and extinguishes at 1624 K. At 1519 K, the pilots of HE3 are the central flamelet plus a first ring that is about to lift. This ring is still partially lifted, but HE3 cannot endure under more extreme circumstances, which explains why 1484 K (30% OF) is not possible. Thus, increasing the jet separation has a negative impact on this burner technology's stabilization mechanism. As a result, closer spacing is advised, which is in line with Veetil et al. [49] and Hindasageri et al. [50] findings. Keep in mind that HE2 and HE3 have identical jet diameters, jet counts, and flamelet ring counts (three).

The same result of decreasing AFT (through φ) at fixed OF = 60% is presented in Fig. 5.20 for the three headends, but not until full blowout. Once more, it can be shown that lowering AFT makes the flame weaker and lengthens it for every headend due to the slower reaction kinetics and slower burning velocity [65]. All of the flamelets are lighted and anchored at 1900 K, and all three headends display powerful, stable flames. At lower AFT, the impact of headend geometry is more noticeable. The majority of the HE2 flamelets are steady at 1600 K, with the outer ring just beginning to rise. In contrast, the third ring is elevated and the peripheral ring in HE1 is quenched. This once more demonstrates the advantage of larger flamelets in HE2. It is demonstrated that a greater jet spacing in HE3 versus HE2 is detrimental to flame stability because the outer ring of HE3 extinguishes and the second ring lifts at 1600 K. Thus, it may be inferred that flamelets' ability to effectively stabilize one another is negatively impacted by increasing their distance from one another. A point when each flamelet is essentially alone and seldom interacts with its neighbors could be attained if the distance were further increased in this scenario, the burner would lose its built-in stabilizing mechanism.

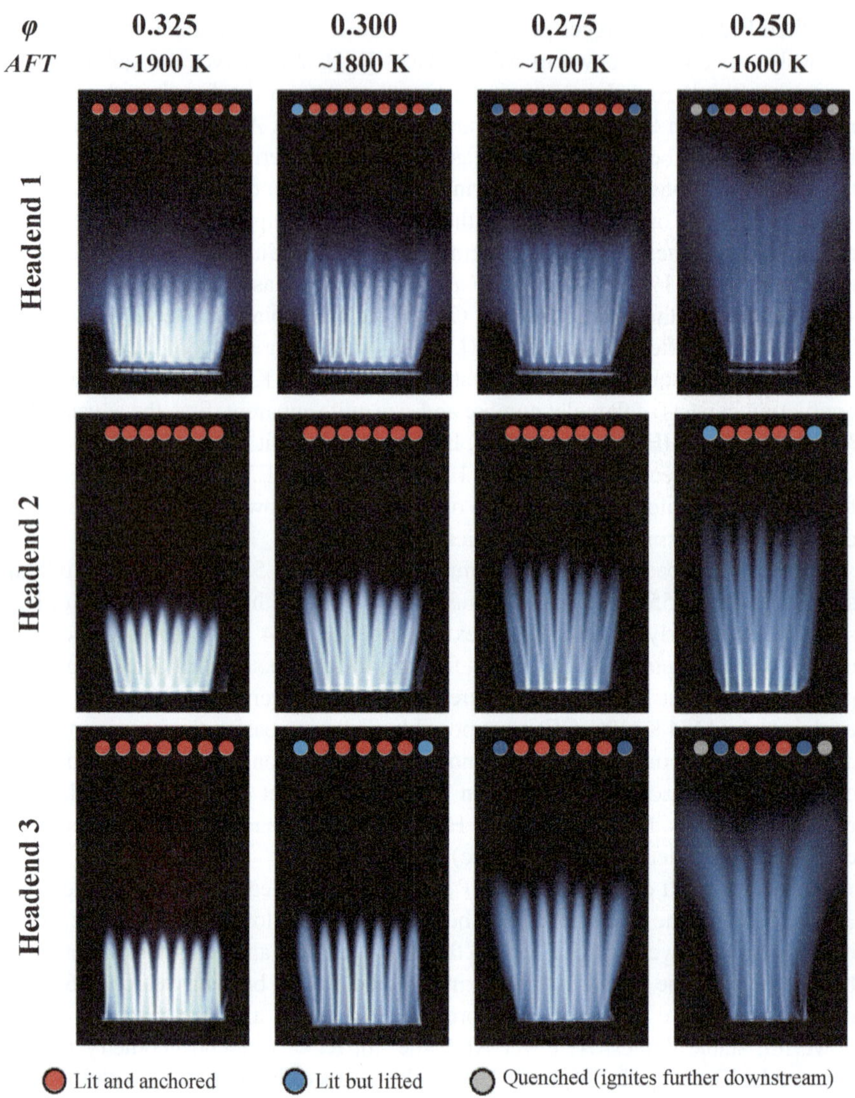

Fig. 5.20 Effect of φ on flame morphology at constant $OF = 60\%$. The HE1 images (top row) are from the work of Aliyu et al. [55]

5.2.4   *Dominating Effects of Adiabatic Flame Temperature on Flame Stability*

The investigation of Fig. 5.18 showed that, even when OF and are different, maintaining AFT constant results in almost comparable flame shapes. Figure 5.21, which shows five flames with widely varying φ (0.325–0.500) and OF (42–60%), but a

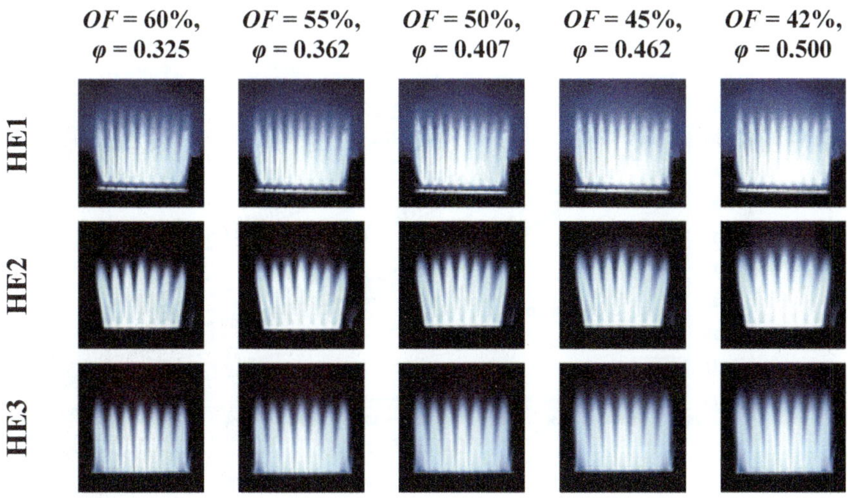

Fig. 5.21 Flame morphology at constant $AFT \approx 1900$ K. The HE1 images (top row) are from the work of Aliyu et al. [55].

mutual AFT of 1900 K for all the three headends, further supports this conclusion. It is evident that, despite the varying OF and φ, the common AFT results in almost identical flame macrostructures for each headend, i.e., flame size, shape, brightness, and the number of stable (non-lifted) flamelets are all rather constant. Thus, it can be concluded that for a variety of burner technologies/geometries, AFT is the main factor influencing the morphology of premixed CH_4 flames.

The stability map of the combustor in Fig. 5.22 provides more evidence of the predominance of AFT. The blowout (i.e., lower flammability limit) and the beginning of dynamic combustion instabilities serve as the stability limitations of each headend in this definition (upper operability limit). The blowout limit was established by gradually lowering φ while maintaining a fixed OF until flame blowout occurred. On the other hand, increasing φ at constant OF led to an upper operability limit where the dynamic instabilities typical of premixed combustion caused loud high-pitched noise at high AFTs; this limit is therefore referred to as the "acoustic limit" in this context. It is not be confused with the flame flashback limit [65, 74] because flashback was prevented by the flame arrestors integrated in each burner headend. The present study is not intended to quantify the frequencies/magnitude or the onset mechanism of this audible noise; as a result, the acoustic limit is shown in Fig. 5.22 as a dotted line, as sensed by the operator's hearing. However, the main objective of this study is to investigate the impact of burner geometry on the blowout limit and the morphology of stable (noise-free) flames.

The combustor stability map was defined within the φ-OF space, as shown in Fig. 5.22, with φ ranging from 0.10 to 0.70 and OF ranging from 25 to 70%. To investigate whether there is any relationship between the stability limits and AFT, it was calculated over the ranges of Fig. 5.22, and the stability map was overlaid

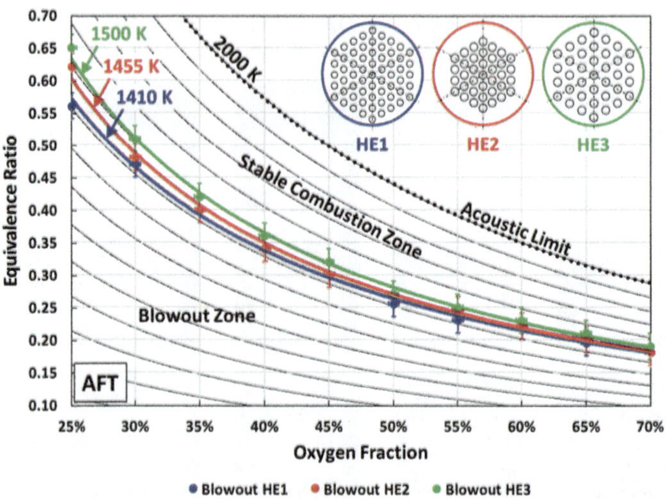

Fig. 5.22 Blowout limits of HE2 and HE3 (current study) in contrast to HE1 (Aliyu et al. [55]), overlaid on the contours of fixed *AFT*. The bulk jet velocity is fixed at 5.2 m/s

on the contours of fixed AFT. Three times each data point was collected. Several conclusions can be drawn from Fig. 5.22 in this instance:

- Constant AFT contours are followed by all three headends' blowout limitations. Therefore, if the jet velocity is maintained, premixed methane combustion, independent of burner geometry, is predominantly controlled by AFT. Thus, for different flame-stabilization techniques and burner geometries, AFT plays a crucial regulatory function in premixed methane combustion. This work suggests using AFT alone rather than the traditional CO_2 dilution ratio to design, control, and define the operation modes and corresponding flame morphologies in premixed oxy-fuel combustors in natural-gas-fired gas turbines. This makes operational convenience possible because only one parameter (AFT) needs to be regulated or monitored rather than several parameters being targeted at once.
- Although HE2 has a 25% larger jet diameter than HE1, and HE3 has a 27% greater jet spacing than HE2, the AFT difference between one blowout limit and the other is only about 45 K. Thus, the technology of gas-turbine micromixers demonstrates its geometry-flexibility. Therefore, they are suggested as a completely versatile technique for oxy-fuel combustion in this study. Micromixers' improved stability and flexibility with fuel and oxidizers have already been discussed in the Introduction section.
- The AFT contours are farther apart at smaller OF and gradually closer together at greater OF in terms of φ. This is due to the fact that at smaller OF (greater CO_2 dilution), the detrimental effects of CO_2 dilution are increased, while at higher OF (abundant oxygen), they are diminished. This also justifies why the three blowout limits are farther apart at smaller OF and almost coincide at greater OF. In other words, the impact of headend geometry is negligible at greater OF.

- For all three headends, the acoustic limit was found to occur nearly at the same value of AFT of 2000 K. However, it depends on the unique perspectives of the test operators. Interestingly, the operator of the HE2 and HE3 tests was not the same person who tested HE1 in the study of Aliyu et al.; yet both operators perceived the acoustic limit almost at the same value of AFT. However, additional research is required to evaluate the precise relationship between premixed methane combustion dynamic instabilities and AFT and burner geometry.
- Early blowout occurs when the diameter of the jets is increased (from HE1 to HE2) while the jet count is decreased (to maintain same combustor power), i.e., the advantage of higher flamelet count in HE1 overcomes the benefit of bigger flamelets in HE2.
- Even when the jet diameter and combustor power remain constant, greater jet spacing (in HE3 versus HE2) causes the blowout to occur earlier. A wider gap between the flamelets makes it harder for them to stabilize one another effectively.

5.2.5 Recommended Burner Design

Part-load operation is given a lot of attention because turndown (i.e., the ratio of full to minimum loads) is one of the important operability characteristics of gas turbines. This emphasizes the significance of the investigations of the burner extinction limit and blowout mechanism discussed above. Oxy-fuel combustion presents an extra difficulty since, as was previously discussed, because it inherently has inferior flame stability compared to traditional air–fuel combustion. Therefore, it is crucial to find burner technology that delivers improved stability while optimizing the burner geometry for wider blowout-side operability. Based on the findings and data presented here, micromixers are recommended as a flexible and remarkably reliable technology for oxy-fuel combustors in zero-emission power plants.

Since HE1 can keep flames burning down to lower AFT before burnout, it has been demonstrated that it performs better than HE2 and HE3. HE1 also more effectively realizes the lean-direct-injection (LDI) concept, which aims to provide rapid reactant mixing through the injection of fuel from multiple sites inside the combustion area (see Fig. 5.23 [75]). The enhanced resilience to lean extinction of LDI air–fuel gas turbines increases the operability window of the combustor on its blowout side [35, 76]. As a result, it is suggested that oxy-fuel micromixers use several jets with smaller diameters and larger inter-jet spacing. Naturally, the ease of fabrication and manufacturing costs also have an effect.

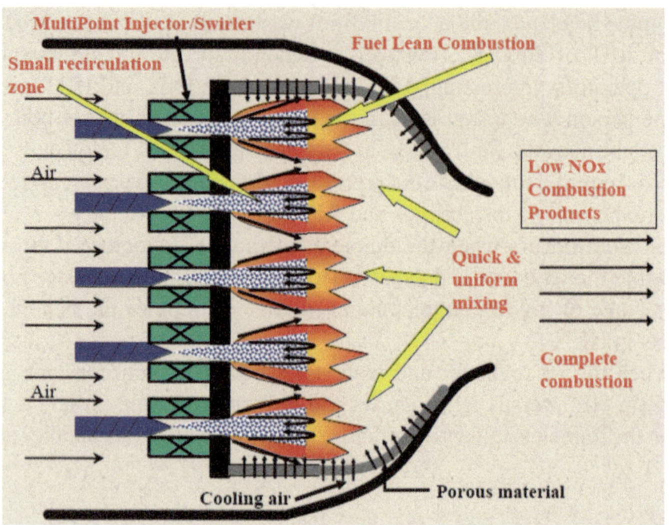

Fig. 5.23 Lean-direct-injection (LDI) concept as implemented in current air–fuel gas turbines [75]

5.3   Characteristics of Lean Premixed Oxy-combustion in Perforated Plate Burner

Global warming brought on by greenhouse gases has a considerable impact on climate change [77–80]. Rapid action is necessary to halt global warming and keep the rate of economic expansion constant. By reducing greenhouse gas emissions, primarily CO_2 emissions, the Paris Agreement 2015 seeks to keep the temperature increase to less than 2 °C above pre-industrial levels (1861–1880). There are practical solutions to reduce CO_2 emissions, such as the adoption of CO_2 capture and storage (CCS) technology and the usage of non-carbon generating technologies. However, just 4.4% of primary energy is used by zero-carbon technologies or renewable energy sources (apart from biomass) [81], as opposed to 85% in 2015 and 78% in 2035 for fossil fuels [82–84]. Therefore, it is essential to implement carbon capture, which might reduce global CO_2 by 15–55% until 2050 [85]. Oxy-fuel combustion, which burns hydrocarbon fuel with pure O_2 rather than air and generates flue gas largely comprised of CO_2 plus H_2O, is the most promising CO_2 capture method. Simple condensation can be used to separate H_2O for cost-effective CO_2 capture. Some CO_2 is recirculated throughout this procedure to maintain control over flame temperature [81]. The characteristics and operability ranges of the resulting flames are significantly altered when CO_2 is used as diluent in an oxy-combustion process instead of N_2 because the two gases have substantially different thermos-physical, chemical kinetics, and radiative properties.

5.3.1 *Characteristics of Fuel/Oxidizer-Flexible Combustion*

The thermo-acoustic stability characteristics of premixed $CH_4/O_2/CO_2$ and $CH_4/O_2/N_2$ flames in a swirl-stabilized combustor was investigated by Shroll et al. [86]. To maintain a specific AFT as the air-flame, stoichiometric oxy-flame conditions were used, and the level CO_2 dilution was adjusted accordingly. The findings showed that AFT is mostly responsible for the alterations in thermos-acoustic modes. Under lean premixed oxy-combustion circumstances, Kutne et al. [87] studied the flame morphology and static stability limits on a swirl combustor. Due to the higher laminar burning velocity, the result demonstrated greater stability with a rise in the oxygen fraction (OF) in the O_2/CO_2 oxidizer. Hu et al. [88] examined the $CH_4/O_2/CO_2$ and $CH_4/O_2/N_2$ laminar flame speeds under atmospheric conditions to look into the impacts of the equivalence ratio (Φ), OF, and the effect of CO_2 in comparison to N_2. It was discovered that the laminar burning velocity exhibited a quadratic relationship to the OF. While maintaining the other parameters, the CO_2-diluted flame speed was also only a sixth as fast as the N_2-diluted one. According to Jerzak and Kuznia [89], premixed $CH_4/O_2/CO_2$ and $CH_4/O_2/N_2$ flames have different blowout processes in a swirl combustor. The findings showed that the operability windows for oxy-flames were much smaller than those for air-flames. Due to the hindered kinetics of oxy-methane combustion, Amato et al. [90] discovered that the operability window of CO_2-diluted flames in a premixed swirl combustor is significantly narrower in contrast to that of N_2-diluted flames. Rashwan et al. [91] studied partially premixed $CH_4/O_2/CO_2$ and CH_4-air flames experimentally on a perforated plate burner. With an increase in the premixing ratio, it has been seen that CO and NOx emissions decline. Increasing the premixing ratio induced a change in the flame color from reddish to pure blue. High soot production levels were found outside of diffusion fires. For a premixed swirl-stabilized gas-turbine model combustor, Abdelhafez et al. [92] showed that, even under stoichiometric circumstances, stable oxy-fuel flames could be attained only if the OF stayed above 22%. The outcomes demonstrated that OF had a significantly bigger impact on flame stability than that of Φ. As a result, steady flames might perhaps be generated at considerably leaner conditions with a higher OF. Rashwan et al. [54] reported similar results, finding that stable oxy-flames required OF in the range of 29–40%. This was related to the detrimental effects of CO_2 on the combustion kinetics, which resulted in limited operability ranges of the combustor.

Hydrogen enrichment is one of the high-potential methods for improving reaction kinetics, particularly close to the blowout limit, and extending the combustor operability limits under premixed oxy-fuel combustion conditions. Given that it doesn't release any damaging poisonous or greenhouse gases during combustion, hydrogen is regarded as one of the cleanest sources of energy. Hydrogen combines with oxygen to form water vapor and thermal energy during burning. Despite these benefits, hydrogen combustion is still not widely used because of its greater price, complicated storage and transportation requirements, and extremely flammable makeup [93]. Recent combustion research, however, suggests that using hydrogen in addition to conventional fuel in combustion can help overcome combustion instabilities, raise operability limits, and reduce NOx emission [94]. Syngas and other hydrogen-rich blends are being studied as alternatives to natural gas because of their improved combustion performance [95, 96]. Hydrogen-enriched fuels can have a significant positive impact on oxy-combustion under premixed conditions [97, 98]. H_2 injection has been shown in several experiments to minimize soot production and CO generation during combustion [99–102]. It can also improve the turbulent flame speed [103], broaden the operability window [104], and boost laminar flame speed [105]. The stability map of hydrogen-enriched oxy-CH_4 flames in a swirl gas-turbine model combustor was established in the work by Imteyaz et al. [97] throughout a range of hydrogen fractions (HF). The findings showed that the blowout and flashback limits lines for static stability follow routes with constant Reynolds number (Re), indicating that Re plays a crucial role in regulating the blowout limit of hydrogen-enriched flames. Other studies [92, 106–108] confirmed that AFT had a significant impact on regulating the blowout limits for oxy-CH_4 and oxy-C_3H_8 flames without hydrogen-enrichment. Various experimental studies [109, 110] also looked at the effects of H_2 content in the syngas fuels. According to a report, lean blowout in syngas combustion is controlled by the mole percentage of hydrogen in the gas.

In a small number of experiments, the effects of hydrogen enrichment for premixed combustors under oxy-fuel combustion circumstances have been well-reported. Examining the accuracy of such results for partly premixed flames anchored over a perforated plate burner, however, is extremely important. The present work intends to examine the effects of fuel and oxidizer flexibility on the operability of combustor over a variety of oxidizer and fuel combination compositions. In order to accomplish this, three different sets of experiments on a partially premixed burner holding flames of hydrogen-enriched compressed natural gas (CNG) under oxy-fuel combustion conditions were conducted over ranges of HF, OF, φ, and inlet flow Re. The findings were contrasted with those from other burners holding flames with various stabilization techniques. In order to explore the involvement of AFT in regulating the combustor stability limits, the operability of the combustor was evaluated under conditions that were close to stoichiometry, and the stability maps were plotted against the contours of AFT.

5.3.2 *Features of Perforated Plate Burner*

The input of fuel and the oxidizer mixture to the burner is depicted in the test rig flow diagram in Fig. 5.24 for assessing the functionality of partially premixed fuel/oxidizer-flexible combustor. Depending on the necessary loading circumstances, equivalence ratio (Φ), and hydrogen fraction, compressed natural gas (CNG) and hydrogen are injected at varying concentrations (HF: volumetric concentration of H_2 in a fuel mixture of H_2 plus CNG). The oxidizer mixture consists of $O_2 + CO_2$ at a high degree of premixing with changing proportions, depending on the operation's required oxygen fraction (OF: volumetric concentration of O_2 in an oxidizer mixture of O_2 and CO_2). A primary burner with an exhaust system, a pilot flame burner, mass flow rate controllers, measuring devices, and oxidizer supply systems make up the test rig.

The perforated plate burner is schematically depicted in Fig. 5.25, which also shows the degree of oxidizer and fuel premixing upstream of the burner. After being thoroughly premixed, H_2 and CNG are transported in a mixture along the fuel flow line to the burner. O_2 and CO_2 from two different vessels that contain oxidizer gases are thoroughly mixed before being fed through the oxidizer flow line. Each gas is measured and recorded independently for its pressure, temperature, and mass flow rate. In a mixing pipe located before the burner inlet portion, the two major flow lines are combined. The partially premixed combustion chamber has a degree of premixing of 7.0 (L/D = 210/30) due to the mixing pipe's dimensions of 210 mm in length and 30 mm in diameter. The perforated plate burner is affixed to the top of the combustor test rig as illustrated in Fig. 5.25 because of the increased flame stability characteristics of this type of burner. The burner has 22 holes, each with a

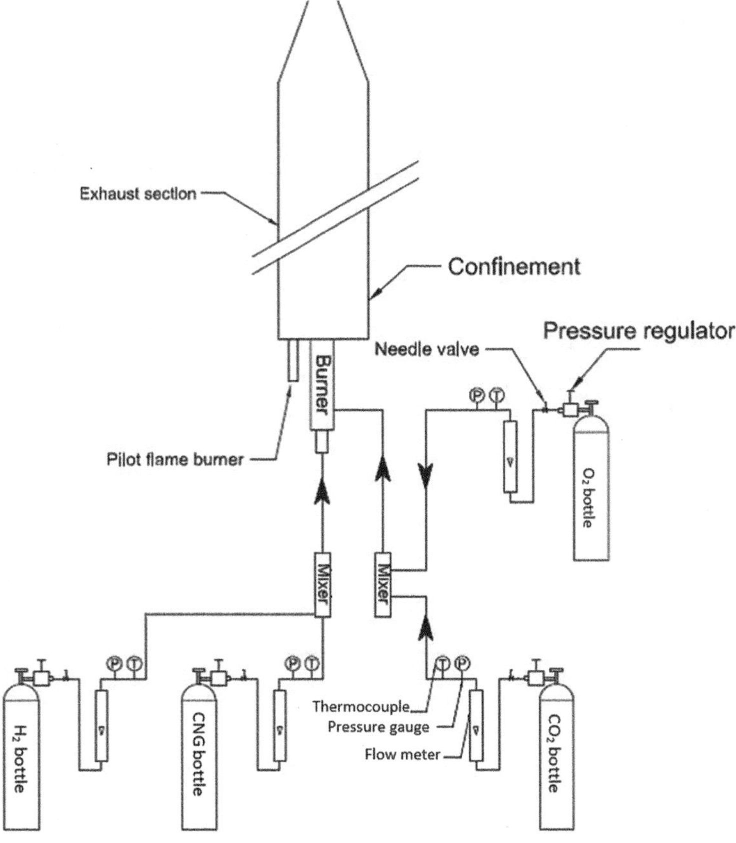

Fig. 5.24   Test rig flow diagram of the partially premixed fuel/oxidizer-flexible combustor

diameter of 4.0 mm, and an exterior diameter of 30 mm. It is 3.0 mm thick. In order to prevent early flame flashback and improve interactions between the flamelets for improved flame stability, the distribution of the holes over the plate was determined to act as an initial flame arrestor. The iron confinement for the combustion chamber, which ensures confined flame operation, has a diameter of 150 mm and a length of 500 mm. To have optical access to the created flames and enable flame observation, a quartz optical window (50 mm × 500 mm) was cut in the confinement. Due to the design of the confinement, heat transmission to the wall only very slightly affects the stability of the created flame, but in a combustor with thermally conductive iron walls, these interactions cannot be entirely ignored [111]. In the event of a flame flashback, multiple flame arrestors were installed in the piping system upstream of the burner to stop upstream flame propagation in the flammable mixture. A top-notch Canon EOS 700D camera with a resolution of 21 Mega Pixels was used to take digital pictures of the numerous flames. A frame rate of 25 frames per second is used by the camera. All photos were captured in night vision mode with a 1/8 s exposure period in a darkened lab space.

Pressurized gas bottles were used to provide all of the gases. As a fuel source, a 200 bar CNG bottle was employed. A 100 bar bottle of hydrogen was used to supply the gas. The oxygen supply system comprises of 120 bar, 99.5% pure oxygen bottles. A bottle with a 99.5% purity rating at 70 bar was utilized for CO_2 supply in a similar supply system. Each bottle's exit is equipped with a pressure regulator to lower the output gas pressure. Each stream then travels via a needle valve and a flow rate

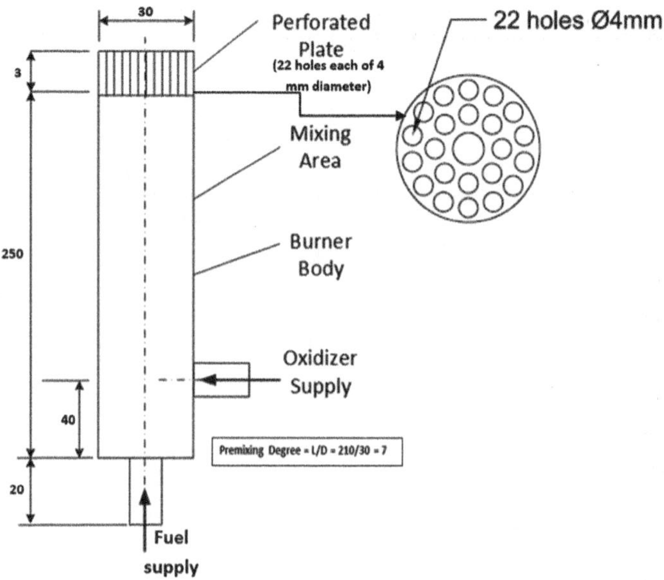

Fig. 5.25 Schematic representation of the perforated plate burner along with its design details (all dimensions are in mm)

controller. The flow rate of CO_2 was regulated using an Omega engineering fl-1448-c mass flow controller with a range of 0–60 L/min and an accuracy of 5%. The flow rate of CNG was managed by a Dwyer Instruments mass flow controller with a range of 0–10 SCFH (standard cubic feet per hour) and an accuracy of 4%. The flow rate of H_2 was managed by a Dwyer Instruments mass flow controller with a range of 0–1.0 SCFH and an accuracy of 4%. The flow rate of oxygen was managed using a Brooks r-6-15-a mass flow controller with a range of 0–10.6 L/min and an accuracy of 1%. Pressure gauges with an accuracy of 2.5% and K-type thermocouples, respectively, were used to measure the pressure and temperature at the exit of each flow controller upstream of the mixing tube. A pipe with a 1.25 cm diameter is used to supply each of the O_2 and CO_2 streams. Before being fed to the main mixing pipe to be combined with the fuel upstream of the burner, both the O_2 and CO_2 streams are mixed in the oxidizer mixer pipe, which has a 0.5-inch diameter and 2.5 m in length. Similar to how the oxidizer mixture is mixed, the CNG and H_2 streams are combined in a separate conduit before being introduced to the main mixing line upstream of the burner. The estimated parameters, which include HF, OF, Re, and φ, have average percentage uncertainties that are, respectively, 8, 9, 12, and 13. Based on a standard uncertainty multiplied by a coverage factor of 2 and a coverage probability of around 95%, the reported enlarged uncertainty is calculated.

This research aims to investigate the impacts of fuel/oxidizer flexibility on the operability of a partially premixed combustor with a premixing degree of 7.0. Under oxy-combustion circumstances, H_2-enriched CNG flames are stabilized across a perforated plate burner in the combustor. Three sets of experiments under various operating settings were carried out to help attain the suggested goal. The first set was carried out to find out how fuel flexibility affected combustor operability and flame macrostructure. The set utilized a steady OF of 29% with a variety of HF, including 0% (pure CNG), 10, 20, and 30%. For each operational HF, the combustor stability map in terms of flashback and blowout constraints was obtained and shown on the φ-Re stability chart. The effect of H_2-enrichment across a range of φ on flame macrostructure was also investigated by contrasting the shapes of the various flames utilizing the flame images taken by a high-speed camera. The effect of OF on combustor operability and flame shape was investigated in the second series of trials. The first set of studies in this collection had their measurements carried out twice, but with different operating factors (32% and 36%, respectively). Measurements were taken in the most recent series of experiments to see if the partially premixed combustor with flames of CNG, H_2, O_2, and CO_2 could function at close to stoichiometry. The set was performed at HF frequencies of 0, 10, 20, and 30% at a constant value of $\varphi = 0.85$. The set measures the flammability thresholds for the oxidizers at various mass flow rates and in terms of OF. As a function of OF and oxidizer mass flow rate, measurements were taken for each operating HF to describe the stability of the flame in terms of flashback and blowout limits, and the findings were displayed on the stability chart. A visual analysis of the flame's appearance was conducted using the acquired images of the flame taken under various operating conditions. The operating conditions for the three sets of experiments are summarized in Table 5.4.

Table 5.4 Summary of the operating conditions for the three sets of experiments

Set number	Φ	OF (% by vol.)	HF (% by vol.)
1	0.2–1.4	29	0
	0.2–1.4	29	10
	0.2–1.4	29	20
	0.2–1.4	29	30
2	0.2–1.6	32	0
	0.2–1.6	32	10
	0.2–1.6	32	20
	0.2–1.6	32	30
	0.2–1.7	36	0
	0.2–1.7	36	10
	0.2–1.7	36	20
3	0.85	25–50	0
	0.85	25–50	10
	0.85	25–50	20
	0.85	25–50	30

To determine the flammability limitations of a partially premixed fuel/oxidizer-flexible combustor retaining H_2-enriched oxy-CNG flames over a range of operating circumstances, all tests in this work were conducted at a fixed degree of premixing of seven. First, an automatic igniter was used to start a small pilot flame. The main burner is assisted in starting safely by the pilot flame burner, which also prevents gas buildup inside the enclosure and any kind of uncontrolled combustion. Then, by regulating the desired values of OF and φ, O_2 and CO_2 gases are introduced to the burner entrance. Then, to achieve the necessary combustion conditions at specified HF and φ, the flow rates of CNG and H_2 are also controlled by the flow controls. Fuel to diluted oxidizer mass-based stoichiometric ratio is written as:

$$\left(\frac{F}{O}\right)_{stoichiometric} = \frac{16 + 2\left(\frac{HF}{1-HF}\right)}{\frac{1}{2}\left(\frac{4-3HF}{1-HF}\right)\left[32 + 44\left(\frac{1}{OF} - 1\right)\right]} = \frac{8 + \left(\frac{HF}{1-HF}\right)}{\left(\frac{4-3HF}{1-HF}\right)\left[8 + 11\left(\frac{1}{OF} - 1\right)\right]}$$

$$(5.12)$$

The following expression is inferred from the implementation of the definition of equivalence ratio as follows:

$$\varphi = \frac{\left(\frac{F}{O}\right)_{actual}}{\left(\frac{F}{O}\right)_{stoichiometric}} = \left(\frac{\dot{m}_{CH_4} + \dot{m}_{H_2}}{\dot{m}_{O_2} + \dot{m}_{CO_2}}\right)\left\{\frac{\left(\frac{4-3HF}{1-HF}\right)\left[8 + 11\left(\frac{1}{OF} - 1\right)\right]}{8 + \left(\frac{HF}{1-HF}\right)}\right\}$$

$$(5.13)$$

OF was maintained constant for each of the first two sets of studies at a specific value. Based on the dimensions and flow characteristics of the mixing pipe before the burner inlet section, the Reynolds number (Re) of the reacting mixture for the various flames was estimated. The mixing pipe has a 210 mm length and 30 mm diameter, which decides how much the combustor is premixed (L/D = 210/30). The upper flammability limit was recorded after a stable flame was created at a particular Re of the reacting mixture, and it was gradually increased by increasing the flow rates of CNG and H_2 at a given HF until the flashback occurred. The same process being repeated while gradually diminishing through lowering.

The operating was held constant for the third series of trials at a value of $\varphi = 0.85$. At the required HF, the mass flow rates of CNG and H_2 were maintained constant. By adjusting the CO_2 flow rate while maintaining a constant mass flow rate of O_2, the OF was changed. Once a steady flame had been established, the mass flow rate of CO_2 was gradually decreased until the flashback occurred, and the flashback circumstances were recorded. Raising the oxidizer flow rate until the blow-off happened, then repeating the preceding operation while recording the blowout conditions. The flammability limit between the two recorded points can be calculated.

5.3.3 Effect of Fuel-Flexibility on Combustor Stability Map

The results under this section quantify the combustor flammability constraints in terms of φ and over a range of the oxidizer mass flow rate based on the information acquired from the first series of experiments as indicated in Table 5.5. At a fixed OF of 29%, the operability restrictions of the combustor were computed for a range of H_2-enrichment values. The restrictions are presented in terms of the φ-Re stability map and contrasted for the various levels of H_2-enrichment in order to quantify the effect of fuel flexibility on the combustor operability. The data also show how the flame appears visually when H_2 is added at a specific Re in the flow of the oxidizer mixture.

Figure 5.26 displays the φ-Re combustor stability map based on experimental results. This equation, $y = 7E\text{-}07x^2 - 0.0013x + 1.4062$, R-squared value is 0.976, where y is the equivalence ratio and x is the Reynolds number, can be used to represent the curve fitting equation for all of the curves in this figure. The plot shows that increasing the HF from 0.0% (pure CNG) to 20% has no discernible impact on the highest flammability limit. But as HF is increased further, to 30%, the upper flammability limit is shifted downward to leaner circumstances at lower Φ. Similar findings were also found in Imteyaz et al.'s [97] examination of a model swirl gas turbine combustor holding fully premixed H_2-enriched oxy-methane flames. The main parameters governing the flashback limit and mechanism, respectively, are the Damkohler number [90, 112] and, for a fixed intake flow Re, the reaction timescale, which is a direct representation of the flame speed. The early flame flashback at a lower φ for the same input flow Re and OF may be explained by the increase in flame speed brought on by the improvement in fuel diffusivity and reactivity brought on by the injection of H_2.

Due to the high ratio of the oxidizer flow rate to the fuel flow rate under part load conditions, where the flame is more likely to blow out, the effect of H_2 addition is

Table 5.5 Experimental operating conditions for the first series of experiments

Set number	OF (% by vol.)	Φ	HF (% by vol.)
1	29	0.2–1.4	0
	29	0.2–1.4	10
	29	0.2–1.4	20
	29	0.2–1.4	30
2	32	0.2–1.6	0
	32	0.2–1.6	10
	32	0.2–1.6	20
	32	0.2–1.6	30
	36	0.2–1.7	0
	36	0.2–1.7	10
	36	0.2–1.7	20
3	25–50	0.85	0
	25–50	0.85	10
	25–50	0.85	20
	25–50	0.85	30

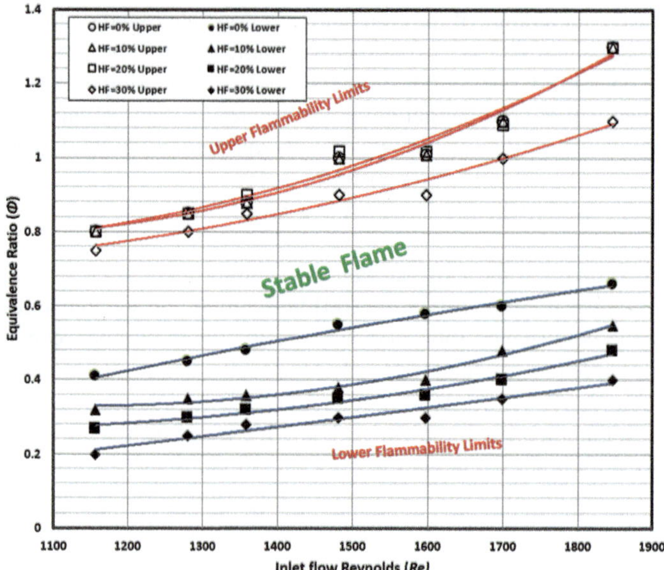

Fig. 5.26 Effect of H_2 enrichment on the Φ-Re flammability limits of the partially premixed combustor holding $CNG/H_2/O_2/CO_2$ flames at OF of 29%

more noticeable. As contrast to pure CNG operability, the inclusion of hydrogen in such circumstances increases the fuel mixture's reactivity and extends the flame's operability to leaner settings. When operating with 30% H_2 addition, the value of φ fell from 0.55 for pure CNG operation to 0.3 at the fixed Re of 1481 when the blowout occurred. The virtually linear trend of the line showing the blowout limit with growing flow Re indicates that the flame blows out at a higher φ for the same HF due to the increased total flow rate and flow velocity. As seen in Fig. 5.26, the flow velocity is greater for the same HF at lower φ values of than the flame speed, delaying the flame flashback to higher values of Φ.

Based on the discussion above, H_2 addition widens the stability region between the flashback and blowout operability constraints of the CNG partially premixed flames at the same inlet flow Re. In order to clearly depict the impact of flow Re on the combustor operability restrictions, the φ-Re combustor stability map in Fig. 5.26 is rearranged to be displayed in the φ HF domain for varied inlet flow Re as shown in Fig. 5.27. In this work, a greater total flow rate that produced a higher inlet flow velocity was used to raise the flow Re at a fixed OF. The flame speed at smaller values of φ is overcome by this flow velocity, which delays flashback to φ greater levels with larger inlet flow Re. The improvement in flame speed caused by the addition of more HF to the fuel mixture reduces this retardation. On the other hand, as inlet flow Re increases, the flame blowout limit shifts to richer circumstances. This can also be ascribed to the fact that raising the inlet flow Re is related to an increase in the total flow rate and inlet flow velocity. Due to the improved fuel diffusivity, mixing, and reactivity in the mixture, this shift is likewise decreased with H_2 addition.

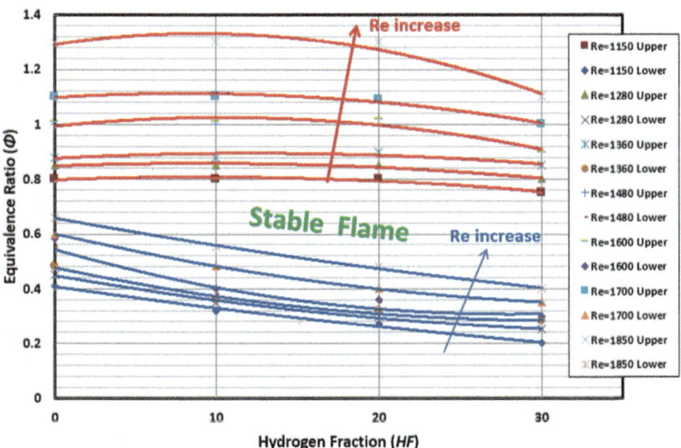

Fig. 5.27 Effect of inlet flow Re on the Φ-HF flammability limits of the partially premixed combustor holding CNG/H_2/O_2/CO_2 flames at OF of 29

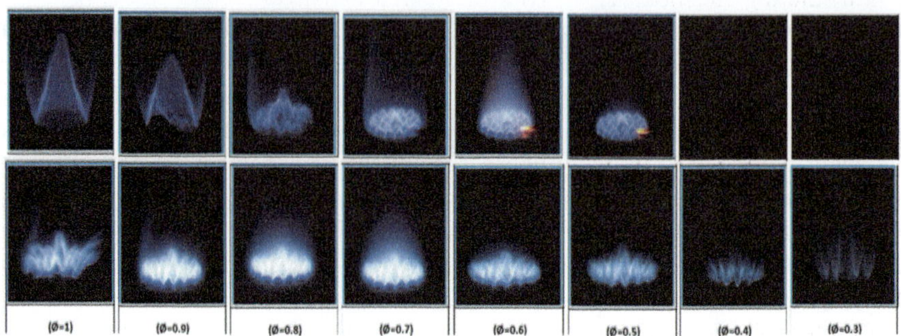

Fig. 5.28 Effect of equivalence ratio, from blowout to flashback at Re = 1481, on the inner cone structure of the CNG/H_2/O_2/CO_2 flames at OF = 29% and for HFs of 0.0% (top row) and 20% (bottom row)

5.3.4 Effect of Fuel-Flexibility on Flame Macrostructure

To evaluate the effect of fuel flexibility on flame structure, digital photos of the various flames were taken at high speed. The visual flame length was calculated by comparing the flame length in the photograph to a reference length scale in the experimental apparatus. To show the real length of the flame in the taken image, a simple ruler that is mounted normally to the burner plate serves as the reference length scale. The distance between the burner plate and the outer cone's tip was used to compute the flame length. A set of flame images that were captured between the blowout and flashback limits at a fixed OF of 29% and a fixed input flow Re of 1481 are shown in Fig. 5.28. Images of pure CNG flames and 20% H_2-enriched flames show how H_2 enrichment affects the color, shape, and appearance of the flames. The H_2-enriched flames are more compact, more stable, and have a smaller outside flame cone as compared to pure CNG flames that are not H_2-enriched. This could be attributed to the enhanced reaction kinetics and elevated diffusivity of the fuel mixture brought about by the addition of H_2. The flame becomes whiter and more stable, with a strong attachment to the plate, when compared to the pure CNG flames that are not H_2 enriched at the same time. The images clearly demonstrate how H_2 enrichment is essential for enhancing flame stability close to the blowout limit. The H_2-enriched flame continues to burn while the pure CNG flame extinguishes at a lower value of $\varphi = 0.4$.

5.3.5 Effect of Oxidizer-Flexibility on Combustor Stability Map

In order to compare two sets of flammability limitations and examine the impact of OF on flame stability under H_2 enrichment, two oxidizer mixture compositions with OF of 32% and 36% are taken into consideration. On the-Re stability map, the flammability limitations for each set are shown taking into account a range of H_2 enrichment, 0.0, 10, 20, and 30%. The results of raising the OF from 32 to 36% are shown in Fig. 5.29. The stability limits of the combustor are wider with the addition of hydrogen when the OF is increased from 29 to 32%, according to a comparison of the flammability limits in Fig. 5.29. The flashback limit will be reached sooner at a lower φ, however, as increasing the OF causes the combustion temperature and flame speed to increase. Because of this, the combustor was unable to maintain stable flames at an OF of 36% and HF of 30%. As illustrated in Fig. 5.29, increasing the OF can expand the flashback limit, causing it to occur at larger values of φ but when the inlet flow is low. This could be explained by the insufficient fuel mixing that occurs in the reacting mixture at lower inlet flow Re values. Raising the OF when the inlet flow is low According to Fig. 5.29, the flame flashback limit is shifted at low Re values to occur at higher at OF = 36% as opposed to the cases at OF = 29% and at OF = 32%. Re improves the flow mixing and results in greater flame anchoring and stabilization. According to Fig. 5.29, with increasing OFs, the lines denoting the stability limits for various HFs converge toward one another. This shows that increasing the OF reduces the impact of HF; in other words, the impact of HF is stronger at lower OF. Additionally, due to the increased reaction rates, increasing OF improves flame stabilization close to the blowout limit.

To clearly depict the effect of flow dynamics on flame stability, the stability limits are shown on the φ-HF map as illustrated in Fig. 5.30. By raising the inlet flow Re at fixed OF and fixed HF, the flashback limit is postponed until higher values of φ where the flame speed approaches the value of the inlet flow velocity. When the input flow Re is increased close to the blowout limit with the response rates reduced, the flame blows out at noticeably higher values of φ when this happens. Figure 5.30 demonstrates that the increase in HF up to 10% at the blowout and flashback flammability limit has no discernible impact. However, because to the higher reaction rates and flame speed,

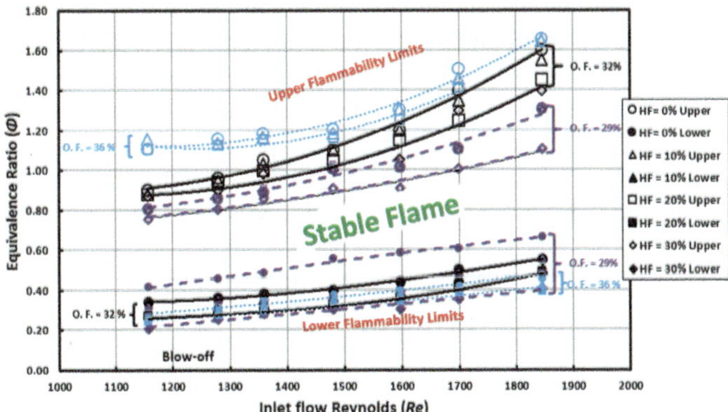

Fig. 5.29 Effect of H_2 enrichment on the Φ-Re flammability limits of the partially premixed combustor holding CNG/H_2/O_2/CO_2 flames at OFs of 29, 32, and 36%

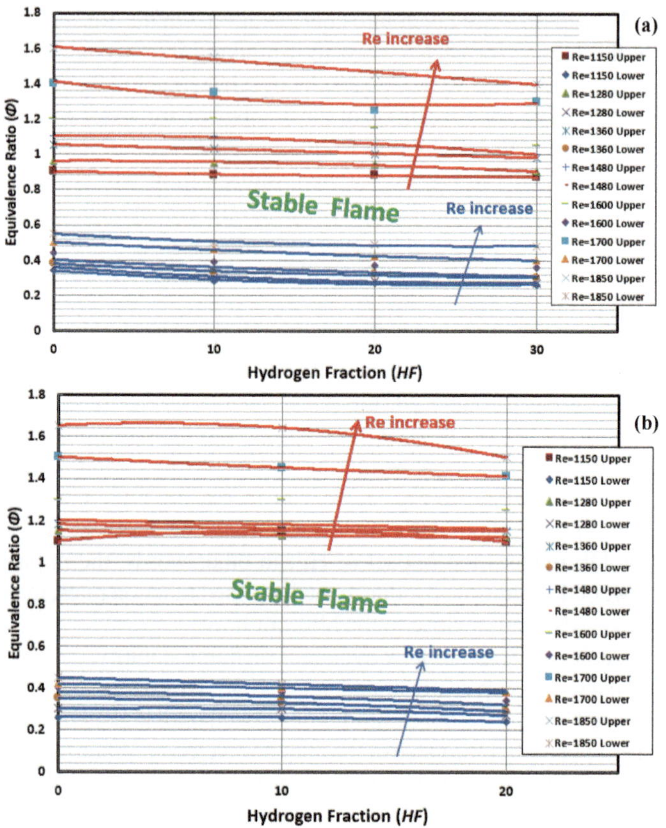

Fig. 5.30 Effect of inlet flow Re on the Φ-HF flammability limits of the partially premixed combustor holding $CNG/H_2/O_2/CO_2$ flames at: **a** OF = 32% and **b** OF = 36%

both flammability limits shift somewhat toward the leaner side with the increase of HF from 20 to 30%. When the OF grows by more than 30%, the advantages of hydrogen enrichment gradually diminish, as shown by a comparison of the stability limits in Figs. 5.27 and 5.30. Because of the faster flame propagation, the flame flash-back was achieved when the OF increased from 32 to 36% and the HF increased beyond 20%, preventing the stability limits at HF = 30% and OF = 36% from being reached.

5.3.6   *Effect of Oxidizer-Flexibility on Flame Macrostructure*

Figure 5.31 compares flame images from near flashback to near blowout at a fixed Re of 1481 to examine how the macrostructure of $CNG/H_2/O_2/CO_2$ flames is affected by the flexibility of the oxidizer (i.e., varying OF). To examine the impact of H_2

enrichment on flame structure, the flames with HF of 0.0 and 20% for each OF are examined. In the rich combustion region, at greater φ than one, the operation of the H_2-enriched flame was not possible, and the flame went out while the non-enriched flame was still burning. This can be explained by the fact that the flame flashback was reached early at lower φ due to the moved upper flame flammability limit to the leaner area with H_2 addition. When compared to pure CH_4 flames, the H_2-enriched flames are brighter, have a more compact inner cone, and are more intense. Compared to pure CH_4 flames, the addition of H_2 increased the operability of the flames towards the blowout limit to leaner circumstances. As the fuel's hydrogen level rises, the reaction mixture's chemical kinetics are accelerated, which explains this. Due to the little flames that lifted off from the perforated plate burner's outside holes as the value of decreased toward flame blowout, the size of the flame cone also decreased. When the OF was raised from 32 to 36% at the same φ, there were no discernible variations in the visual flame length; however, when the OF was raised from 29 to 36%, there are discernible variances in flame morphologies (see Fig. 5.32 and compare Figs. 5.28 and 5.31). In a number of recent studies on swirl-based combustors [92, 106–113] as well as a micromixer-based combustor [107], it has been noted that flames of similar and OF (i.e., same AFT) at the same inlet flow velocity result in very similar flame macrostructure. This indicates that similar flames can be obtained while fixing one of the flame characteristic parameters, which is AFT, and one of the flow. This may explain why the flame shapes are comparable at fixed when variations in OF are not appreciably different, such as when OF was increased from 32 to 36%. However, because of the large change in the AFT, increasing the OF from 29 to 36% led to major changes in flame form. Higher OF causes larger, more stable flames with better flame anchoring to the perforated plate and faster burning characteristics. Due to the greater flame temperature and radiation intensity inside the combustor, the overall flame length was prolonged at higher OFs with longer reddish outer cone. Figure 5.33 compares the visual total length of 20% H_2-enriched and non-enriched CNG flames as a function of for OFs of 29, 32, and 36% at Re = 1481. Because there was less oxidizer available for combustion, all flames increased in overall length while decreasing in φ. Due to the increased flame burning velocity caused by the injection of H_2, the flames within the inner cone grow shorter and more intense, which causes the flame core to move toward the base. By making more oxygen available, increasing the OF in the oxidizer mixture lengthened the reaction region downstream. Due to the greater diffusivity of H_2 in the reacting mixture, which prolongs the reactions downstream of the burner plate, the flame extension was further improved with H_2 enrichment in comparison to the non-enriched flame. This may explain why the flame length was increased when the OF was increased from 32 to 36% at a constant HF of 20% (see Fig. 5.33) because the downstream processes will be enhanced by the additional oxygen in the presence of H_2.

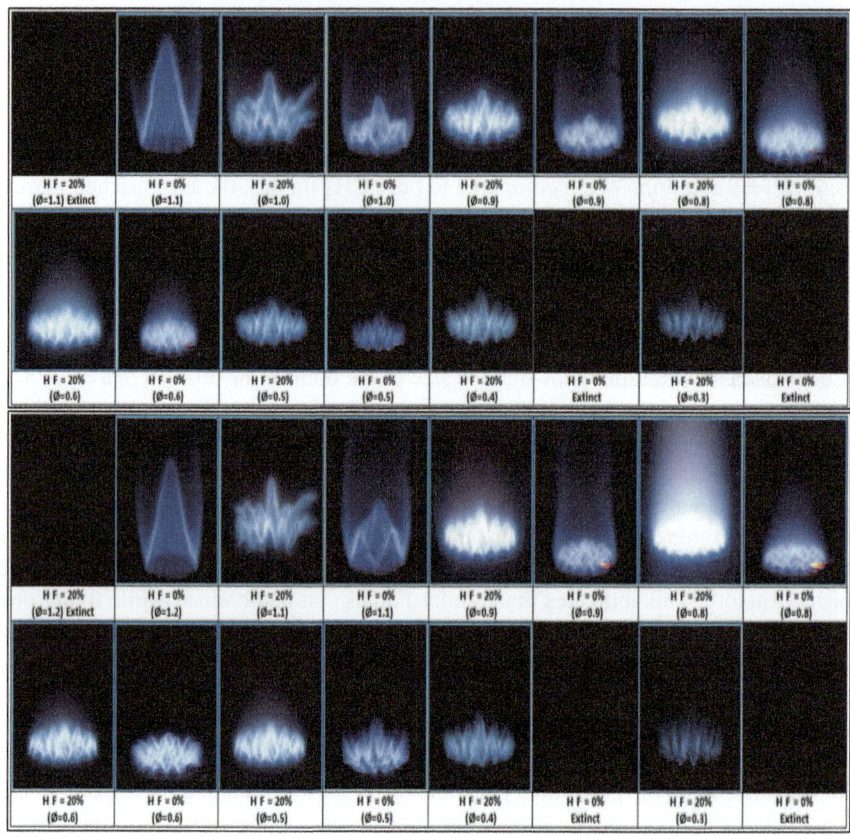

Fig. 5.31 Effect of equivalence ratio, from blowout to flashback at Re = 1481, on the inner cone structure of the CNG/H$_2$/O$_2$/CO$_2$ flames for HFs of 0.0 and 20% and at: (upper two rows of flames) OF = 32%, and (lower two rows of flames) OF = 36%

5.3.7   Near-Stoichiometry Operability of the Combustor

Oxy-fuel combustion systems are advised from an economic standpoint to burn fuel close to stoichiometric conditions. Given that a sizeable percentage of the power produced is utilized to separate oxygen, this is done to fully use the oxygen that is available for combustion and prevent it from being lost in the exhaust stream. In order to characterize the flammability limits and flame morphologies of CNG/H$_2$/O$_2$/CO$_2$ at $\varphi = 0.85$, a series of tests were carried out. Figure 5.34's stability map illustrates the combustor flammability limitations for a range of H$_2$-enrichment levels in terms of OF and oxidizer flow rate. The findings demonstrated that the perforated plate burner combustor could not operate for all flames that were close to stoichiometric conditions at OFs below 25% owing to flame blowout and over 46% due to flame flashback. H$_2$ addition can increase the operability of oxy-fuel combustion systems

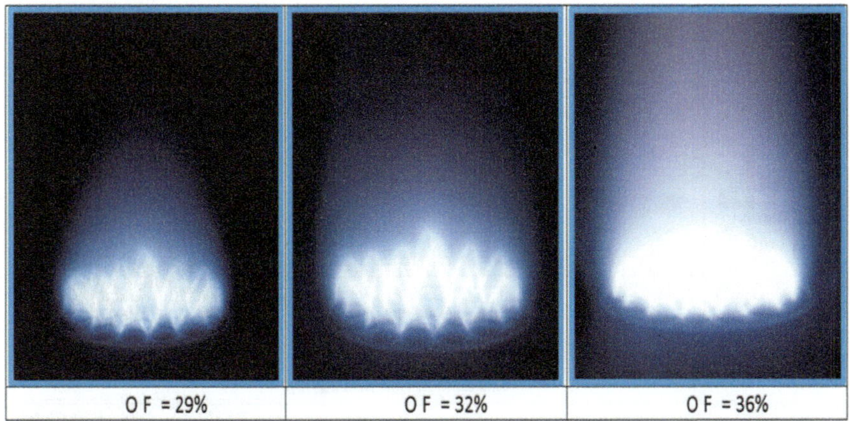

Fig. 5.32 Effect of oxygen fraction on visual CNG/H$_2$/O$_2$/CO$_2$ flame appearance at $\Phi = 0.8$, Re = 1481, and HF = 20%

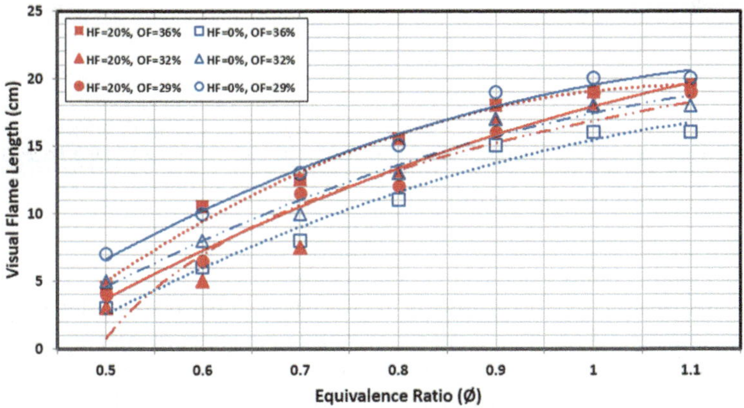

Fig. 5.33 Effect of equivalence ratio on the visual flame length of the CNG/H$_2$/O$_2$/CO$_2$ flames at Re = 1481 for HFs of 0.0 and 20% and at OFs of 29, 32, and 36%

to lower OFs due to its strong reactivity and diffusivity. When the HF was increased from 0.0% to 30% at Re of 1481, the combustor was able to decrease with the OF from 28 to 25%. Due to the large increase in flame speed brought on by the addition of H$_2$, however, flame flashback occurred at lower OFs. When HF was increased from 0.0 to 30%, the value of OF at which flashback was achieved at Re of 1481 decreased from 42 to 38%. In order to prevent flame flashback, it is advised to maintain OF below 36% for the CNG/H$_2$/O$_2$/CO$_2$ flames with HF of 30%. The findings support the notion that H$_2$ addition is recommended during part-load operation but not favored under full load conditions from an economic and combustor operability perspective.

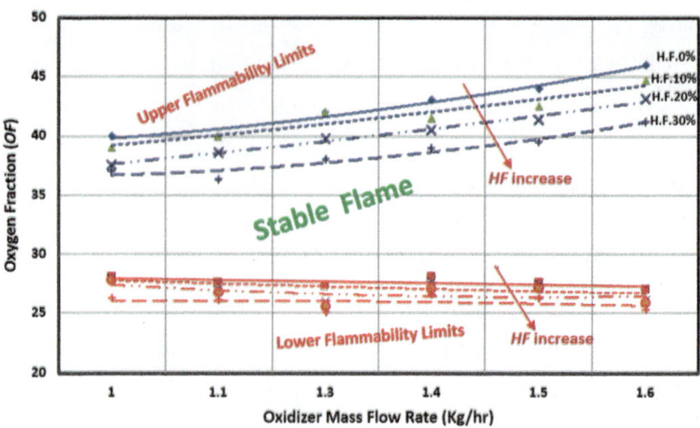

Fig. 5.34 Effect of H_2 enrichment on the near stoichiometry flammability limits, in terms of OF and oxidizer mass flow rate, of the partially premixed combustor holding $CNG/H_2/O_2/CO_2$ flames at $\Phi = 0.85$

Flashback and blowout limitations did not change noticeably with the addition of H_2 up to HF of 10%.

5.3.8   The Role Adiabatic Flame Temperature on Controlling Combustor Operability

The flammability limits obtained from the first two sets of trials were reorganized to be exhibited on the φ-HF static stability map on background contours of the matching AFT as shown in Fig. 5.35 in order to investigate the effect of AFT in influencing combustor operability limitations. To do that, a code known as the Engineering Equation Solver (EES) was created to calculate the contours of AFT at a given flow rate of 1481 and over the ranges of HF and that were taken into consideration. The graphs demonstrate that there is no relationship between the AFT contours and the line denoting the combustor blowout limit. In addition, the flow Re emerges as the key factor dictating the combustor blowout limit based on the stability maps shown in Figs. 5.27 and 5.23. This indicates that the flow characteristics (in terms of Re) rather than the flame characteristics have a greater impact on the combustor operability close to the blowout limit (in terms of AFT). On the other hand, regardless of the operational values of, HF, and OF, the line indicating the combustor flashback limit at Re of 1481 coincides, but not very well, with a line of constant φ, AFT = 2200 K. This indicates that the combustor AFT plays a part in regulating the combustor's operability close to the flashback limit in addition to the flow Re's primary function. In order to prevent flame flashback while the combustion system is operating at medium to full load conditions, the parameter AFT must be taken into account. This is a brand-new discovery for the current perforated plate burner combustor containing $CNG/H_2/O_2/$

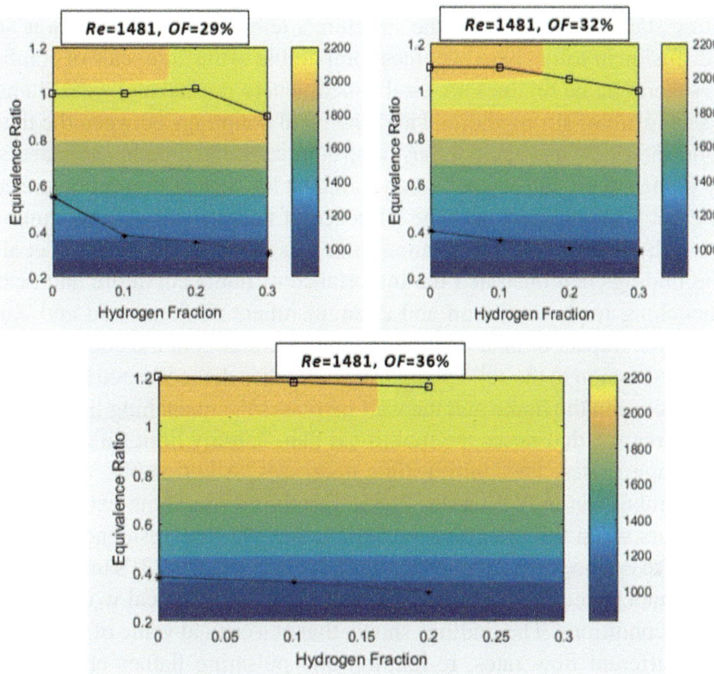

Fig. 5.35 Combustor Φ-HF static stability map plotted on the contours of adiabatic flame temperature, in K, for different oxygen fractions at fixed Re of 1481

CO_2 flames that have been partially premixed. However, recent investigations for swirl-based [92, 106, 108, 113] and micromixer-based combustors [107] at fixed inlet flow velocity, but without H_2 enrichment, revealed a similar outcome. This suggests that AFT, when having similar intake flow characteristics, i.e. at fixed inlet flow Re, plays a function in controlling the combustor flashback limit regardless of the flame type (partially or totally premixed flames).

5.4   Characteristics of Lean Premixed Oxy-combustion in Micro-structured Catalytic Honeycomb Burner

Recent advancements in nanotechnology significantly reduce the need for micro electromechanical devices (MEMS). Compact, long-lasting, and quickly recharged power supply became essential. Utilizing fuels with high power densities will thereby increase the operational lifetime of these micro-power generators [114, 115]. As a result, the development of MEMS shows prospective interest in micro-scale combustion. As the combustor size is decreased, the surface to volume ratio rises, the flame temperature starts to couple with the structure's temperature, and the time scale of thermal diffusion in solid phase becomes comparable to the time scale of combustion.

Thus, several flame regimes are established, where normal and weak flames may coexist, as a result of strong thermal and chemical couplings between the flame, the flame-solid interface, and the combustor structure. Isolated flame cells and steady/unsteady flame streets can be seen when wall heat loss and auto-ignition are coupled at discrete hot places [116–118]. The influence of flow velocity, Lewis number, and channel width on the flame propagation limit was investigated by Daou et al. [119, 120]. The findings demonstrated the importance of flame curvature and near wall-flame quenching in the initiation and extinguishment of flames. Ju and Xu [121] investigated the impact of heat loss, taking into account both the coupling of flames and heat transmission in the solid phase. The outcomes demonstrated that the thermal interaction between the flame and the wall increases the quenching limit and creates a new flame regime that raises the traditional flammability limit. In actuality, as the wall temperature rises, the igniting time decreases. Additionally, weak flames or flameless combustion may be seen. Flame instability via flame extinction and re-ignition occurs when the ignition time approaches the flow residence time and the combustor size is close to the quenching limit. Jackson et al. [122]'s investigation into the physics underlying the flame instabilities involved numerical work at fixed wall temperature conditions. The findings shown that at a critical value of the Damköhler number at different flow rates, re-ignition and pulsating flames characterized by severe oscillations occur.

5.4.1   Characteristics of Micro-scale Combustion

Micro-combustors have different benefits over traditional thrusters. The formation of sustainable combustion and the ignition of the propellants become more difficult as the combustor's size decreases. The increased wall heat and radical losses are to blame for this. On the other hand, when the temperature of the flame rises, the rise in material strength at high temperatures and pressure becomes another challenging issue. To address these issues, a variety of micro-thruster types with various length scales, materials, propellants, and combustor ignition techniques have recently been thought about and researched. Numerous concerns become more significant as combustion volumes shrink. These include gas-phase reaction wall quenching, fluid mixing, thermal control, and residence duration. The typical lengths of the microdevices are brief. Small Peclet and Reynolds numbers are the outcome of this. Because of this, the flow is primarily laminar, and as a result, the importance of viscous effects and diffusive transport of mass and heat increases. Low Reynolds numbers also make reactant mixing a major issue in micro systems. Since turbulence mixing is ineffective, species mixing mostly occurs by diffusion. For a chemical reaction to start, nearby laminar streams must quickly and thoroughly mix with one another. The surface-to-volume ratio rises when the combustor scale is reduced, which causes a large heat loss to the chamber wall. The temperature gradient within the solid wall also lessens as a result of the lower Biot number. The flow residence time must be longer than the time needed for the chemical processes in order to have a full combustion. For thorough mixing in non-premixed combustion, more space and time are required. The flame quenches if the total power produced inside the combustor is less than the loss to the wall.

As the combustor size lowers, the wall temperature rises. The significant thermal connection between the flame and wall is to blame for this. The square of the combustor length scale is also related to the diffusion time scale. As a result, as the combustor length scale is decreased, the diffusion time of chemical radicals from the flame region to the wall will equal the distinctive production time of radicals. As a result, the surface of the wall experiences a strong kinetic radical quenching. Models of micro-scale combustion will no longer be relevant unless radical recombination in the wall, particularly at high wall temperatures, is taken into account since radical quenching will be paired with thermal quenching. Micro-scale combustion can be a fresh platform for new thinking study and development on efficient energy conversion arrangements, together with the current worries of climate change and energy security. This is applicable to vehicles that need tiny, effective, and low burn propulsion engines as well as power generation applications. Thus, further research on scale-down engines like micro-gas turbines, reciprocating engines with spark and compression ignition, Wankel engines, and free-piston engines is still encouraged. The innovative concept basics, such as specialized instabilities/dynamics, weak flame in micro-scale systems, interaction between homogeneous and heterogeneous reactions, and non-equilibrium effects, are on the horizon. Wall/flame influences as well as thermal/chemical interactions are crucial. The small volume/surface ratio is to blamed for this. There may be new discoveries and fundamentals as a result of research into other scaling issues in fluid dynamics, thermodynamics, heat transfer, and chemical combustion/reaction.

Hua et al. [123] and Norton and Vlachos [124] have researched the effects of combustor size and operating parameters, particularly the heat transfer conditions at chamber wall (e.g., adiabatic wall, with heat loss and heat conduction within the wall). The findings suggested that stable combustion may be sustained by maintaining suitable thermal conditions at particular dimensions and increasing the ratio of flow residence time in chamber to chemical reaction time. For premixed CH_4-air mixtures, Li and Choua [125] investigated the impact of combustor size and shape, input velocity profile, and slip-wall boundary condition on the flame temperature. In the same velocity range and for the same hydrodynamic diameter, the results favor the planar combustor over the cylindrical one. If combustion could be maintained in tight spaces with gaps as tiny as 0.3 mm, Karagiannidis et al. Additionally, it was shown that the slip-wall boundary's influence on these micro combustors is minimal in comparison to the effects of bulk velocity and gas temperature. In order to examine the flammability thresholds, quenching diameters, and temperature distributions of three hydrocarbon fuels, methane, propane, and hydrogen. Tang et al. [126] investigated experimentally the combustion characteristics of the fuels. In comparison to propane and methane/air flames, the results demonstrated wider stability limitations for hydrogen/air flames.

In a micro-planar combustor, Tang et al. [127] investigated the impact of hydrogen addition on the combustion properties of premixed methane/air. The outcomes demonstrated that, depending on the mass flow rate, the flame generated by employing pure methane is pushed away from the entrance. In addition, the flame was gradually moved toward the inlet part by the addition of hydrogen. The impact of hydrogen feed splitting on combustion stability and exit gas temperature was investigated by Shabanian and Reza [128]. To learn how to reduce the distortion of hot entrance sites, hydrogen is injected in small amounts. As a result of the increased number of input dividing sections, which decreased heat losses and raised output temperature, the findings indicated uniform temperature distribution. Experimental recordings of the transient combustion characteristics in a cylindrical micro-combustor were made by Li and Choua [129]. The findings give stability maps for various combustor lengths defined by equivalence ratio and flow velocity using recorded acoustic emissions. The outcomes demonstrated that the flame stabilizing properties of the backward-facing step combustor design.

Utilizing catalytic combustion to prevent combustion instabilities under these micro-scale settings is the current prevalent approach [130, 131]. However, a number of limitations, such as durability and maximum operating temperature, restrict the use of catalytic combustion in the production of electricity. Additionally, catalytic combustion is a water-shed remedy to deal with flame instabilities brought on by the heat losses produced in such micro burners. The flammability limitations can be expanded and safety privileges granted because surface catalytic processes can be sustained at low temperatures and significantly lower fuel–air ratios. Through the use of temperature and concentration slips, Xu and Ju [132, 133] and Aghalayam et al. [134] investigated the non-equilibrium effects on the wall and radical quenching in a micro-scale combustor. The findings demonstrated how the slip effect affects catalytic reactions by dominating the effects of both reaction rate and diffusion transport, which causes the spread of a nonlinear reaction rate distribution at higher Damköhler numbers. The stability limitations of heterogeneous/homogeneous combustion for methane-fueled channel-flow catalytic platinum-coated micro reactors were investigated experimentally and numerically by Karagiannidis et al. [130]. A fully elliptic two-dimensional (2-D) model was used to conduct the numerical investigation. The gas-phase chemistry expanded the low-velocity stability limits due to the established intense flames, which was a significant discovery. Due to the heat release, which was predominantly caused by the incomplete oxidation of methane, the high-velocity blowout limitations were also increased.

Low activation energy from the catalytic burning of hydrogen fuel over the catalysts results in extremely quick reactions. Under fuel-leaner conditions, hydrogen-air mixtures were discovered to self-ignite across platinum wires [135]. The mixes with more fuel, however, ignite at a temperature higher than room temperature [136]. In order to enable micro-scale combustion without an ignition source, hydrogen is utilized as an exciting helper, as it encourages the creation of radicals that attack both CH_4 and H_2. The impact of hydrogen addition on the propane-air mixtures' ignition properties over platinum-coated micro-channels was examined quantitatively by Chen and Xu [137]. The simulation findings demonstrated hydrogen-aided auto ignition of propane-air mixtures at warmed feed temperature. The outcomes demonstrated that the equivalence ratio and the thermal conductivity of the wall both affect where the localized ignition occurs. With consideration for the coupling phenomenon between heat and mass transport in these micro burners, it was discovered that the minimum hydrogen content for self-ignition ranged from 0.8 to 2.8% depending on the wall conductivity and the inlet velocity. By adjusting the gas flow rate, reactant concentrations, and composition, Barbato et al. [138] investigated experimentally the auto-ignition of methane and methane-hydrogen mixtures over a Pt-LaMnO$_3$ catalytic honeycomb. The findings showed that fuel hydrogen enrichment can greatly lower the amount of heat supply required to start combustion. Zimont [139] used the zero-dimensional (0-D) methodology created by Semenov et al. [140] and Zel'dovich et al. [141] to analyze the auto ignition characteristics over the same reactor as Barbato et al. [134]. In addition to providing sensitivity analyses for the adjustment of each parameter and its impact on the ignition characteristics, this study provides operation limits that were not examined in the tests. The findings demonstrated that whereas the self-ignition temperatures for CH4 and CH4/H$_2$ fuels are very different, the quenching temperatures are identical. In their study of methane transient catalytic combustion, Cimino and Di Benedetto [142] demonstrated that the flame front moves from the reactor's outlet to its inlet during startup and shutdown conditions, converting the methane-air mixture to CO$_2$ and H$_2$O due to a pre-heating temperature that is specified and unaffected by the flow rate or reactor length. In a micro-planar combustor with and without a cross plate, Tang et al. [143] studied the combustion characteristics of premixed propane/air experimentally and numerically. The flame shape and temperature distribution of the external wall are significantly influenced by the length of the cross-plate. When compared to a straight-channel combustor, it improves the heat transfer between the mixture and the inner wall of the combustor with an external wall temperature rise of at least 90 K. By utilizing a dimensionless plate length of 5/9 for the new combustor, the flame stability is improved; at an equivalence ratio of 0.7, the blowout limit can be raised by 0.4 m/s.

In the current study, computational fluid dynamics (CFD) analysis is done to confirm the auto ignition behavior of CH$_4$-Air mixtures across a 3-D micro catalytic honeycomb reactor with and without hydrogen addition (Unlike the 0-D approach used by Zimont [139]). The current study's objective is to show how the flow rate, equivalence ratio, and reactant composition affect the minimum ignition temperature control during steady-state circumstances to preserve the reactor's stable functioning. Analysis of the role of H$_2$ addition in relation to total flow rate. To identify the positions of the flame during the startup conditions, transient simulations are also carried out, taking into account the 3-D description of the gas and solid phases, heat conduction through the solid wall, and gas and surface reactions in the reactor.

5.4.2 *Features of Catalytic Honeycomb Reactor*

Figure 5.36a depicts a schematic representation of the catalytic honeycomb reactor's mixed input, preheating, and exhaust sections. To prevent heat loss, the honeycomb monolith is covered in an alumina blanket. The bypass of the reacting flow is then inserted in a quartz tube that is placed within a tubular electric furnace. The lab-scale reactor constructed by Barbato et al. [138] has the exact same planned domain. Since the surface temperature is thought to be the lowest ignition temperature, the electric furnace allows for a step control on that temperature. In addition to preheating the introduced mixture, this temperature also supplies the heat required to continue combustion while overcoming significant heat losses brought on by the high surface to volume ratio of such a micro reactor. According to Fig. 5.36b, the micro-reactor is modeled as a honeycomb sliced into a disk with dimensions of 17 mm in diameter and 12 mm in length. The honeycomb has a solid wall thickness of 0.051 mm and 900 cell per square inch. When these values are substituted in Eqs. 5.14 and 5.15, 317 cells with 0.85 mm cell diameter are produced, and the volume of the catalyst in the honeycomb is roughly 27.2% of the total volume [144].

$$D_{cell} = 25.4 \left((cpsi)^{-\frac{1}{2}} - \frac{t_w}{1000} \right) \text{mm} \tag{5.14}$$

$$A_{flow} = \frac{cpsi}{(2.54)^2} \frac{\pi}{4} D_r^2 D_{cell}^2 \ \text{mm}^2 \tag{5.15}$$

Platinum catalyst with surface site density of 2.7×10^{-8} kg mol/m^2 and 1.1 g weight is coated on the inner surfaces of the reactor cells.

In order to relate their properties and calculate the heat loss during combustion, three engineering materials are defined for the quartz tube, alumina blanket, and honeycomb monolith zones. Table 5.6 lists the engineering characteristics of various materials.

The localized self-ignition temperature is expected to be between 40 and 160 slph for two different compositions of gas mixtures that are injected into the reactor.

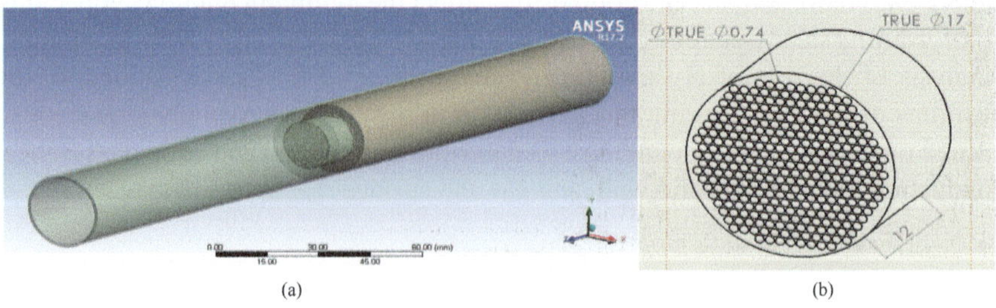

(a)   (b)

Fig. 5.36 Representation of the catalytic honeycomb reactor: **a** full 3-D representation and **b** honeycomb cut

Table 5.6 Characteristic properties of zone materials

Material	$k \left(\frac{W}{m.k}\right)$	$\rho \left(\frac{kg}{m3}\right)$	$C_p \left(\frac{J}{kg.k}\right)$	Zone
Quartz glass	0.71	2500	1045	Quartz tube
Alumina	0.125	64	871	Alumina blanket
Cordierite	2	2600	1464	Honeycomb Monolith

Table 5.7 Characteristics of the considered fuel mixtures in the present study

	Fuel	
	CH_4	CH_4–H_2
	Mix1	*Mix2*
H_2 (%)	–	2.1
CH_4 (%)	2.8	2.2
O_2 (%)	10	10
N_2 (%)	87.2	85.7
Equivalence ratio (Φ)	0.56	0.54
Input heat flux, kJ N l^{-1} (W/m²)	0.9	0.9
Thermal power, W	10–35	10–35
Volume flow rate range 11–44.5 cm³/s		

Despite having differing compositions, Mix1 and Mix2 get the same amount of heat input, as shown in Table 5.7.

5.4.3 Combustion Modeling in Catalytic Honeycomb Reactor

The fundamental governing equations for catalytic combustion read without taking into account radiation and body force terms, in particular:

$$\rho u \frac{dA_c}{dx} + \rho A_c \frac{du}{dx} + u A_c \frac{d\rho}{dx} = \sum_{m=1}^{M} a_{i,m} \sum_{k=1}^{K_g} \dot{S}_{k,m} W_k \qquad (5.16)$$

The gas, which consists of kg species, has an axial velocity of u and a (mass) density of. W k is the species k's molecular weight, and (S k) is the rate at which this species is produced molar by all surface processes. The cross sectional (flow) area and the effective internal surface area per unit length of material, respectively, in the reactor are represented by the numbers Ac and a (i, m). Ac and a (i, m) are both flexible functions of x.

The balance between pressure forces, inertia, viscous drag, and momentum added to the flow by surface reactions is expressed in the momentum equation for the gas.

Thus:

$$A_c \frac{dP}{dx} + \rho u A_c \frac{du}{dx} + \frac{dF}{dx} + u \sum_{m=1}^{M} a_{i,m} \sum_{k=1}^{K_g} \dot{S}_{k,m} W_k = 0 \tag{5.17}$$

The balance between pressure forces, inertia, viscous drag, and momentum added to the flow by surface reactions is expressed in the momentum equation for the gas. Thus:

$$\rho u A_c \left(\sum_{k=1}^{k_g} h_k \frac{d Y_k}{dx} + \overline{c_p} \frac{dT}{dx} + u \frac{du}{dx} \right) + \left(\sum_{k=1}^{k_g} h_k Y_k + \frac{1}{2} u^2 \right) \sum_{m=1}^{M} a_{i,m} \sum_{k=1}^{K_g} \dot{S}_{k,m} W_k$$

$$= a_e Q_e - \sum_{m=1}^{M} a_{i,m} \sum_{k=1}^{K_g} \dot{S}_{k,m} W_k h_k \tag{5.18}$$

$$\dot{S}_k = 0 \quad k = k_s^f, \ldots k_s^l \tag{5.19}$$

where $\dot{S}_{k,m}$ is the rate of heterogeneous generation, which typically varies on both the surface and gas compositions. The steady-state conservation equations are easily expressed in equation under the assumption that these speciee immobile (5.19). Every species in every surface phase (n) present on every surface material is subjected to the surface species conservation equation (m).

ANSYS CHEMKIN-PRO17.2 is used in the current work to model the combustion of a surface honeycomb monolith reactor. The combustion process was predicted using Deutschmann et al. [145]'s basic heterogeneous kinetics mechanism. The model takes into account 17 gas-phase species (including CH_4, CH_3, CH_2, CH, CH_2O, HCO, CO_2, CO, H_2, H, O_2, O, OH, HO_2, H_2O_2, N_2) and 11 site species (Pt(S), H(S), H_2O(S), OH(S), CO(S), CO_2(S), CH_3(S), CH_2(S), CH(S), C(S), and O(S)) with 58 volumetric reactions. Through importing reaction mechanisms from CHEMKIN for CH_4 combustion over Pt catalyst, including backward reaction, third body efficiency, and coverage-dependent reaction, both volumetric and surface elementary reaction rates are assessed.

The inlet border of the reactor is known as the mass flow-inlet, and its mass is computed for various fuel flow rates at ambient temperature. At the reactor's exit portion, a pressure-outlet boundary is used. Equations (5.20 and 5.21) regulate coupled heat transport at all interfaces between the solid and fluid phases under the no-slip condition [146]:

$$-\lambda_s \frac{\partial T_s}{\partial r} + \sum_{k=1}^{K_j} M_k \dot{S}_k h_k = -\lambda_g \frac{\partial T_g}{\partial r} \tag{5.20}$$

$$\dot{S}_k M_k + \rho D_k \frac{\partial Y_k}{\partial r} = 0 (k = 1, 2, \ldots, k_i) \tag{5.21}$$

In this equation, (S_k) is defined as the heterogeneous generation or depletion rate of the gaseous or surface species k. (5.22):

$$\dot{S}_k = \sum_{m=1}^{N_j} k_{f,m} v_{k,m} m \prod_{S=1}^{K_i+k_j} [X_s] v'_{s,m} (k = 1, 2, \ldots, K_i + K_j) \tag{5.22}$$

where X_s is the molar concentration of species S, $K_{f,m}$ is the Arrhenius rate of the mth heterogeneous reaction, and $v_{k,m}$ and $v'_{s,m}$ are the stoichiometric coefficients:

$$k_{f,m} = A_m T^{\beta m} \exp\left(\frac{-E}{RT_s}\right) \prod_{S=1}^{k_j} \Theta_s^{\mu_{s,m}} \exp\left(\frac{\varepsilon_{s,m} \Theta_s}{RT_s}\right) \tag{5.23}$$

where $\mu_{s,m}$ and $\varepsilon_{s,m}$ are the coverage dependence parameters. However, the rate coefficient, K_{fm}, is calculated for the adsorption reaction by the sticking coefficient, S_0:

$$k_{f,m} = \frac{S_{0,m}}{\Gamma^\tau} \sqrt{\frac{RT_s}{2\pi M_m}} \tag{5.24}$$

The molar concentration of surface species k, X_s, is defined by:

$$X_s = \Gamma \Theta_s \tag{5.25}$$

where Γ is the surface site density of Pt catalyst with the value of 2.7×10^{-9} mol/cm^2, and Θ_s is the coverage fraction calculated by:

$$\frac{d\Theta_s}{dt} = \frac{S_s}{\Gamma} \tag{5.26}$$

The CHEMKIN program, which is integrated into the honeycomb monolith reactor, uses the aforementioned geometric characteristics to create a complete description of the catalytic reactor. The 250 mm long and 25.4 mm wide domain consists of a honeycomb monolith, an alumina blanket, and a quartz tube. According to Fig. 5.37, this domain is discretized into around one million cells. The CHEMKIN program, which is integrated into the honeycomb monolith reactor, uses the aforementioned geometric characteristics to create a complete description of the catalytic reactor. The 250 mm long and 25.4 mm wide domain consists of a honeycomb monolith, an alumina blanket, and a quartz tube. According to Fig. 5.37, this domain is discretized into around one million cells.

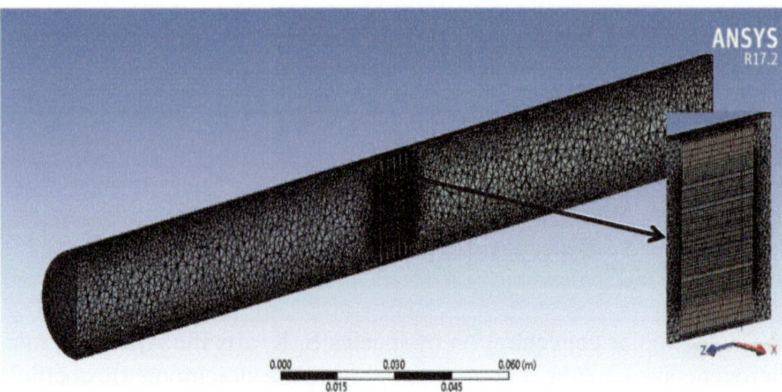

Fig. 5.37 Representative mesh of the developed model

The quartz tube's surface has been tuned to have a constant wall temperature that is the same as the temperature required for localized self-ignition. On the honeycomb monolith's internal surfaces, surface reactions are permitted. Laminar models are used to simulate the flow in these honeycomb shapes under steady-state circumstances. For these micro burners, the turbulence-chemistry interaction is simulated using a finite-rate model. To accurately model the entire chemical process, a sophisticated mechanism with 58 volumetric reactions and a 23-step heterogeneous mechanism is chosen. By importing CHEMKIN files into FLUENT and turning on wall surface reactions, these reactions are defined. To simulate the experimental technique, a coupled pressure–velocity solver and pseudo transient mode were activated. Only the honeycomb reactor zone is discretized to 3 million cells for transient simulations, with the mesh aspect ratio conveniently chosen to capture the localized ignition position. Symmetry plane is used for the full-size model to cut the solution time in half and hence, the calculation time. The semi-implicit technique for pressure linked equations (SIMPLE) algorithm with double precision solver is used to model the coupling between pressure and velocity.

Results are shown under pseudo-auto-thermal circumstances, where some heat is supplied from outside the reactor and some heat should be generated throughout the reaction to keep the combustion going. The system is kept at a controlled temperature using an electric furnace since the heat losses are too great to allow thermal auto sustainability with a reaction heat source alone. Since it may be assumed that the temperature of the incoming gas to the reactor after passing through the preheating section, the controlled temperature of the complete external system is known as the inlet mixture temperature. To establish the minimum ignition temperature (MIT) for Mix1 and Mix2, a parametric analysis of the input temperature is conducted. To evaluate the model at the reference case study, the present simulation uses similar experimental settings to Barbato et al. (using Mix1 composition at equivalence ratio of 0.56). The influence of the equivalence ratio, flow rate, and addition of H_2 to CH_4 on the ignition properties of the micro-structured catalytic combustor is investigated through a number of parametric simulations.

5.4.4   Auto-Ignition Characteristics

For both Mix1 and Mix2, Fig. 5.38 shows the fluctuation of adiabatic flame temperature with minimum ignition temperature. At the same MIT, Mix2 had a little higher adiabatic flame temperature than Mix1, while having the same heating input. This could be explained by Mix2 containing a sizable amount of hydrogen. The projected adiabatic flame temperature and Boehman's measurements of temperatures showed excellent agreement [147].

The inlet temperature was step-controlled until auto-ignition occurred and the flame spread in order to calculate the MIT for this micro-catalytic reactor under these operating conditions. The furnace keeps working to ensure sustainable combustion while overcoming the considerable heat losses brought on by the high surface-to-volume ratio. Figure 5.39 depicts the change in maximum flame temperature along the reactor for various inlet mixture temperatures of Mix1 and Mix2 at a volumetric flow rate of 22 cm^3/s (80 slph). The plot demonstrates that combustion begins at MIT of 500 °C for Mix1 and 472 °C for Mix2, respectively. Above these values, there are no appreciable changes in flame temperature. Additionally, it seems that mixing H$_2$ with CH$_4$ advances the kinetics mechanism and raises the auto-ignition temperature to about 30 °C. Due to the large surface-to-volume ratio of the reactor, there is only a little difference between MFT and AFT in the ignited zone. Other factors, which have a negligible effect, are therefore neglected.

Figure 5.40 shows the axial distribution of the flame temperature along the burner and compares it to experimental data from [138, 146] recorded at three different points inside the reactor (3 mm from the inlet, at the middle and 3 mm before the outlet). The findings demonstrate that the current numerical model is capable of producing

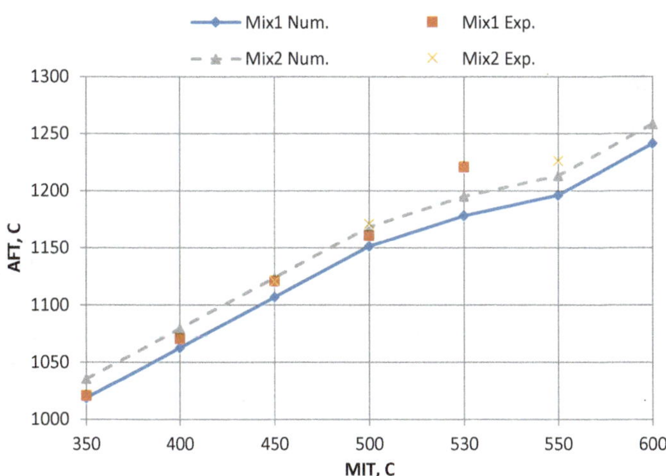

Fig. 5.38 Variation of adiabatic flame temperature versus pre-determined MIT for both *Mix1* and *Mix2*

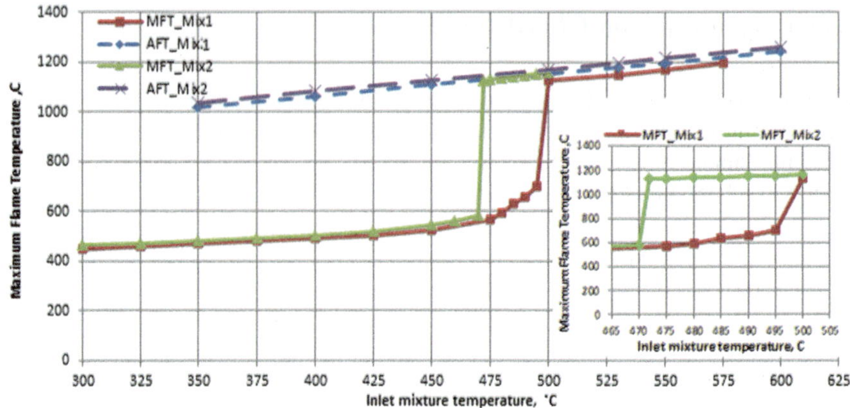

Fig. 5.39 Variation of maximum flame temperature with different inlet mixture temperatures for *Mix1 and Mix2*

an accurate prediction of flame temperature distribution. The differences between the experimental data and the numerical results, however, can be attributed to the model's neglect for the radiation impact or to mistakes in temperature measurement brought on by the thermocouple's placement inside the reactor. The catalytic honeycomb has thermocouples with 0.5 mm tips that are arranged in close proximity to the reactor's solid wall. Because of the substantial reduction in gas flow caused by the thermocouples, even though the temperature measurement in the honeycomb catalyst is a weighted average of the gas temperature, it is thought to be more representative of the surface temperature.

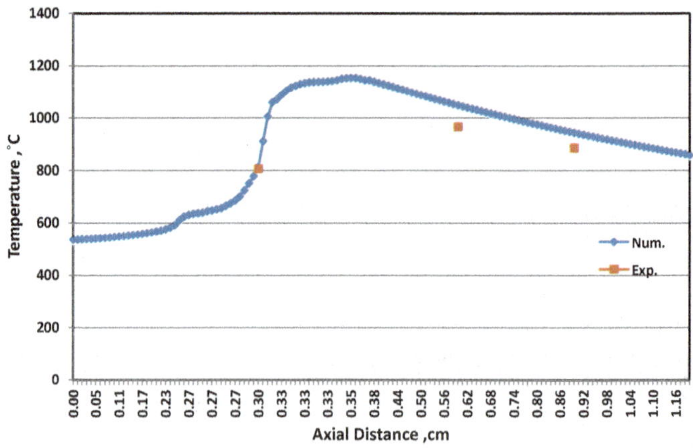

Fig. 5.40 Variation of the flame temperature along the micro reactor in comparison with the experimental data by [138, 146]

The statistics from MIT, which were acquired for comparable catalytic micro reactors [142, 147], are in good agreement with the current numerical findings for the identical gas mixture. Figures 5.39 and 5.40 show that ignoring radiation and other operational issues, such as the method for inlet temperature monitoring, had a small negative impact on the present work's agreement with the experimental results. Furthermore, the measured findings of Karagiannidis et al. [130] at pressure of 1 bar are convenient with the current numerical results of MIT. However, due to the reactivity of both the catalytic and gaseous pathways at these high pressures, a significant divergence becomes visible (up to 16 bar).

It was specified that the outer wall's weighted-average total heat transfer coefficient would be 17 $W/m^2/k$. This figure differs from the theoretically determined value of 48.6 $W/m^2/k$ by Zimont [139], which is a localized value. Figures 5.41 and 5.43 show the heat generated during the combustion process, which is analyzed in order to anticipate the auto-ignition characteristics. The surface heat generation rate and the volumetric heat production rate are the two primary divisions of the heat production rate. The investigation shows that the catalytic surface heat production rate plays a significant role in boosting activation energy and starting the combustion process. The surface heat production rate profile, shown in Fig. 5.41, features two spikes that advance the single spike of the volumetric heat production rate profile. This approach increases the activation energy of intermediate species and quickens the kinetics mechanism's reaction rates. The distributions of species concentrations along the reactor, as depicted in Fig. 5.42, can be used to determine the locations of both surface and volumetric heat production rates. The position of its peak and the fluctuation in H_2 mole fraction found indicate that the catalyst's surface reactions have a significant impact. The creation of OH and the location of its peak, however, are fundamentally where volumetric reactions appear to have an impact.

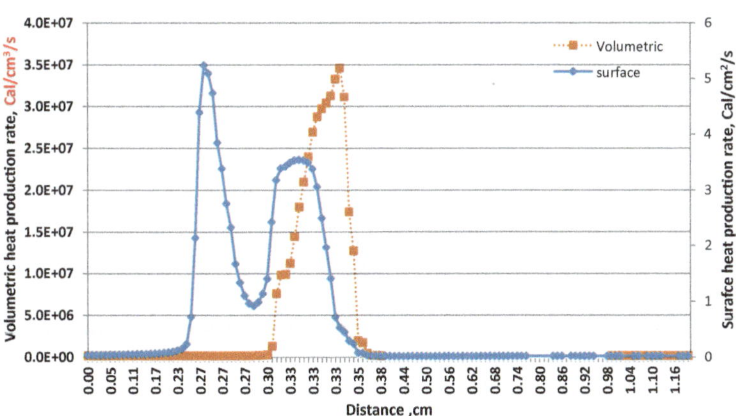

Fig. 5.41 Distributions of the axial surface and volumetric heat production rates (Cal/cm^2/s) along the micro reactor

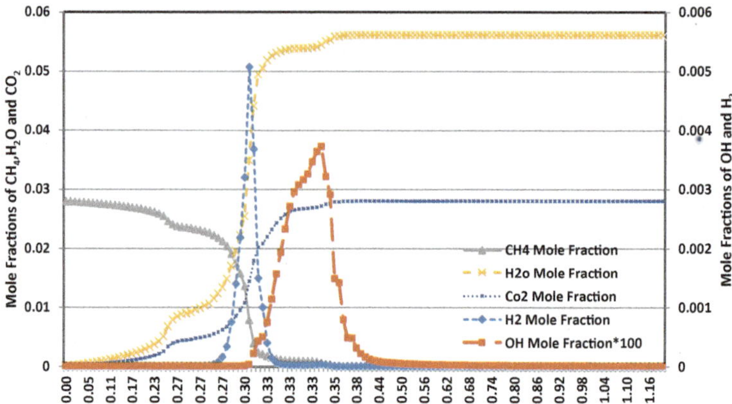

Fig. 5.42 Axial distributions of species mole fractions along the micro reactor

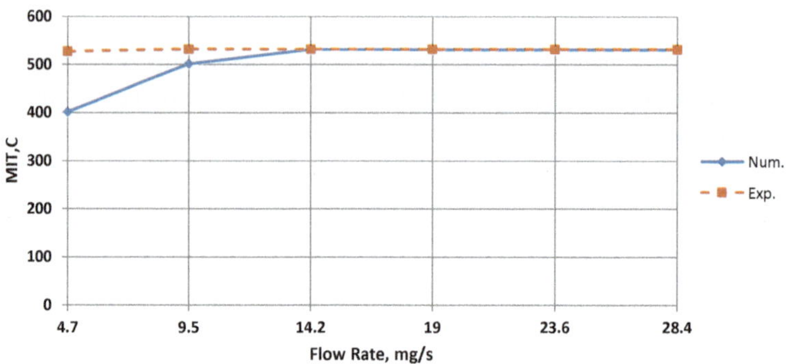

Fig. 5.43 Effect of flow rate of *Mix1* on the *MIT* and comparison with the experimental data by Chen and Xu [138]

5.4.5   Effect of Flow Rate on Auto-Ignition Characteristics

Mix1's MIT is examined in relation to the flow rate through the micro reactor, and the findings are compared to the experimental data [138] shown in Fig. 5.43. At greater flow rates (from 13 mg/s to 28.4 mg/s), it is demonstrated that the numerical and experimental results of MIT totally agree. At lower flow rates (4.7 mg/s), there is a sizable difference between the numerical and experimental results of MIT, and this difference gets smaller the higher the flow rates go. Figure 5.44 displays the maximum flame temperature for Mix1 at various flow rates. It is demonstrated that the influence of altering flow rate on minimum ignition temperature substantially depends on the balance between volumetric heat generation and volumetric heat losses for lower flow rates (4.7 mg/s and 9.5 mg/s). Conversely, when the minimum ignition temperature is held constant, this impact disappears for larger flow rates (23.6 mg/s and 28.4 mg/s).

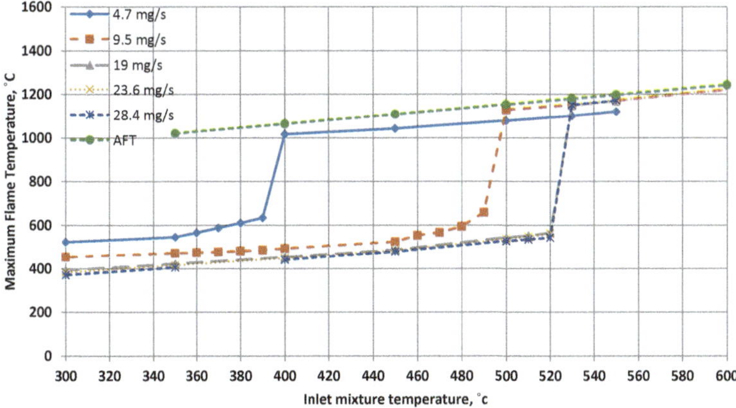

Fig. 5.44 Variation of the maximum flame temperature versus inlet mixture temperature at different flow rates for *Mix1*

Figure 5.45 illustrates the flammable range of equivalence ratios required for methane combustion in the micro-catalytic combustor at 9.5 g/s flow rate. It is demonstrated that the limits of flammability under lean mixture conditions vary from 0.45 to 0.75 and occur often. The results indicate experimentally novel operational restrictions for methane combustion over micro catalytic reactors that had not previously been documented. The experimental results by Dogwiler et al. [146] show excellent consistency with this range of equivalence ratios and guarantee catalytically stabilized combustion (CST) of a lean methane-air mixture at an equivalence ratio of 0.4. The results match those obtained by Reinke et al. [148] in terms of numbers.

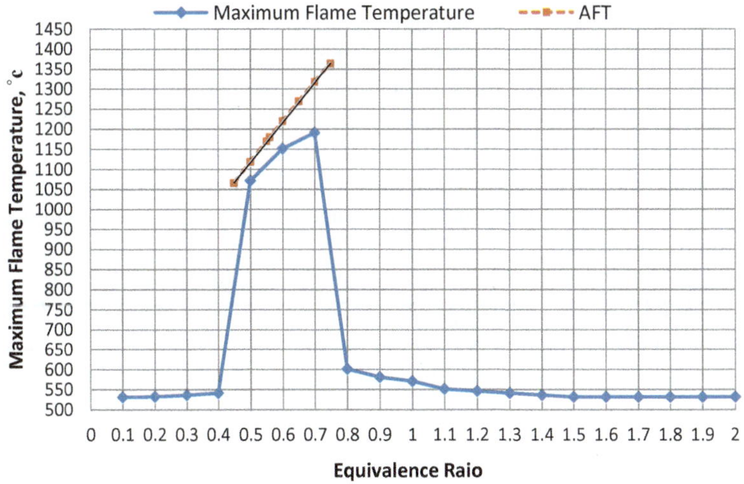

Fig. 5.45 Variation of the maximum flame temperature versus equivalence ratio

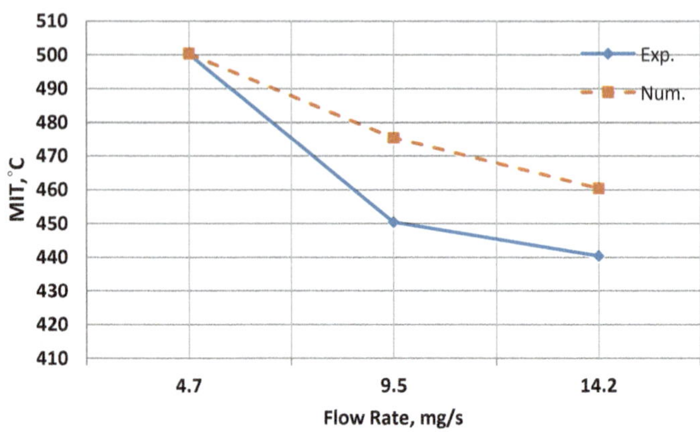

Fig. 5.46 Variation of numerical and experimental results [138] of *MIT* versus flow rate for *Mix2*

The change of the MIT's numerical and experimental results [138] for flow rate for H_2–CH_4 fuel composition is shown in Fig. 5.46. (Mix2). The fuel reactivity is increased and the mechanism of CH_4 ignition is promoted, as seen in Fig. 5.46, when part of the CH_4 is replaced with H_2. Each case's variation in the heat transfer coefficient is not taken into account. Higher flow rates cause a greater difference between the calculated and experimental data because the constant heat transfer coefficient assumption is false.

The steady state temperature contours along the electric furnace lit at MIT and run by Barbato et al. are shown in Fig. 5.47 [138]. To gain heat till achieving the MIT, which is nearly equivalent to the controlled electric surface temperature, Mix1 is introduced to the preheating section at a temperature of 27 °C. In Fig. 5.47, the temperature contours inside the outer radial micro-tubes and along the tubes of the micro-burners are magnified. The temperature on the outer radial tubes is lower than that on the inner tubes despite having the same Reynolds number (Re) due to high heat losses at the outer surface. The link between combustion and heat losses inside these micro-reactors is solidified by these temperature contours. The velocity contours along the reactor domain are depicted in Fig. 5.48. Despite the fact that the velocity inside the micro-tubes is discovered to be around 72 cm/s, a low value of Re is obtained because of the honeycomb structure, where the flow is assumed to be in the laminar domain.

Figure 5.49a shows the methane consumption together with the location of combustion. CO_2 is one of the byproducts of combustion, as seen in Fig. 5.49b. The reactor exit CO concentration is found to be less than 3 ppm, which is in excellent agreement with the measurements made using experimental data from [138].

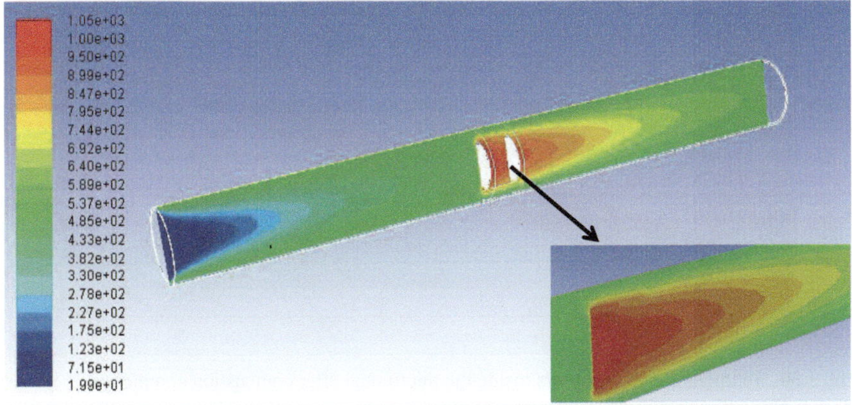

Fig. 5.47 Temperature contours inside the electric furnace domain

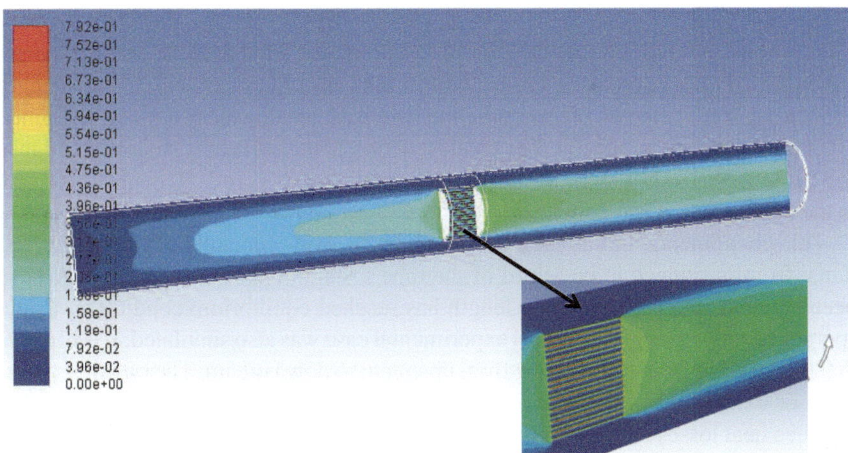

Fig. 5.48 Velocity contours along the electric furnace domain

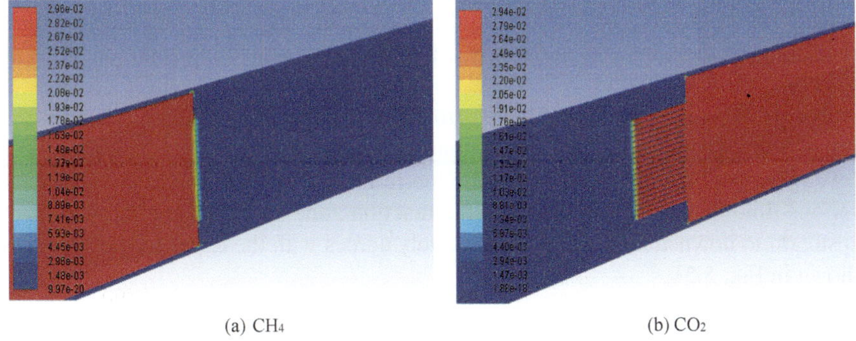

(a) CH₄   (b) CO₂

Fig. 5.49 Contours of CH_4 and CO_2 mole fractions along the electric furnace domain

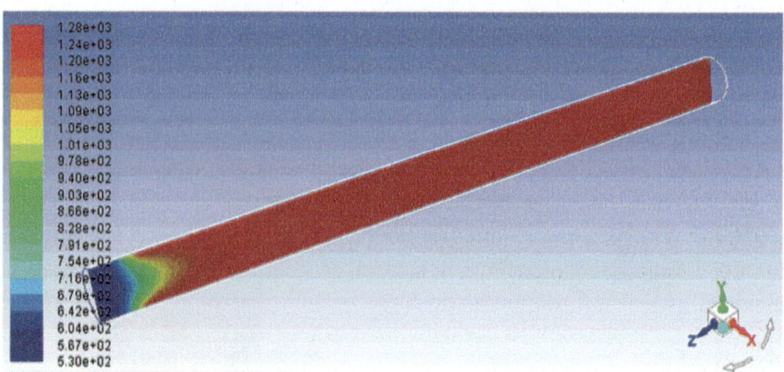

Fig. 5.50 Temperature (K) contours inside the micro tube after combustion completion

5.4.6 Unsteady/Transient Operation

The combustion inside a tiny tube burner is simulated adiabatically under transient conditions. For a micro-tube burner with a 0.85 mm diameter and 11 mm length simulation, a time step of 1.0 microsecond was used, and the mesh domain contained 40,000 total numbers of cells. The catalyst action at the surface is observed to cause the flame front to shift from upstream to downstream. Figure 5.50 depicts one shot at the conclusion of the combustion process.

This phenomenon has already been seen in the experimental study of [142]. The flame front is situated at one-third of the tube's length once stable conditions have been reached, and the remaining length has reached equilibrium conditions. Under sporadic circumstances, the actual experimental case was also simulated. It is obvious that the reaction front is travelling from upstream to downstream. The position (internally or externally) of the micro-burner, which heavily influences the relationship between heat losses and heat generated from combustion, determines the stable location of self-ignition. The response front shifts backward, as illustrated in Fig. 5.51, according to an experimental conclusion made by Cimino and Di Benedetto [142], and temperature profiles change their maximum from the outflow portion to the intake section. Additionally, it was said that at steady state circumstances, the first zone converts methane almost entirely, while the entire remaining zone functions as a heat source.

For the axisymmetric part of the reactor, which included 3 million mesh cells, the simulation was done with 1.0 microsecond time steps. Figure 5.52, which shows the development of the auto-ignition inside the reactor, shows temperature contours at various time intervals. It is seen that the reaction zone appears to be moving from upstream to downstream, which qualitatively agrees with the experimental findings shown in Fig. 5.51

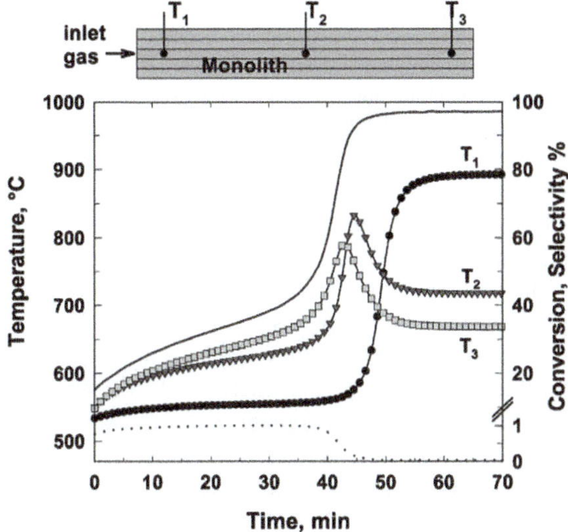

Fig. 5.51 Transient temperature profiles reported experimentally by Cimino and Di Benedetto [142]

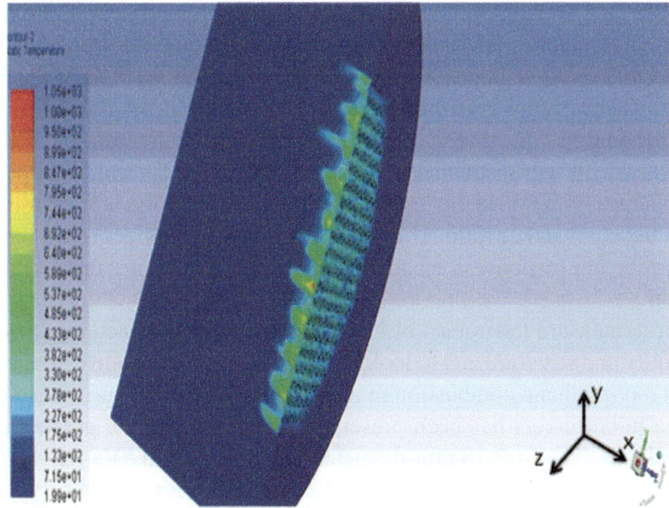

Fig. 5.52 Temperature contours at about 0.5 s at a central cross-sectional plane of the reactor

5.5   Concluding Remarks

This chapter presents the state and in-depth analysis of the investigations on the flow-field, combustion, and stability properties of premixed flames while taking into account various burner designs for a wide range of industrial applications. For many industrial applications, the properties of lean premixed (LPM) combustion have been explored, and a number of in-depth studies on the topic have been given. First, a study on premixed $CH_4/O_2/CO_2$ flames in a swirl-stabilized combustor has been presented. The entering combustible mixture's bulk throat velocity was held constant at 5.2 m/s. The O_2/CO_2 oxidizer's oxygen content was maintained at 60% vol. while the impact of the equivalence ratio was investigated from the blowout limit to the flashback limit. The mesh was constructed and LES simulations were run in the complete three-dimensional (3-D) domain of the combustor using the commercial Ansys Fluent 19.2 software. Based on the probability density function (PDF) model, the reaction rate and thermochemical characteristics were averaged. Through comparisons of predicted OH* concentration maps with visual flame appearance and actual data on axial and radial temperature profiles, the computational model was successfully confirmed. While adjusting the equivalence ratio, three distinct flame structures—(I) double conical flame, (II) corner-stabilized flame, and (III) swirl-stabilized (V-shaped) flame—were seen. Near the blowout limit, the twin conical flame is seen at low equivalence ratios. With two unique reaction zones and two inner recirculation zones (IRZ) separated by a cooler, reaction-free corner recirculation zone, this flame stabilizes within both the inner and outer shear layers (CRZ). Higher equivalence ratios reveal the corner-stabilized flame. A flame that stabilizes inside a responding CRZ is created when the downstream reaction zone shrinks and combines with the upstream one. A transition to the typical V-shape of swirl-stabilized flames with noticeably greater IRZ and weaker CRZ is further induced by increasing the equivalence ratio. Thus, this flame's IRZ stabilizes it. Up until a flashback, when the hot response zone advances closer to the burner throat, the V-shape is dominant.

In order to measure the impact of burner geometry (jet diameter and spacing) on flame stability and structure and to look into the geometry flexibility of micromixers, a study on oxy-methane combustion in a premixed multi-hole burner that replicates gas-turbine micromixers has been presented. We compared the geometry of three burner headends. The 61 3.18 mm-diameter jets on Headend #1's (HE1) hexagonal matrix are evenly spaced at 5.5 mm apart. With a 5.5 mm separation between each of its 37 larger-diameter (3.97 mm) jets, Headend #2 (HE2) is also organized hexagonally. In order to maintain the same combustor power, HE2 has fewer jets than HE1. In comparison to Headend #2 (HE2), Headend #3 (HE3) has jet spacing that is 7.0 mm wider (27% more). Visual imaging and temperature measurements were made in a few different flames for each headend, and the stability maps of the three headends were also established. To preserve comparable cold flow conditions and turbulence intensities, the entire investigation was carried out at a fixed flame-base jet velocity of 5.2 m/s. Despite the significant variations in jet diameter and headend spacing, the three blowout limits were determined to be only about 45 K apart. Gas-turbine micromixers are thus advocated here as a really versatile (fuel/oxidizer/geometry) and superiorly robust technology for oxy-fuel combustion in zero-emission power plants because this amply proves the geometry flexibility of these devices. HE1 fared better than HE2 and HE3. Larger flamelets in HE2 that are more stable than the smaller ones in HE1 are produced when the jet diameter is increased (from HE1 to HE2). This is true even at adiabatic flame temperatures (AFT) as low as 1658 K. However, as blowout drew near, the advantages of having more flamelets in HE1 outweighed those of larger flamelets in HE2; as a result, HE1 persisted down to 1410 K, compared to 1455 K in HE2. The lean-direct-injection (LDI) idea, which has demonstrated that the use of numerous small-scale flamelets is best for extending the combustor operability window in air–fuel gas turbines, is interestingly closer to HE1 than HE2. Increasing the jet separation was also found to be undesirable since it makes it more difficult for them to effectively stabilize one another; HE3 blew out at around 1500 K, compared to 1455 K in HE2. Additionally, it was discovered that, for fixed flame-base velocities, AFT is the main factor affecting the stability and structure of premixed methane flames. It was discovered that flames with the same AFT had identical macrostructures, and the blowout limit of each headend followed an AFT contour. This discovery is consistent with earlier research on swirl-stabilized $CH_4/O_2/CO_2$ and $CH_4/O_2/N_2$ flames, although none of those studies looked at how this finding is influenced by geometry. The originality of the current study is thus in showing that this holds true for a range of burner geometries.

Third, a thorough experimental inquiry has been offered above to look at how the operability of the partially premixed combustor stabilizing flames of $CNG/H_2/O_2/CO_2$ over a perforated plate burner is affected by the flexibility of the fuel and oxidizer. Three sets of studies covering a range of hydrogen fraction (HF), oxygen fraction (OF), equivalence ratio (Φ), and Reynolds of the inlet flow have produced the results that have been reported (Re). The initial set was conducted to explore the impacts of fuel flexibility on flame stability and macrostructure over a range of HF, from 0.0 to 30%, and at a fixed OF of 29.0%. The second set was carried out to determine the impact of oxidizer flexibility for OF between 29.0 and 36.0% on flame stability and shape. To ascertain the combustor operability close to the stoichiometric state ($\Phi = 0.85$), the third set of tests was conducted. Due to its strong reactivity and diffusivity, the results indicated that the blowout border with H_2 addition was expanded. However, due to the faster flame, the flashback border with H_2 addition was reduced near the leaner state. With the adding H_2, the flame's stability, intensity, and anchoring all enhanced while its length shrank. Similar-AFT flames displayed comparable flame shapes. The combustor's operability was restricted in the range of OF from 25% (at blowout) to 46% at $\Phi = 0.85$, close to stoichiometry (at flashback). The close stoichiometry process demonstrated that the H_2 adding can decrease with the operating value of OF at immovable inlet stream Re; however, the value of OF must not surpass 36% at HF of 30% to prevent flashback. The findings demonstrated that at the blowout limit and near the flashback limit, respectively, the inlet flow Re and reaction kinetics rates determine the flame stability. The findings indicated that AFT plays a part in regulating the combustor operability close to the flashback limit when the inlet flow Re is fixed. Whatever the flame type, partially or totally premixed, whatever the burner kinds, swirl or perforated plate, this finding was reported. Based on combustor operability and economic considerations, it is determined that H_2 addition is recommended for part-load operation but not favored under full load conditions.

Finally, a study on the auto-ignition properties of micro-structured catalytic honeycomb reactors for mixtures of CH_4 and CH_4-H_2 has been reported. ANSYS 17.2, a commercial program, has been used to examine the coupling of heat transmission and combustion in such a micro-burner under pseudo-auto-thermal steady state settings and transient instances. For the computations of species concentrations and reaction rates, a comprehensive CHEMKIN reaction kinetics mechanism has been taken into account. Utilizing the information found in the literature for a similar rector design, numerical validation of the minimum ignition temperature (MIT) values of CH_4-air mixes was done. Investigations have been done into how flow rate affects the MIT of CH_4-air mixes. Investigations have also been done into how hydrogen enrichment affects the MIT of CH_4-air mixtures. The addition of H_2 to CH_4 has been found to enhance the kinetics process and raise the auto-ignition temperature by roughly 30 °C. The findings demonstrated that for various mass flow rates of CH_4-air mixtures, ignition takes place at the same temperature. However, as the flow rate was increased for CH_4–H_2-air combinations, the ignition temperature dropped. Under the conditions of a lean mixture, the flammability limitations of methane combustion in a micro catalytic reactor have been established (equivalence ratio from 0.45 up to 0.75). The findings revealed previously unreported experimentally new operational limitations for methane combustion in micro catalytic reactors. It has been seen that the response front is obviously travelling from upstream to downstream under transitory conditions, mirroring the stated experimental behavior.

Acknowledgements The authors appreciate the support received for the preparation of this book from the Deanship of Research Oversight and Coordination (DROC) at King Fahd University of Petroleum & Minerals (KFUPM). The support provided by the Interdisciplinary Research Center for Hydrogen Technologies and Carbon Management (IRC-HTCM) on project number INHE2308 is highly appreciated. Also, the support received through the KFUPM Consortium for Hydrogen Future through the projects numbered H2FC2309 and H2FC2315 is highly appreciated.

Nomenclatures

A	Total stream area of each burner headend
A	Collision factor
Ac	Cross sectional (flow) area
AFT	Adiabatic flame temperature
$a_{i,m}$	Effective internal surface area per unit length
a_e	Heat flux surface area per unit length
CH_4	Methane gas
CO_2	Carbon dioxide

CRZ	Corner recirculation zone
$\overline{c_p}$	Mean heat capacity per gas unit mass
D	Diameter of one jet
D_k	Species diffusion coefficient
E	Activation energy
F	Exerted drag force on the gas by the tube wall
H_2	Hydrogen
HE1	Headend #1
HE2	Headend #2
HE3	Headend #3
H_2O	Water vapor
h_k	Specific enthalpy of species k
IRZ	Inner recirculation zone
$K_{f,m}$	Arrhenius rate of the mth heterogeneous reaction
K_g	Total number of gaseous species
LES	Large eddy simulation
Mix1	Mixture 1
Mix2	Mixture 2
MM	Micromixer
M_s	Number of surface species
MIT	Ignition Minimum Temperature (°C)
MFT	Flame Maximum Temperature (°C)
N	Number of jets of each burner headend
N_2	Nitrogen
NOx	Nitrogen oxides ($NO + NO_2$)
O_2	Oxygen
O.F.	Oxygen fraction, volumetric percentage of oxygen in the O_2/CO_2 oxidizer mixture
P	Absolute pressure
PF	Combustor profile factor
PPB	Perforated-plate burner
PVC	Processing vortex core
Q_e	Heat flux
S_k	Heterogeneous molar production rate of kth species
$Slph$	Standard litre per hour (L/hr)
SLPM	Standard liters per minute
$S_{0,m}$	Sticking coefficient
T	Absolute gas temperature
T_s	Surface temperature
u	Axial velocity of the gas
$v_{k,m}, v'_{s,m}$	Stoichiometric coefficients of reactions
W_k	Gas-phase species molecular weight
$\overline{X_s}$	Molar concentration of species S
X_k	Molar concentration of surface species k
Y_k	Gas-phase species mass fraction

Z	Axial location inside the combustor
z, y, x	Streamwise, transverse, and lateral physical coordinates

Subscripts

k, s	Indices for gas-phase and surface species
G	Gas

Greek Symbols

Φ	Equivalence ratio
θ_s	Surface species coverage
Γ	Surface site density
λ_s	Surface thermal conductivity
λ_g	Thermal conductivity of gas
μ	Viscosity
ρ	Density
τ	Reactor residence time
ρ	Density of reactant mixture
μ	Dynamic viscosity of reactant mixture
$\mu_{s,m}, \varepsilon_{s,m}$	Coverage dependence parameters

References

1. R.E. Anderson, S. MacAdam, F.V. Viteri, D.O. Davies, J.P. Downs, A. Paliszewski, ASME turbo expo, power land, sea air 781–791 (2008)
2. D. Cieutat, I. Sanchez-Molinero, R. Tsiava, P. Recourt, N. Aimard, C. Prébendé, Energy Procedia **1**, 519–526 (2009)
3. N. Perrin, C. Paufique, M. Leclerc, Energy Procedia **63**, 524–531 (2014)
4. B.J.P. Buhre, L.K. Elliott, C.D. Sheng, R.P. Gupta, T.F. Wall, Prog. Energy Combust. Sci. **31**, 283–307 (2005)
5. A.M. Gamal, A.H. Ibrahim, E.M.M. Ali, F.M. Elmahallawy, A. Abdelhafez, M.A. Nemitallah et al., Energy Fuels (2017). https://doi.org/10.1021/acs.energyfuels.6b02874
6. A. Amato, B. Hudak, P. D'Carlo, D. Nobl, D. Scarborough, J. Seitzman et al., J. Eng. Gas Turbines Power (2011). https://doi.org/10.1115/1.4002296
7. P. Kutne, B. K. Kapadia, W. Meier, M. Aigner, Proc. Combust. Inst. 3383–3390 (2011)
8. R. Marsh, J. Runyon, A. Giles, S. Morris, D. Pugh, A. Valera-Medina et al., Proc. Combust. Inst. **36**, 3949–3958 (2017)
9. P. Jourdaine, C. Mirat, J. Caudal, A. Lo, T. Schuller, Fuel **201**, 156–164 (2017)
10. H.G. Watanabe, Proc. ASME Turbo Expo **2015**, 1–10 (2015)

11. S.S. Rashwan, A.H. Ibrahim, T.W. Abou-Arab, M.A. Nemitallah, M.A. Habib, Appl. Energy **169**, 126–137 (2016)
12. B. Shi, J. Hu, S. Ishizuka, Combust. Flame **162**, 420–430 (2015)
13. C.Y. Liu, G. Chen, N. Sipöcz, M. Assadi, X.S. Bai, Appl. Energy **89**, 387–394 (2012)
14. M.A. Habib, S.S. Rashwan, M.A. Nemitallah, A. Abdelhafez, Appl. Energy **189**, 177–186 (2017)
15. A. Abdelhafez, S.S. Rashwan, M.A. Nemitallah, M.A. Habib, Appl. Energy (2018). https://doi.org/10.1016/j.apenergy.2018.01.097
16. A. Abdelhafez, M.A. Nemitallah, S.S. Rashwan, M.A. Habib, Energy and Fuels **32** (2018). https://doi.org/10.1021/acs.energyfuels.8b01133
17. P. Weigand, W. Meier, X.R. Duan, W. Stricker, M. Aigner, Combust. Flame (2006). https://doi.org/10.1016/j.combustflame.2005.07.010
18. W. Meier, X.R. Duan, P. Weigand, Combust. Flame (2006). https://doi.org/10.1016/j.combustflame.2005.07.009
19. M.A. Nemitallah, G. Kewlani, S. Hong, S.J. Shanbhogue, M.A. Habib, A.F. Ghoniem, Energy (2016). https://doi.org/10.1016/j.energy.2015.12.010
20. S.S. Rashwan, A.H. Ibrahim, T.W. Abou-Arab, M.A. Nemitallah, M.A. Habib, Appl. Energy (2016). https://doi.org/10.1016/j.apenergy.2016.02.047
21. S.S. Rashwan, A.H. Ibrahim, T.W. Abou-Arab, M.A. Nemitallah, M.A. Habib, Energy **122**, 159–167 (2017)
22. S. Brohez, C. Delvosalle, G. Marlair, Fire Saf. J. **39**(5), 399–411 (2004)
23. M.A. Nemitallah, M.A. Habib, Appl. Energy **111**, 401–415 (2013)
24. D. Veynante, L. Vervisch, Prog. Energy Combust. Sci. **28**, 193–266 (2002)
25. N. Peters, Meas. Sci. Technol. (2001). https://doi.org/10.1088/0957-0233/12/11/708
26. G. Kewlani, T. Supervisor, *Large Eddy Simulations of Premixed Turbulent Flame Dynamics : Combustion Modeling, Validation and Analysis* (2014). http://hdl.handle.net/1721.1/93863
27. F. Emanuel, H. Pérez, Subfilter scale modelling for large eddy simulation of lean hydrogen-enriched turbulent premixed combustion. Thesis 1–186 (2011)
28. F.E. Herná, C.P.T. Groth, L. Gü, Int. J. Hydrogen Energy **39**, 7147–7157 (2014)
29. S. Taamallah, Z.A. LaBry, S.J. Shanbhogue, M.A. Habib, A.F. Ghoniem, J. Eng. Gas Turbines Power **137**, 071505 (2015). https://doi.org/10.1115/1.4029173
30. S. Taamallah, Z.A. Labry, S.J. Shanbhogue, A.F. Ghoniem, Proc. Combust. Inst. **35**, 3273–3282 (2015)
31. S. Taamallah, S.J. Shanbhogue, A.F. Ghoniem, Combust. Flame **166**, 19–33 (2016)
32. W. Jerzak, M. Kuznia, J Nat Gas Sci Eng **29**, 46–54 (2016)
33. N. Shelil, A.J. Griffiths, N. Syred, Numerical study of stability limits of premixed-swirl flames, in *45th AIAA/ASME/SAE/ASEE Joint Propulsion Conference and Exhibit*, 2–5 August 2009, Denver, Colorado, (2009), pp. 1–15. https://doi.org/10.2514/6.2009-4926
34. N. Syred, A. Khalatov, *Advanced Combustion and Aerothermal Technologies* (2007). https://doi.org/10.1007/978-1-4020-6515-6
35. M.A. Nemitallah, A.A. Abdelhafez, A. Ali, I. Mansir, M.A. Habib, Frontiers in combustion techniques and burner designs for emissions control and CO_2 capture: a review. Int. J. Energy Res. 4730 (2019)
36. M.A. Haque, M.A. Nemitallah, A. Abdelhafez, I.B. Mansir, M.A. Habib, Review of fuel/oxidizer-flexible combustion in gas turbines. Energy Fuels **34**, 10459–10485 (2020)
37. Ali, M.A. Nemitallah, A. Abdelhafez, B. Imteyaz, M.M. Kamal, M.A. Habib, Numerical and experimental study of swirl premixed $CH_4/H_2/O_2/CO_2$ flames for controlled-emissions gas turbines. Int. J. Hydrogen Energy **45**, 29616–29629 (2020)
38. Ali, M.A. Nemitallah, A. Abdelhafez, M. Hussain, M.M. Kamal, M.A. Habib, Comparative analysis of the stability and structure of premixed $C_3H_8/O_2/CO_2$ and $C_3H_8/O_2/N_2$ flames for clean flexible energy production. Energy **214**, 118887 (2021)
39. Y.H. Li, G.B. Chen, F.H. Wu, H.F. Hsieh, Y.C. Chao, Effects of carbon dioxide in oxy-fuel atmosphere on catalytic combustion in a small-scale channel. Energy **94**, 766–774 (2016)

40. Y. Song, C. Zou, Y. He, C. Zheng, The chemical mechanism of the effect of CO_2 on the temperature in methane oxy-fuel combustion. Int. J. Heat Mass Transf. **86**, 622–628 (2015)

41. Y.H. Li, G.B. Chen, Y.C. Chao, Effects of flue gas addition on the premixed oxy-methane flames in atmospheric condition. Energy Procedia **75**, 3054–3059 (2015)

42. Amato, B. Hudak, P. D'Carlo, D. Noble, D. Scarborough, J. Seitzman, T. Lieuwen, Methane oxycombustion for low CO_2 cycles: blowoff measurements and analysis. J. Eng. Gas Turbines Power **133**(6), 1–9 (2011)

43. Y. Xie, J. Wang, M. Zhang, J. Gong, W. Jin, Z. Huang, Experimental and numerical study on laminar flame characteristics of methane oxy-fuel mixtures highly diluted with CO_2. Energy Fuels **27**(10), 6231–6237 (2013)

44. P.H. Joo, M.R.J. Charest, C.P.T. Groth, Ö.L. Gülder, Comparison of structures of laminar methane-oxygen and methane-air diffusion flames from atmospheric to 60atm. Combust. Flame **160**(10), 1990–1998 (2013)

45. M. Ditaranto, J. Hals, Combustion instabilities in sudden expansion oxy-fuel flames. Combust. Flame **146**(3), 493–512 (2006)

46. M. Ditaranto, R. Anantharaman, T. Weydahl, Performance and NOx emissions of refinery fired heaters retrofitted to hydrogen combustion. Energy Procedia **37**(1876), 7214–7220 (2013)

47. J. Oh, D. Noh, Laminar burning velocity of oxy-methane flames in atmospheric condition. Energy **45**(1), 669–675 (2012)

48. P. Kutne, B.K. Kapadia, W. Meier, M. Aigner, Experimental analysis of the combustion behaviour of oxyfuel flames in a gas turbine model combustor. Proc. Combust. Inst. **33**(2), 3383–3390 (2011)

49. J. Edacheri Veetil, B. Aravind, A. Mohammad, S. Kumar, R.K. Velamati, Effect of hole pattern on the structure of small scale perorated plate burner flames. Fuel **216**(December 2017), 722–733 (2018)

50. V. Hindasageri, P. Kuntikana, R.P. Vedula, S.V. Prabhu, An experimental and numerical investigation of heat transfer distribution of perforated plate burner flames impinging on a flat plate. Int. J. Therm. Sci. **94**, 156–169 (2015)

51. K.S. Kedia, A.F. Ghoniem, Mechanisms of stabilization and blowoff of a premixed flame downstream of a heat-conducting perforated plate. Combust. Flame **159**(3), 1055–1069 (2012)

52. E.V. Jithin, V.R. Kishore, R.J. Varghese, Three-dimensional simulations of steady perforated-plate stabilized propane-air premixed flames. Energy Fuels **28**(8), 5415–5425 (2014)

53. M. Gamal, A.H. Ibrahim, E.M. Ali, F.M. Elmahallawy, A. Abdelhafez, M.A. Nemitallah, S.S. Rashwan, M.A. Habib, Structure and lean extinction of premixed flames stabilized on conductive perforated plates. Energy Fuels **31**(2), 1980–1992 (2017)

54. S.S. Rashwan, A.H. Ibrahim, T.W. Abou-Arab, M.A. Nemitallah, M.A. Habib, Experimental study of atmospheric partially premixed oxy-combustion flames anchored over a perforated plate burner. Energy **122**, 159–167 (2017)

55. M. Aliyu, A.A. Abdelhafez, S.A. Said, A. Mohamed, Characteristics of oxy-fuel combustion in lean pre-mixed multi-hole burners. Energy and Fuels (2019)

56. M.H. du Toit, A.V. Avdeenkov, D. Bessarabov, Reviewing H_2 combustion: a case study for non-fuel-cell power systems and safety in passive autocatalytic recombiners. Energy Fuels **32**, 6401–6422 (2018)

57. H.H.W. Funke, N. Beckmann, J. Keinz, S. Abanteriba, Numerical and experimental evaluation of a dual-fuel dry-low-NOx micromix combustor for industrial gas turbine applications **50855**, V04BT04A045 (2017)

58. T. Asai, S. Dodo, M. Karishuku, N. Yagi, Y. Akiyama, A. Hayashi, Performance of multiple-injection dry low-NOx combustors on hydrogen-rich syngas fuel in an IGCC pilot plant. J. Eng. Gas Turbines Power **137**(9), 091504 (2015)

59. W.D. York, W.S. Ziminsky, E. Yilmaz, Development and testing of a low NOx hydrogen combustion system for heavy-duty gas turbines. J. Eng. Gas Turbines Power **135**(2), 022001 (2013)

60. H. Lee, S. Hernandez, V. McDonell, E. Steinthorsson, A. Mansour, B. Hollon, Development of flashback resistant low-emission micro-mixing fuel injector for 100% hydrogen and syngas fuels 411–419 (2009)

61. T. Asai, S. Dodo, H. Koizumi, H. Takahashi, S. Yoshida, H. Inoue, Effects of multiple-injection-burner configurations on combustion characteristics for dry low-NOx combustion of hydrogen-rich fuels **54624**, 311–320 (2011)
62. H.H.W. Funke, N. Beckmann, S. Abanteriba, An overview on dry low NO x micromix combustor development for hydrogen-rich gas turbine applications. Int. J. Hydrogen Energy **44**(13), 6978–6990 (2019)
63. H.H.W. Funke et al., Numerical and experimental characterization of low NOx micromix combustion principle for industrial hydrogen gas turbine applications **44687**, 1069–1079 (2012)
64. S. Dodo, T. Asai, H. Koizumi, H. Takahashi, S. Yoshida, H. Inoue, Combustion characteristics of a multiple-injection combustor for dry low-NOx combustion of hydrogen-rich fuels under medium pressure **54624**, 467–476 (2011)
65. Abdelhafez, S.S. Rashwan, M.A. Nemitallah, M.A. Habib, Stability map and shape of premixed $CH_4/O_2/CO_2$ flames in a model gas-turbine combustor. Appl. Energy **215**(December 2017), 63–74 (2018)
66. S. Taamallah, N.W. Chakroun, H. Watanabe, S.J. Shanbhogue, A.F. Ghoniem, On the characteristic flow and flame times for scaling oxy and air flame stabilization modes in premixed swirl combustion. Proc. Combust. Inst. **36**(3), 3799–3807 (2017)
67. S. Abdelwahid, M. Nemitallah, B. Imteyaz, A. Abdelhafez, M. Habib, Effects of H_2 enrichment and inlet velocity on stability limits and shape of $CH_4/H_2–O_2/CO_2$ flames in a premixed swirl combustor. Energy Fuels **32**(9), 9916–9925 (2018)
68. S. Brohez, C. Delvosalle, G. Marlair, A two-thermocouples probe for radiation corrections of measured temperatures in compartment fires. Fire Saf. J. **39**(5), 399–411 (2004)
69. Y.A. Cengel, M.A. Boles, *Thermodynamics: An Engineering Approach*, 8th edn. (McGraw-Hill Education, New York, 2015)
70. M. Li, Y. Tong, M. Thern, J. Klingmann, Investigation of methane oxy-fuel combustion in a swirl-stabilised gas turbine model combustor. Energies **10**(5) (2017)
71. R. Marsh et al., Premixed methane oxycombustion in nitrogen and carbon dioxide atmospheres: measurement of operating limits, flame location and emissions, in *Proceedings of the Combustion Institute*. Proc. Combust. Inst. **36**(3), 3949–3958 (2017)
72. H.K. Kayadelen, Effect of natural gas components on its flame temperature, equilibrium combustion products and thermodynamic properties. J. Nat. Gas Sci. Eng. **45**(November), 456–473 (2017)
73. H. Lefebvre, D.R. Ballal, *Gas Turbine Combustion: Alternative Fuels and Emissions*, 3rd edn. (CRC Press, ISBN 9781420086058, 2010), p. 134
74. Abdelhafez, M.A. Nemitallah, S.S. Rashwan, M.A. Habib, Adiabatic flame temperature for controlling the macrostructures and stabilization modes of premixed methane flames in a model gas-turbine combustor. Energy and Fuels **32**(7), 7868–7877 (2018)
75. J. Cai, Aerodynamics of lean direct injection combustor with multi-swirler arrays. Ph.D. thesis, University of Cincinnati, 2006. http://rave.ohiolink.edu/etdc/view?acc_num=ucin11 48233034
76. N.T. Weiland, T.G. Sidwell, P.A. Strakey, Testing of a hydrogen diffusion flame array injector at gas turbine conditions. Combust. Sci. Technol. **185**(7), 1132–1150 (2013)
77. A. Dai, T. Zhao, J. Chen, Climate change and drought: a precipitation and evaporation perspective. Curr. Clim. Chang. Reports **4**(3), 301–312 (2018)
78. S. Lawal, C. Lennard, B. Hewitson, Response of southern African vegetation to climate change at 1.5 and 2.0° global warming above the pre-industrial level. Clim. Serv. **16**, 100134 (2019)
79. Y. Liu et al., Impacts of 1.5 and 2.0 °C global warming on rice production across China. Agric. For. Meteorol. **284**, 107900 (2020)
80. Q. Sun et al., Global heat stress on health, wildfires, and agricultural crops under different levels of climate warming. Environ. Int. **128**, 125–136 (2019)
81. M.A. Habib, H.M. Badr, S.F. Ahmed, R. Ben-Mansour, K. Mezghani, S. Imashuku, G.J. la O', Y. Shao-Horn, N.D. Mancini, A. Mitsos, P. Kirchen, A.F. Ghoneim, A review of recent developments in carbon capture utilizing oxy-fuel combustion in conventional and ion transport membrane systems. Int. J. Energy Res. **35**, 741–764 (2010)

82. S. Nader, Paths to a low-carbon economy-the Masdar example. Energy Procedia **1**, 3951–3958 (2009)
83. J. Goldemberg, The promise of clean energy. Energy Policy **34**, 2185–2190 (2006)
84. L. Yildiz, Fossil fuels. Comprehensive Energy Systems (2018)
85. IPCC Fourth Assessment Report: Climate Change 2007 (Geneva, Switzerland). Clim. Chang. **1340** (2007)
86. A.P. Shroll, S.J. Shanbhogue, A.F. Ghoniem, Dynamic-stability characteristics of premixed methane oxy-combustion. J. Eng. Gas Turbines Power **134**(5), 051504 (2012)
87. P. Kutne, B.K. Kapadia, W. Meier, M. Aigner, Experimental analysis of the combustion behaviour of oxyfuel flames in a gas turbine model combustor. Proc. Combust. Inst. **33**, 3383–3390 (2011)
88. X. Hu, Q. Yu, J. Liu, N. Sun, Investigation of laminar flame speeds of $CH_4/O_2/CO_2$ mixtures at ordinary pressure and kinetic simulation. Energy **70**, 626–634 (2014)
89. W. Jerzak, M. Kuznia, Experimental study of impact of swirl number as well as oxygen and carbon dioxide content in natural gas combustion air on flame flashback and blow-off. J. Nat. Gas Sci. Eng. **29**, 46–54 (2016)
90. A. Amato et al., Methane oxycombustion for low CO_2 cycles: blowoff measurements and analysis. J. Eng. Gas Turbines Power **133**(6), 061503 (2011)
91. S.S. Rashwan, A.H. Ibrahim, T.W. Abou-arab, M.A. Nemitallah, M.A. Habib, Experimental investigation of partially premixed methane–air and methane–oxygen flames stabilized over a perforated-plate burner. Appl. Energy **169**, 126–137 (2016)
92. A. Abdelhafez, M.A. Nemitallah, S.S. Rashwan, M.A. Habib, Adiabatic flame temperature for controlling the macrostructures and stabilization modes of premixed methane flames in a model gas-turbine combustor. Energy Fuels **32**, 7868–7877 (2018)
93. Conserve Future Energy. [Online]. Available: https://www.conserve-energy-future.com/adv antages_disadvantages_hydrogenenergy.php
94. H.S. Kim, V.K. Arghode, A.K. Gupta, Flame characteristics of hydrogen-enriched methane-air premixed swirling flames. Int. J. Hydrogen Energy **34**, 1063–1073 (2009)
95. M. Ball, M. Wietschel, The future of hydrogen—opportunities and challenges. Int. J. Hydrogen Energy **34**, 615–627 (2009)
96. P.P. Edwards, V.L. Kuznetsov, W.I.F. David, N.P. Brandon, Hydrogen and fuel cells: towards a sustainable energy future. Energy Policy **36**, 4356–4362 (2008)
97. B.A. Imteyaz, M.A. Nemitallah, A.A. Abdelhafez, M.A. Habib, Combustion behavior and stability map of hydrogen-enriched oxy-methane premixed flames in a model gas turbine combustor. Int. J. Hydrogen Energy **43**, 16652–16666 (2018)
98. S. Taamallah, K. Vogiatzaki, F.M. Alzahrani, E.M.A. Mokheimer, M.A. Habib, A.F. Ghoniem, Fuel flexibility, stability and emissions in premixed hydrogen-rich gas turbine combustion: technology, fundamentals, and numerical simulations. Appl. Energy **154**, 1020–1047 (2015)
99. F. Delattin, G. Di Lorenzo, S. Rizzo, S. Bram, J. De Ruyck, Combustion of syngas in a pressurized microturbine-like combustor: experimental results. Appl. Energy **87**, 1441–1452 (2010)
100. A.E.E. Khalil, A.K. Gupta, Hydrogen addition effects on high intensity distributed combustion. Appl. Energy **104**, 71–78 (2013)
101. M. Gu, H. Chu, F. Liu, Effects of simultaneous hydrogen enrichment and carbon dioxide dilution of fuel on soot formation in an axisymmetric coflow laminar ethylene/air diffusion flame. Combust. Flame **166**, 216–228 (2016)
102. S.H. Park, K.M. Lee, C.H. Hwang, Effects of hydrogen addition on soot formation and oxidation in laminar premixed C_2H_2/air flames. Int. J. Hydrogen Energy **36**, 9304–9311 (2011)
103. F. Halter, C. Chauveau, I. Gökalp, Characterization of the effects of hydrogen addition in premixed methane/air flames. Int. J. Hydrogen Energy **32**, 2585–2592 (2007)
104. S. Abdelwahid, M.A. Nemitallah, B. Imteyaz, A. Abdelhafez, M.A. Habib, Effects of H2 enrichment and inlet velocity on stability limits and shape of $CH_4/H_2–O_2/CO_2$ flames in a premixed swirl combustor. Energy Fuels **32**, 9916–9925 (2018)

105. M. Sardarabadi, M. Hosseinzadeh, A. Kazemian, M. Passandideh-Fard, Experimental investigation of the effects of using metal-oxides/water nanofluids on a photovoltaic thermal system (PVT) from energy and exergy viewpoints. Energy **138**, 682–695 (2017)

106. A. Abdelhafez, S.S. Rashwan, M.A. Nemitallah, M.A. Habib, Stability map and shape of premixed $CH_4/O_2/CO_2$ flames in a model gas-turbine combustor. Appl. Energy **215**, 63–74 (2018)

107. M. Aliyu, A. Abdelhafez, S.A.M. Said, M.A. Habib, M.A. Nemitallah, I.B. Mansir, Characteristics of oxyfuel combustion in lean-premixed multihole burners. Energy Fuels **33**(11), 11948–11958 (2019)

108. A. Ali, M.A. Nemitallah, A. Abdelhafez, I.G. Alsakhawy, M.M. Kamal, M.A. Habib, Static stability and combustion characteristics of oxy-propane flames in a premixed fuel-flexible swirl combustor. Energy Fuels **33**(11), 11996–12007 (2019)

109. M.A. Habib, E.M.A. Mokheimer, S.Y. Sanusi, M.A. Nemitallah, Numerical investigations of combustion and emissions of syngas as compared to methane in a 200 MW package boiler. Energy Convers. Manag. **83**, 296–305 (2014)

110. Q. Zhang, D.R. Noble, T. Lieuwen, Characterization of fuel composition effects in $H_2/CO/CH_4$ mixtures upon lean blowout. J. Eng. Gas Turbines Power **129**, 688–694 (2007)

111. I.A. Ibrahim, T.W. Abou-Arab, S.S. Rashwan, M.A. Nemitallah, M.A. Habib, Effects of oxidizer flexibility and bluff-body blockage ratio on flammability limits of diffusion flames. Appl. Energy **178**, 19–28 (2016)

112. J.F. Driscoll, Turbulent premixed combustion: flamelet structure and its effect on turbulent burning velocities. Prog. Energy Combust. Sci. **34**, 91–134 (2008)

113. M.A. Nemitallah, S. Alkhaldi, A. Abdelhafez, M.A. Habib, Effect analysis on the macrostructure and static stability limits of oxy-methane flames in a premixed swirl combustor. Energy **159**, 86–96 (2018)

114. Y. Yan, Y. Liu, H. Li, W. Huang, Y. Chen, L. Li, Z. Yang, Effect of cavity coupling factors of opposed counter-flow micro-combustor on the methane-fueled catalytic combustion characteristics. J. Energy Resour. Technol. JERT-18-1199 (2018)

115. V. Shilapuram, B. Bagchi, O. Ozalp, R. Davis, Statistical modeling of hydrogen production via carbonaceous catalytic methane decomposition. J. Energy Resour. Technol. **140**(7), 072006–072006–8 (2018)

116. O. Askari, M. Elia, M. Ferrari, H. Metghalchi, Auto-ignition characteristics study of gas-to-liquid fuel at high pressures and low temperatures. J. Energy Resour. Technol. **139**(1), 012204–012204–6 (2016)

117. G. Amador, J. D. Forero, A. Rincon, A. Fontalvo, A. Bula, R. Vasquez Padilla, W. Orozco, Characteristics of auto-ignition in internal combustion engines operated with gaseous fuels of variable methane number. J. Energy Resour. Technol. **139**(4), 042205–042205–8 (2017)

118. Y. Ju, K. Maruta, Microscale combustion: technology development and fundamental research. Prog. Energy Combust. Sci. **37**(6), 669–715 (2011)

119. J. Daou, M. Matalon, Influence of conductive heat-losses on the propagation of premixed flames in channels. Combust. Flame **39**, 128–321 (2002)

120. J. Daou, M. Matalon, Flame propagation in Poiseuille flow under adiabatic conditions. Combust. Flame **49**, 124–337 (2001)

121. Y. Ju, B. Xu, Effects of channel width and Lewis number on the multiple flame regimes and propagation limits in mesoscale. Combust. Sci. Technol. **53**, 178–189 (2006)

122. T.L. Jackson, J. Buckmaster, Z. Lu, D.C. Kyritsis, L. Massa, Flames in narrow circular tubes. Proc. Combust. Inst. **31**(1), 955–962 (2007)

123. J. Hua, M. Wu, K. Kumar, Numerical simulation of the combustion of hydrogen–air mixture in micro-scaled chambers. Chem. Eng. Sci. **60**, 3497–3506 (2005)

124. D.G. Norton, D.G. Vlachos, Combustion characteristics and flame stability at the microscale: a CFD study of premixed methane/air mixtures. Chem. Eng. Sci. **58**, 4871–4882 (2003)

125. J. Li, S.K. Choua, A numerical study on premixed micro-combustion of CH_4–air mixture: effects of combustor size, geometry and boundary conditions on flame temperature. Chem. Eng. Sci. **150**, 213–222 (2009)

126. A. Tang, Y. Xu, J. Pan, A comparative study on combustion characteristics of methane, propane and hydrogen fuels in a micro-combustor. Int. J. Hydrogen Energy **40**, 16587–16596 (2015)

127. A. Tang, Y. Xu, J. Pan, Combustion characteristics and performance evaluation of premixed methane/air with hydrogen addition in amicro-planar combustor. Chem. Eng. Sci. **131**, 235–242 (2015)

128. S.R. Shabanian, S. Reza, CFD study on hydrogen-air premixed combustion in a micro scale chamber. Iran. J. Chem. Chem. Eng. **29**(4) (2010)

129. J. Li, S.K. Choua, Study on premixed combustion in cylindrical micro combustors: transient flame behavior and wall heat flux. Exp. Thermal Fluid Sci. **33**, 764–773 (2009)

130. S. Karagiannidis, J. Mantzaras, G. Jackson, Hetero-/homogeneous combustion and stability maps in methane-fueled catalytic micro reactors. Proc. Combust. Inst. **31**, 3309–3317 (2007)

131. R. Sui, J. Mantzaras, R. Bombach, Hetero-/homogeneous combustion of fuel-lean methane/oxygen/nitrogen mixtures over rhodium at pressures up to 12 bar. Proc. Combust. Inst. **36**, 4321–4328 (2017)

132. B. Xu, Y. Ju, Concentration slip and its impact on heterogeneous combustion in a micro scale chemical reactor. Chem. Eng. Sci. **72**, 60–73 (2005)

133. B. Xu, Y. Ju, Theoretical and numerical studies of non-equilibrium slip effects on a catalytic surface. Combust. Theor. Model. **79**, 10–16 (2006)

134. P. Aghalayam, P.A. Bui, D.G. Vlachos, The role of radical wall quenching in flame stability and wall heat flux: hydrogen-air mixtures. Combust. Theor. Model. **30**, 2–6 (1998)

135. D.G. Norton, D.G. Vlachos, Hydrogen-assisted self-ignition of propane/air mixtures in catalytic micro-burners. Proc. Combust. Inst. **30**, 473–2430 (2005)

136. R. Sui, J. Mantzaras, R. Bombach, Homogeneous ignition during fuel-rich $H_2/O_2/N_2$ combustion in platinum-coated channels at elevated pressures. Combust. Flame **180**, 184–195 (2017)

137. J. Chen, D. Xu, Transient simulation of the hydrogen-assisted self-ignition of fuel-lean propane-air mixtures in platinum-coated micro-channels using reduced-order kinetics. J. Chem. Technol. Metall. **51**, 90–98 (2016)

138. P.S. Barbato, G. Landi, R. Pirone, G. Russo, A. Scarpa, Auto-thermal combustion of CH_4 and CH_4–H_2 mixtures over bi-functional Pt-LaMnO$_3$ catalytic honeycomb. Catal. Today **147S**, 271–278 (2009)

139. V.L. Zimont, Theoretical study of self-ignition and quenching limits in a catalytic micro-structured burner and their sensitivity analysis. Chem. Eng. Sci. **134**, 800–812 (2015)

140. N.N. Semenov, Zur Theorie des Verbrennungsprozesses Z. Angew. Phys. **48**(7–8), 571–582 (1928)

141. Y.B. Zel'dovich, G.I. Barenblatt, V.B. Librovich, G.M. Machviladze, *The Mathematical Theory of Combustion and Explosions Plenum Publishing Corporation.* (New York, 1985)

142. S. Cimino, A. Di-Benedetto, Transient behaviour of perovskite-based monolithic reactors in the catalytic combustion of methane. Catal. Today **69**, 95–103 (2001)

143. A. Tang, J. Deng, Y. Xu, J. Pan, T. Cai, Experimental and numerical study of premixed propane/air combustion in the micro-planar combustor with a cross-plate insert. Appl. Therm. Eng. **136**, 177–184 (2018)

144. ANSYS Chemkin-Pro Theory Manual 17.2. ANSYS, Inc. San Diego (2016)

145. O. Deutschmann, R. Schmidt, F. Behrendt, J. Wamat, Numerical modeling of catalytic ignition. Sym. (Int.) Combust. **26**, 1747–1754 (1996)

146. U. Dogwiler, J. Mantzaras, P. Benz, Two-dimensional modelling for catalytically stabilized combustion of a lean methane-air mixture with elementary homogeneous and heterogeneous chemical reactions. Combust. Flame **116**, 243–258 (1999)

147. A. Boehman, Radiation heat transfer in catalytic monoliths. AIChE J. **44**, 2745–1755 (1998)

148. M. Reinke, J. Mantzaras, R. Bombach, Gas phase chemistry in catalytic combustion of methane/air mixtures over platinum at pressures of 1–16 bar. Combust. Flame **141**, 448–468 (2005)

Chapter 6
Applications of Fuel/Oxidizer-Flexible Premixed Combustion in Gas Turbines

6.1 Hydrogen-Enriched Combustion for Gas Turbine Applications

In the last few decades, carbon capture technologies have been the subject of significant research as a strategy to counteract the dangers of global warming. Although these technologies have a bright future, they nevertheless face a number of obstacles, such as energy costs associated with carbon-free emissions, retrofitting with prevailing plants, and combustion features [1, 2]. Oxyfuel combustion, in which hydrocarbon fuel gases are burned in nitrogen-depleted air, is recognized as one of the efficient carbon capture systems. By condensing the water vapour out, this technology produces flue gas that is mostly composed of CO_2 and water vapour, which can be easily separated. To reduce the undesirable high temperature, some flue gases are recycled into the combustion chamber. Due to the possibility of some combustion instability being introduced on account of the greater proportion of CO_2 in the combustion atmosphere, further research into the foundations and modeling of oxyfuel combustion is necessary [3, 4].

According to studies, adding CO_2 to the combustion environment has a considerable impact on rates of reaction and the laminar flame speed [5, 6]. Additionally, in a CO_2 rich environment, heat transfer and flow conduct are likewise changed [7, 8]. It has been discovered that blending fuels with greater burning speeds can improve the stability properties of premixed oxy-flames [9]. Hydrogen is regarded as a leading alternate clean energy means for energy-mix fuel mixes for gas turbine combustion and other applications due to its inherent advantages. The first and second laws efficiency of the scheme have also been reported to rise with hydrogen addition [10]. It is recognized as a top-notch addition with the benefits of reduced NOx, a good combustion temperature, and broader operability limits in ultra-lean premixed combustion [11]. Additionally, the accumulation of hydrogen reduces the rate of soot initiation and surface growth, which lowers the rate of soot production during combustion [12, 13]. A lean turbulent premixed flame's static stability has been shown to be greatly

© The Author(s), under exclusive license to Springer Nature Singapore Pte Ltd. 2024
M. A. Nemitallah et al., *Hydrogen for Clean Energy Production: Combustion Fundamentals and Applications*, https://doi.org/10.1007/978-981-97-7925-3_6

improved by even a little amount of hydrogen enrichment in the hydrocarbon fuel mixes [14, 15].

According to Gersen et al. [16], the ignition delay time decreases as the hydrogen content in the fuel blends rises. The trend may be correlated using the mixing law put forth by Cheng and Oppenheim [17]. Similar to this, the laminar flame speed exhibits a linear relationship with the fuel mixture's hydrogen content [18, 19]. According to Tang et al. [20], when the hydrogen proportion in the fuel mixture is large, the relationship between the unscratched flame promulgation speed and the unstretched laminar flame speed is considerably more strong. The increased molecular diffusivity of hydrogen as well as the enhanced production of H, O, and OH radicals may be responsible for the rise in the laminar burning velocity of the hydrogen-rich flames [21]. Although the turbulent flame speed is monotonically increased by hydrogen enrichment in lean mixtures, the effect is primarily determined by the type of hydrocarbon fuel used in rich regimes [22]. Halter et al.'s [14] discovery that the combustion intensity S_T/S_L is improved by growing the hydrogen portion in the fuel blend suggests that the influence is extra pronounced on the turbulent speed of the flame (S_T) than the laminar flame speed (S_L). The observation might be explicated by the fact that increasing the curvature of the flame front causes more small-scale wrinkle formation, which in turn increases the turbulent flame speed [23, 24].

Other reports, supported by experimental research [13] and numerical analysis done with GRI-Mech 3.0 [20], imply that the acceleration of burning velocity is connected to the acceleration of radical generation in the flame. Even though the adiabatic flame temperature (Tad) may fall with hydrogen merger at a certain thermal charging due to its lesser heating value, higher combustibility of hydrogen raises the local flame temperature. The reaction zone expands more quickly as a result of the local flame temperature increase, which prevents cooler flow recirculation [11]. In their study of the thermoacoustic instability of propane oxy-combustion, Abubakar and Mokheimer [25] found that when the hydrogen content of the fuel blend increases, the acoustic amplitude decreases.

The static stability of the flames is impacted by the fuel blend's addition of hydrogen. It was noted that flash-back occurs through the outer thin boundary layer at significantly lower flow rates when swirl numbers are smaller and the center recirculation zone is more constrained. Thus, it was determined that a variety of hydrogen fuel blends can be used under premixed swirl combustion to reduce the flashback [26]. Additionally, it has been discovered that adding hydrogen to hydrocarbons increases the lean blowout limit and lowers NOx emission in gas turbines [27]. According to Schefer [15], increasing the hydrogen component in the fuel mixture to 20% causes the equivalency ratio at which lean blowout (LBO) was found to drop by about 15%. The Damkohler number and flame thickness in oxy-methane flames tend to grow with increased hydrogen and oxygen fractions [28]. When Zhang et al. [29] looked into the relationship between LBO and Damkohler number, they discovered that for different mixes of $H_2/CO/CH_4$ combination, the blowout was observed at a consistent value of Damkohler number (0.4). For both methane/air and oxy-methane flames at fixed bulk mean throat velocity, the flashback was reported to occur at fixed Tad consistently in earlier investigations by the authors of the current work on the identical premixed swirl combustor [30, 31]. Due to CO_2's greater heat capacity than N_2, the CO_2 flames displayed lower Tad than the analogous N_2 flames with the same oxygen percent and equivalency ratio. Flames from the same Tad also have the similar shapes. The effects of adding hydrogen to methane on the static stability and macrostructure of oxy-flames were examined in a prior work on the identical premixed swirl combustor [9]. It was discovered that the fuel mixture's operability window is the widest at a hydrogen mole fraction of roughly ($50\%H_2$–$50\%CH_4$).

6.2   Premixed Swirl Combustors Holding Hydrogen Enriched Oxy-methane Flames

In this section, two studies based on numerical simulations and experimental measurements of premixed swirl gas turbine models carrying hydrogen-rich oxy-methane flames under various operating circumstances are discussed in detail. The goal of the first study (Sect. 6.2.1) is to experimentally and numerically analyze the stability, structure, and emissions of premixed CO_2-diluted hydrogen-enriched oxyfuel fires in a dry low emission (DLE) swirl combustor, a topic that is becoming more and more interesting but has not yet been well studied. The experiment was carried out in stoichiometric circumstances, as is customary for oxyfuel applications. The study investigates the flame properties for a variety of reactant combination HF, OF, and bulk throat velocities (Uin). To the best of the authors' knowledge, the function of the OF in hydrogen-enriched oxyfuel fires has not yet been studied. Since flame features are a fundamental indication of a reaction's level of oxidation (combustion), the impact of the OF on these characteristics is significant. On the other hand, the flame dynamics are greatly impacted by the presence of diluent (CO_2) in various concentrations. Gaining knowledge of the OF for premixed hydrogen-enriched flames will help gas turbine designers, manufacturers, and operators switch from the current air-based gas turbines to premixed hydrogen-enriched oxy-fuel turbines that are more effective and environmentally beneficial. In terms of the combustor efficiency and lower emissions, it would also assist them in analyzing and selecting the best OF for the needed loading conditions. The effect of the OF, which was not covered in the investigations by Suliman et al. [32] and Nemitallah et al. [33], is now well understood thanks to the current work. Additionally, the emissions properties caused by modifications to HF, OF, and Uin are described here, highlighting the originality of the current study. The current statistics and subsequent analysis offer helpful knowledge for reducing the adaptation problems when existing air–fuel gas turbines are converted to operate on hydrogen-enriched oxy-fuel. In the second research (Sect. 6.2.2), the identical combustor is used with fuel that is 50% H_2 and 50% CH_4, and the effects of inlet Reynolds number, equivalency ratio, and swirl number on the stability and combustion behavior of H_2/CH_4 oxy-flames were further examined experimentally and statistically. The research advances our understanding of the behavior of hydrogen/methane flames in an O_2/CO_2 oxidizer stream under a variety of operating conditions.

6.2.1 Swirl Premixed $CH_4/H_2/O_2/CO_2$ Flames for Emission Control

The test scenarios explored in this study are shown in Table 6.1, and the representative diagram of the lab-scale vertical gas-turbine model combustor employed in this study is shown in Fig. 6.1. In Ref. [34], the test rig has been thoroughly described. The mixing plenum on the burner has an approximate length-to-diameter (L/D) ratio of 20. Different gases are introduced into the plenum, which is upstream of the burner, by various inlets. Before the gases reach the burner's throat, the mixing area makes sure they are correctly mixed. An open-ended cylindrical quartz tube is positioned at the top of the burner to prevent outside air from being drawn into the combustion area. The flow rates are managed by Aalborg Inc. mass flow controllers, which have a full-scale uncertainty of 0.5%. To add swirl to the premixed stream, a 550 swirler is employed. Following is the equation used to compute the swirl number [35, 36].

$$SN = \frac{2}{3}\left\{\frac{1 - \left(D_{cb}/D_{in}\right)^3}{1 - \left(D_{cb}/D_{in}\right)^2}\right\} \tan(\alpha_{sw}) = 0.98 \tag{6.1}$$

where D_{cb} represents the diameter of swirler center body, D_{in} is the inner diameter of mixing plenum, and α_{sw} is the swirler blade angle. After passing via a converging–diverging nozzle with a throat diameter of 20 mm, the whirling gas mixture enters the combustor. The diverging part prevents rapid expansion of the mixture into the combustor, while the converging part gently accelerates the mixture while guaranteeing minimal loss to the swirl.

Table 6.1 Test matrix for premixed H_2-enriched oxy-methane combustion

Case	U_{in} (m/s)	OF (%)	Thermal loading (kW)	HF (%)
1	5.2	30	3.88	20
2			3.82	40
3			3.74	60
4			3.88	20
5		34	4.31	
6		38	4.72	
7	4.4	30	3.28	
8	5.2		3.88	
9	6.0		4.47	

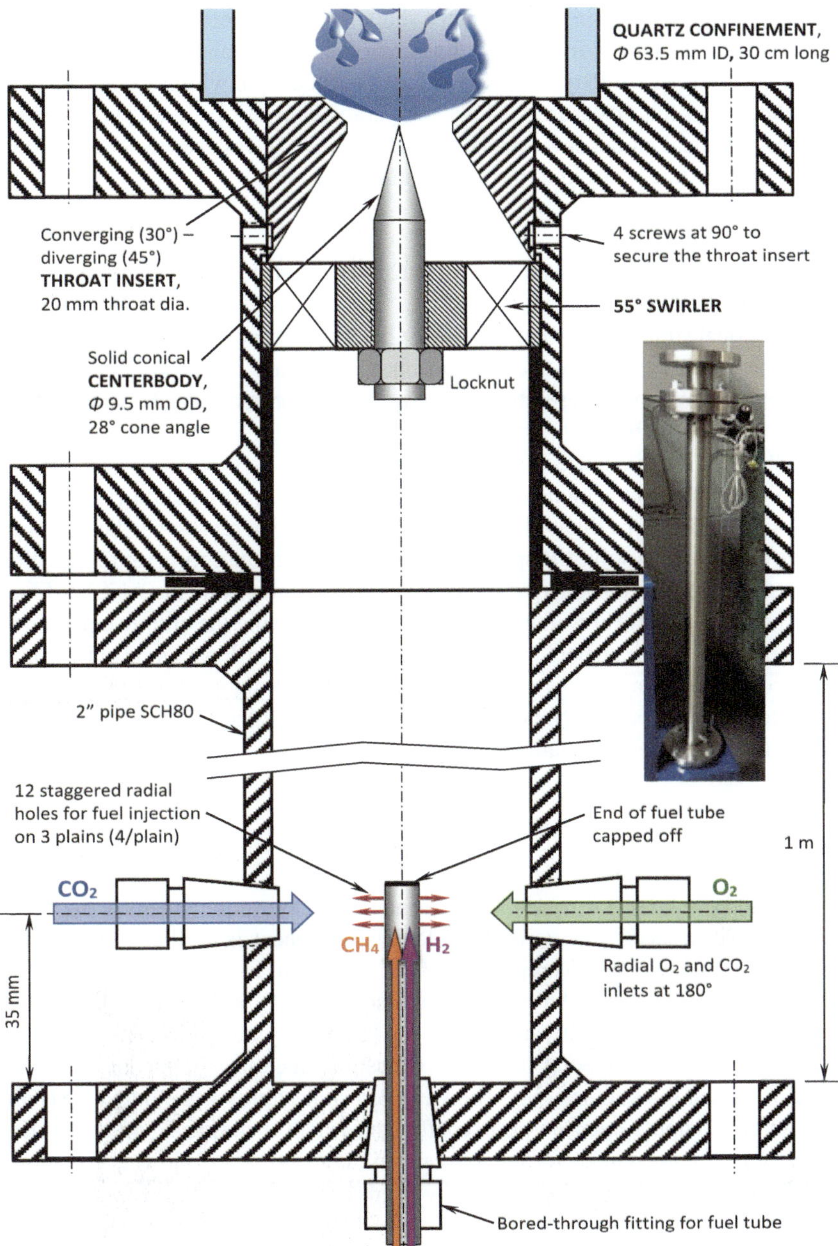

Fig. 6.1   The premixed swirl model gas-turbine combustor schematic

Using a high-resolution camera with a shutter speed of 1/60 s, an f-stop setting of f/5.6, and an ISO sensitivity of 1600, flame photos were obtained through the transparent quartz tube. An R-type (PtRh13%-Pt) thermocouple with a 1.0 mm junction diameter was used to measure the local temperatures inside the quartz confinement. The mathematical model developed by Brohez et al. [37] was used to account for the radiation inaccuracy related to the temperature data.

6.2.1.1   Modelling Premixed H_2-Enriched Oxy-methane Combustion

The numerical simulations were run using CH_4/H_2 fuel and O_2/CO_2 oxidizer mixes at stoichiometric conditions and at atmospheric pressure. The power density of the model combustor closely resembles the industrial gas turbines' usual operating range (3.5 to 20 MW/m^3/bar). The test scenarios include HF (20–60% by volume), OF (30–38% by volume), and Uin (4.4–6.0 m/s) ranges.

The integrated mathematical model uses a large eddy simulation (LES) model for scalar variables in a three-dimensional (3D) region that represents the model gas turbine combustor to solve the species transport equation and conservation equations of mass, momentum, and energy. All simulations were run using the ANSYS-Fluent 19.2 commercial software suite. By simultaneously solving the elliptic equations in the 3D domain, heat transport, emission characteristics, and flow regime of hydrogen-enriched oxy-methane flames are anticipated. The following general equation serves as the foundation for the species transport equation and the conservation of mass, momentum, and energy equations:

$$\frac{\partial}{\partial x_J}\left(\overline{\rho U_j}\emptyset + \overline{\rho u_j \varphi}\right) = \frac{\partial}{\partial x_J}\left[\Gamma_\varphi \frac{\partial \emptyset}{\partial x_J}\right] + \rho \overline{S_\emptyset} \tag{6.2}$$

where $\emptyset$ and φ represent Reynold's average and fluctuating components of the dependent variable, respectively, u_J is the velocity component in J direction, Γ_φ is the diffusion component, and $S_\emptyset$ represents the source term.

The radiative heat transfer equation was solved using a discrete ordinate (DO) model, with one radiative iteration for every five energy iterations. This model has been suggested for modeling oxyfuel combustion since it is applicable to a wide range of optical thicknesses [38]. The weighted sum of grey gas model (WSGGM) was used to compute the absorption coefficient, which is confirmed under a variety of operating circumstances for the gaseous mixture [39]. The following equation is used to determine the overall emissivity () in terms of the weighted sum of grey gases:

$$\varepsilon = \sum_{i=0}^{i=I} \alpha_{\varepsilon,i}(T)\left[1 - e^{-\kappa_i\, PL}\right] \tag{6.3}$$

$\alpha_{\varepsilon,i}$ indicates the emissivity weighting factor for the grey gas i at temperature T. κ_i represents the adsorption coefficient with P being the partial pressure of absorbing gases and L being the gas layer thickness (or path length).

While small eddies dissipate the kinetic energy of the turbulence, large eddies represent the typical length of the mean flow, and together they roughly describe flow turbulence. The LES modeling technique in this study was used to address turbulence. In LES, large eddies are kept, and smaller eddies are filtered out since they are too small to have any major impact on the flow field due to their energy. Particularly when they are adequately close to Kolmogorov scale, these filter scales are thought to be estimators in general. For the purpose of researching impacts below filter scales, sub-filter scale models are created using the LES technique. In order to assess the correctness of the numerical model, the current study applies the LES model to a model gas turbine combustor and compares the projected results with the existing experimental data. The produced simulation results are examined following validation.

The following equation for a quantity was used to perform the low-pass filtering operation [33]:

$$\overline{\varphi}(x, t) = \int G(r, x)\varphi(x - r, t)dr \tag{6.4}$$

For the full domain, integration is solved by this equation. The G filter function decides the size of the resolved eddies. Continuity, Navier–Stokes equations, and the filtering function are used to produce the general equations in the following forms:

$$\frac{\partial \rho}{\partial t} + \frac{\partial}{\partial x_i}(\rho \overline{u}_i) = 0 \tag{6.5}$$

$$\frac{\partial}{\partial t}(\rho \overline{u}_i) + \frac{\partial}{\partial x_j}(\rho \overline{u}_i \overline{u}_j) = \frac{\partial}{\partial x_j}\left(\mu \frac{\partial \sigma_{ij}}{\partial x_j}\right) - \frac{\partial \overline{p}}{\partial x_i} - \frac{\partial \tau_{ij}}{\partial x_j} \tag{6.6}$$

where σ_{ij} represents the stress tensor (because of the molecular viscosity), defined by:

$$\sigma_{ij} \equiv \left[\mu\left(\frac{\partial \overline{u}_i}{\partial x_j} + \frac{\partial \overline{u}_j}{\partial x_i}\right)\right] - \frac{2}{3}\mu \frac{\partial \overline{u}_i}{\partial x_j}\delta_{ij} \tag{6.7}$$

The subgrid-scale stress τ_{ij} is given by:

$$\tau_{ij} \equiv \rho \overline{u_i u_j} - \rho \overline{u}_i \overline{u}_j \tag{6.8}$$

Subgrid-scale stresses are produced by the filtering procedure and require modeling. Because of this, the isotropic portion of the subgrid-scale stress and filtered static pressure terms are combined, and the deviator portion is modeled using turbulent viscosity:

$$\tau_{ij} = -2\mu_t \overline{S}_{ij} \tag{6.9}$$

where μ_t and $\overline{S}_{ij}$ are the turbulent viscosity and rate of stain tensor, respectively. Wall-Adapting Local Eddy-viscosity (WALE) model was used for solving the turbulent viscosity, and is given by:

$$\mu_t = \rho L_s^2 \frac{\left(S_{ij}^d S_{ji}^d\right)^{3/2}}{\left(\overline{S}_{ij}\overline{S}_{ji}\right)^{5/2} + \left(S_{ij}^d S_{ji}^d\right)^{5/4}} \tag{6.10}$$

where L_s symbolizes mixing length for subgrid-scale. Both L_s and S_{ij}^d are given as:

$$L_s = \min\left(\kappa d, C_w V^{1/3}\right) \tag{6.11}$$

$$S_{ij}^d = \frac{1}{2}\left(\overline{g}_{ij}^2 \overline{g}_{ji}^2\right) - \frac{1}{3}\delta_{ij}\overline{g}_{kk}^2 \tag{6.12}$$

where κ denotes von Karman constant. WALE constant is given by C_w, d represents distant to the closest wall, and V is the volume of computational cell. $\overline{g}_{ij}$ is defined as:

$$\overline{g}_{ij} = \frac{\partial \overline{u}_i}{\partial x_j} \tag{6.13}$$

For hydrogen-enriched oxy-methane simulations, the partially premixed combustion model was changed to accommodate the fully premixed combustion situation. For the purpose of obtaining additional scalar variables, the transport equations for the mean mixture fraction $\overline{f}$, mixture fraction variance $\overline{f'^2}$ and mean reaction progress variable (c) were solved. The mass fraction of a particular species in the mixture is what is meant by the reaction progress variable, which has a value of 1.0 for burnt gas mixtures and 0 for unburnt mixtures. The given mixture fraction is:

$$f = \frac{sY_{fu} - Y_{ox} + Y_{ox,0}}{sY_{fu,1} + Y_{ox,0}} \tag{6.14}$$

The basis for computing species fractions is the equation shown below:

$$\overline{\emptyset} = \overline{c} \int_0^1 \emptyset_b(f)p(f)df + (1 - \overline{c}) \int_0^1 \emptyset_u(f)p(f)df \tag{6.15}$$

Reaction progress variable, c, is given by:

$$c = \left(\frac{\sum_{i=1}^{n} Y_i}{\sum_{i=1}^{n} Y_{i,ad}} \right) \tag{6.16}$$

Following correlation was used for estimating the flame thickness:

$$\delta = \frac{\lambda_u}{\rho_u c_p S_L} \tag{6.17}$$

where ρ_u and λ_u represent thermal density and conductivity, respectively.

Damköhler number is given by the ratio of the reaction rate to the diffusion rate:

$$Da = \frac{\left(l_t / u' \right)}{\left(\delta / S_L \right)} \tag{6.18}$$

Since the combustor is axis-symmetric, only one quadrant of the modeled reactor (Fig. 6.2) was taken into account for all simulations in order to shorten the computation time. A mesh independence investigation was conducted, and 40,000 finite volume cells were chosen. The burner region produced finer meshes, whilst the combustor's outlet zone produced coarser meshes. Through the intake, different $CH_4/H_2/O_2/CO_2$ compositions with constant yet distinct throat velocities (U_{in}) were fed into the combustor. For the introduction of a swirl flow with a 0.98 swirl number, tangential and axial flow components were utilised [33]. The vortex technique algorithm was employed to find a solution for the varying velocity. The mean mixture fraction was calculated using the predetermined values of the equivalence ratio and mixture composition. At the combustor wall, a mixed thermal boundary condition was established with both convection and external radiation as heat transfer mechanisms, a free stream temperature of 300 K, and a convective heat transfer coefficient of 20 W/m^2/K. The 6.0 mm wall thickness was chosen for the semi-transparent border type. For computing the radiative heat transfer through the quartz tube, the non-gray band model with 4.0 wavelength bands was employed. Periodic boundary conditions were applied to both symmetric faces of the quadrant portion of the combustor under consideration in order to handle the tangential components of whirling flow. At a fixed equivalence ratio of 1.0 (i.e., stoichiometric condition), all simulations were run while altering the values of OF, HF, and U_{in}.

6.2.1.2 Global Behavior of $CH_4/H_2/O_2/CO_2$ Flames

An essential component of analysis for gaining a complete grasp of flame characteristics is the visualization of flame form. The evaluation of various flame shapes also sheds light on the extinction mechanisms for flames. The outlines of OH radicals used to construct flame forms are compared to photos of natural luminosity in Fig. 6.3. It depicts the overall size and shape of the flames for a variety of flammable mixes examined in the current study. The impact of changing HF on the flame form is depicted in Fig. 6.3a. While the flame has a spread-out shape for fuel mixtures

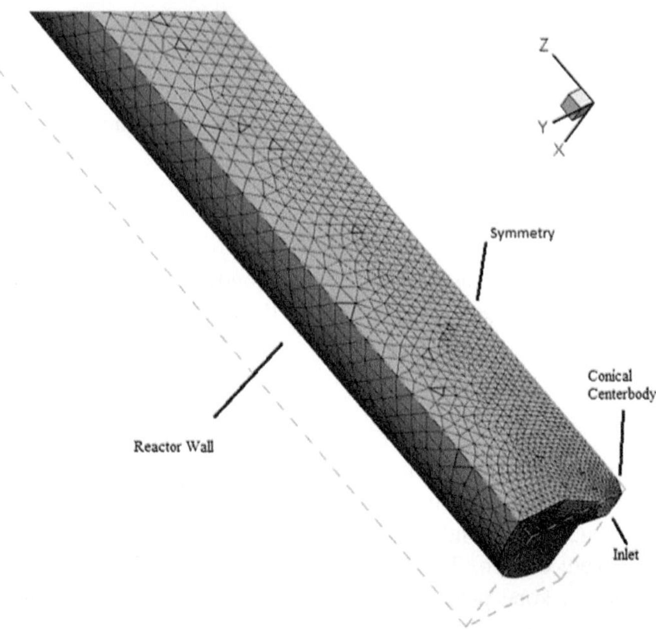

Fig. 6.2 Combustor quadrant mesh applied throughout the computational processes

with lower hydrogen fractions (HF = 40%), it becomes compact and perturbed for fuel mixtures with higher hydrogen fractions (HF = 60%), indicating improved reactivity of the mixture, which is also indicated by the higher OH concentration seen in the numerical flame shape. This conclusion is consistent with other research indicating that hydrogen enrichment increases laminar flame speeds [40, 41]. Due to local quenching brought on by the near-wall cooling effect, no reactions take place in the outer recirculation zone (i.e., the shadowy area beneath the flame beside the burner wall).

Figure 6.3b illustrates how changing OF has an impact on the flame form. The experimental photographs display a brighter, more compact flame with higher OF, indicating that oxygen enrichment has increased the combustible mixture's reactivity. More oxidizer being available to the fuel causes higher reactive species concentrations, speeding up the reaction. The numerical flame morphologies also demonstrate that with increasing oxygen fractions (OF = 38%) in the oxidizer mixture, the concentration of OH rises noticeably.

The impact of Uin on the combustion behavior is examined in Fig. 6.3c. As opposed to greater velocity (Uin = 6.0 m/s), low velocities (Uin = 4.4 m/s) produce stronger flames. This results in improved combustion efficiency at lower bulk velocities due to the radicals' extended residence time at lower velocities to interact and react with other chemical species. This effect is further supported by numerical flame forms, which have larger OH concentrations at slower speeds. The numerical findings in all three of the comparisons shown in Fig. 6.3 correspond with the experimental

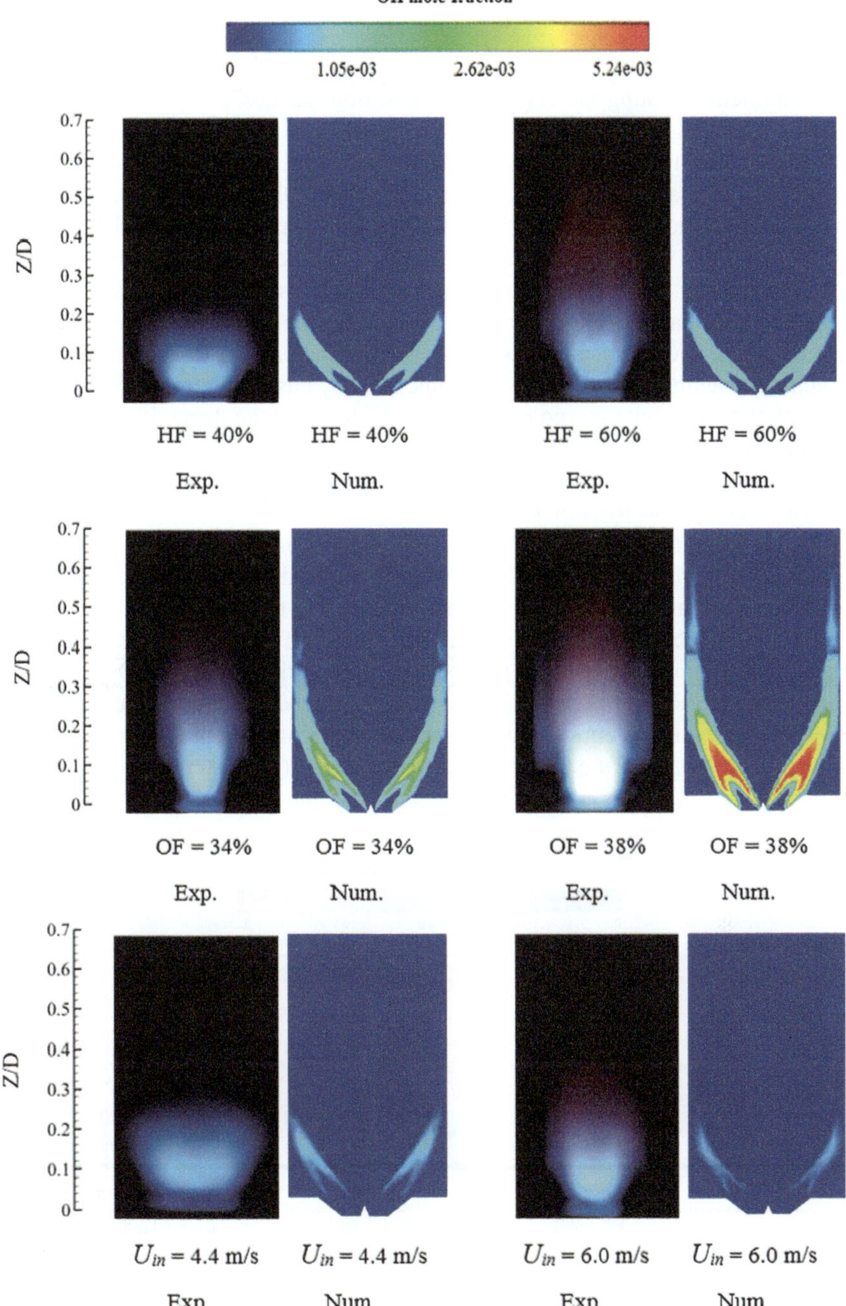

Fig. 6.3 Comparison of calculated (OH contours) and actual (camera photos) flame forms under stoichiometric circumstances. Top: effect of HF at OF = 30% and U_{in} = 5.2 m/s; middle: effect of OF at HF = 20% and U_{in} = 5.2 m/s; bottom: effect of U_{in} at HF = 20% and OF = 30%

flame shapes, demonstrating the accuracy of the numerical code applied to these simulations.

The reaction progress variable (c) is a crucial measure for describing flame thickness and demonstrating the degree of combustion reaction completion by demonstrating the change from the unburned reactant mixture to the fully consumed products. As it is customary for premixed flames, the value of c in this study is dependent on the concentration of OH. The effects of HF, OF, and Uin on the reaction's progress are depicted by the contours of the progress variable in Fig. 6.4. The figure rarely shows any variations, especially close to the burner exit where the flame stabilizes, suggesting that the individual impacts of increasing hydrogen- and oxygen-enrichment on flame thickness are not likely to be substantial. A slight shortening of the flame is seen, indicating faster reflexes as anticipated. Despite the increase in angular momentum that displaces the flame front by pushing it outwards, increasing the bulk throat velocity has no effect on the thickness of the flame. It is also important to notice that the resulting increase in turbulence rarely has an impact on the flame thickness, indicating that chemical kinetics plays a dominating role here and is unaffected by flow dynamics.

Figure 6.5 shows the impact of HF, OF, and U_{in} on overall flame thickness (δ preheat zone + reaction zone). Thinner flames are produced with higher H_2 and O_2 concentrations, and they seem to be more responsive to OF than HF. The derived Damköhler number (Da) is displayed in Fig. 6.6 as a function of HF, OF, and U_{in}. Figure 6.6a, b demonstrate a large rise in Da with the addition of H_2 and O_2, indicating quicker reactions at greater HF and/or OF. Decreased residence time causes the transport time scale to outweigh the chemical time scale, which further contributes to decreased Da when the inlet velocity is increased (Fig. 6.6c). Da is more than unity for every scenario in Fig. 6.6, indicating the importance of chemical kinetics

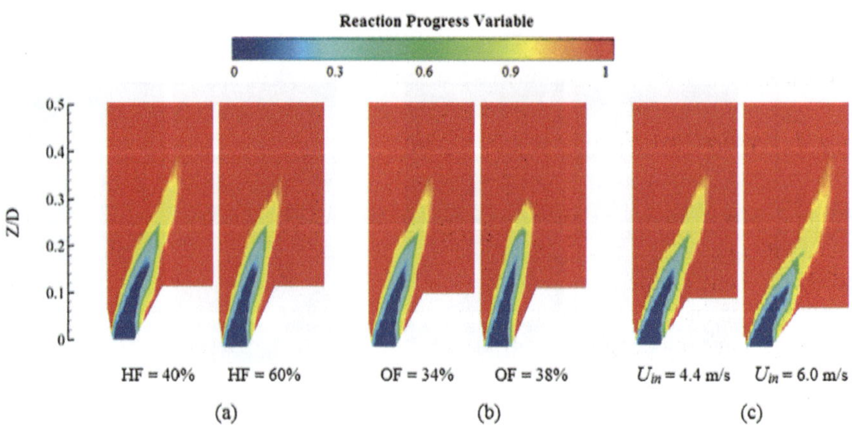

Fig. 6.4 Progress variable contours calculated under stoichiometric circumstances. **a** Effect of HF at OF = 30% and U_{in} = 5.2 m/s, **b** Effect of OF at HF = 20% and U_{in} = 5.2 m/s, and **c** Effect of U_{in} at HF = 20% and OF = 30%

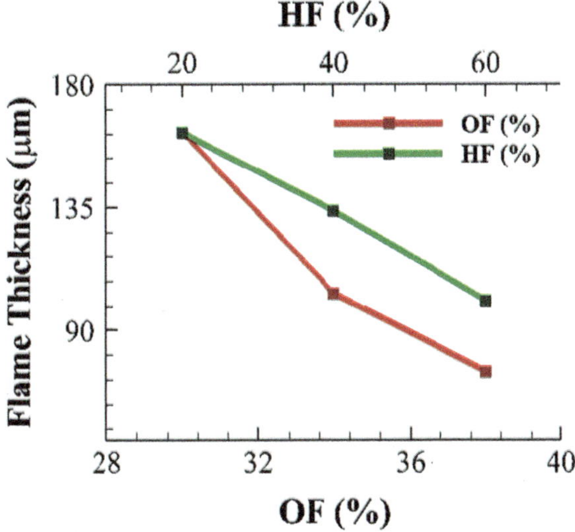

Fig. 6.5   Effects of HF and OF (both at $U_{in} = 5.2$ m/s) on flame thickness at stoichiometric combustion

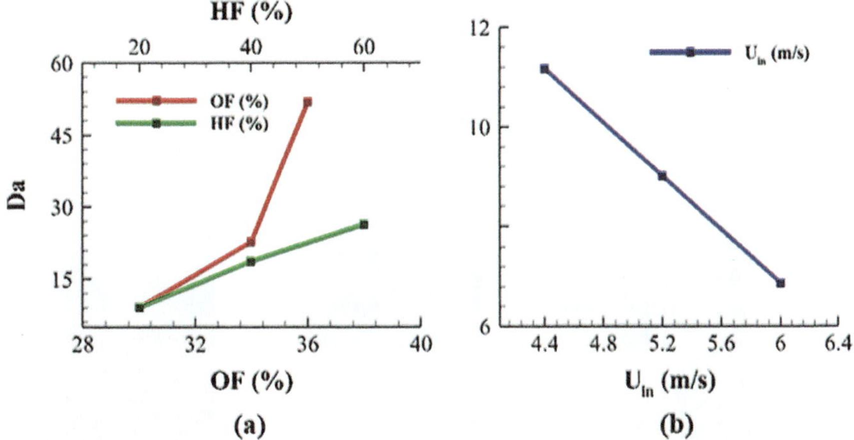

Fig. 6.6   Variation of Da at stoichiometric combustion in terms of: **a** HF and OF at $U_{in} = 5.2$ m/s and **b** U_{in} at HF = 20% and OF = 30%

in the observed flames. Turbulence affects the general outside structure of the flame by causing it to become wrinkled and spread out, but it has no effect on the flame's inner structure.

6.2.1.3   Flow-Field Structure of $CH_4/H_2/O_2/CO_2$ Flames

The numerical data is further investigated by mapping the velocity vectors onto static temperature contours as illustrated in Fig. 6.7 in order to evaluate flow-flame interactions in the current test conditions. The formation of the inner and outer recirculation zones (IRZ and ORZ, respectively) identifies the general characteristics of the turbulent reactive flow field. In all circumstances, the temperature rises to levels that are appropriate and safe for the gas turbine blades in the upstream region, where it hits around 2300 K. In the same current combustor, Suliman et al. [32] conducted an experimental investigation of the flame stability and reported the stability map plotted against the contours of AFT, as shown in Fig. 6.8. The results validate the present numerical temperature data, showing a maximum temperature of roughly 2400 K for the steady flame scenario of HF = 20%, OF = 30%, and U_{in} = 6.0 m/s.

The impact of changing HF on the flow and temperature fields is examined in Fig. 6.7a. A huge primary eddy (PE) is seen at lower HF = 40%; it is located at Z/D = 0.6 and is preceded by a minor recirculation zone just upstream that denotes the presence of a smaller secondary eddy (SE). The PE moves toward the burner exit, which is centered at Z/D = 0.2, as the reactant mixture's hydrogen concentration rises (HF = 60%), while two SEs are formed upstream of it and begin rotating in opposition to one another. Eddy formation increases the mixing caused by turbulent motion, which boosts the flow reactivity.

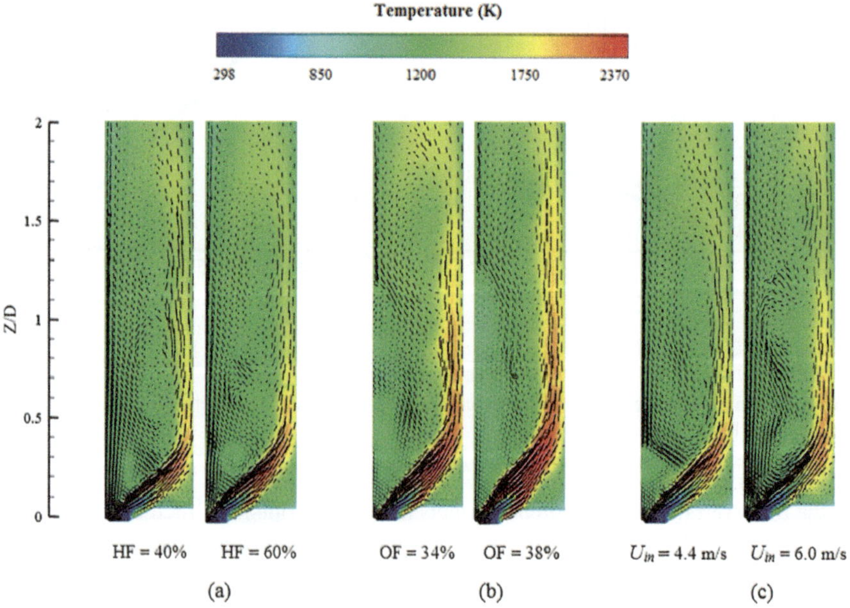

Fig. 6.7 Velocity vectors over temperature plots at stoichiometric combustion. **a** Effect of HF at OF = 30% and U_{in} = 5.2 m/s, **b** Effect of OF at HF = 20% and U_{in} = 5.2 m/s, and **c** Effect of U_{in} at HF = 20% and OF = 30%

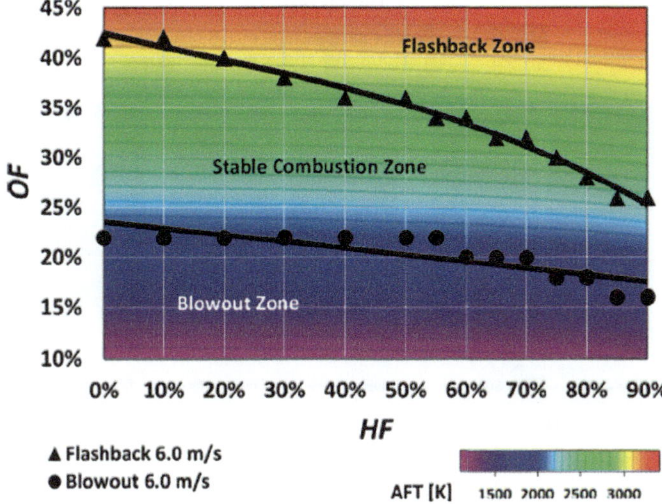

Fig. 6.8 Flame stability map over the plots of AFT at $U_{\text{in}} = 6.0$ m/s [32]

Figure 6.7b illustrates how the oxygen fraction affects the structure of the thermal flame. Because there are more O radicals available in the combustible mixture, increasing the OF causes greater temperatures. As multiple eddies arise as a result of these radicals' interactions with the flammable mixture, reactivity is further increased. A PE appears to be turning counterclockwise at lower oxygen concentrations (OF = 34%), followed by a SE rotating clockwise. Further upstream, another smaller SE may be seen rotating in the clockwise direction. The secondary eddies split into multiple smaller, counter-rotating eddies as the oxygen concentration rises (OF = 38%), demonstrating stronger flame-vortex interactions caused by turbulence-induced coarse mixing. As seen in the experimental flame forms depicted in Fig. 6.3a and b, increasing the hydrogen and/or oxygen concentrations generally accelerates the reaction speeds and produces intense, compact flames.

Figure 6.7c shows the physical impact of U_{in} on the reactant combination. One PE and one SE are generated, spinning counter-clockwise, at lower input velocities ($U_{\text{in}} = 4.4$ m/s), where reactant species have more opportunity to interact. On the other hand, with a higher input velocity ($U_{\text{in}} = 6.0$ m/s), just one PE is seen rotating counter-clockwise, indicating that there was less coarse mixing caused by turbulence in the mixture. A smaller ORZ is seen at the bottom corner of the combustor in all three comparisons (of HF, OF, and U_{in}), which generally remains unaffected, most likely because it is farther away from the flame.

The tendency for rotation in the flow field, which is also strongly related to the development of turbulent eddies, is adequately represented by vorticity. The magnitude of the vorticity for various mixture compositions is depicted in Fig. 6.9. The main reaction zone, or stabilized flame, is indicated by the space between the outer and inner shear layers. According to Fig. 6.9a, the inner shear layer thickens as the

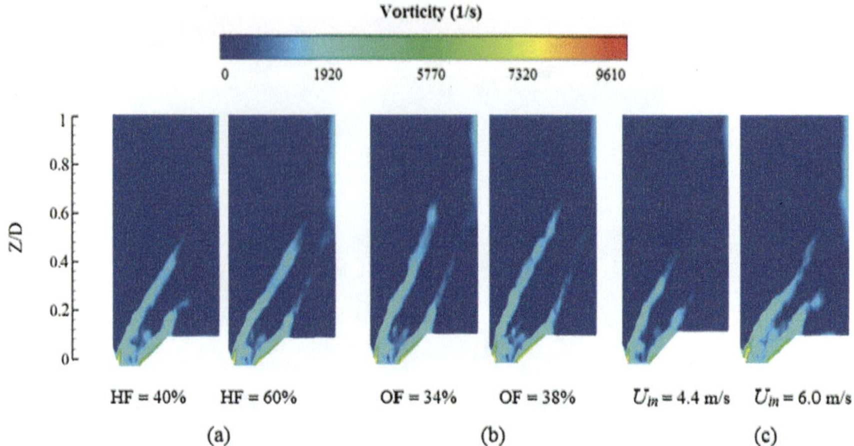

Fig. 6.9 Plots of vorticity (1/s) at stoichiometric combustion. **a** Effect of HF at OF = 30% and U_{in} = 5.2 m/s, **b** Effect of OF at HF = 20% and U_{in} = 5.2 m/s, and **c** Effect of U_{in} at HF = 20% and OF = 30%

HF rises. The increased levels of vorticity close to the central axis are to blame for this. Remember that the two flames' overall static temperature fields are identical (Fig. 6.7a), but that the flame with enhanced HF = 60% has a zone of localized high temperature. Increased HF comes from the buoyant forces being driven by this "hot spot". Similar to the increase in vorticity, the inner shear layer thickness increases with increasing oxygen concentration in the mixture (Fig. 6.9b). Figure 6.7b also shows the elevated local static temperatures. As seen in Fig. 6.9c, vorticity has a high magnitude when U_{in} is higher, but as inlet velocity rises, the lengths of the inner and outer shear layers shorten. This indicates that the ORZ predominates at higher inlet velocities (Uin = 6.0 m/s); the flame stabilizes at the corner region of the combustor and the outer shear layer is extended. The local zone temperatures are dominant with U_{in} = 6.0 m/s (Fig. 6.7c), which would have contributed to ORZ dominance and increased vortex concentration.

6.2.1.4 Emission Attributes of $CH_4/H_2/O_2/CO_2$ Flames

Here, the impacts of HF, OF, and U_{in} on the product formation rate (PFR) are examined in order to study the emission characteristics of the flames, as shown in Fig. 6.10. With the addition of hydrogen (Fig. 6.10a) and/or oxygen (Fig. 6.10b), the PFR in the comparatively thin inner shear layer increases, resulting in compact flames. However, compared to HF, the OF has a far higher impact on PFR. PFR is increased by increasing the amount of oxygen in the oxidizer mixture, which promotes the oxidation of the CH_4 and H_2 fuel species. Due to higher concentrations of smaller H_2 molecules, the rise in HF also raises the PFR. However, in terms of PFR, the effect of higher oxygen concentration outweighs the effect of more hydrogen in

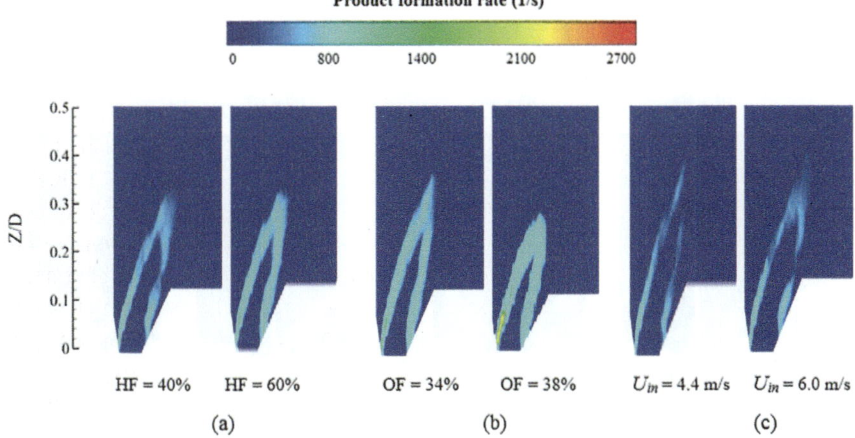

Fig. 6.10 Plots of product formation rate (1/s) at stoichiometric combustion. **a** Effect of HF at OF = 30% and U_{in} = 5.2 m/s, **b** Effect of OF at HF = 20% and U_{in} = 5.2 m/s, and **c** Effect of U_{in} at HF = 20% and OF = 30%

the fuel mixture. It happens because there are more O_2 molecules available, which oxidize more fuel molecules and lead to efficient combustion. This shows that the OF dominates the properties and pace of combustion.

Shorter flame structure is produced with lower U_{in} due to an increase in the PFR in the inner shear layer (Fig. 6.10c). However, the impact of input velocity on PFR is negligible, suggesting that the chemical rather than the physical characteristics of the flame determine combustion rate. The stability transition of the flames from the ORZ to the IRZ is not captured by the computational model, but the trend of PFR provides some insight. It is possible to recollect the flame images from Fig. 6.3 for corner flame stabilization at lower HF and OF and greater U_{in}, revealing the ORZ's major influence on flame stability; however, as HF and OF increase and U_{in} decreases, the ORZ's influence lessens.

Figure 6.11 illustrates the analysis of the mole fraction distributions of CO, OH, CO_2, and H_2O species to determine the emission characteristics of the current flame instances. The fuel mixture produces higher CO and lower CO_2 concentrations when the HF content is raised. One of the potential causes of this is that the addition of hydrogen increases the number of H species that are available and compete with CO for the reaction of atomic O, delaying the creation of CO_2 and promoting the formation of OH. Atomic O and H tend to interact more quickly than CO does when it oxidizes to CO_2. Additionally, higher HF causes the premixed stoichiometric oxy-methane flames' combustor temperatures to rise, which promotes CO_2 dissociation into CO and O species. The elevated CO levels would have also been a result of this. Furthermore, the generation of H_2O via the OH reaction pathway is facilitated by the availability of more H species as a result of hydrogen enrichment. Figure 6.12 illustrates the rise in effluent CO emissions at the combustor exit from 140 ppm at HF = 20% to 279 ppm at HF = 60%.

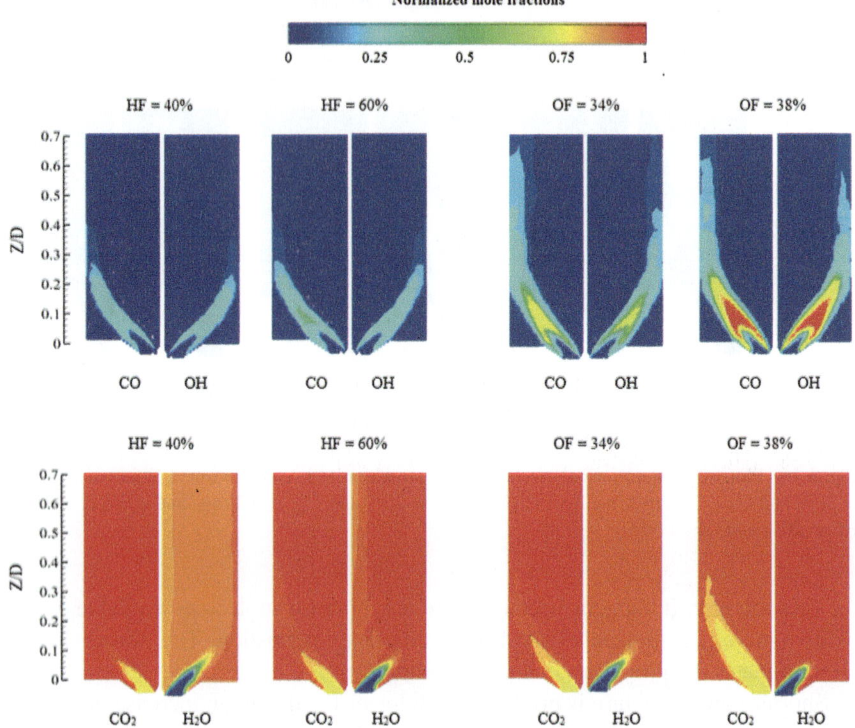

Fig. 6.11 Mole fractions of CO, OH, CO_2, and H_2O at stoichiometric combustion with varying HF and OF

Fig. 6.12 CO emissions at combustor exit at stoichiometric combustion with varying HF and OF at $U_{in} = 5.2$ m/s

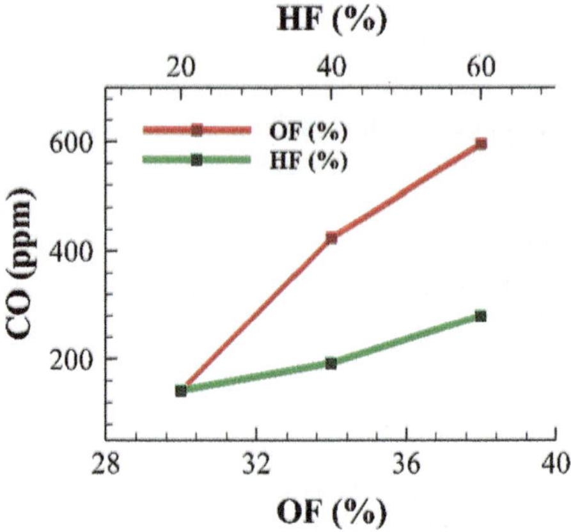

With rising OF, the mole fractions of CO and OH species improve. In the O_2/CO_2 oxidizer combination, higher O_2 concentrations speed up the rate at which CH_4 and H_2 molecules break down into smaller species like CO and OH. CO_2 is a component of the reactant oxidizer combination as well as one of the complete combustion products. On one side, CO is oxidizing to CO_2, but due to the high temperatures, CO_2 dissociates into CO and O species, which dominates. As a result, raising OF raises CO and lowers CO_2 concentrations. Additionally, raising the OF promotes OH oxidation, which leads to higher H_2O concentrations. At OF = 38%, the CO emissions rise from 140 to 594 ppm (Fig. 6.12). The role of the OF on combustion and emissions characteristics is more dominating than that of the HF, as was previously mentioned, and Fig. 6.12 furthers this knowledge.

6.2.2 Operability of Premixed Swirl Combustor Under Hydrogen-Enrichment

The same combustor described in the preceding section is used in this study to assess the combustor operability under hydrogen enrichment using a standard numerical approach. The oxidizer stream and fuel stream compositions were fixed for each experiment. The hydrogen fraction in the fuel mixture (CH_4/H_2) was preserved at 50% (by vol.), while the oxygen fraction in the oxidizer stream (O_2/CO_2) was set at 30% (by vol. The chosen oxidizer composition was made in light of earlier research [3, 19, 42] that indicates the flame characteristics at this composition are comparable to those of traditional air combustion. As a result, it is crucial to conduct further research in order to integrate oxy-combustion technology into the current setup. The operating power density of the experimental testing ranges from 3.5 to 20 MW/m^3/bar, which is typical of gas turbine operation on an industrial scale. This study's base case test was run at 1.0 equivalency ratio, 5.2 m/s throat velocity, and 0.98 swirl number. To create a comparison study based on the impacts of these parameters on the stability and combustion properties of the created flames, these parameters were adjusted separately. The document [31] contains the specifics of the test protocol used in this investigation, which is identical to our prior study.

6.2.2.1 Effect of Equivalence Ratio

The effects of the equivalency ratio on the stability and flame morphology of premixed H_2–CH_4–O_2–CO_2 flames have been investigated. The fuel mixtures' hydrogen content was fixed at 50%, while the composition of the oxidizers was set to 30%O_2–70%CO_2. By altering the flow rates of both the fuel and the oxidizer, the equivalency ratio was changed while maintaining a consistent volumetric flow rate across all cases. Stoichiometric (near flashback limit), $\varphi = 0.75$ (middle stable combustion), and $\varphi = 0.5$ (near blowout limit) are the three situations that have been taken into consideration. Figure 6.13 compares photos taken with a high-definition camera, which show the flame shape that was measured experimentally, with numerical contours of OH mole fractions on the vertical plane that passes through the center of the combustor, which show the flame shape that was estimated numerically. The extension of our prior work [9] is shown in the figure with an equivalency ratio of 1. The experimental and numerical images of the flames' forms reveal that the numerical model is effective in predicting the flame structure seen in the figure. Compared to flames at lower equivalence ratios, the flame at $\varphi = 1.0$ is compact, intense, bright, violent, and characterized by a high OH concentration. As seen in Fig. 6.14, the flame exhibits a typical V-shape that is stabilized at both the inner and outer shear layers (ISL and OSL). The flame stabilization transitions from swirl stabilized to corner stabilized as the equivalence ratio is decreased, and this change is visible at $\varphi = 0.75$. When the equivalency ratio is lower ($\varphi = 0.5$), the flame is completely stabilized at the inner shear layer and adopts the confinement-like shape that corner stabilized flames are known for. Due to decreased temperature and reaction rates, the numerical findings demonstrate lower OH concentrations. A single big eddy is seen in the inner recirculation zone (IRZ) with lower equivalence ratios; there is no outside recirculation. A tiny outer recirculation begins to form as the equivalency ratio is raised (to 0.75), and the flame also begins to stabilize near the OSL. Additionally, a second, smaller eddy that is moving in a clockwise manner forms, and the center of the huge eddy in the IRZ slips downstream away from the burner. This can be explained by the powerful reaction occurring at greater equivalency ratios, which raises the temperature and speeds up the burning of gases. The expansion of burnt gases, which increases disturbance and creates different shear layers, also contributes to the formation of smaller eddies and a distinct outer recirculation zone. When the equivalency ratio is increased further (to 1.0), the main eddy moves farther away from the burner and two smaller eddies form in the IRZ, revolving in opposite directions. The flame is firmly stabilized between the two shear layers and a clearly defined outer recirculation zone (ORZ) is also seen to emerge.

Figure 6.15 shows the temperature profiles in the radial direction at heights of $Z/D = 0.1$ and 0.5 from the burner. When the combustion advances towards the leaner regimes, the peak at the height of $Z/D = 0.1$ becomes diffused and lower and is farther from the center of the burner. The burnt-hot gases are represented by the peaks on either side of the valley, which stands in for the jet of cooled reactive gases. The wall where the flame is stabilized without any external recirculation is attached to the valley at $\varphi = 0.5$. Due to the presence of a unique ORZ, the temperature level drops at 1.0 less than it does in the other two circumstances. Similar patterns can be seen in the temperature profiles at $Z/D = 0.5$, as the temperature level drops as the combustion becomes leaner. Furthermore, due to radiative and convective losses to the environment, the temperature level drops towards the combustor wall. Figure 6.16 shows the rate of product formation (PFR) in the radial direction at heights of $Z/D = 0.1$ and 0.5 from the burner. PFR features peaks closer to the burner center at the stoichiometric combustion, with a valley between the two peaks signifying the unburnt jet. Peaks move farther from the burner center as combustion becomes leaner, and the valley deepens. Additionally, variations in the PFR profile

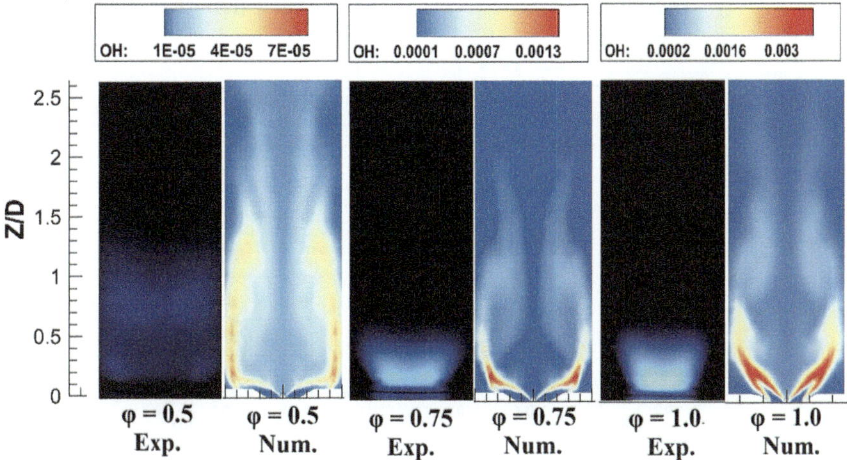

Fig. 6.13 Experimental and numerical flame macrostructure at different equivalence ratios (case $\varphi = 1.0$ [9])

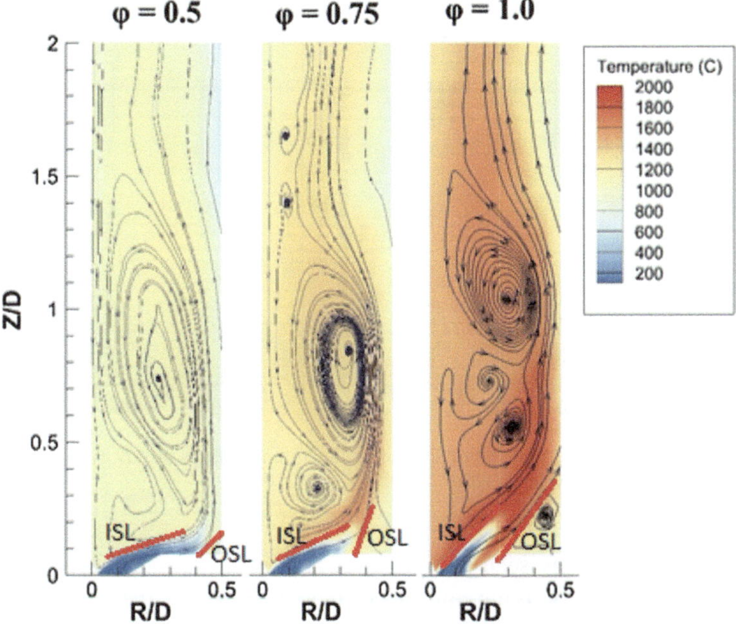

Fig. 6.14 Velocity field over temperature plots at different operating equivalence ratios (case $\varphi = 1.0$ [9])

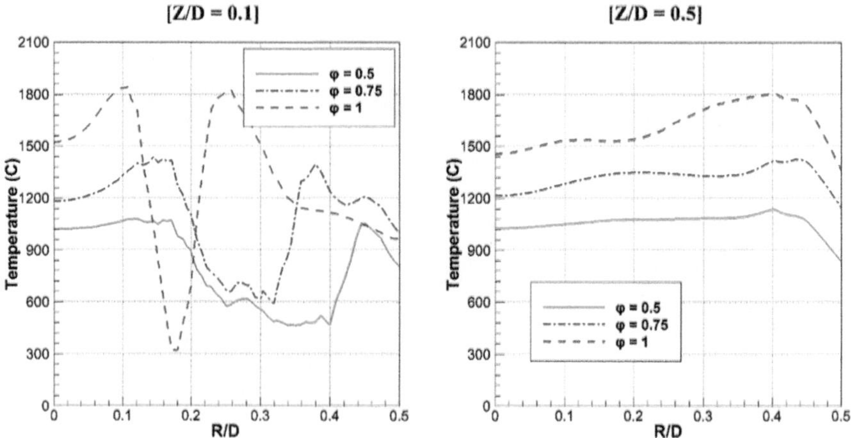

Fig. 6.15 Radial temperature profiles at heights of $Z/D = 0.1$ and 0.5 from the burner at different equivalence ratios

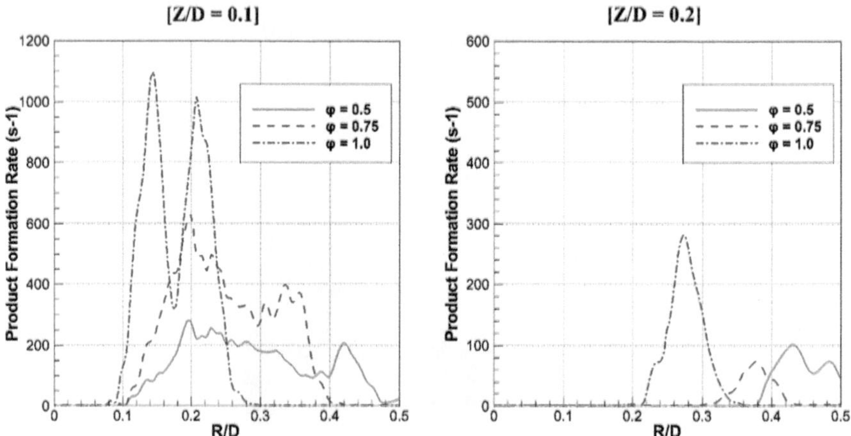

Fig. 6.16 Product formation rate along the radial direction at heights of $Z/D = 0.1$ and 0.5 from the burner at different equivalence ratios

at leaner combustion are seen as a result of high strain rates and localized flame quenching. Additionally, as the equivalence ratio is decreased, the PFR's magnitude lowers, indicating slower reaction speeds.

Figure 6.17 shows how the equivalency ratio affects the flame thickness. Increasing from 31.9 m at $\varphi = 1.0$ to 55.5 m at $\varphi = 0.75$, the flame thickness. The flame thickness rises to more than three times the stoichiometric flame thickness with a further decrease in the equivalency ratio ($\varphi = 0.5$). As a result, the flame thickness approaches the Kolmogorov scales at lower equivalence ratios. The fact that the Karlovitz number rises from 0.24 to 1.82 when the equivalence ratio falls from 1.0 to

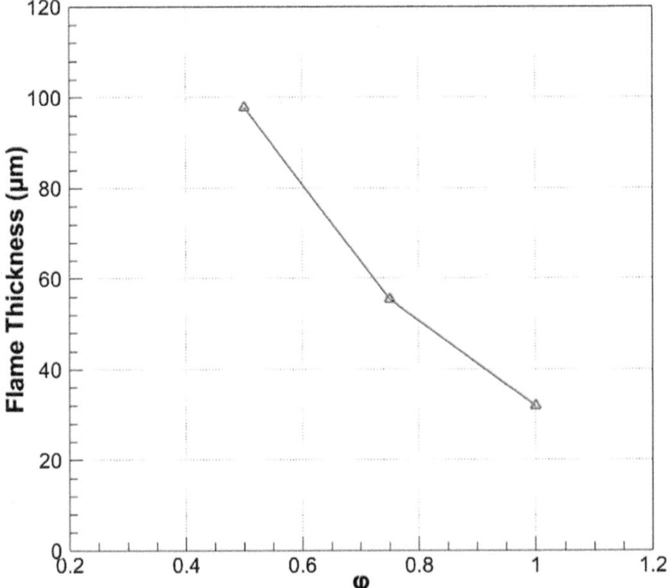

Fig. 6.17 Influence of operating equivalence ratio on flame thickness

0.5, respectively, makes this clear from Fig. 6.18. The integral time scale is therefore still longer than the chemical time scale, but the smallest eddies are equivalent to the flame thickness and are capable of interacting with the inner flame. The Damkohler number has dropped from 1672 (at 1.0) to 159 (at 0.5), demonstrating that chemical kinetics continues to dominate the reaction rate. The thickened flame regime is a term used to describe this type of flame. Additionally, the Kolmogorov eddies' stretch brings the crucial flame stretch even closer, raising the likelihood of flame quenching. This occurrence can be predicted by the model, as shown in Fig. 6.19. As the equivalency ratio drops, so does the critical rate of strain as well as the flame strain rate. However, until both values are virtually equal at $\varphi = 0.5$ and the effect of flame quenching becomes apparent, the critical strain rate considerably declines relative to the flame strain rate. The experimental findings, which showed that any attempt to lower the equivalency ratio lower than 0.5 was ineffective and that the flame blowout was observed, also confirm this.

6.2.2.2 Effect of Inlet Reynolds Number

This section examines how the throat Reynolds number affects the stability and flame morphology of premixed H_2–CH_4–O_2–CO_2 flames. Based on the hydraulic throat diameter, the volumetric flow rates of the reactant gases were changed to obtain the Reynolds numbers of 7696, 9095, and 10,494. The reactant gases' mass flow rates

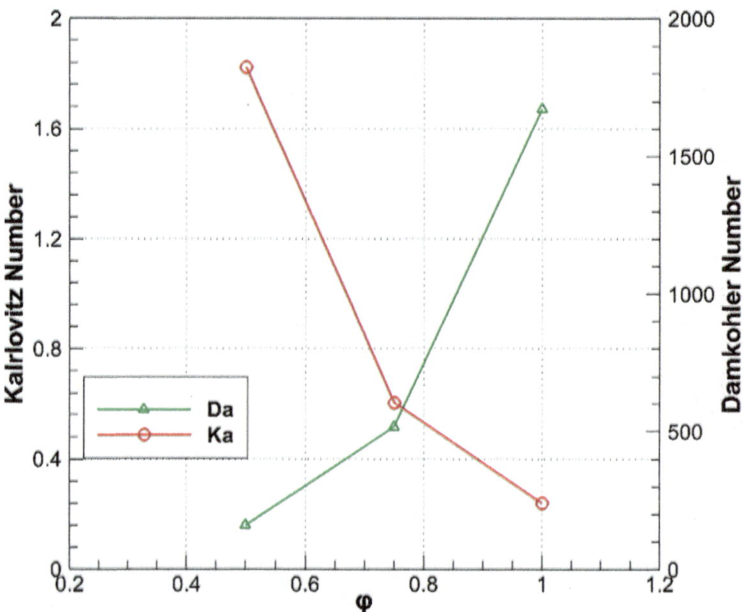

Fig. 6.18 Influence of operating equivalence ratio on the Karlovitz and the Damkohler numbers

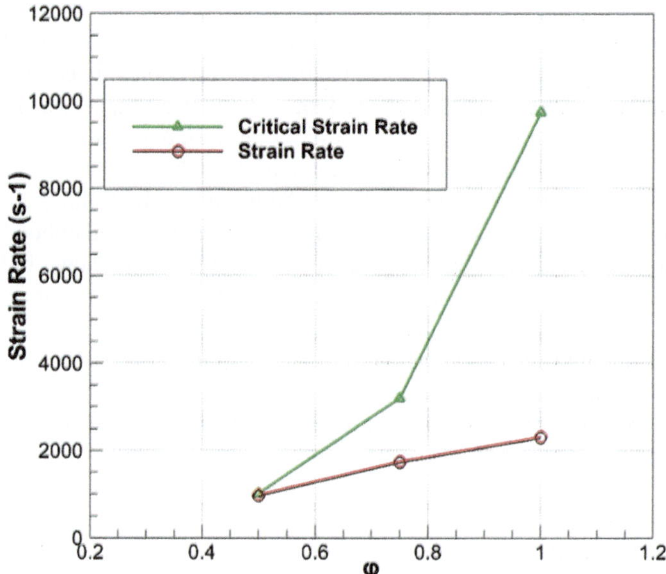

Fig. 6.19 Critical and normal strain rates at different equivalence ratios

were changed to reach the Reynolds numbers, and the corresponding burner exit velocities in the three cases were 4.4, 5.2, and 6.0 m/s. The hydrogen fraction and equivalency ratio were set at 50% and 1.0, respectively. Figure 6.20 displays the high-definition photos of an experimental flame along with contours of the OH radical concentration. The findings show that the flame form is unaffected by Reynolds number, with the exception of a minor lengthening of the flame, which is represented by both the experimental and numerical flame shapes. Higher burner axial exit velocities can be the cause of the flame elongation with Reynolds number that has been seen in prior studies [43]. The flames are typically V-shaped and stabilized between the inner and outer shear layers at all of the tested Reynolds values. Figure 6.21, which displays the velocity flow-field over the three examples' temperature contours, helps to better visualize the regime. A noticeable ORZ and a sizable IRZ define the flames. The center of the huge eddy in the IRZ shifts downstream away from the burner as the Reynolds number rises. Additionally, the secondary smaller eddies, which are distinguishable at the lower Reynolds number, are engulfed by the principal eddy at the higher Reynolds number as it lengthens and expands. Localized temperature distribution characterizes the smaller primary eddy, however as the Reynolds number rises, the primary eddy lengthens and improves the temperature dispersion. But the recirculation zones were only supposed to have a minor influence on the structure of the flame.

When the Reynolds number goes further, as shown in Fig. 6.22, no discernible change is seen in the Damkohler number, which was found to rise from 1050 at Re = 7596 to 1670 at Re = 9095. Beyond 9000, it was seen that the effect of Reynolds number on the Karlovitz number began to decline. This asymptotic pattern might be explained by the close proximity of the flashback point to Re = 9000 [9]. As the Reynolds number rises, the power density increases, and the flame becomes extremely unstable. Figure 6.23 shows how the throat Reynolds number affects the strain rate. As the Reynolds number rises from 7596 to 10,494, the strain rate was observed to rise from 2033 to 2665 s^{-1}. However, this slight increase in strain rate

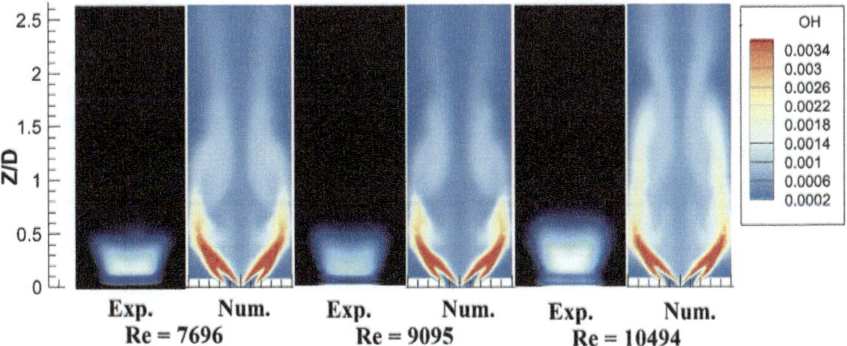

Fig. 6.20 Experimental and numerical flame macrostructure at different Res (Case Re = 9095 [9])

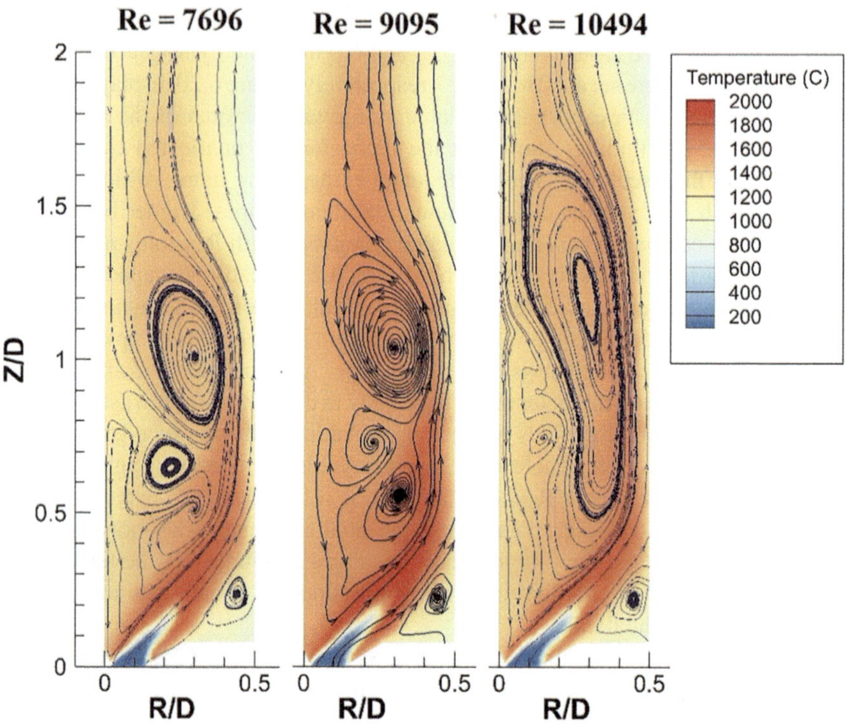

Fig. 6.21 Velocity field over temperature plots at different Res (Case Re = 9095 [9])

is still well below the critical rate of strain in these circumstances, therefore no appreciable impact on the flame is seen.

6.2.2.3   Effect of Flow Swirl

This section investigates numerically the impact of the swirl number on the stability and flame properties of H_2–CH_4–O_2–CO_2. All examples had constant volumetric flow rates, but the swirl number was altered by varying the velocity's tangential component. In the oxidizer stream, the hydrogen proportion was fixed at 50%, while the oxygen fraction was set at 30%. At stoichiometric circumstances, all the flames were examined. Figure 6.24 shows the OH mole fraction contours for the three situations that were analyzed. It was discovered that the flame with a lower swirl number (SN = 0.69) had a longer length and distinctive "V" form. When can be observed from the graph, when the SN increases, the flame length decreases and the flame widens. Both the inner and outer shear layers stabilize the flames at SN = 0.69 and 0.98. However, like the flames at lower equivalency ratios mentioned previously, the flame at a greater swirl number (SN = 1.48) tends to stabilize at the inner shear layer.

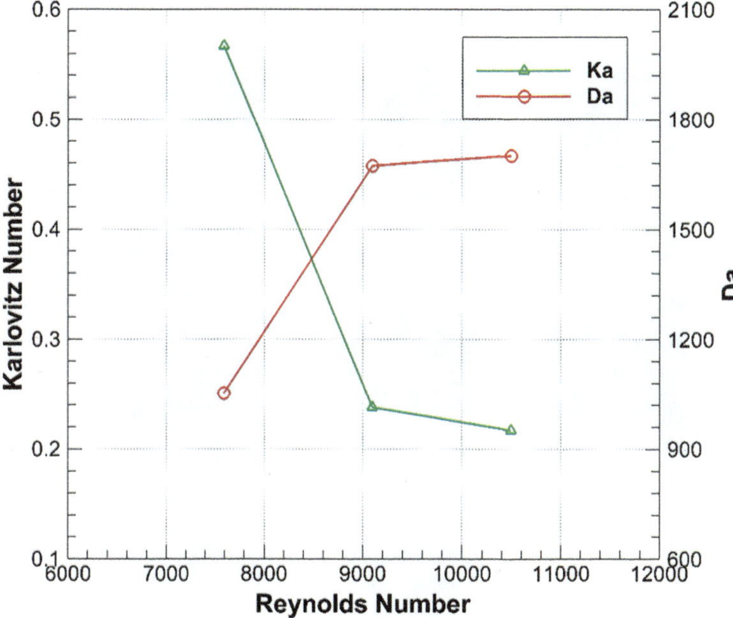

Fig. 6.22   Effect of Reynolds number on Ka and Da numbers

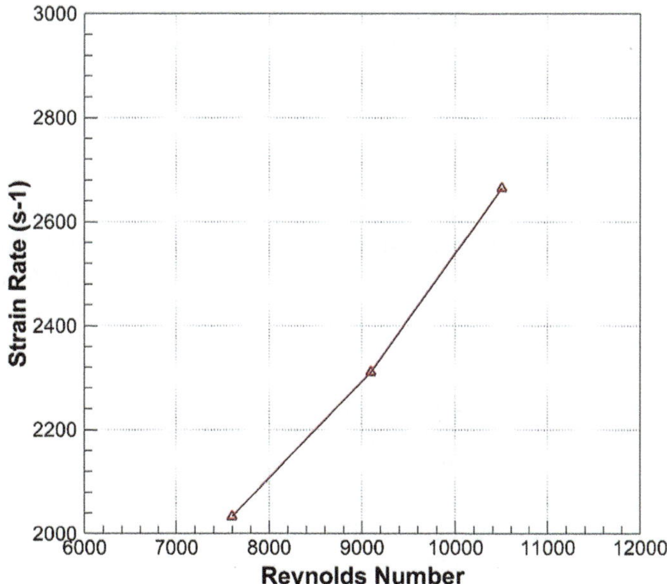

Fig. 6.23   Strain rate at different Res

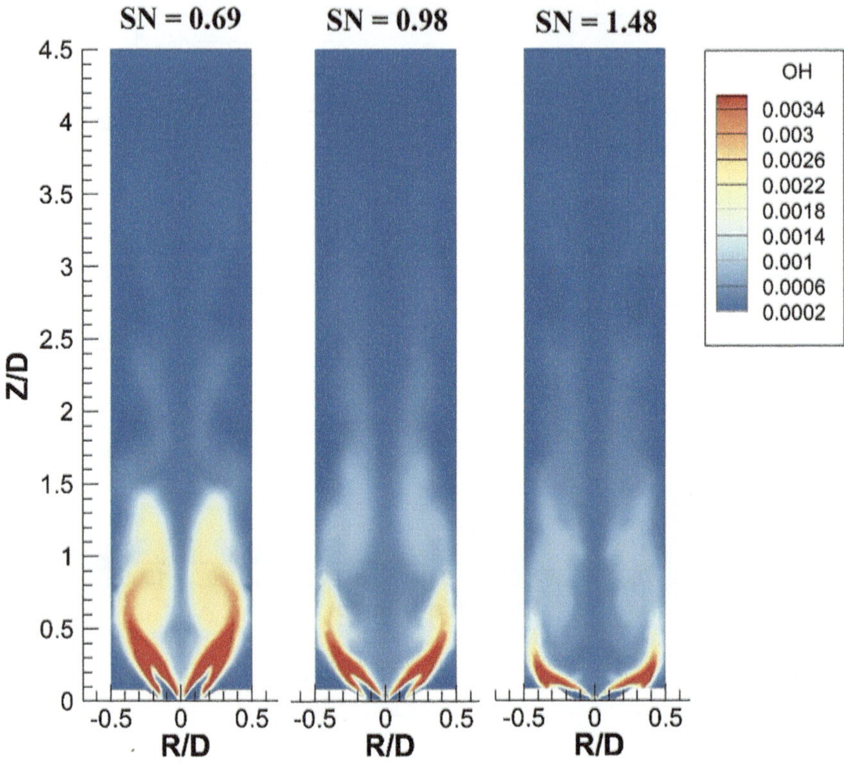

Fig. 6.24 Calculated flame macrostructure at different swirl numbers (Case SN = 0.98 [9])

Figure 6.25 shows the velocity flow-field over the various SN's temperature contours. At the lower swirl number (SN = 0.69), a small secondary eddy revolving in the opposite direction from the first large eddy is shown in the IRZ. In this regime, a unique ORZ is also seen. More secondary eddies form as the swirl number rises to 0.98, and the ORZ becomes constricted yet still recognizable. The principal eddy in the IRZ moves farther away from the burner downstream. Higher swirl number flames (SN = 1.48) exhibit a massive primary eddy in the IRZ that engulfs all lesser eddies. Additionally, a second recirculation zone that also rotates counterclockwise seems to form further downwind of the combustor. The flame tends to stabilize at the inner shear layer as the ORZ decreases. The reaction becomes more severe as the tangential component of the velocity increases, and the temperature distribution concentrates close to the burner base. The vortex continues to exist further downstream of the combustor when the swirl number is raised. When the vortex reaches the combustor's exit in the scenario with a larger swirl number (SN = 1.48), a low-pressure zone is created in the middle. As a result, a recirculation zone is seen close to the combustor output, where it entrains ambient air and raises the combustion's

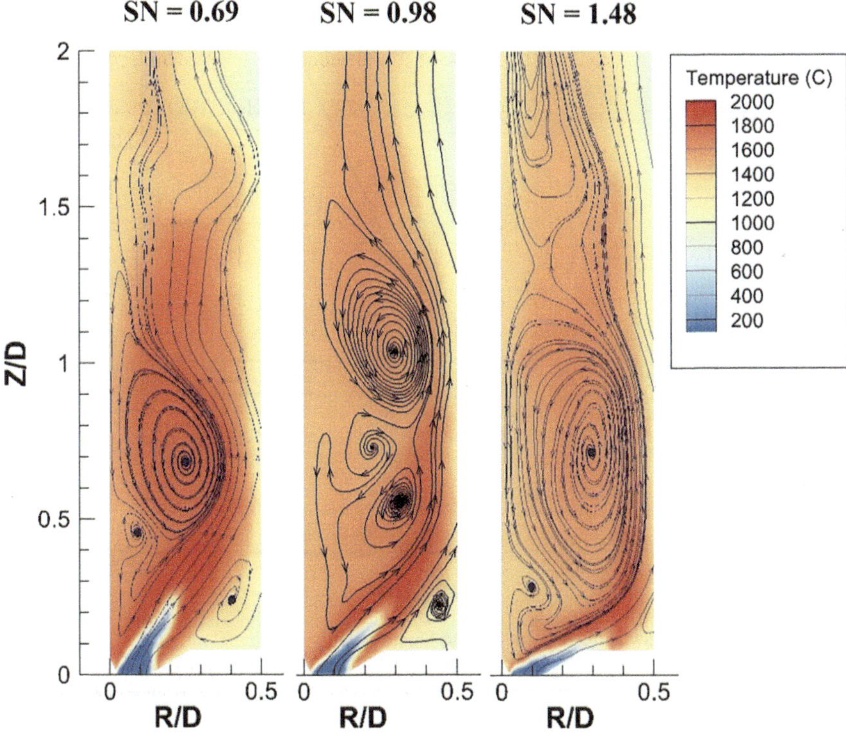

Fig. 6.25 Velocity field over temperature plots at different swirl numbers (Case SN = 0.98 [9])

overall stoichiometry. But given that the reaction zone virtually reaches its conclusion before the recirculation zone (Fig. 6.24), it scarcely has an impact on combustion.

Figure 6.26 shows the impact of the swirl number on the Karlovitz number and Damkohler number. Both of the non-dimensional numbers rise with the swirl number, albeit Ka just little, from 0.22 to 0.29, and SN slightly, from 0.69 to 1.48, respectively. This shows that increased turbulence is the cause of the Kolmogorov scale declining with swirl number. However, a significant increase in the Da (from 1300 to 2140) is shown along with the swirl number shift, indicating a longer integral time scale than the chemical time scale as a result of the enlargement of the massive eddy as previously illustrated in Fig. 6.25. The impact of the swirl number on the strain rate is depicted in Fig. 6.27. When the swirl number was increased from 0.60 to 1.48, respectively, it was discovered that the strain rate grew monotonically from 1540 to $2917 \, s^{-1}$.

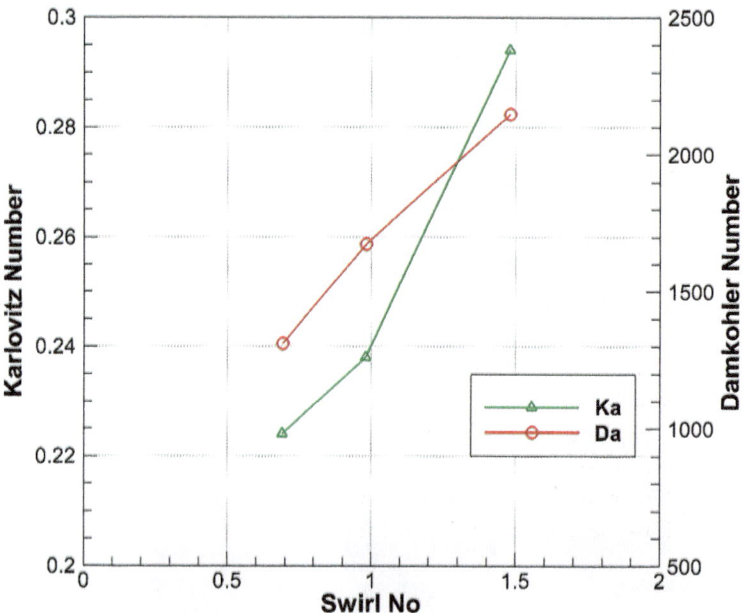

Fig. 6.26 Influence of swirl number on Ka and Da numbers

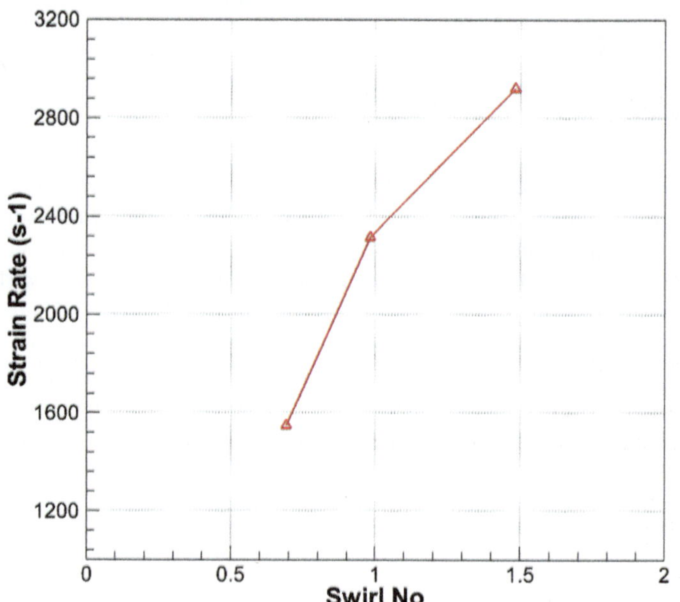

Fig. 6.27 Strain rate over range of swirl numbers

6.3 Premixed Swirl Combustor Holding Air- and Oxy-propane Flames

Due to rising fuel consumption and greenhouse gas (GHG) emissions, fossil fuel combustion is receiving more attention as the world's energy needs grow. To address this energy demand, large-scale energy generation typically uses gas turbines powered by fossil fuels, making future development in gas turbines essential for improving fuel efficiency and lowering emissions [44]. Air-combustion is used in the majority of those gas turbines, which causes the production of dangerous GHG gases including CO_2, CO, NOx, SOx, etc., which are mostly to blame for global warming and air pollution. Even though there are several methods available and in use today to capture combustion exhaust gases, these methods are still too expensive and sophisticated to be used in every gas turbine application. Among those methods, oxyfuel combustion has the potential to be a workable replacement for traditional air–fuel combustion [45, 46]. In contrast to air combustion, oxy-combustion uses pure O_2 as the oxidizer. There is no chance of thermal NOx (by the Zeldovich mechanism) creation during the combustion process because only O_2 is delivered to the combustor. This is true unless impure fuel is used or there is an air leak into the combustor. The lack of nitrogen results in a lower volume of the oxidizer, which is compensated by recirculating some CO_2 from the exhaust gas. Since only O_2 is utilized as an oxidizer, the main components of the exhaust stream are CO_2 and H_2O. H_2O may be readily separated from CO_2 by condensation, and the leftover CO_2 can be recycled or recovered for use elsewhere [47]. A further benefit of oxyfuel combustion is that it generates approximately 75% less flue gas than the conventional air-combustion system [48]. As a result, implementing oxyfuel combustion may result in the use of zero-emission power plants (ZEPPs).

Lean premixed (LPM) combustion, which lowers the combustor's operating temperature and NOx emissions, is the typical combustion condition in air-combustion. Alternatively, since producing oxygen from air is an expensive and energy-intensive process, oxyfuel combustion operates close to stoichiometric state to lower the oxygen consumption. By reducing the intensity of the flame using recirculated CO_2 from the exhaust, the high temperature produced by oxyfuel combustion, which involves burning fuel in pure oxygen, is decreased. Oxyfuel combustion is the ideal replacement for air–fuel combustion due to its higher flame temperature, greater emission management, and less hardware modifications [49, 50]. Furthermore, the move from the traditional air-combustion to the cutting-edge oxy-combustion technology is motivated by the strict pollution restrictions and the depleting fossil fuel reserves. However, switching to oxy-combustion necessitates a thorough understanding of the behavior of combustion with various fuel/oxidizer mixtures.

Due to its inherent advantage in flexible fuel/oxidizer combustion, oxy-combustion technology has received a lot of attention recently [51–54] in a number of review works on gas turbine combustion. This is also reflected in the authors' most recent review article [55], which singles out swirl-stabilized burners as one of the combustors that is particularly well suited for the use of oxy-combustion technology. The research on oxyfuel combustion in various burners, such as dry low NOx (DLN) burners, swirl-stabilized enhanced vortex (EV) burners, perforated plate (PPB) burners, and micromixer (MM) burners, has been supplemented throughout the last years by a number of studies that have been published in the open literature. The swirl-stabilized burners have received the most oxy-combustion application research out of the group. In addition to burners, research has also been done on oxy-combustion with different fuels (such as methane, propane, syngas, ammonia, and hydrogen) at various combustion parameters, such as equivalence ratio (), oxygen fraction (OF), adiabatic flame temperature (Tad), and flame speed (FS). Methane has been used as the fuel in the bulk of investigations on oxyfuel combustion in swirl-stabilized combustion chambers [4, 9, 10, 56–64], while propane has been the subject of far fewer studies [25, 65–72] than methane. In a model swirl-stabilized non-premixed gas turbine combustor, Abubakar et al. [65, 66] examined the features of oxy-propane combustion and contrasted the findings with those of propane-air combustion. When the swirl number is 1.0, the results indicated a stable oxy-propane flame that was almost stoichiometric. They noted that the critical velocity of oxygen-propane combustion was six times greater than that of air-propane combustion, which controls the transition from an attached flame to a lifted flame. According to one study, both flame and flow dynamics affect the oxy-propane flame's blowout characteristics. Additionally, compared to those at swirl numbers of 0.6 and 1.5, propane oxyfuel combustion showed larger stability limits at 1.0 [66]. Abubakar et al. [68] looked at the features of non-premixed oxy-propane combustion with hydrogen enrichment in a different study. When hydrogen enrichment was increased by 30% (by volume), the oxy-propane flame's blowout limit increased by 2.5%. Contrarily, adding hydrogen increased the amount of carbon monoxide (CO) that was released into the exhaust, reaching a high of 1.93% CO at a 40% (by volume) hydrogen concentration in the feeding fuel mixture. The line of lean blowout limit is found to follow the constant adiabatic flame temperature contour, according to research by Ali et al. [67, 70] on the CO_2-dialuted oxy-propane flame in premixed combustion settings. At the same adiabatic flame temperature, it was also seen that the flame forms were identical. Recent experimental research [25] on the thermoacoustic instability of oxy-propane flames found that vortex-flame interactions may contribute to the thermoacoustic instability. Up to 60% CO_2 content, an increase in auditory amplification was seen, after which the oscillations decreased. Alternately, when the hydrogen content in the feeding fuel mixture increased, the acoustic amplitude decreased, which can be explained by the weakening of vortex-flame interaction at higher hydrogen concentrations [25].

Propane has the potential to be employed as an alternative fuel to methane-based premixed combustion systems since it has a higher carbon content, higher adiabatic flame temperature, and a higher Wobbe index value [55]. Therefore, for practical application, a thorough grasp of the oxy-propane combustion is essential. To the best of our knowledge, the literature has rarely taken into account the specific properties of oxy-propane combustion in premixed gas turbine combustors. Two research on air- and oxy-propane combustion for gas turbine combustion applications are described in this section. In the first study, combustion, exhaust gas concentrations, and stability characteristics of CO_2-diluted oxy-propane combustion in a model premixed swirl-stabilized gas turbine combustor are all covered in detail using large eddy simulations (LES) and experimental measurements. The study presents the stability and combustion properties of the $C_3H_8/O_2/CO_2$ flame over a wide range of operating circumstances in an effort to enrich the oxy-propane studies for gas turbine application. In order to provide designers and operators of gas turbines with knowledge of the stable flame behavior, the second study (Sect. 6.3.2) examines the combustion performance of these flames in a dry low emission (DLE) lean premixed (LPM) swirl-stabilized model gas-turbine combustor. Rarely is information on the static stability and macrostructure of oxy-propane flames found in the literature that is currently available. The goal would be to replace the current air-based combustors with more environmentally friendly oxy-fuel combustors. According to the authors' best knowledge, this study fills this gap by empirically examining the behavior of premixed $C_3H_8/O_2/N_2$ and $C_3H_8/O_2/CO_2$ flames. By using fuel-flexible gas turbines, the benefits of O_2-enriched air-propane and oxy-propane combustion can be taken advantage of with little retrofitting. In this study, the combustion stability and structure of these flames are examined throughout a range of oxygen fractions (21–70%) and equivalency ratios (0.1–1.0) at a fixed flame-base bulk throat velocity of 5.2 m/s.

6.3.1 Characteristics of Oxy-propane Flames in Swirl Combustor

In this work, a swirl-stabilized dry low emission (DLE) model gas turbine combustor that adopts the lean-premixed (LPM) combustion approach is used to explore the properties of propane-fueled oxy-combustion. A quartz tube with an inner diameter of 76 mm and a length of 30 cm surrounds the combustion zone. The base plate is made of steel and is 10 mm thick. The center of the combustor headend is occupied by a solid conical center body with a 9.5 mm outside diameter and a 28° cone angle. 5.2 m/s bulk throat velocity and a swirler with 8 vanes at a 55° angle are used to fully premix the fuel-oxidizer mixture before it enters the combustion chamber. Equation (6.19), which calculates the swirl number (SN), [73, 74]:

$$SN = \frac{2}{3}\left[\frac{1 - \left(\frac{D_{cb}}{D_{in}}\right)^3}{1 - \left(\frac{D_{cb}}{D_{in}}\right)^2}\right] \tan(\alpha_{sw}) = 0.98 \tag{6.19}$$

where D_{cb}, D_{in}, and α_{sw} refer to diameter of swirler center body, inner diameter of mixing plenum, and swirler blade angle, respectively.

While the exhaust gas combination mostly consists of CO_2 and H_2O, the incoming gas mixture also contains C_3H_8, O_2, and CO_2. The presence of carbon dioxide at the entrance indicates the recirculation of CO_2 from the exhaust stream following CO_2

Table 6.2 The conditions of the different flames considered in the present numerical study identified by the case number

Case number	Oxygen fraction (OF) (%)	Adiabatic flame temperature (T_{ad}), K	Equivalence ratio (Φ)	Remarks
1	60.0	1681	0.26	Fixed OF
2		2006	0.34	
3		2303	0.42	
4	45.0	2302	0.597	Fixed T_{ad}
5	35.0		0.80	
6	48.0	2156	0.50	Fixed Φ
7	39.0	1837		
8	35.0	1694		

collection and H_2O condensation. This work examines experimentally and numerically the stability and flame characteristics of oxy-propane combustion over a range of operational equivalency ratios and oxygen fractions (φ: 0.26–0.80, OF: 35%–60%). Our earlier experimental study [67] has more information regarding the experimental setup, measurement methods, and testing process.

Based on the experimental findings, some oxy-propane flames from the stable flame operation zone were chosen in order to conduct a thorough numerical investigation. In Table 6.2, the chosen flames of various OF, φ, and Tad are listed. The parameters of combustion and emissions, as well as flame form, temperature distribution, streamlines, and vorticity, were compared for various flames. The results are shown in a dimensionless way, regardless of the combustor dimensions, with the axial and radial distances expressed as Z/D and R/D, respectively, in relation to the combustor diameter (Fig. 6.28).

6.3.1.1 Modelling Premixed H_2-enriched Oxy-propane Combustion

The LES model in ANSYS Fluent 19R3 was used to solve the conversation equations of mass, momentum, and energy as well as the transport equation for a scalar variable in the three-dimensional (3D) computational domain. In order to anticipate the properties of heat transfer, flow, and emission, the 3D domain elliptic equations have to be simultaneously solved. Equation (6.20) [75] contains the general conservation equation for mass, momentum, energy, as well as species transport. Here, u_j is the velocity component along the j direction, _ is the diffusion component, S_ is the source term, and are the dependent variable's Reynolds averaged and fluctuating components, respectively.

$$\frac{\partial}{\partial x_j}\left(\overline{\rho U}_j \Phi + \overline{\rho u_j \emptyset}\right) = \frac{\partial}{\partial x_j}\left[\Gamma_\emptyset \frac{\partial \Phi}{\partial x_j}\right] + \rho \overline{S}_\Phi \qquad (6.20)$$

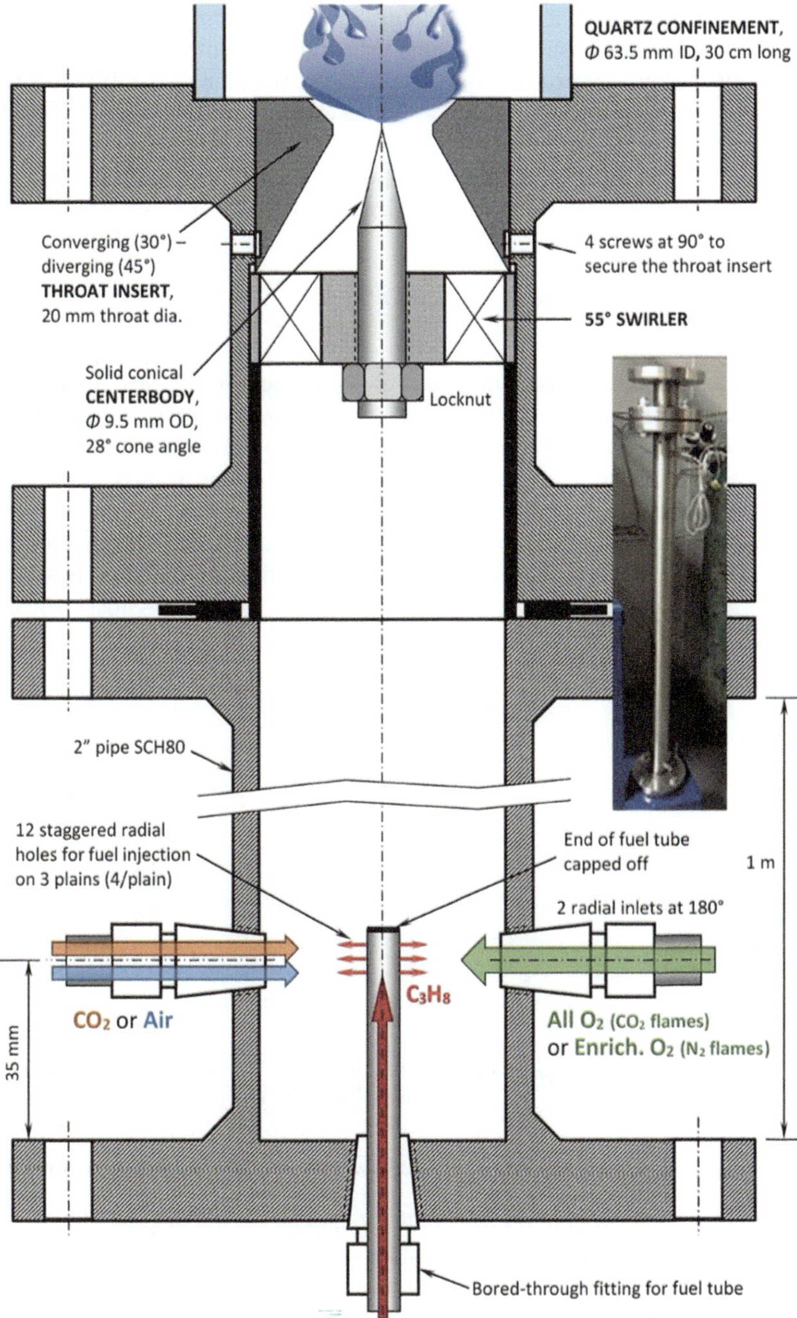

Fig. 6.28 Diagrammatic depiction of the swirl-stabilized model gas turbine combustor experimental setup

The discrete ordinate (DO) modeling approach, which is recommended for modeling oxyfuel combustion and is valid for a wide range of optical thickness, was used to forecast radiant heat transfer. One radiative iteration of the DO model was performed for every 10 energy iterations in the current investigation. The weighted sum of gray gas model, or WSGGM, was used to compute the absorption coefficient of the gaseous mixture under a variety of operating conditions [76]. Equation (6.21) shows the total emissivity (ε) in terms of the weighted sum of gray gases.

$$\varepsilon = \sum_{i=0}^{i=I} \alpha_{\varepsilon,i}(T)\left[1 - e^{-\kappa_i PL}\right] \tag{6.21}$$

The two forms of eddies that make up flow turbulence are large and minor eddies. While small eddies result in the kinetic energy of turbulence being dissipated, huge eddies refer to the typical length of the mean flow. In this work, turbulence was reduced by filtering out smaller eddies using the LES modeling method. The low-pass filtering procedure for a quantity can be performed using Eq. (6.22) [9], where G is the filter function used to determine the scale of resolved eddies and the integration covers the full domain. As a result, after incorporating the filtering function, Eqs. (6.23) and (6.24) display the expression for continuity and the Navier–Stokes equation, respectively.

$$\overline{\varphi}(x, t) = \int G(r, x)\varphi(x - r, t)dr \tag{6.22}$$

$$\frac{\partial \rho}{\partial t} + \frac{\partial}{\partial x_i}(\rho \overline{u}_i) = 0 \tag{6.23}$$

$$\frac{\partial}{\partial t}(\rho \overline{u}_i) + \frac{\partial}{\partial x_j}(\rho \overline{u}_i \overline{u}_j) = \frac{\partial}{\partial x_j}\left(\mu \frac{\partial \sigma_{ij}}{\partial x_j}\right) - \frac{\partial \overline{p}}{\partial x_i} - \frac{\partial \tau_{ij}}{\partial x_j} \tag{6.24}$$

Here σ_{ij} and τ_{ij}, corresponding to stress tensor from molecular viscosity and subgrid-scale stress, are expressed by Eqs. (6.25) and (6.26), respectively.

$$\sigma_{ij} \equiv \left[\mu\left(\frac{\partial \overline{u}_i}{\partial x_j} + \frac{\partial \overline{u}_j}{\partial x_i}\right)\right] - \frac{2}{3}\mu \frac{\partial \overline{u}_i}{\partial x_j}\delta_{ij} \tag{6.25}$$

$$\tau_{ij} \equiv \rho \overline{u_i u_j} - \rho \overline{u}_i \overline{u}_j \tag{6.26}$$

Additionally, the isotropic portion of the subgrid-scale stresses and filtered static pressure terms were combined to simulate the subgrid-stresses that emerged from the filtering operation. Equation (6.27) illustrates the modeling of the deviator portion using turbulent viscosity.

$$\tau_{ij} = -2\mu_t \overline{S}_{ij} \tag{6.27}$$

Turbulent viscosity (μ_t) is determined as:

$$\mu_t = \rho L_s^2 \frac{\left(S_{ij}^d S_{ij}^d\right)^{3/2}}{\left(\overline{S}_{ij}\overline{S}_{ij}\right)^{5/2} + \left(S_{ij}^d S_{ij}^d\right)^{5/4}} \tag{6.28}$$

Here, L_s represents the mixing length for subgrid-scale as expressed in Eq. (6.29), while the term S_{ij}^d can be evaluated from Eq. (6.30). The terms κ, d, C_w, and V in Eq. (6.29) symbolize the Karman constant, distant to the closest wall, WALE constant, and volume of the computational cell, respectively.

$$L_s = \min\left(\kappa d, C_w V^{1/3}\right) \tag{6.29}$$

$$S_{ij}^d = \frac{1}{2}\left(\overline{g}_{ij}^2\overline{g}_{ij}^2\right) - \frac{1}{3}\delta_{ij}\overline{g}_{kk}^2; \quad \overline{g}_{ij} = \frac{\partial \overline{u}_i}{\partial x_j} \tag{6.30}$$

The partially premixed combustion model, which was altered to account for fully premixed combustion, was used to model the burning of CO_2-diluted oxy-propane. To obtain additional scalar variables, transport equations for the mean reaction progress variable (c), mean mixture fraction (f), and mixture fraction variance ((f')^2) were solved. In this case, the mass fraction of a suitable species, whose value is zero in the unburned mixture and 1.0 in the entirely burned mixture, corresponds to the reaction progress variable. Equation (6.31), where s and Y are the oxygen to fuel ratio and mass fraction, can be used to calculate the mixture fraction. The terms Y_{ox} and Y_{fu} correspond to the mass fractions of oxygen and fuel, respectively, whereas the subscripts 0 and 1 denote the incoming stream of oxidizer and fuel, respectively.

$$f = \frac{sY_{fu} - Y_{ox} + Y_{ox,0}}{sY_{fu,1} + Y_{ox,0}} \tag{6.31}$$

Equation (6.32) can be used to calculate temperature and species fractions, or density-weighted average scalars, from the probability density function (PDF) and progress variable. In this statement, the subscripts u and b stand for unburnt and burned mixes, respectively.

$$\overline{\emptyset} = \overline{c}\int_0^1 \emptyset_b(f)p(f)df + (1 - \overline{c})\int_0^1 \emptyset_u(f)p(f)df \tag{6.32}$$

On the other hand, the progress variable can be stated in terms of the mass fraction of the species and is computed using Eq. (6.33), where Y_i and $Y_{(i,ad)}$ are the mass fraction of the species as a product and the mass fraction of the species after complete combustion, respectively. The progress variables' values vary from 0 to 1, with 0 denoting an unburnt mixture and 1 denoting an already-burnt mixture.

$$c = \left(\frac{\sum_{i=1}^{n} Y_i}{\sum_{i=1}^{n} Y_{i,ad}} \right) \tag{6.33}$$

Equation (6.34) can be used to calculate the heat release factor (), which is a non-dimensional metric used to quantify the heat released during combustion. Here, T_{ad} is the temperature of the adiabatic flame, and T_0 is the temperature of the reactant mixture that has not yet burned.

$$\gamma = \left(\frac{T_{ad} - T_0}{T_{ad}} \right) \tag{6.34}$$

Flame thickness (δ) can be calculated from the correlation between unburnt thermal conductivity (λ_u), density (ρ_u), specific heat capacity (c_p), and laminar flame speed (S_L), as indicated in Eq. (6.35).

$$\delta = \frac{\lambda_u}{\rho_u c_p S_L} \tag{6.35}$$

The Damkohler number (Da), evaluated from Eq. (6.36), refers to the ratio of reaction rate to diffusion rate. The terms l_t and u' indicate the integral length scale and root mean square (RMS) of velocity fluctuation.

$$Da = \frac{l_t/u'}{\delta/S_L} \tag{6.36}$$

The computational domain was determined to be an axis-symmetric quadrant of the tubular shape, as illustrated in Fig. 6.29, because the model gas turbine combustor was housed in a cylindrical quartz tube. To assess the resulting meshes' mesh independence, a grid test was run. Finally, a mesh of around 40,000 cells was chosen for the numerical investigation, with fine meshes near the combustor input and slightly coarser meshes near the outlet zone. A prior numerical analysis by the authors of the current work on the identical combustor contains specifics of the grid independence test for the combustor model [9].

Different fuel-oxidizer mixes were taken into account when a series of numerical simulation cases was run at a constant bulk throat velocity of 5.2 m/s and delivered into the combustion zone through the combustor headend. To generate a swirl flow with SN = 0.98, the axial and tangential components of the reactant mixture were specified in the inlet boundary condition. For species modeling, partially premixed combustion was used, and the vortex method technique was used to solve for the changing velocity. For various reactant inlet compositions, the mean mixture fraction was computed, and the associated laminar flame speeds were determined from ANSYS Chemkin using the GRI-Mech 3.0 mechanism [77].

It was decided to use both external-radiation heat transfer and mixed thermal boundary conditions for the combustor wall. The wall's convective heat transfer coefficient was set at 20 W/m2K, while the temperatures for the free-stream and

Fig. 6.29 Combustor mesh adopted in the developed numerical model

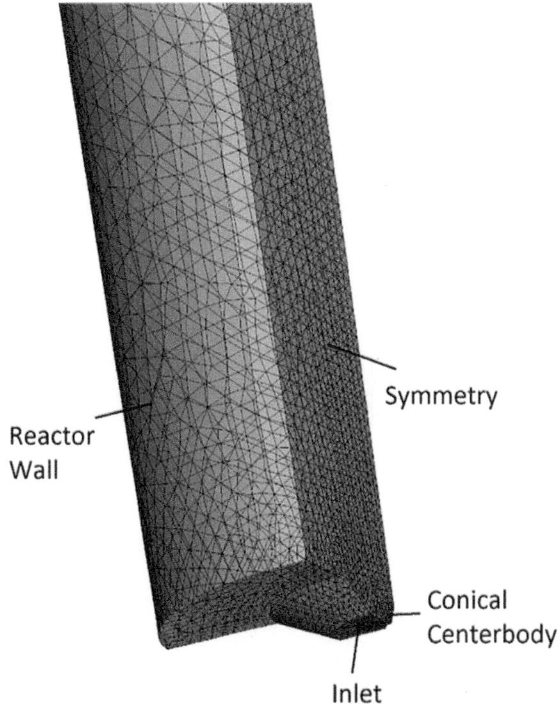

external radiation were 300 K and 290 K, respectively. In addition, a quartz wall with a semi-transparent type border was taken into consideration, with a wall thickness of 6 mm. The quadrant's symmetry faces are taken into consideration as periodic boundary conditions in order to appropriately compute for the tangential component of the whirling flow.

The experimental data from the authors' earlier study on oxy-propane combustion were used to validate the numerical model that was constructed in this study [67]. The validation example has the mixing conditions OF = 0.30 and Φ =89%. As shown in Fig. 6.30, the calculated radial temperature distributions made with the use of the current numerical model were compared to those obtained experimentally at an axial distance of z = 10 cm, OF = 0.30, and Φ =89%. The computed and observed data indicated a respectable degree of consistency in temperature distributions. In the following section of this study, additional comparisons of the model results and the experimental data in terms of flame macrostructure are offered. The comparisons revealed high levels of agreement.

6.3.1.2   Global Flame Behavior

Experimental research on pre-mixed oxy-propane flames was conducted over the ranges of OF and Φ to identify the combustor operability ranges between the blowout

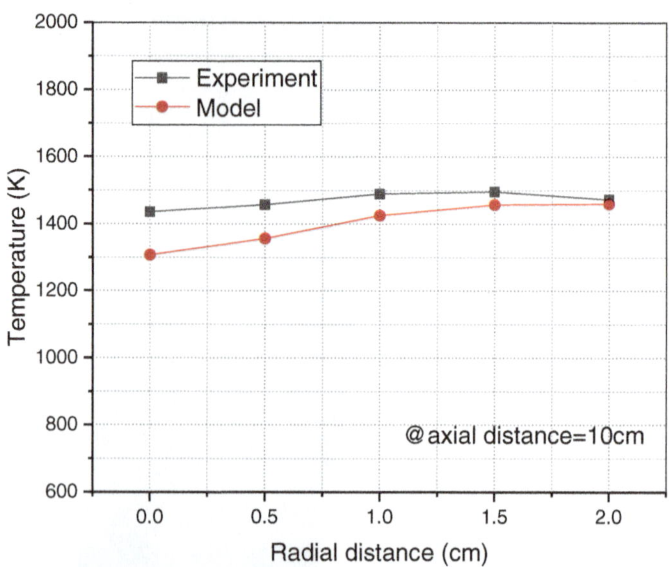

Fig. 6.30 Numerical versus experimental radial temperature profiles at an axial location of $z = 10$ cm, OF $= 0.30$, and $\Phi = 89\%$ [67]

and flashback flammability limitations. Tad and FS were found to be correlated, with the FS changing only in response to Tad, independent of OF and Φ. Additionally, neither the combustor power density (PD) nor the input Reynolds Number (Re) showed any discernible influence on the flame speed. Since the line exhibiting the flame blowout limit follows the contours of constant Tad, as illustrated in Fig. 6.31, the adiabatic flame temperature was identified as a regulating parameter for flame stability. The flashback curve, in contrast to the blowout, does not follow the Tad contour; instead, the pattern is determined by the reaction rate. In the experimental study, flame forms at various OF and Φ, and consequently at various Tad, were also studied. The $C_3H_8/O_2/CO_2$ flames tend to take on comparable forms and have similar temperature profiles close to the flame center at the same Tad. Additionally, in earlier studies on oxy-methane combustion in a swirl-stabilized burner as well as in a micromixer (MM) burner [78, 79], the influence of Tad on flame macrostructure was seen.

Understanding combustion behavior requires understanding flame perception, which can also shed light on flame extinction mechanisms. As shown in Fig. 6.32, the flame morphologies from the experimental inquiry [67] and the created model for the aforementioned circumstances are contrasted. The numerical flame-shapes were expressed from the outlines of OH radicals at the center-plane of the combustor, and experimental flames were photographed using a 24 MP camera (shutter speed: 1/60, f-stop: 5.6, ISO: 1600). It should be noted that not all simulation instances considered the quenching impact of the cool combustor bottom wall, which led to modest variations in the captured flame forms from the estimated ones as shown in Fig. 6.32.

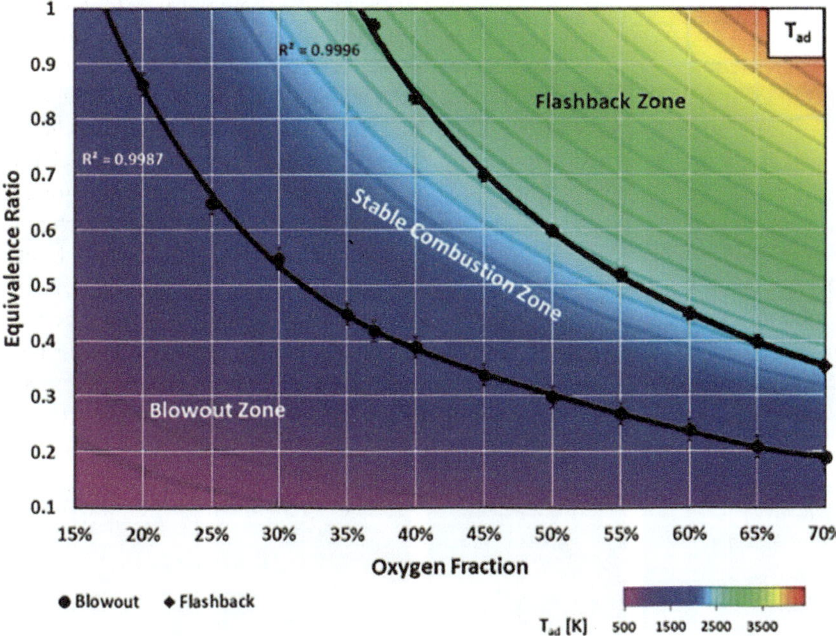

Fig. 6.31 Stability map of the model gas turbine combustor holding $C_3H_8/O_2/CO_2$ flames, plotted over the contours of T_{ad} at fixed bulk inlet velocity of 5.2 m/s [67]

Low reaction rates brought on by this quenching effect of the cool combustor base wall can be explained by the existence of dark narrow zones next to the combustor wall and ORZ (near burner base). The computational flames, on the other hand, are clearly extended to the burner wall buster.

The flame forms in Fig. 6.32 illustrate how changing at a fixed OF affects instances 1–3. The created flame is towering and weak, as well as having a spread-out form, for example Φ, case-1. As Φ rises, Tad rises as well, strengthening the flame with accelerated flame pace. Figure 6.33 shows the rise in flame speed (FS). As a result of this rise in Tad and FS, the flames in cases 2 and 3 become noticeably shorter and more compact. The numerical flame morphologies in Fig. 6.32 show that the increased reactivity of the combustible mixture, which also increases the concentrations of OH radicals inside the combustor, is indicated by the shorter flame length. A spike in OH was caused by increasing Φ from case-1 to case-3. Both models and testing produced nearly identical flame forms in examples 4 and 5. This is explained by the fact that these two flames share the same Tad, which is consistent with the experiment's results [67]. Similar findings from earlier research [30, 56, 67, 70, 78, 80] were reported. Due to the different OF values required for the two flames to have the same Tad, even though the two cases are of different, their flame lengths are the same. Additionally, at higher Tad, $C_3H_8/O_2/CO_2$ flames in swirl-stabilized combustor develop the V-shape characteristic. By contrasting the flame shapes for examples 6 through 8, it

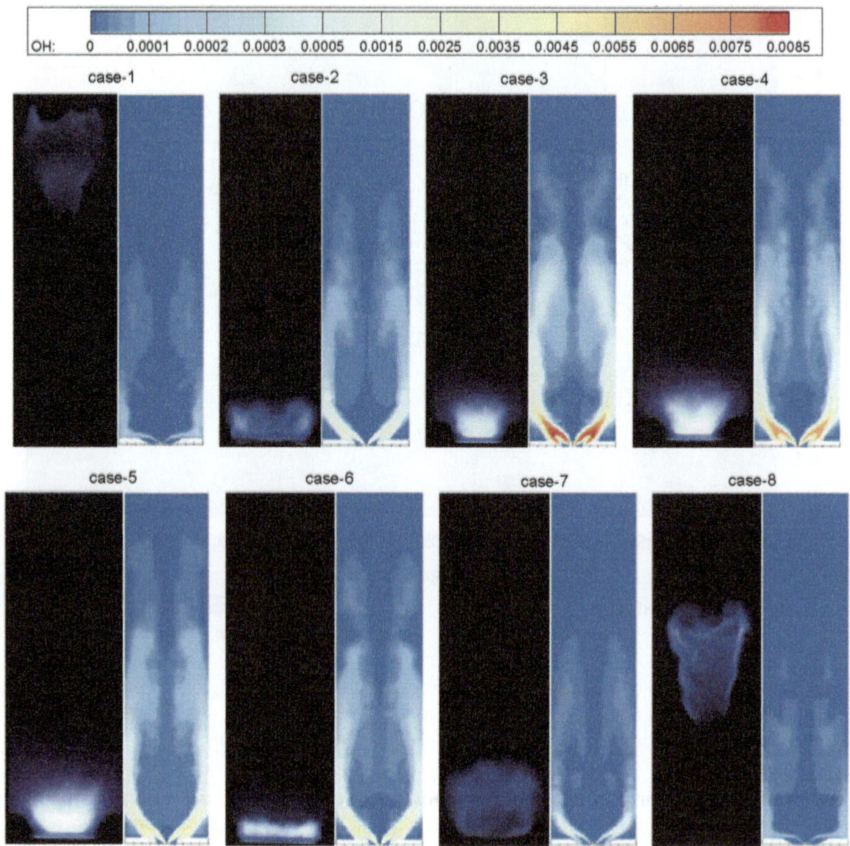

Fig. 6.32 Comparison of experimental and numerical flame shapes for the cases 1–8

is possible to see how changing the OF at affects the oxy-propane flame shapes. In this instance, the flame in case-6 is brief and powerful, the flame in case-7 is higher and relatively weaker, and the flame in case-8 is both the tallest and weakest of the three. As a result of the accelerated combustion chemistry's improved flame speed and increased reactivity of the mixture in oxygen-rich environments (see Fig. 6.33), the flame weakens when switching from case-6 to case-8 due to the combustor's decreased oxygen availability. The numerical counterpart of these three flames shows that the lowered chemistry also impacts the species concentration and lowers the OH concentrations. The spread-out nature of the flame at lower OF is also indicated by computational flame structures.

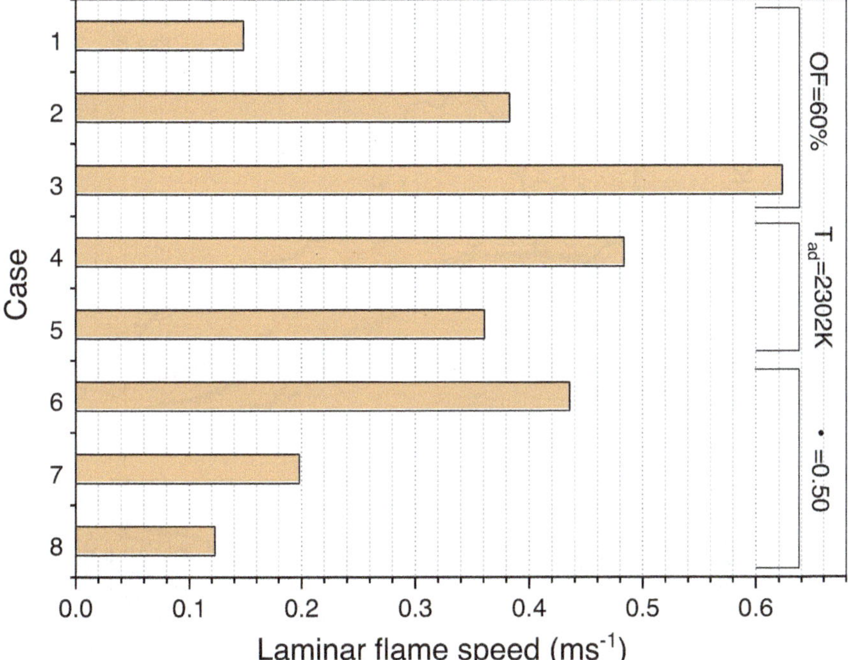

Fig. 6.33   Laminar flame speed over the different cases under investigation

6.3.1.3   Temperature Profiles and Flow-Field Structure

One of the most crucial factors in the design and production of any combustion system is the combustion temperature. The anticipated combustion temperature needs to be as high as is feasible given the combustor material's limit. As a result, this study looks closely at the combustor temperature. The temperature distributions and their normalized ($1000/T$) representation at the combustor centerline are shown in Fig. 6.33. The temperature of the combustor rises from the inlet to $Z/D = 1.1$ in all circumstances, as depicted in Fig. 6.34, and then it begins to fall r OF. For examples 3, 4, 5, and 6, however, the temperature initially falls and then rises to $Z/D = 1.1$. This is due to the creation of eddies in the inner recirculation zone (IRZ), which, as will be discussed in a later section of this paper, circulates convective heat in this zone. It is important to note that case-3, with its higher values of OF and Tad, obtained the highest axial temperature. The $1000/T$ temperature distribution likewise undergoes a different transition, with the value decreasing until $Z/D = 1.1$ and then rising from that point to the combustor outlet.

Figure 6.35 shows the radial temperature distributions and its normalized distributions at various axial distances (at $Z/D = 0.5, 1.5, 2.5,$ and 3.5) for the investigated cases. The temperature rises only slightly at the axial region of $Z/D = 0.5$, up to $R/D = 0.28$. The temperatures then rapidly rise until $R/D = 0.44$, after which they drop

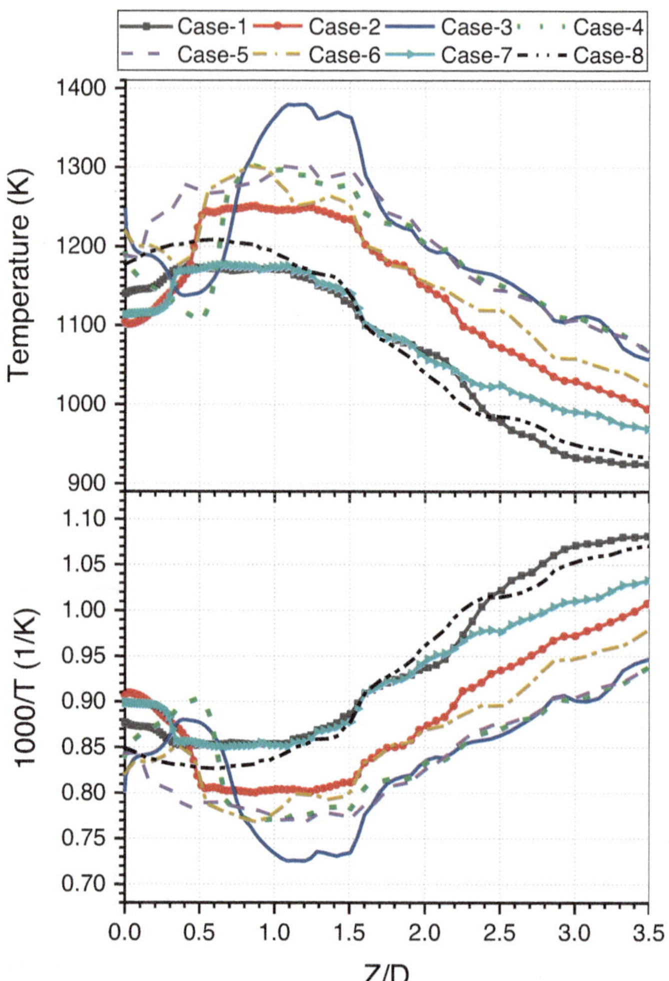

Fig. 6.34 Axial and normalized distributions of temperature along the centerline of the combustor for the different cases under investigation

again as a result of heat transfer to the combustor wall. Temperatures rise until R/D = 0.38 at Z/D = 1.5 before beginning to fall. This axial point experiences temperatures between 1100 and 1650 K. The temperature, which ranges from 720 to 1400 K at Z/D = 2.5, rises from the center point to R/D = 0.18 (or R/D = 0.3 in some circumstances), then falls. The radial temperatures neat the exit portion (Z/D = 3.5) for all examples range from 700 to 1200 K. The temperature rises up to R/D = 0.36 and then decreases to R/D = 0.5 near the combustor wall. At various axial points, the temperature profiles (measured in 1000/T) shift in opposition to the corresponding temperatures.

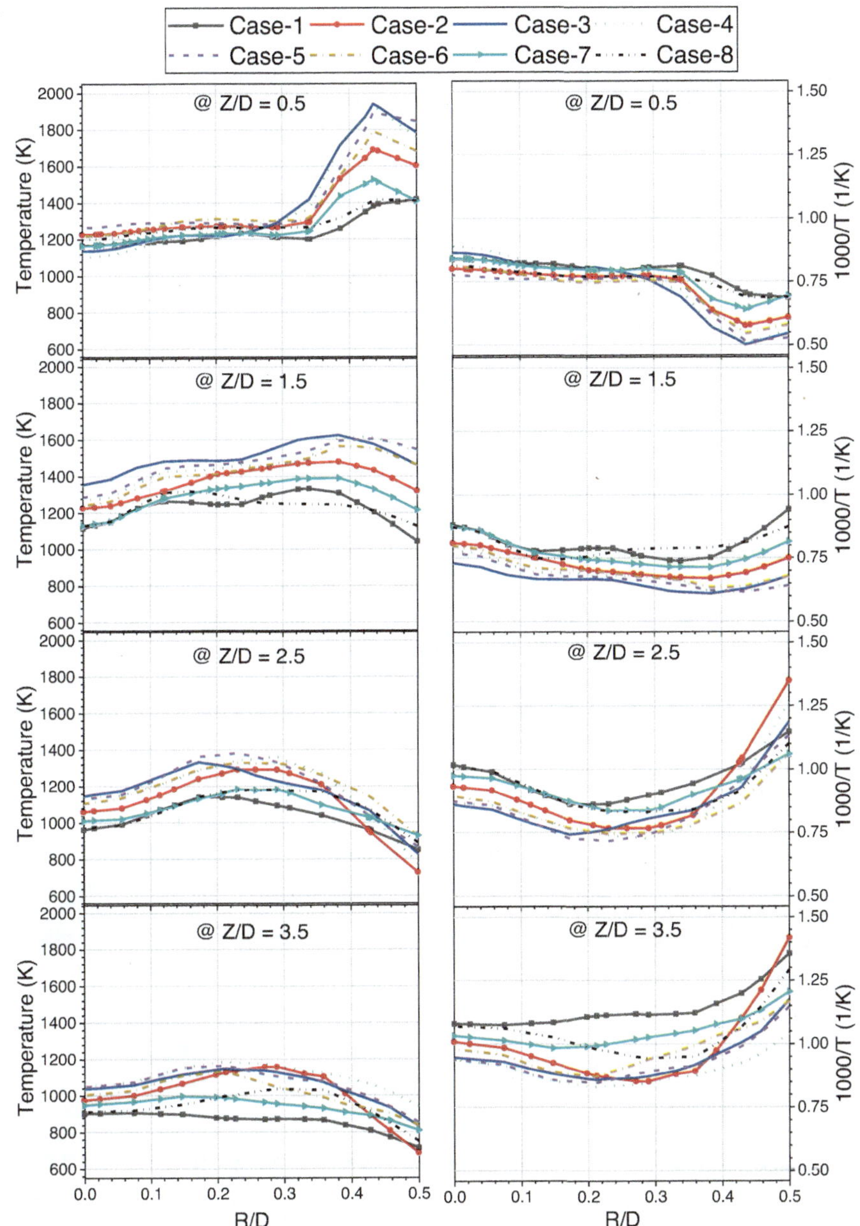

Fig. 6.35 Radial and normalized distributions of temperature within the combustion zone at axial locations of Z/D of 0.5, 1.5, 2.5 and 3.5 for the different cases under investigation

The temperature contours provide a clearer picture of the temperature distributions along radial directions. Figure 6.36 displays the radial temperature contours for flames with constant Tad of 2302 K at various axial locations. As seen in Fig. 6.36, while having varied OF and Φ, the distributions of temperature for the different cases are quite comparable. The similar Tad level in both situations is responsible for the similar temperature profile. The experimental investigation [67] reported findings that were similar. High temperature zone (1600.0–1900.0 K) close to the wall of the combustor and comparatively low temperature sector (1200–1500 K, 1300–1500 K) in the central region are the two temperature zones that can be distinguished at Z/D of 0.5. Temperature profiles for Z/D of 1.5 reveal that temperatures for cases 4 and 5 assortment from 1200 to 1500 K and 1300 to 1500 K, correspondingly, relatively evenly around the radius. Three temperature zones are shown for the two examples for Z/D = 2.5: an inner and an outer zone that are both "relatively low temperature zones," as well as a third zone that is higher in temperature and situated in the middle of the two. The calculated temperatures in the exterior and inner regions are practically identical, ranging from 1200 to 1300 K and 800 to 1200 K, respectively, while the third zone in between has temperatures between 1300 and 1400 K. Similar temperature zones to those found at Z/D = 2.5 may be seen in the calculated temperature profiles at Z/D of 3.5. For examples 4 and 5, the calculated temperatures in the inside, outside, and third in-between regions are approximately 1100 K, (900.0–1100.0 K), and 1200.0 K, correspondingly, at this axial distance, Z/D = 2.5.

Understanding the flow behavior and temperature fluctuations of the numerous combustion flames in the combustor is made possible by flow-field structure. Based

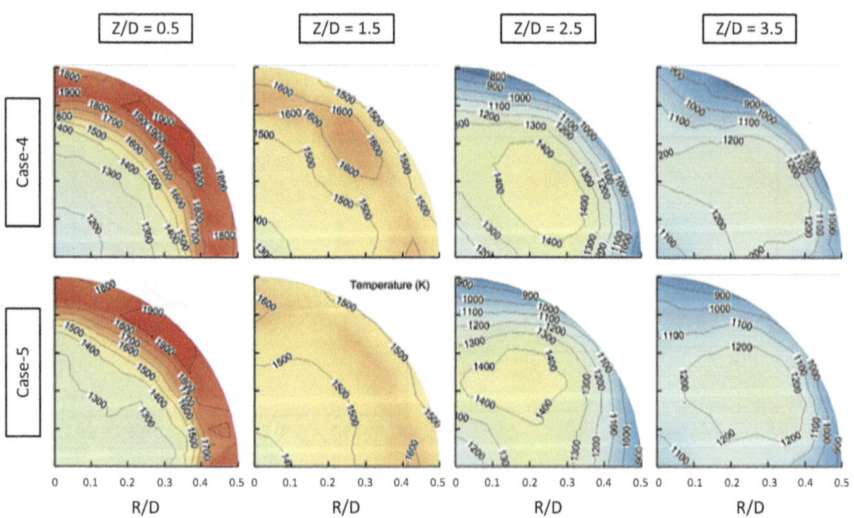

Fig. 6.36 Plots of temperature contour for flames of cases 4 and 5 for the same adiabatic flame temperature (2302 K) at dissimilar axial locatiosn of Z/D = 0.5, 1.5, 2.5 and 3.5

on the estimated profiles of velocity streamlines and vorticity magnitude, the flow-regions for the analyzed flame instances are inspected. The creation of the external recirculation region (ORZ) and internal recirculation region (IRZ) in a turbulent reactive flow is better understood with the aid of velocity streamlines. As well, vorticity demonstrates the flow field's trend to rotate and is crucial for the development of turbulent eddies. Temperature profiles are given alongside velocity streamlines (shown in Fig. 6.38) and vorticity magnitudes (shown in Fig. 6.39) for improved flow-field visualization.

The velocity streamlines along the center-plane contours of the combustion temperature are shown in Fig. 6.37. The produced eddies in both IRZ and ORZ set the examined cases' flow-field apart. Two IRZ are always visible: one near the combustor's tidy entrance and another close to the exit. By contrast, the ORZ is only visible in powerful, stable flames. From instance 1–3, it is possible to see how at fixed OF affects the IRZ and ORZ. In comparison to the flames produced at equivalency ratios of 0.34 and 0.42, the one at 0.24 produces a dimmer and more dispersed flame. The flame's wide distribution prevented an ORZ from forming. A little clockwise eddy and the ORZ begin to develop as the rises, and at $\Phi = 0.42$, the spinning of the ORZ begins. Since the FS grows at higher Φ, the abrupt expansion of the reacting mixture into the bigger combustor volume causes the creation of a more intense ORZ. Therefore, by adjusting the value of operating Φ, the eddy in ORZ may be controlled, which is comparable to what was found from a study of air-propane combustion in a backward-facing step combustor [81].

The IRZ's turbulent eddies are also changing concurrently with the rise in. From the combustor intake to the axial location of Z/D = 1.5 at $\Phi = 0.26$, two counter-clockwise-rotating eddies can be visible. These eddies combine into a single, bigger eddy with two epicentres that is shifted downstream when is increased to 0.34. The

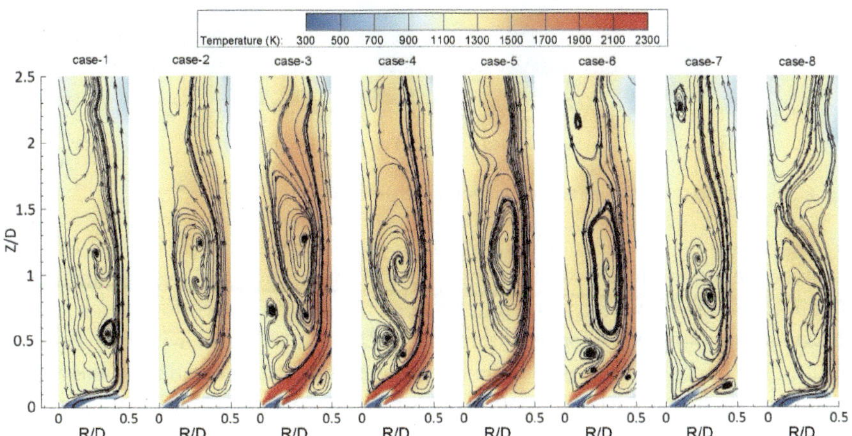

Fig. 6.37 Velocity streamlines over the plots of temperature for the different cases

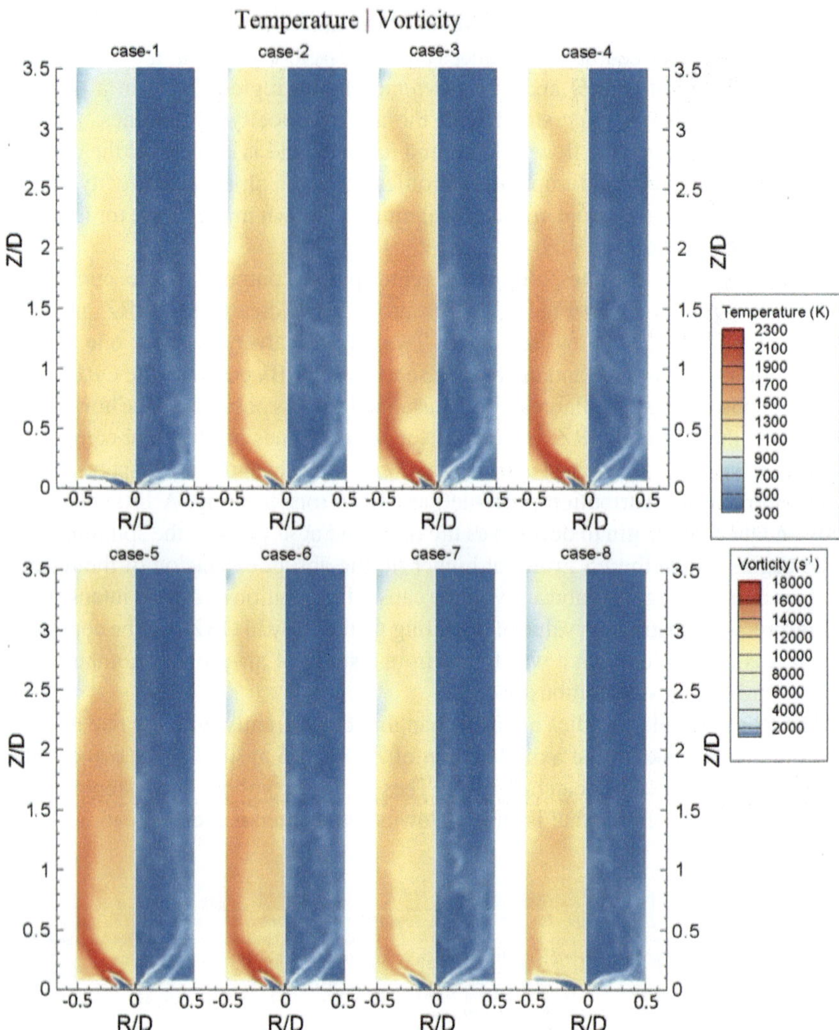

Fig. 6.38 Plots of temperature and vorticity for the different cases

combined counter-clockwise spinning eddy strengthens and flows further downstream at $\Phi = 0.42$, where it is found between 0.5 and $Z/D = 2$. However, a smaller clockwise-spinning eddy forms within $0.5 \leq Z/D \leq 1$ at higher with this displacement of the larger primary eddy. The interaction of turbulence and reactions within the flame is indicated by the formation of a smaller secondary eddy. The higher flame speed at altitude demonstrates a firm anchoring of the flame to the burner inlet, which is what is responsible for the eddies migrating away from the combustor intake. This phenomenon was also noted in a recent study on hydrogen-enhanced oxy-methane combustion [9], where the production of secondary eddies and their relocation to the

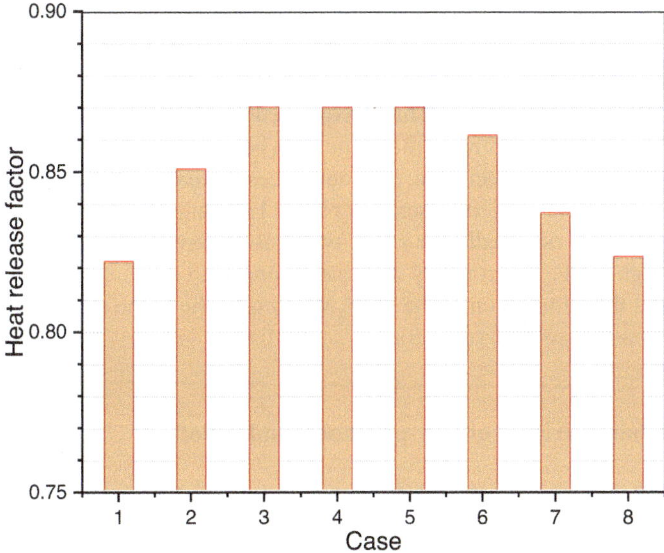

Fig. 6.39 Heat release factor for the different flame cases

downstream of the combustor were noted with increasing hydrogen addition. The effect on the IRZ was identical in our case of oxy-propane combustion to that in the case of hydrogen-enriched methane combustion since the fuel's hydrogen content rises with an increase in Φ.

According to velocity streamlines in the circumstances of fixed Tad, each flame has a sizable primary eddy spinning in the region of $0.5 \leq Z/D \leq 2$. For case-4, however, there are two more secondary eddies in the IRZ close to the burner inlet that are spinning in opposition to one another. In contrast to example 5, the circulation in the ORZ is also more obvious in case 4. Higher OF and lower Φ significantly contribute to better combustion chemistry, which in turn causes the development of these extra eddies in case-4. In examples 6–8 operated at fixed Φ, the impact of OF on the development of eddies is more evident. At higher OF, case-6, two opposing spinning secondary eddies are observed between the combustor intake and $Z/D = 0.5$ within the IRZ, and a primary eddy arises and rotates counter-clockwise within the zone $0.5 \leq Z/D \leq 1.5$. On top of that, higher OFs can also observe such strong rotation in the ORZ. Secondary eddies in the IRZ vanish at OF = 39%, whilst the primary eddy grows and moves upstream. The flame spreads out more, shrinking the ORZ. The major eddy in the IRZ weakens and drifts even more in the direction of the combustor inlet at the lowest OF, OF = 35%. The ORZ is removed and the flame is fully spread out in this instance.

The vorticity magnitude and related temperature contours for various combination compositions are shown in Fig. 6.38. According to Fig. 6.38, where the flame is stabilized and temperatures can reach as high as 2300 K, reaction takes place between the inner shear layer (ISL) and the outer shear layer (OSL). The temperature decreases to a more acceptable level required for safe gas turbine blade operation as the flame spreads towards the exit area. The ISL's thickness grows slightly as $0.5 \leq Z/D \leq 1$ at fixed OF condition increases. This is explained by the increased vorticity close to the central axis. Additionally, as the flame core between the ISL and OSL thickens and reaches a greater temperature, buoyant forces increase, the IRZ eddies migrate downstream, and the vorticity rises as a result. For flame cases with constant Tad, vorticity levels are essentially the same, with case-4 having slightly more vorticity due to its higher OF. The vorticity decreases along with the thickness and length of the OSL and the elimination of the ORZ, weakening the flame as the OF decreases from 48% (case-6) to 35% (case-8).

6.3.1.4 Characteristics of Oxy-propane Combustion

This section discusses the heat release factor (γ), product formation rate (PFR), progress variable, flame thickness (δ), and Damkohler number (Da) for some of the $C_3H_8/O_2/CO_2$ flame examples now under examination. The heat released during a combustion process is represented by the heat release factor. PFR, as its name suggests, describes the rate at which products are created, whereas progress variables show how quickly a mixture goes from being unburned to being totally burned. The term "flame thickness" describes the thickness at which the reaction zone heats the unburnt mixture to ignition temperature. The proportion between the reaction rate and the diffusion rate is represented by the Damkohler number. The heat release factor (γ) for CO_2 diluted oxy-propane combustion for the various flames of various mixture compositions is shown in Fig. 6.39. This factor rises with the flame's Tad at a constant temperature for the unburned mixture. As seen in Fig. 6.39, for examples 1–3, γ increases with Φ because the rise in Φ raises the Tad of the flame at a fixed OF, whereas for cases 4 and 5, is similar since they have the same Tad. The flame's Tad was lowered by reducing OF at fixed Φ, which also reduced the heat release factor γ for flame cases 6–8.

The center-plane contours of the PRF (left side) and progress variable (right side) close to the combustor intake are shown in Fig. 6.40. Because of the spread-out character of the flame at lower, Φ case -1, the PFR can only be observed on the ISL. The PFR rises on both ISL and OSL as Φ grows, from case-1 to case-3, and its distribution becomes more compact, indicating a shorter, more intense flame. Similar trends are seen in the progress variable, and the transition zone from unburned mixture to fully burned products likewise shortens. It's crucial to remember that even though the transfer from the ORZ to the IRZ stabilization modes couldn't be made [82], PFR trends might still be used to make the change. When comparing flame examples 4 and 5 of the same Tad (2302 K), PFR is stronger in case 4 and the unburned mixture zone is bigger. Due to the fact that the chemical element of flame determines the rate of combustion, case-4's lower Φ and higher OF allow the fuel-oxidizer mixture burn more quickly and produce a more compact flame than case-5's [83]. Cases 6

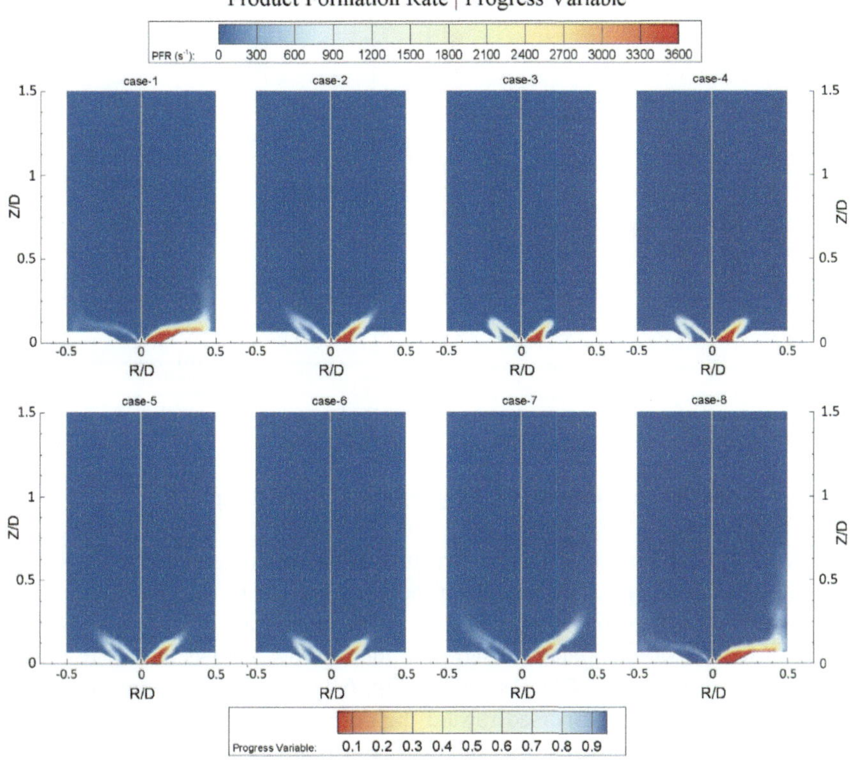

Fig. 6.40 Plots of product formation rate and reaction progress variable for $C_3H_8/O_2/CO_2$ flames at different flow conditions

through 8 at fixed Φ show an increase in flame length as the OF decreases, with case 8 showing a tendency for the flame to stabilize at the corner region of the combustor.

The flame thickness and Damkohler number for the various flame situations are shown in Fig. 6.41. As seen in Fig. 6.41a, the more compact flame that is seen at higher Φ causes the flame thickness to decrease with a rise in at fixed OF. Additionally, as shown in Fig. 6.41b, the flame compact structure denotes a stronger flame with noticeably faster reactions in the core, which is indicated by the noticeably higher Damkholer number. Flame thickness measurements for flames of $\Phi = 0.26$, 0.34, and 0.42 at OF $= 60\%$ are 222.7 m, 85.5 m, and 52.1 m, respectively, whereas the Da values are 5.3, 30.5, and 80.9, respectively. Due to the more compact nature of the flame in case-4 compared to case-5, δ is higher (84.7 m) with relatively lower Da (26.3) in case-4 and lower (64.7 m) with increased Da (45.1) in case-5. Last but not least, for examples 6 through 8, the flame thickness grows and Da drops when OF is reduced at a fixed Φ. At $\Phi = 0.5$, the flame thickness measurements are 72.7 m, 158 m, and 254.3 m, respectively, with Da values of 35.7, 8.6, and 4.1 m at OF $= 48\%$, 39%, and 35%.

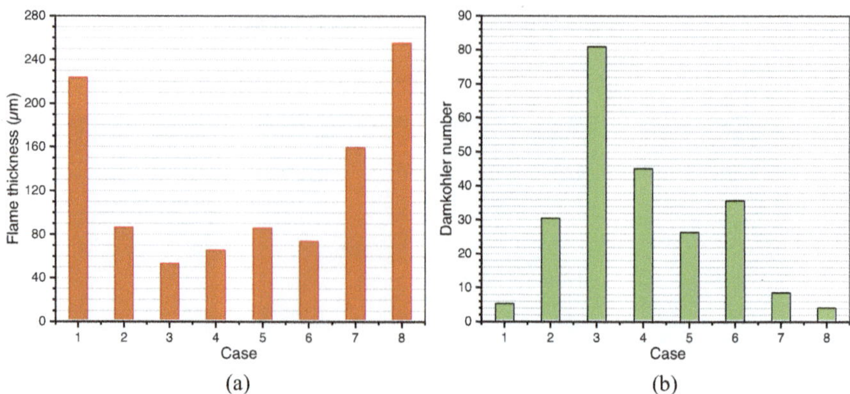

Fig. 6.41 Impact of mixture composition on: **a** flame thickness and **b** Da number

6.3.1.5   Emission Attributes of Oxy-propane Combustion

A fuller understanding of the flame structure and emission characteristics can be gained by examining the distributions of mole fractions of C_3H_8, O_2, CO, and CO_2 within the combustion zone. Figure 6.43 shows the mole fraction distributions of O_2 and CO_2 for the various flame scenarios, while Fig. 6.42 shows the mole fraction distributions of C_3H_8 and CO. As seen in Fig. 6.42, CO is barely present in case 1 and propane is more dispersed toward the corner of the combustor. The fuel's supply of oxidizer is sufficient for both the C and H atoms at a lower (OF = 60%). The concentration of C_3H_8 decreases as Φ rises, whereas the concentration of CO rises. This results in the full combustion of C_3H_8, which can be attributed to the increased flame speed and response rates at higher Φ. The presence of CO, on the other hand, increases at greater (OF = 60%) because the H and O atoms tend to react more quickly, and C must compete with H for O. This results in an increase in the mole fraction of CO in the combustor due to the lower oxidation of CO for CO_2 production. The significantly more fuel-rich atmosphere in example 3 (see Fig. 6.43) can also be attributed to the rise in CO_2. Cases 4 and 5 can also be explained by the same process. Due to the greater and lower OF in case-5 compared to case-5, the concentrations of C_3H_8 and CO are higher. In examples 6 through 8, the CO mole percentage falls along with the drop in OF at a constant Φ of 0.5. O_2 concentration falls as CO_2 concentration rises at lower OF. Figures 6.42 and 6.43 show that when the CO_2 concentration in the oxidizer mixer rises, the residence time increases at the same bulk velocity [84], enhancing the interaction between the radicals and causing lower CO and higher CO_2 concentrations inside the combustor.

Figure 6.44 displays the CO concentrations in parts per million (ppm) at the combustor center line and at the combustor outflow. The majority of CO emission is generated within the axial range, 0Z/D2.5, as illustrated in Fig. 6.44a. This is due to the impact of the IRZ being present in the same axial range. Some of the CO combines with the available oxygen and turns into CO_2 as the flow moves towards the exit.

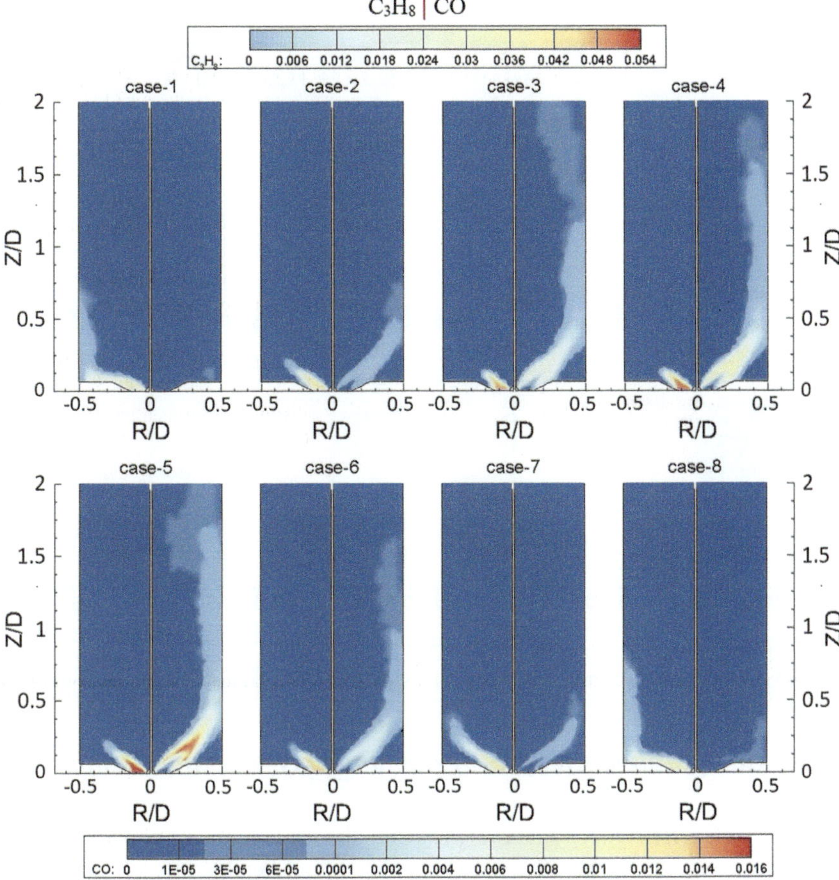

Fig. 6.42 Contours of mole fractions of propane (left) and CO (right) for the different flame cases

This results in a substantially lower CO emission at the combustor outlet than the maximum value of its axial dispersion. Case-5 ($\Phi = 0.8$ and $OF = 35\%$) showed the highest CO emission among the various mixture compositions both axially within the combustor and at the combustor output. Due to its greater Φ and lower OF, case-5 contains proportionally more fuel and less oxygen than the other instances, as previously shown in Figs. 6.42 and 6.43. Due to this relatively fuel-rich environment, the fuel does not receive enough oxygen to react with, cannot completely convert to CO_2, and as a result emits more CO than in the comparison cases.

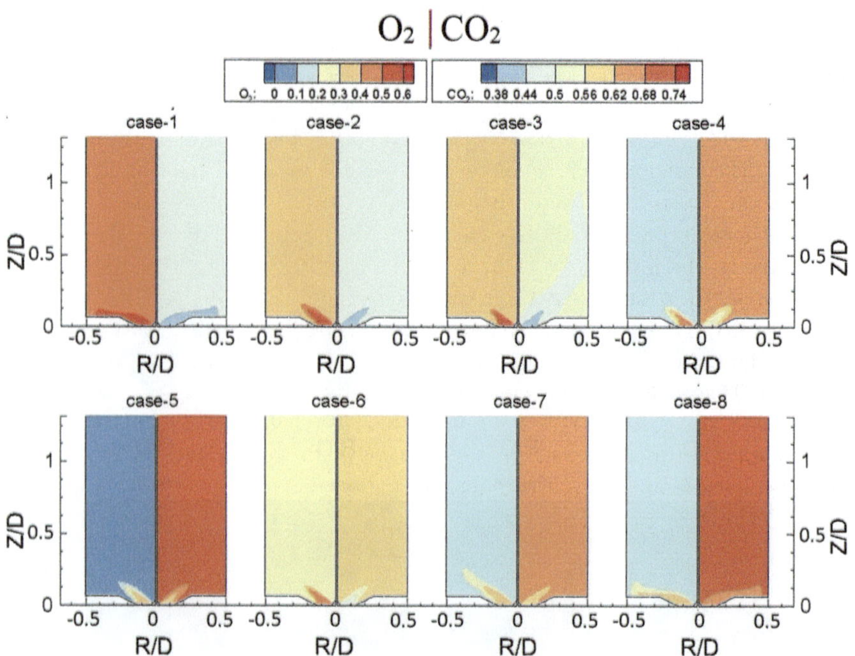

Fig. 6.43 Contours of mole fractions of O_2 (left) and CO_2 (right) for the different flame cases

6.3.2 Comparative Analysis of Air- and Oxy-propane Flames

The premixed $C_3H_8/O_2/N_2$ and $C_3H_8/O_2/CO_2$ flames at ambient conditions were examined using the LPM DLE swirl-stabilized fuel-flexible model gas-turbine combustor, which was discussed in the section above. This combustor's power density (3.5–20 MW/m^3/bar) is comparable to that of running gas turbines. Blowing out and flashback limits are two flame extinction limits that are used to quantify the flame's static stability. The same flame-base bulk throat velocity (5.2 m/s) is used for all the flames tested here to maintain comparable cold-flow conditions and turbulence intensities. In order to create swirling flow in the combustor, a 55° geometric swirler was added in the headend. Through the use of a converging–diverging insert with a 2.0 cm throat diameter, the flow past the swirler was quickened. The insert prevented any rapid expansion in the divergent region while allowing for steady convergence of the flow in the convergent part. After the burner headend, a quartz tube was put in place to keep the flames contained by atmospheric air. For the purpose of preventing vortex breakdown at the flow centerline, which could cause early flashback, the swirler was designed with a conical center body that extended within the throat insert. A premixed reactant mixture was produced upstream of the swirler by injecting air, CO_2, C_3H_8, and O_2 gases in a mixing plenum with a length of 1.0 m and a diameter of 2.0 in. The gas flow rates were managed by thermal mass flow controllers from Aalborg Inc.,

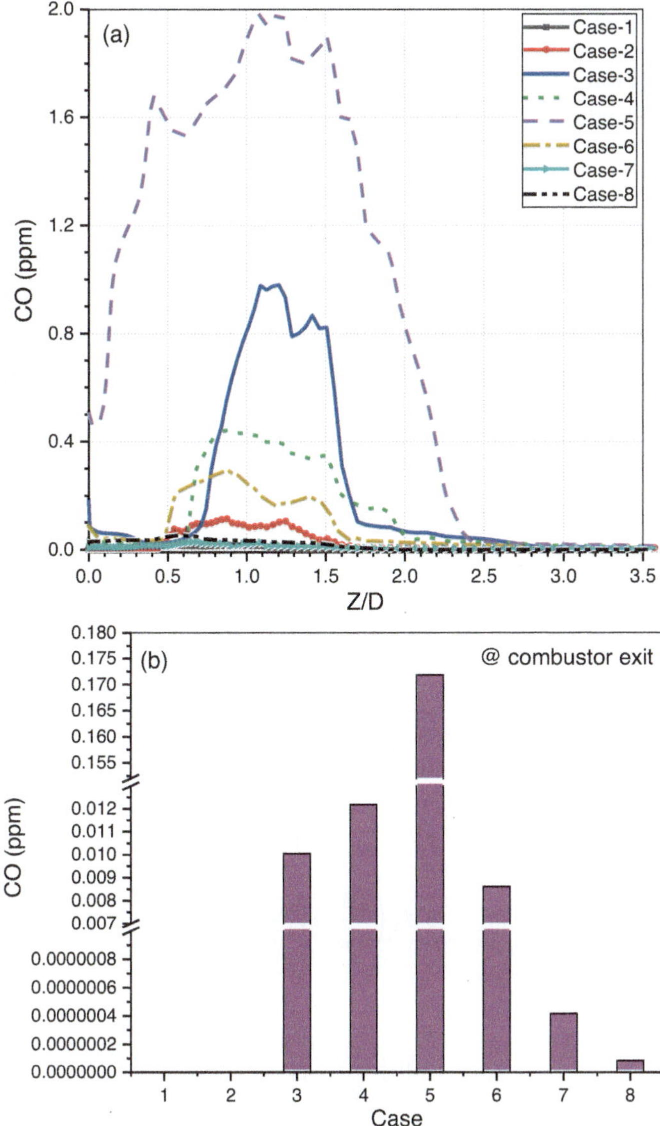

Fig. 6.44 CO concentrations in ppm for the different flame cases: **a** along the axial direction, and **b** at combustor outlet

with full-scale accuracy of 0.5%. Calculations showed that the resulting errors in OF and were 0.4% and 0.02 respectively. The injection of the oxidizer gases, namely air/ CO_2 and O_2, was done from opposing sides, and the injection of the C_3H_8 fuel was done through radial holes on a coaxial tube that was sealed at the top. To investigate the effectiveness of premixing, various stable flames with various compositions were used. Photographic photos of flames obtained with a 24 MP camera at 1600 ISO, 1/ 60-s shutter speed, and 5.6 f-stop settings were used for the analysis of flame shapes.

The bulk flame-base velocity of all the examined flames was 5.2 m/s. The method for maintaining the bulk throat velocity fixed at 5.2 m/s has been thoroughly described in Ref. [85]. The mass flow rates of the various reactants were simultaneously changed to reach a certain data point (defined by OF and φ) at this given bulk velocity. In the current study, ranged from 0.1 to 1.0, whereas the oxygen fraction ranged from 21 to 70%.

Richer φ was sought for a fixed OF if the flashback extinction limit was desired, and slimmer was sought to observe flame blowout. It is important to note that, prior to approaching the flashback limit, higher auditory loudness and a compact flame form (connected to the combustor throat) were seen. Weaker raised flames indicated an extinction limit close to a breakout.

The following equations, which were obtained from the generalized method of Ref. [85], were used to compute the species flow rates for C_3H_8, O_2, and diluent X (either N_2 or CO_2).

$$\dot{m}_{C_3H_8} = 44\left(\frac{pAv_a}{R_uT}\right)\left(1 + \frac{5}{\varphi OF}\right)^{-1} \tag{6.37}$$

$$\dot{m}_{O_2} = 160\left(\frac{pAv_a}{R_uT}\right)\left(\varphi + \frac{5}{OF}\right)^{-1} \tag{6.38}$$

$$\dot{m}_X = 5M_X\left(\frac{pAv_a}{R_uT}\right)\left(\frac{1-OF}{\varphi OF + 5}\right) \tag{6.39}$$

Therefore,

$$\dot{m}_{N_2} = 140\left(\frac{pAv_a}{R_uT}\right)\left(\frac{1-OF}{\varphi OF + 5}\right) \tag{6.39a}$$

and

$$\dot{m}_{CO_2} = 220\left(\frac{pAv_a}{R_uT}\right)\left(\frac{1-OF}{\varphi OF + 5}\right) \tag{6.39b}$$

The required airflow rate was calculated by using $\dot{m}_{N_2}$ from Eq. 3a and the fact that air consists of 76.7% nitrogen by mass, i.e.,

$$\dot{m}_{Air} = \frac{\dot{m}_{N_2}}{0.767} \tag{6.40}$$

Depending on Eqs. (6.38), (6.39a), and (6.40), the flow rate of enrichment oxygen was determined as

$$\dot{m}_{enO_2} = \dot{m}_{N_2} + \dot{m}_{O_2} - \dot{m}_{Air} \tag{6.41}$$

The bulk dynamic viscosity (μ_{mix}) of the reactant mixture was calculated as follows [86]:

$$\mu_{mix} = \frac{\sum y_i \mu_i \sqrt{M_i}}{\sum y_i \sqrt{M_i}} \tag{6.42}$$

Likewise, the bulk density (ρ_{mix}) was calculated using the following formula:

$$\rho_{mix} = \frac{P \sum \dot{m}_i}{R_u T \sum (\dot{m}_i / M_i)} \tag{6.43}$$

The Reynolds number at the throat of the burner was calculated as

$$Re = vD\rho_{mix}/\mu_{mix} \tag{6.44}$$

The mass flow rate of propane was used to define the combustor power density (PD) as:

$$PD = \frac{\dot{m}_{C_3H_8} CV_{C_3H_8}}{p \times combustor\ volume} \tag{6.45}$$

where $CV_{C_3H_8}$ is the tabularised propane calorific value, and the combustor size is the internal volume of the quartz imprisonment.

A thermodynamic analysis was conducted for estimating the adiabatic flame temperature (AFT) based on the tabulated enthalpies of formation of C_3H_8, H_2O, and CO_2 and the sensible (temperature dependent) enthalpies of CO_2, O_2, N_2, and H_2O [87]. This was done because the exhaust-gas composition and AFT are dependent on both OF and φ of the reactant mixture. Table 6.3 provides an overview of the operational circumstances for the current investigation.

Table 6.3 Operating conditions

Parameter	Range in $C_3H_8/O_2/CO_2$ flames	Range in $C_3H_8/O_2/N_2$ flames
Oxygen fraction (%)	21–70	21–70
Equivalence ratio	0.1–1.0	0.1–1.0
Power density (MW/m³/bar)	0.34–10.61	0.34–10.61
Reynolds number	8047–11,089	6573–7493

6.3.2.1 Air-Combustion versus Oxy-combustion

Since the thermophysical qualities of O_2 and N_2 are equivalent, it is commonly believed that changes in the oxygen fraction (OF) of a diluted O_2/N_2 oxidizer combination won't significantly alter its thermophysical behavior. Compared to air–fuel flames, oxy-fuel flames have significantly different features from air–fuel flames, and as a result, OF has a greater impact on key combustion parameters such the adiabatic flame temperature (AFT).

Figure 6.45 shows the combustor stability maps for N_2- and CO_2-diluted flames against backdrop contours of constant AFT, which show the blowout and flashback limitations. The blowout and flashback limits were found to be accurately represented by the exponential fits, with R2 values of 0.9991 and 0.9978 for N_2 flames and 0.9966 and 0.9989 for CO_2 flames, respectively. The stable zone of combustion is represented by the region between the stability limits, whereas the regions below and above are known as the blowout and flashback zones, respectively. Figure 6.45 shows the combustor stability maps, which show the blowout and flashback limits for N_2- and CO_2-diluted flames over constant AFT backdrop contours. With R2 values of 0.9991 and 0.9978 for N_2 flames and 0.9966 and 0.9989 for CO_2 flames, respectively, it was discovered that the exponential fits properly represented the blowout and flashback limitations. The stable zone of combustion is the region between the stability limits, while the blowout and flashback zones, respectively, are the regions above and below.

The modest mistake associated with the current simplified computation of AFT that disregards CO_2 dissociation is responsible for the slight deviation from AFT contours found in the blowout and flashback limits of CO_2 flames. As seen in Fig. 6.46 in terms of the mole percentage of effluent CO_2 in both flame sets, CO_2 flames emit much more effluent CO_2 than N_2 flames. The blowout and flashback limitations of CO_2 flames would have followed constant-AFT contours if CO_2 dissociation had been taken into consideration while calculating AFT. It should be noted that the AFT error is highest where the mole fraction of effluent CO_2 grows along the blowout and flashback limits of CO_2 flames in the direction of lower OF.

In Fig. 6.47, the stability maps of the CO_2 and N_2 flames are combined for comparison. As was previously mentioned, the slopes of the constant AFT contours in CO_2 flames are steeper [$(\partial\varphi/\partial OF)_{AFT}$], showing a higher influence of the OF on AFT in oxy-propane flames. In addition, for any given combination of φ and OF, the AFT of a CO_2 flame is smaller than that of a N_2 flame. The aforementioned observations are the result of N_2 having a lower heat capacity than CO_2. This is a novel observation for C_3H_8 flames and coincides with earlier findings for CH_4 flames [85] that the static stability limits correlate well with AFT for both CO_2 and N_2 flames.

The following are some further important observations:

- For the investigated fixed flame-base bulk velocity of 5.2 m/s, the blowout limits for both CO_2 and N_2 flames were reached at nearly the same AFT of 1580 K.
- For the investigated fixed flame-base bulk velocity of 5.2 m/s, the flashback limits for both CO_2 and N_2 flames were reached at the same AFT of 2350 K.

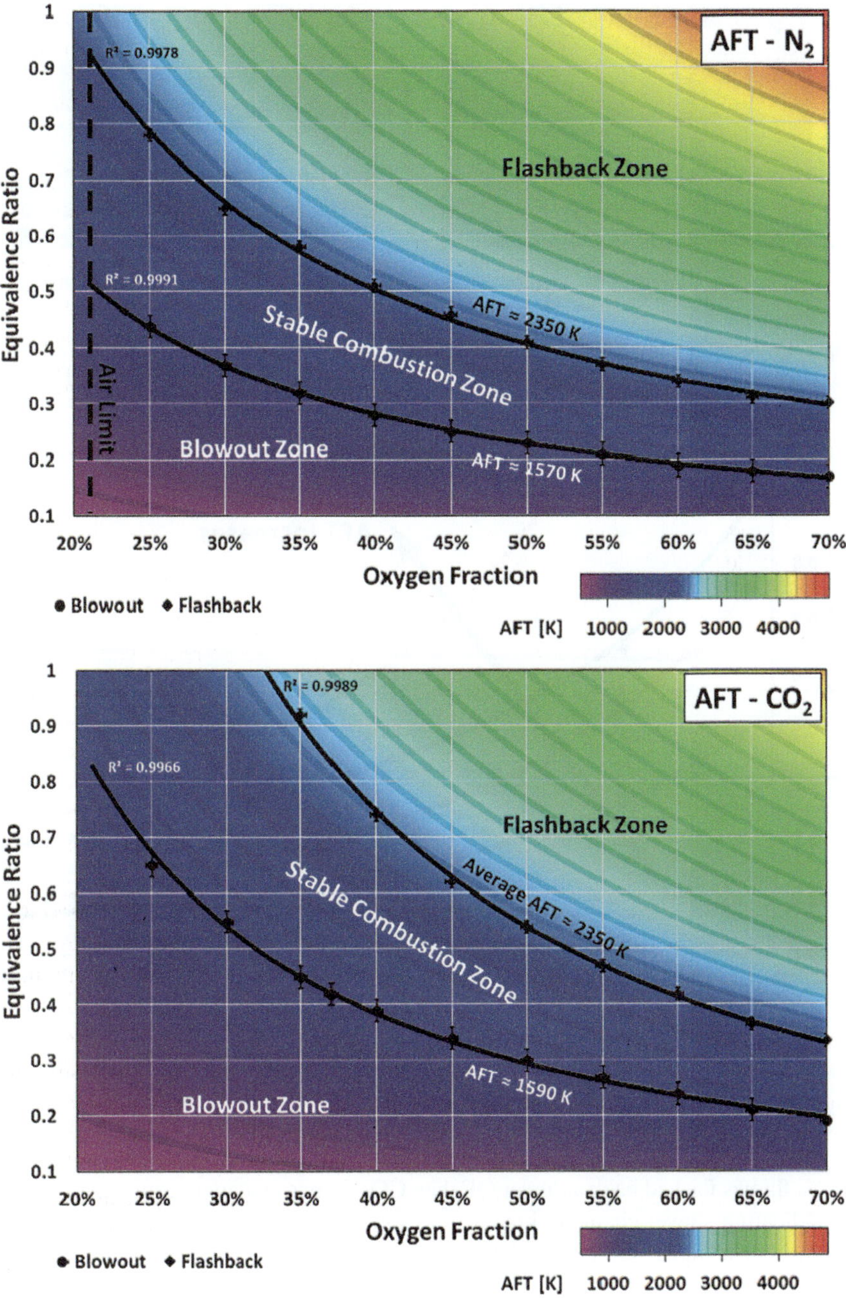

Fig. 6.45 Stability maps plotted against the contours of AFT. Top: $C_3H_8/O_2/N_2$ flames, bottom: $C_3H_8/O_2/CO_2$ flames

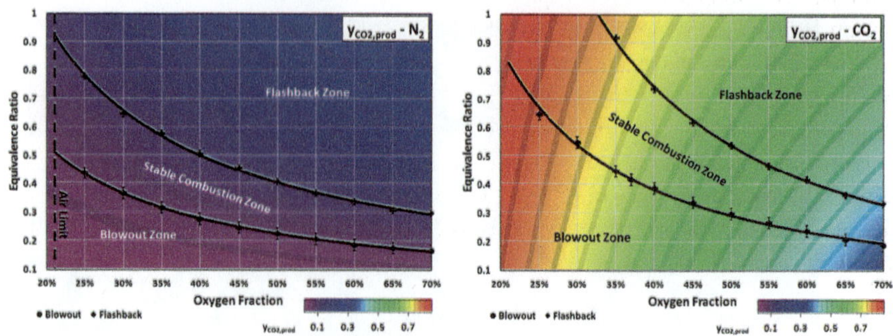

Fig. 6.46 Stability maps of $C_3H_8/O_2/CO_2$ flames (right) and $C_3H_8/O_2/N_2$ flames (left) over the plots of effluent-CO_2 mole fraction

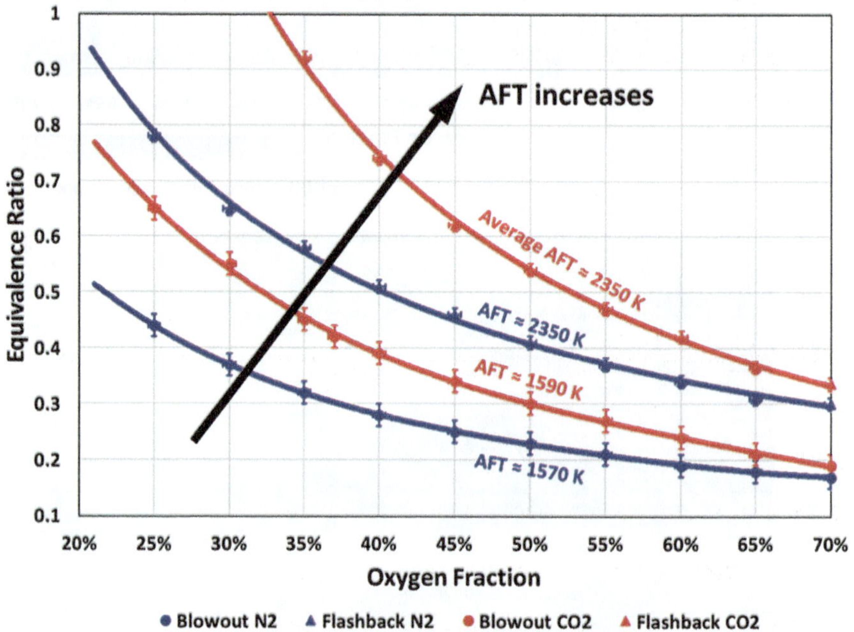

Fig. 6.47 Comparison of the stability maps of $C_3H_8/O_2/N_2$ and $C_3H_8/O_2/CO_2$ flames

- N_2 flames flashed back at lower OF than CO_2 flames at a given equivalency ratio, indicating slower chemical kinetics in CO_2 flames, which need more oxygen to flashback. This result is completely consistent with Ref. [88].

At increasing oxygen concentrations, the stability properties of CO_2 and N_2 flames behave in the same way. This is due to the lower CO_2 and N_2 concentrations in the oxidizer mixture at greater OF, which has a lessened impact on the behavior of the flame stability.

While the authors have previously noted a substantial correlation between $CH_4/O_2/N_2$ and $CH_4/O_2/CO_2$ flames and AFT [85, 89], the current study also detects this correlation for $C_3H_8/O_2/N_2$ and $C_3H_8/O_2/CO_2$ flames. This leads to the conclusion that AFT has a universally dominant role in determining response kinetics. The flame extinction limits are primarily controlled by reaction kinetics when the flame-base bulk velocity is fixed. At the beginning of flashback extinction, the flow velocity and flame speed are delicately balanced. Each flashback limit is attained at a constant flame speed (FS), which is indicated by the use of constant flame-base bulk velocity in this context. This further supports the predominating role of AFT because the direct dependency of FS on AFT shown by the constant AFT and FS along each flashback limit is in line with the findings of Kaskan [90] and Van Maaren et al. [91].

The current analysis and findings make a compelling case for basing premixed oxy-propane combustors' design and operation, particularly for high and medium load configurations, largely on AFT. Existing air–fuel gas turbines already follow this procedure. As a result, even though the latter is employed to control the flame temperature in air–fuel combustors, gas-turbine manufacturers base their design, operation, and evaluation of the combustors on the flame temperature instead of the equivalency ratio (for the aforementioned reasons). By operating the combustor at constant pressure drop across its headend, it is possible to maintain constant flame-base bulk flow velocity in existing LPM air–fuel gas turbines. This procedure keeps a buffer against flashbacks. The stability maps are replotted in Fig. 6.48 against the boundaries of constant Reynolds number, Re, to further highlight the dominant function of AFT in regulating the reaction kinetics of the current flames. The Re maps show the impact of CO_2 having a higher density than O_2 and N_2. As was previously seen with the AFT, the slopes of the Re contours $\left[\left(\frac{\partial \varphi}{\partial OF}\right)_{Re}\right]$ on the CO_2 map are significantly steeper than those on the N_2 one, indicating a higher effect of the OF over the oxy-fuel Re map. The steeper slopes of Re contours are attributed to the increased CO_2 density. Despite the N_2 map appearing to match Re outlines, there is no correlation seen for CO_2 flames. This demonstrates that AFT, not Re, is the main regulating parameter.

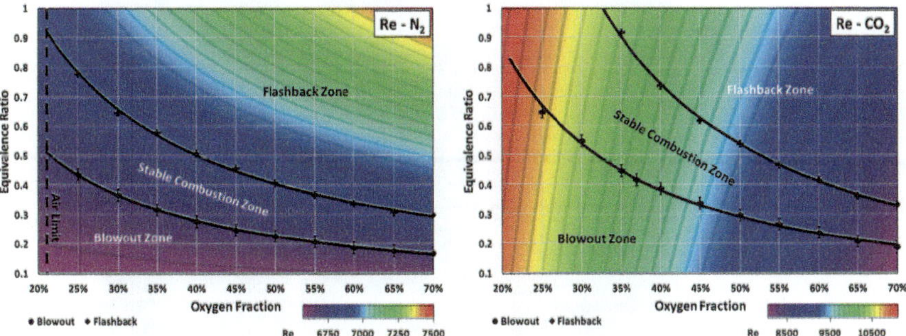

Fig. 6.48 Stability maps of $C_3H_8/O_2/CO_2$ flames (right) and $C_3H_8/O_2/N_2$ flames (left) over the plots of constant bulk throat Re

6.3.2.2   Flame-Structure Analysis Based on Adiabatic Flame Temperature (AFT)

While the preceding section focused on how AFT affects reaction kinetics and, in turn, regulates flame stability, this section investigates how AFT affects flame structure. For both N_2 and CO_2 flames, Fig. 6.49 depicts the effect of on flame structure at constant OF (=60%). A tall weak flame that is stabilized by the outside recirculation zone gives way to a shorter compact flame that is stabilized by the inner recirculation zone as the value rises. The altered flame structure is a result of the higher's increased reaction kinetics. It should be emphasized that the N_2 and CO_2 flames are not being compared at the value of as shown in Fig. 6.49, but rather at ordinary values of AFT It is evident that CO_2 and N_2 flames at a particular AFT have almost comparable structures, despite the differing values in each AFT group. This confirms that AFT alone has a monopolistic influence over the reaction kinetics of the current C_3H_8 flames.

Figure 6.50 illustrates the analysis of natural-luminosity images of several flames at a fixed AFT of 2135 K but differing, OF, and diluent type (N_2 vs. CO_2) in order to focus more closely on the impact of AFT. It is noted that, regardless of, OF, and φ diluent type, maintaining AFT constant often leads to roughly identical structures for both CO_2 and N_2 flames. For CH_4 flames, similar results have already been reported [85]. Again, due to the little mistake associated with the oversimplified computation of AFT, slight shape changes are seen within CO_2 flames. Remember that N_2 flames

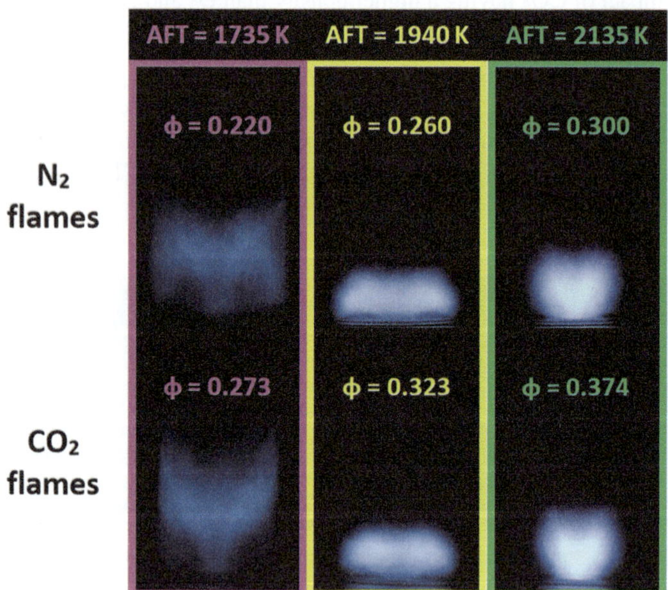

Fig. 6.49 Images of N_2 and CO_2 flames at constant OF = 60%

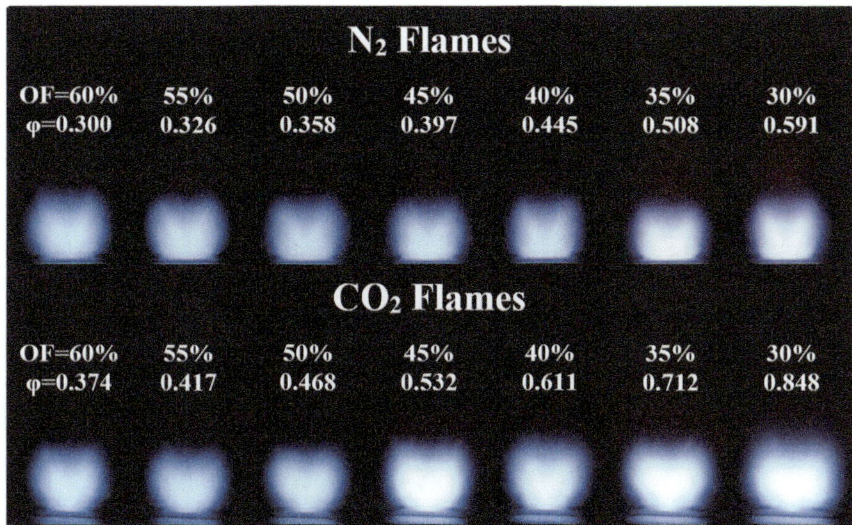

Fig. 6.50 Images of N_2 and CO_2 flames at a fixed AFT of 2135 K but different φ and OF

have less effluent CO_2 than CO_2 flames do (Fig. 6.46), therefore if CO_2 dissociation had been taken into account in the computation of AFT, the CO_2 flames seen in Fig. 6.50 would have been similar. Given that CO_2 has a higher heat capacity than N_2, a higher φ must be present in the CO_2 flame in order to match the N_2 flame's AFT at the same OF in Fig. 6.50. This higher φ can be seen as a penalty to account for CO_2's greater heat capacity.

Figure 6.51, in contrast to Fig. 6.49, focuses on comparing N_2 and CO_2 flames at constant OF and φ but varied AFT. Due to CO_2's larger heat capacity, switching from N_2 to CO_2 reduces the AFT as expected. Lower AFT causes slower reaction kinetics, which worsens flame stability. As a result, unlike the constant-AFT analysis, maintaining φ at any OF does not maintain flame stability and form. The combined examination of Figs. 6.49, 6.50, and 6.51 demonstrates that AFT is essential in regulating the structure and behavior of premixed flames.

The flame shape in both N_2 and CO_2 flames was observed for a range of φ (from near flashback to near blowout) at various oxygen fractions in order to conduct additional analysis of the flame macrostructure and stabilizing behavior. As seen in Fig. 6.52, the stability maps of N_2 and CO_2 flames have been indicated with the value of φ at which flame macrostructure drastically changed. In the near-flashback region, or between the flashback limit and the isotherm of 2100 K, N_2 flames appear to be substantially more compact, indicating faster response kinetics and flame stabilization in the inner recirculation zone (IRZ). As φ is reduced, the flame enlarges and enters the outer-recirculation-zone (ORZ) stability phase, which is indicated by an isotherm of 1950 K. Thus, the grey area denotes the transition from IRZ to ORZ. As the ORZ loses strength, further reduction in (green data points) results in the flame lifting off the burner surface and low-intensity combustion. The flame moves into

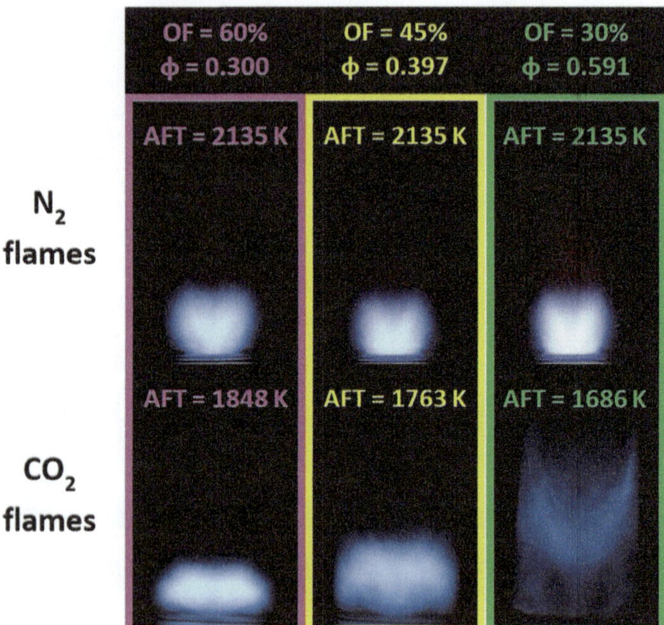

Fig. 6.51 N_2 and CO_2 flames of the same OF and φ but different AFT

the near-blowout region (blue data points) with any additional decrease in φ, where it assumes a conical form without any recirculation zones. Similar transitions were detected in the CO_2 flames, as shown in Fig. 6.52, with the IRZ-to-ORZ transition taking place between 2180 and 2030 K. It is important to note that a major decrease in flame size accompanied the change in flame macrostructure from compact flames to long conical ones.

6.4   Highly Diluted Oxy-fuel Multi-hole Combustor with Hydrogen Enrichment

Between 1990 and 2090, the average global temperature is expected to rise by a maximum of 8 °C [92]. The primary offender, CO_2, is responsible for around 77% of all greenhouse-gas emissions that cause global warming [93]. Fossil fuels, which are expected to continue to be a substantial source of energy for the next fifty years [94], are the primary source of CO_2 emissions. Numerous solutions to this issue have been put forth. A promising technique that can be used in both new and existing power plants [95] to produce an exhaust stream that is primarily made up of H_2O and CO_2, where the latter can be separated through simple condensation of the former, is oxygen-fuel combustion, in which oxygen is separated from air and used

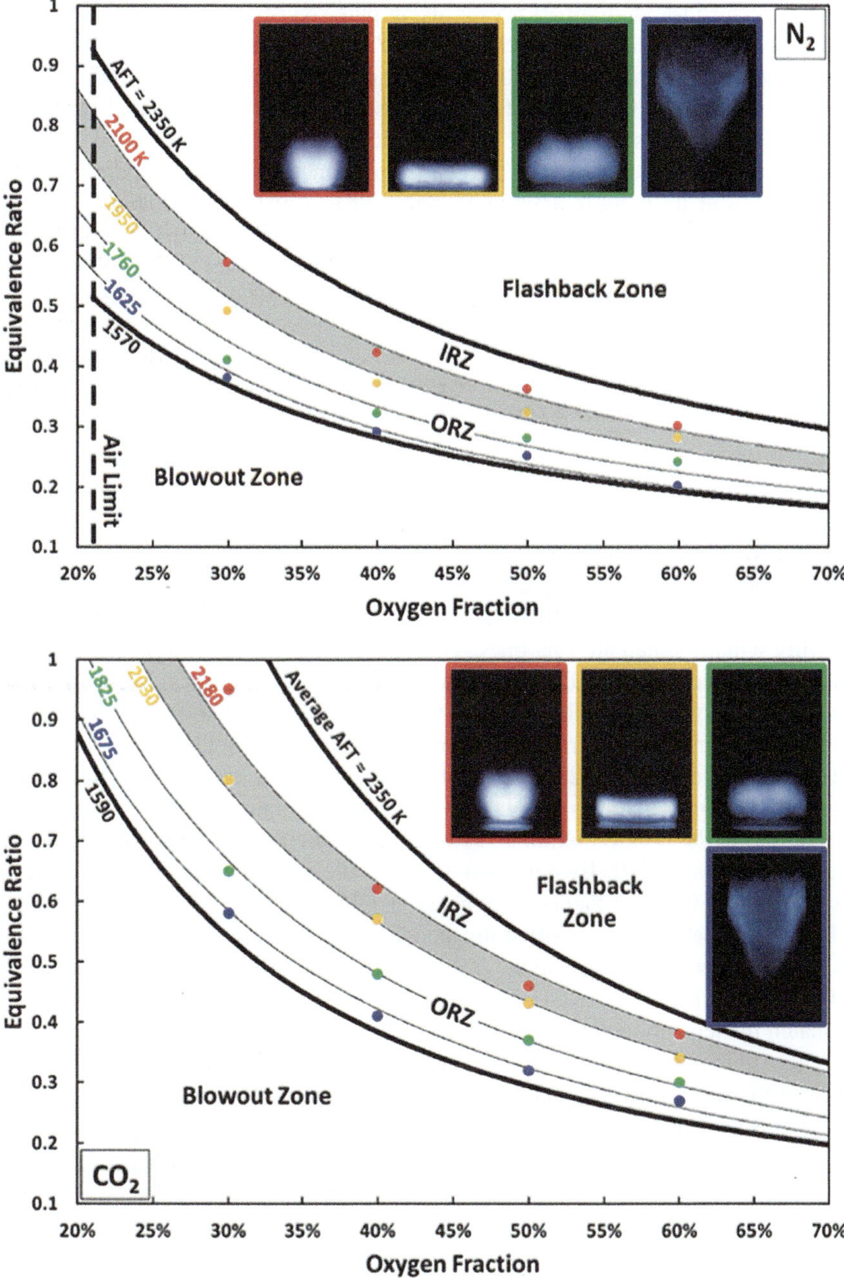

Fig. 6.52 Transition from IRZ to ORZ stabilization in N_2 (top) and CO_2 (bottom) flames

as an oxidizer instead of air. However, using pure oxygen causes abnormally high flame temperatures, endangering the hardware's metallurgical integrity. Exhaust gas recirculation, often known as EGR, lowers the flame temperature by recirculating some of the effluent CO_2 to dilute the fresh stream of oxygen and fuel. The oxy-combustion process should occur near to stoichiometry to rationalize the usage of oxygen [96], as the separation of oxygen from air incurs additional operating costs [97]. As a result, the flame temperature is only managed by EGR. Following a detailed experimental investigation on testing hydrogen-enriched oxy-methane highly CO_2-diluted flames for applications in zero emission Allam power cycles, a review of highly diluted flames for zero-emission power cycles is offered in this part.

6.4.1 Highly Diluted Flames for Zero-Emission Allam Power

The combustion characteristics are impacted by CO_2 dilution from three perspectives: radiation effect [98], thermal effect [99], and kinetic effect [100]. This is because the thermos-physical properties of N_2 and CO_2 differ greatly from one another. The reaction kinetics are slowed by CO_2, which also negatively impacts the flame speed and combustion efficiency [101, 102]. As a result, oxy-fuel combustors have smaller stability windows than air–fuel ones. This is true for premixed, partially premixed, and unpremixed flames [103], as well as nonpremixed [104] ones. Experimentally and numerically analyzing the impact of CO_2 dilution on laminar flame speed, Halter et al. [105] found that CO_2 reduces flame speed more than N_2. Oxy-fuel fires cannot be sustained at the average 21% oxygen fraction (OF) of air, according to Nemitallah and Habib [6] and Ditaranto and Hals [106]. Instead, a minimum OF = 30% (by vol.) is required in the O_2/CO_2 oxidizer, and even then, the flame is still less stable than the one in the air [107]. By demonstrating that the adiabatic flame temperature of stoichiometric CH4/air flames is attained at OF = 30% in $CH_4/O_2/CO_2$ flames, Abdelhafez et al. [108] supported this conclusion. Similar minimum OF limits for oxy-fuel fires were reported in several research, specifically between 27 and 35% [109] and 25% [97, 110]. Even OF = 36% is required, according to Song et al. [111], to match the temperature profiles with air flames.

The Allam power cycle [112, 113] is a high-pressure supercritical CO_2 Brayton cycle used in operating oxy-fuel power plants with carbon capture [114]; see Fig. 6.53 [115]. It is named after Rodney J. Allam. This cycle can significantly reduce the cost of electricity [116], while capturing all CO_2 emissions at the high purity and pressure levels required for downstream CO_2 reuse and/or sequestration [117]. It also matches the typical net efficiencies of existing combined-cycle power plants without carbon capture. It is interesting to note that the Allam cycle's combustor only increases the temperature of the CO_2 working fluid by 400 degrees Celsius, from 770 to 1160. This slight temperature increase implies that the oxygen flame must be significantly diluted by CO_2 that has been cycled. While the secondary CO_2 stream is used for liner cooling and dilutes the combustion products to lower the mean exit temperature to just 1160 °C [118], the primary CO_2 flow is split to create a 23% OF within the

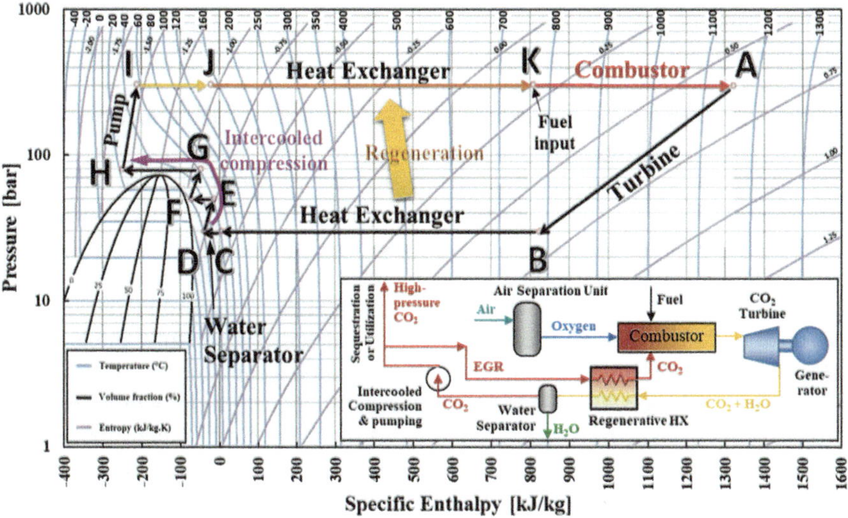

Fig. 6.53 Allam cycle for power generation with CO_2 capture

primary reaction zone. As mentioned above, previous studies were unable to achieve such high levels of dilution, which prompted the current study to look for and analyze an efficient burner technology that can a) support highly diluted oxy-flames, b) is oxidizer-flexible to accommodate a variety of oxygen fractions, and c) is fuel-flexible to tolerate a variety of fuel compositions as well as hydrogen enrichment.

Since H_2 has been found to improve a number of air–fuel combustion characteristics, including reaction kinetics [119], combustion intensity [120], emissions [121], soot formation [122], and resistance to strain-induced extinction at leaner operating conditions [123], hydrogen enrichment is proposed as a way to increase the stability and turndown of oxy-fuel combustors. The disadvantage is early flashback, though, as a result of H_2's faster flame propagation speed due to its higher burning velocity [121]. The production of CO is also favored by the presence of H_2 [124]. Few research have looked into H_2-enriched oxy-fuel fires despite the clear benefits of H_2 enrichment in air–fuel applications and the availability of studies in this field. In a premixed swirl-stabilized combustor, Imteyaz et al. [125] studied CH_4/H_2–O_2/CO_2 flames over a wide range of hydrogen fraction (HF) and equivalency ratio. They claimed that when HF increased, the blowout and flashback limits shifted toward the leaner side as a result of a faster flame. Abdelwahid et al. [126], who investigated premixed CH_4/H_2–O_2/CO_2 flames under stoichiometric circumstances, provided support for this. Due to the improved chemical kinetics and increased burning velocity brought on by the addition of H_2, which generated flashback even at low oxygen percentages, a narrower stable flame zone was seen. At greater HF, smaller, more compact flame sizes were also noted. H_2-enrichment was observed to raise the critical strain rate and decrease the quenching potential in premixed oxy-fuel flames in a related numerical investigation [127]. Additionally, when HF grows, the Karlovitz number (Ka) falls and the Damkohler number (Da) increases. Additionally, it was noted that adiabatic flame temperature has less of an impact on the combustor's static stability limitations than does the reactant mixture's Reynolds number.

The employment of a burner technology that is adaptable enough to support various enrichment levels without experiencing flashback or instabilities is necessary for the successful application of H_2 enrichment. The micromix combustion technology [128] is a strong contender in this situation. A micromixer (MM) is a multi-tube premixing fuel injector that separates the flame into multiple tiny flamelets and combines the fuel and oxidizer at the microscale. A number of manufacturers of air–fuel gas turbines, including Honeywell Garrett [129], Mitsubishi Hitachi Power Systems [130], and General Electric [131], have already successfully implemented MMs, an emerging lean-premixed air–fuel combustion technology that is inherently safe against flashback. But MMs have never before been put through an oxy-fuel combustion test. When H_2-rich fuels were evaluated for air combustion in MM combustors, York et al. [132] and Asai et al. [133] reported steady flashback-free operation for this type of combustor. A MM combustor with about 1600 flamelets was created by Funke et al. [134] to be integrated into an operational gas turbine. They claimed that the MM combustor's use enabled stable engine running during acceleration, idle, and start-up. In an MM air combustor, Dodo et al.'s [135] simulation of Integrated Gasification Combined Cycle (IGCC) conditions included carbon capture. Simulants of syngas fuel with H_2, CH_4, and N_2 were employed, along with three different carbon capture rates of 0, 30, and 50%. The MM burner operated steadily at atmospheric pressure without blowout or flashback.

Three knowledge gaps in the available literature are highlighted in the review above. First off, prior research has shown that the majority of swirl-based burner technologies are unable to operate at the high levels of dilution (exhaust-gas recirculation) needed in the oxy-fuel combustors of zero-emission power cycles. Second, oxy-fuel combustion circumstances have never been tested on the naturally flexible and reliable MM gas-turbine technology. Research on this technology is still in its early phases, and the literature only contains a small amount of data that pertains to air combustion. Third, further research is still required to determine how H_2 enrichment affects oxy-fuel flames, especially when combined with a cutting-edge technology like MMs. As a result, the study covered in the next section adopted a novel strategy to meet the difficult dilution requirements of the oxy-fuel combustors of zero-emission power cycles, particularly the Allam cycle, by using MM technology with H_2 enrichment. In order to increase combustion efficiency as well as simulate the real-world operation of an oxy-fuel combustor, experiments were conducted at an equivalency ratio of 0.9 (near stoichiometry). We looked at the impact of H_2 enrichment on the combustor blowout limit for fuel HF with a range of 0–90% and oxidant OF with a range of 13–40%. To investigate the influence of various parameters on flame macrostructure and stability, specific flames were imaged.

6.4.2 Combustor Operating Conditions

In the current work, premixed CH_4, H_2, O_2, and CO_2 flames in a model micromixer gas turbine combustor are investigated. A detailed description of the experimental facility can be found elsewhere [89, 126]. The architecture of the actual premixing combustion equipment is depicted in Fig. 6.54. In contrast to real micromixers, where fuel is injected and mixed with oxidizer inside the headend, here fuel was added to a mixing plenum together with O_2 and CO_2. This simplification was done in order to prevent any impact of jet-to-jet fluctuation in equivalency ratio on combustor performance by ensuring that an almost completely premixed reactant mixture enters the headend. The mixing plenum is large enough to provide almost flawless premixing, with an aspect ratio L/D of 20. Nevertheless, a quick test was performed to demonstrate this by shifting the fuel tube axially from the nominal position depicted in Fig. 6.54 by a range of -25 to $+ 100$ mm. The combustor stability map and flame shape were unaffected by the position of the fuel tube, confirming that almost perfect reactant premixing is accomplished upstream of the burner headend.

The headend was converted from a micromixer to a multi-hole burner by the use of a mixing plenum. The geometry of the multi-hole headend under examination is shown in Fig. 6.55. Uniform jet spacing is thus maintained throughout the headend by distributing 613.18 mm diameter jets on a hexagonal matrix such that each core jet is surrounded by 6 equally spaced jets with a distance of 5.5 mm. The core jet is encircled by four concentric hexagonal "rings" of jets in this arrangement, which likewise produces hexagonal symmetry. As previously discussed, micromixers are naturally resistant to flashback, but the inlets of the multi-hole burner's holes were purposefully designed smaller in diameter (1.6 mm) to serve as flame arrestors for operator safety. The flow can recover after the arrestors and depart the headend in a fully grown form thanks to the main holes' aspect ratio L/D of $40/3.18 = 12.6$, which is large enough.

In order to enable flame imaging, the experiments were carried out in an optically accessible quartz combustor at atmospheric pressure and temperature (no preheating). Aalborg Inc. thermal mass flow controllers were used to regulate all gas flow rates with an uncertainty of 0.5% of full scale, translating to an inaccuracy of 0.4% in the computed HF and OF. Visual still photographs of selected flames were taken with a high-definition digital camera set at 5.6 f-stop, 1/60 shutter speed, and 3200 ISO sensitivity.

To determine the value of the combustor profile factor, temperature measurements were also taken in the plumes of a few different fires. We used an R-type thermocouple (Platinum Rhodium - 13%/Platinum) with a 1-mm junction, which has a measurement accuracy of 1.5 °C or 0.25% (whichever is higher) and can measure temperatures up to 1480 °C. The National Instruments (NI) computer interface LabVIEW and its data acquisition (DAQmx) card were used to get the temperature values. According to the criterion outlined by Brohez et al. [136], the temperature data were adjusted for radiation error to the surroundings.

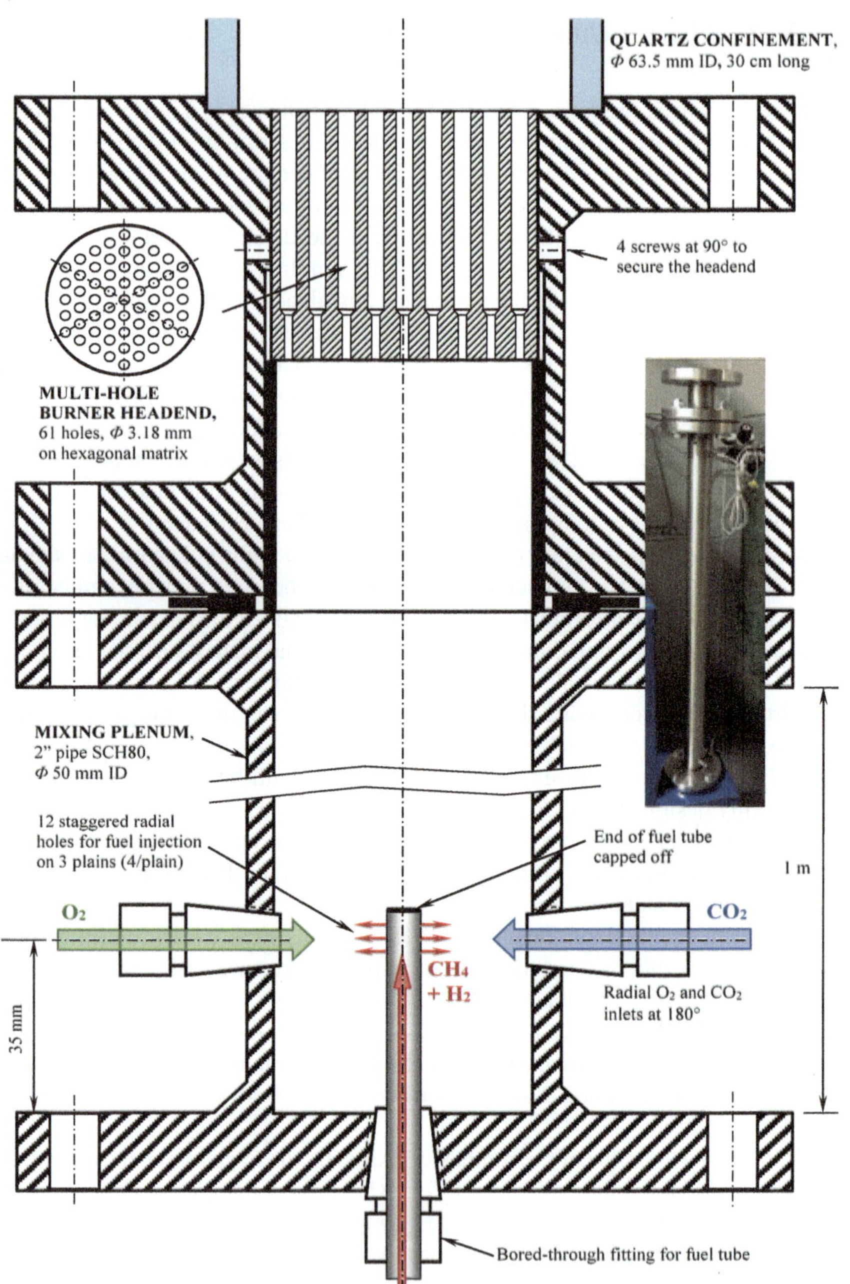

Fig. 6.54 A close-up of the premixing combustor setup with the multi-hole burner headend installed

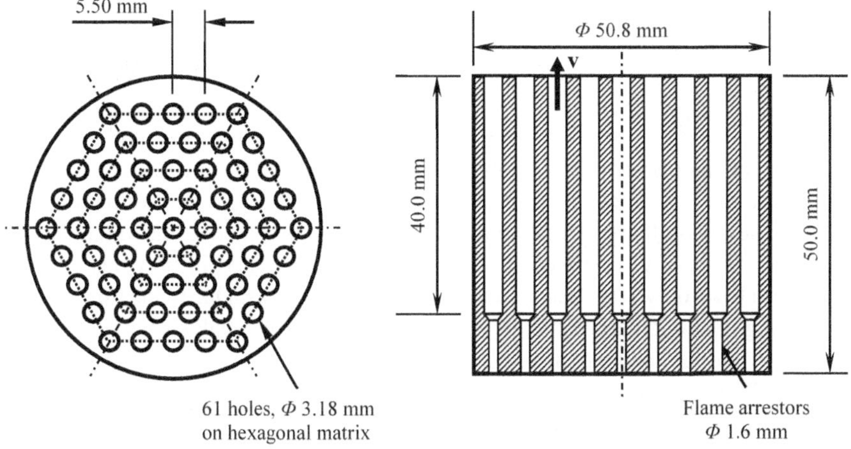

Fig. 6.55 Micromixer burner: sectional elevation (right) and top view (left)

To simulate the actual operation of an oxy-fuel gas turbine and increase combustion efficiency, all experiments were carried out at close to stoichiometric conditions ($\varphi = 0.9$). In order to ensure consistent flow conditions and turbulence intensities across all isothermal flow fields, the input jet velocity, v, was maintained at 5.2 m/s for all fires. As v has been shown to be a more relevant parameter influencing flame stability than intake Reynolds number [89, 137], maintaining v constant also cancels out its effect on flame stability. To keep v constant, the flow rates of CH_4, H_2, O_2, and CO_2 were computed and changed simultaneously. The reader is directed to earlier works by the authors [108, 126] for more information on these computations.

The combustor power density is given as [138]:

$$PD = \frac{\dot{m}_{CH_4} \times CV_{CH_4} + \dot{m}_{H_2} \times CV_{H_2}}{p \times combustor\ volume} \tag{6.46}$$

Reynolds number is given as:

$$Re = \frac{vD\rho_{mix}}{\mu_{mix}} \tag{6.47}$$

The mixture density (ρ_{mix}) and dynamic viscosity (μ_{mix}) are given as [139]:

$$\rho_{mix} = \frac{p \sum \dot{m}_i}{R_u T \sum (\dot{m}_i/M_i)} \tag{6.48}$$

$$\mu_{mix} = \frac{\sum y_i \mu_i \sqrt{M_i}}{\sum y_i \sqrt{M_i}} \tag{6.49}$$

The mixture flow rate ($\dot{m}_{mix}$) and the pressure drop (Δp) are defined as:

$$\dot{m}_{mix} = vA\rho_{mix} \tag{6.50}$$

$$\Delta p = \frac{1}{2}\rho_{mix}v^2 \tag{6.51}$$

Based on tabulated information on the enthalpies of formation and sensible enthalpies, the oxy-combustion process of H_2-enriched CH_4 was examined to estimate the adiabatic flame temperature (AFT). When compared to the value of AFT itself, the effect of CO_2 dissociation on AFT was calculated in previous research [140, 141] to be only 100 K under stoichiometric conditions. Due to the tiny discrepancy being regarded acceptable, it was anticipated that the analysis would be made simpler by the fuel being completely burned.

The following steps were taken to quantify the combustor stability map:

1. Light the combustor at a stable position that was first discovered by trial and error using only $CH_4/O_2/CO_2$ (i.e., no H_2).
2. Aim for stable operation at a specific target OF by gradually adjusting the flow rates and adding H_2 in a strategic manner.
3. After achieving a steady flame at the required OF, the HF is gradually reduced at this OF in order to reach the blowout point.
4. Reignite, then modify (steps 1 and 2). The HF is gradually increased at constant OF to reach the top limit at this OF when a steady flame has been achieved at the desired OF.
5. Repetition of steps 1–3 will help you calculate the blowout (lower) limit at various OF values.
6. Carry out step 4 again to calculate the upper limit for various OF values.

6.4.3 Stability Mapping of the Combustor

In order to create an oxy-fuel combustor that can sustain highly diluted flames, similar to those required in the Allam-cycle combustor, this study studies the application of micromixer gas-turbine technology with H_2 enrichment. The combustor stability map within the HF-OF space is displayed in Fig. 6.56a–d against a background contour of various flow and combustion parameters. The blowout limit is self-explanatory, however the upper limit that defines the stable combustion zone is referred to here as the "acoustic limit." This should not be confused with the anticipated flashback limit because flashback was prevented by the burner headend's built-in flame arrestors. As the name suggests, the acoustic limit refers to the beginning of a loud screech noise that was heard at the top edge of the stable combustion zone. The acoustic limit is thus represented as a dashed line without separate data points, only to illustrate the beginning of audible noise as subjectively experienced by the test operator because the magnitude and frequency of this noise were not measured in this work. This noise denotes limited premixed combustion-specific instability [142]. In order to protect the combustor and increase its operability window, LPM gas-turbine combustors are

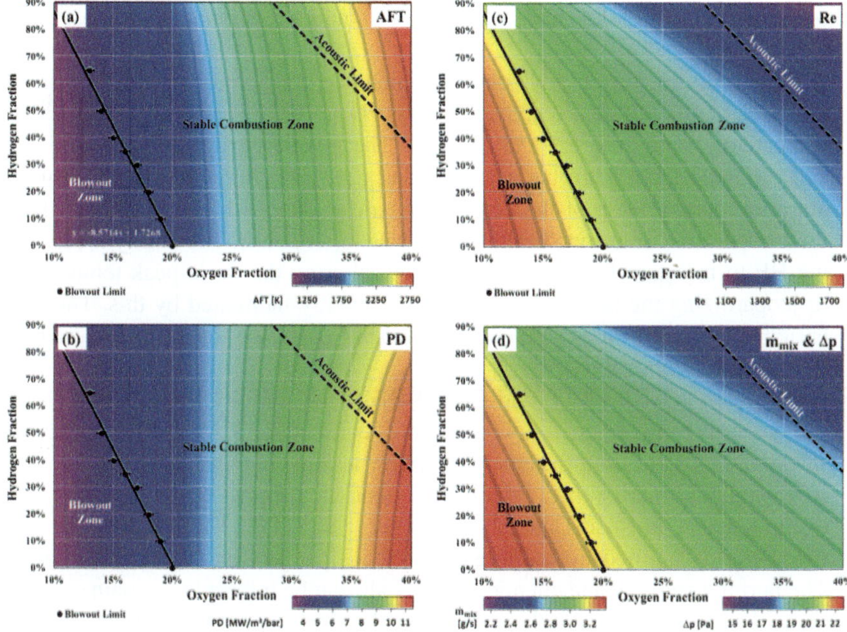

Fig. 6.56 a Stability map over the contours of AFT. **b** Stability map over the contours of PD.
c Stability map over the contours of Re. **d** Stability map over the contours of $\dot{m}_{mix}$, and Δp

generally fitted with active and/or passive methods to bypass or minimize certain instabilities. However, the current model combustor lacks these features, therefore the stable operation zone had to be shut down as soon as instabilities appeared. However, the main goal of this work is to increase the operability window on the blowout side, not this restriction.

Figure 6.56a–d provide several significant new findings:

- The micromixer technology was able to maintain an oxy-methane flame at OF as low as 21%, even without H_2 enrichment (i.e., at HF = 0%). Compared to the 30% minimum threshold mentioned in the literature for swirl-based technologies [97, 106–108], this is a significant improvement. Thus, while constructing new oxy-fuel gas turbines or converting existing gas turbines to oxy-combustion, micromixers should be the technology of choice. The blowout mechanism is examined in more depth in the following section to explain why micromixers have better stability.

- The combustor maintained ignition down to a record-low OF value of 13%, which corresponds to AFT = 1265 K (990 °C, 1800 °F), exceeding the Allam cycle's (1160 °C) dilution criteria. This provides excellent operating flexibility and a ground-breaking boost in the cycle's turndown and low-load capacities. Surprisingly, the lean blowout limit of the majority of existing LPM air–fuel gas-turbine combustor technology is significantly lower than an AFT of 1265 K. If preheating is used, even lower OF and AFT values should be possible with the current combustor.

- HF has little impact on AFT or PD, which are both predominantly under the direction of OF. For instance, the AFT and PD are 1272 K and 4.31 MW/m^3/bar, respectively, with 70% HF and 13% OF. These values are extremely similar to 1221 K and 4.15 MW/m^3/bar at 0% HF and 13% OF. Therefore, the exit temperature and heat release of the combustor barely alter with H_2 enrichment from the perspectives of turbine health and engine input power. The ability to regulate the HF to maintain flame stability without altering the cycle peak temperature or jeopardizing the health of the turbine is greatly facilitated by this. Thus, H_2 enrichment is advised as a practical approach to enhance the limited operability of oxy-fuel combustion systems.

- In contrast to $CH_4/O_2/CO_2$ and $CH_4/O_2/N_2$ flame stability limits, CH_4/H_2–O_2/CO_2 stability limits do not follow contours of constant AFT [89, 108]. Figure 6.56c shows the blowout limit to somewhat follow a shape of constant Re (1660). The following will clarify this: Premixed H_2-enriched flames have been observed to blow out at constant Damkohler numbers (Da) [125, 143]. A constant flow time scale is implied by the usage of fixed input jet velocity in this context. The burning velocity and chemical time scale can be inferred to remain constant along the blowout limit. Furthermore, from the viewpoint of Re, Re regulates flow turbulence and consequently impacts the burning velocity [125]. As previously discussed [101, 102, 125, 126], H_2 addition (greater HF) and CO_2 dilution (lower OF) have opposing impacts on the reaction kinetics and burning velocity. It may be concluded that the burning velocity is about constant along a constant-Re contour since moving along it is expected to maintain a certain amount of turbulence as the effects of HF and OF balance one another out. The blowout limit is seen to occur at roughly constant Re, which is consistent with the Da analysis and explains the observation. Thus, our results are exactly in line with those of Refs. [125, 127].

- The combustor stability map covers the PD range of 4.5–11.5 MW/m^3/bar; at its top end, acoustic dampening can be used to exceed the acoustic limit. This PD range is equivalent to the real operating range of industrial gas turbines, which is normally between 3.5 and 20.0 MW/m^3/bar [97].

Another crucial performance metric is the overall mass flow rate via the combustor, or m_{mix}, which indicates the mass flow rate that the turbine uses to produce electricity. Even though the inlet jet velocity, v, is maintained in this case, Eq. (6.50) still states that m_{mix} changes since mix varies with OF and HF. It may also be inferred from Eqs. (6.50) and (6.51) that m_mix is proportional to the pressure drop over the burner headend, p, for constant v. Thus, m_{mix} and p have a similar plot in Fig. 6.56d. Due to its importance in the design and operation of LPM air–fuel gas turbines, p has been examined in this article. Since air has a stable chemical composition and the density of air is constant at a particular compressor-discharge pressure and temperature, Eq. (6.51) translates constant p to constant v. Operating the air–fuel combustor at constant p means maintaining a constant mass flow rate through the turbine, and the amount of fuel burned in the combustor can be used to regulate the peak cycle

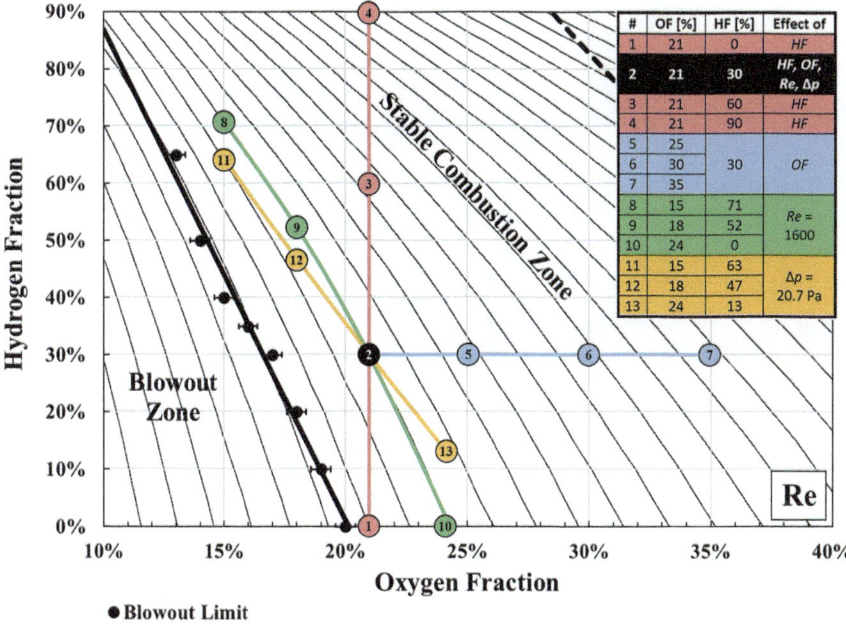

#	OF [%]	HF [%]	Effect of
1	21	0	HF
2	21	30	HF, OF, Re, Δp
3	21	60	HF
4	21	90	HF
5	25		
6	30	30	OF
7	35		
8	15	71	
9	18	52	Re = 1600
10	24	0	
11	15	63	
12	18	47	Δp = 20.7 Pa
13	24	13	

Fig. 6.57 Operating conditions of the selected flames

temperature. However, oxygen-fuel gas turbines are quite different because, even with constant fuel and oxygen flow rates, the peak cycle temperature is regulated by CO_2 recirculation. Mix is therefore a changeable parameter in oxy-fuel gas turbines that also influences the mass flow rate generating power in the turbine.

6.4.4 Flame Macrostructure

A high-definition camera was used to take pictures of the visual flame appearance in order to study the impact of various parameters on flame macrostructure. In order to target the influence of each parameter individually and span the operability window, thirteen different flames were carefully chosen. Figure 6.57 depicts how these flames are distributed throughout the stable combustion zone.

Figure 6.58 shows an example of a flame image. Remember that the reactant combination ($CH_4 + H_2 + O_2 + CO_2$) is discharged through 61 holes in the burner headend in the form of 61 jets. The principal reaction zone consists of 61 flamelets because each jet burns in a little conical flame front known as a "flamelet." A secondary zone in the shape of a common plume is present downstream of the flamelets. The isometric view was unable to distinguish between the several flamelets since the flame was visible through the quartz combustor wall, as shown in Fig. 6.58. The camera had to be positioned so that it was perpendicular to one of the three

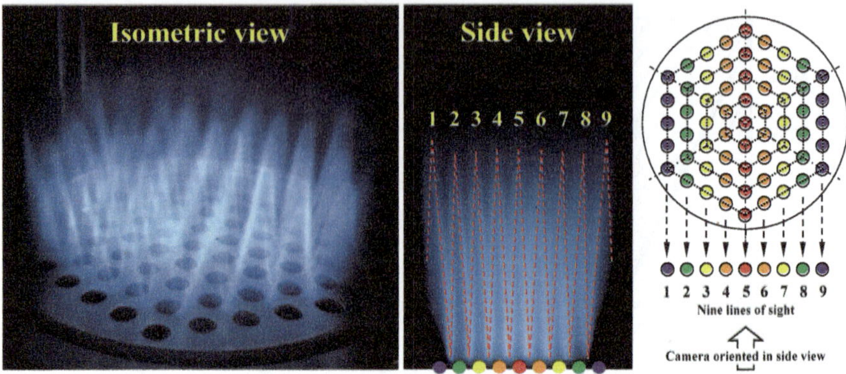

Fig. 6.58 Sample flame image and schematic description of the direction of camera view

axes that make up the headend's hexagonal symmetry. As a result, the camera saw each flamelet in a line of sight as a single flamelet. The 61 flamelets in Fig. 6.58's schematic top view are colored to show how they were reduced to nine lines of sight and only appeared as nine flamelets in the side-view image. Additionally, the camera's alignment with an axis of symmetry enabled the identification of anchored, lifted, and lighted rings in the concentric hexagonal rings surrounding the central jet seen in Fig. 6.55. Lines #1&9 in Fig. 6.58's schematic are deemed to be accurate representations of the fourth (periphery) ring because they stand in for 10 flamelets in this ring. Lines 2 and 8 stand for the third ring's eighth and fourth flamelets, respectively. But if positions #1 and #9 in the side-view flame image are raised and positions #2 and #8 are anchored,

Figure 6.59 displays the photos illustrative of the influence of hydrogen enrichment on flame macrostructure at a fixed OF = 21%. At HF = 0%, the flame is faint and on the verge of blowing out, as can be seen. The second flamelet ring is raised (marked blue), the third and fourth rings are unlit (marked white), and only the central flamelet and the first ring are tethered (marked red). Rings 3 and 4 are less stable because of this flame's low OF and HF, which lower burning velocity in these rings below the jet velocity. However, when the bulk velocity turns favorable, the unburned reactant combination ignites and burns further down in the combustor. This results in the formation of a recirculation zone (see Fig. 6.59), where rings 3 and 4 were seen. Surprisingly, the lit pilot core of these outer rings' 19 flamelets, only 7 of which are anchored, and 12 of which are even lifted, support the 42 unlit jets. In other words, the stabilization mechanism of micromixer technology works so well that the combustor is only stabilized by around 30% of its jets. However, because rings 3 and 4 make up around 70% of the combustor power, their delayed combustion greatly lengthens the entire flame. Additionally, the flame fills the whole cross-section of the quartz container, as can be seen.

Flame stability improves as the HF is raised to 30%, as shown by rings 2 getting anchored and rings 3 successfully igniting although still being lifted. But ring 4 is still unlit. Recirculation zone size shrinks, and the length of the flame as a whole gets

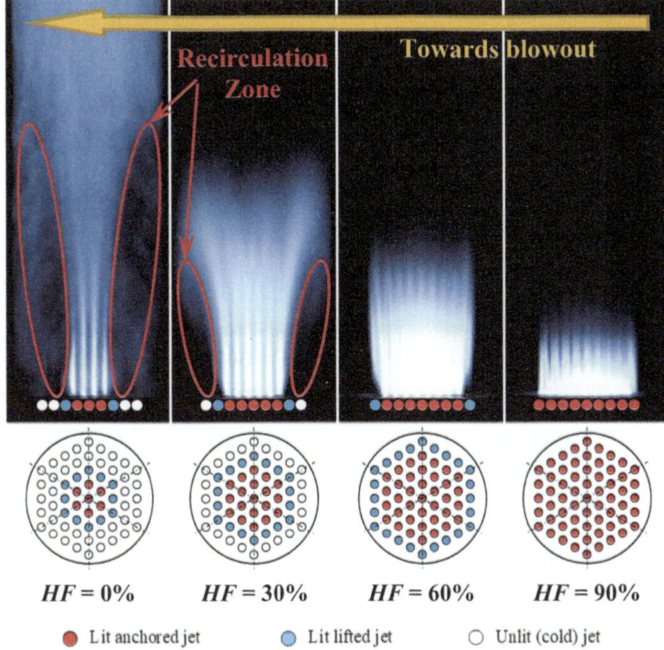

Fig. 6.59 Impact of HF on flame shape at a fixed OF $= 21\%$

noticeably shorter. The overall flame length decreases substantially if the combustion of the 18 flamelets in ring 3 (30% of the combustor power) is no longer delayed. As seen in Fig. 6.59, lowering the HF to 60% completely removes the recirculation zone since the burning velocity is now sufficient to anchor all core flamelets and ignite ring 4 with just slight lifting. The flame now has a greater Da, which causes the flame to be shorter. At 90% HF, all 61 flamelets were visible.

Figure 6.60 at a constant HF of 30% illustrates the impact of changing the OF on the macrostructure of the flame. CO_2 dilution is decreased by raising the OF at given equivalency ratio. Less CO2 causes the flame to become more compact and noticeably brighter, indicating improved reaction kinetics and greater heat release. The adiabatic flame temperature is mostly governed by OF, as was previously described in the analysis of Fig. 6.56a. Therefore, lowering CO_2 dilution raises AFT and subsequently flame brightness. Figure 6.59 shows that the HF had a minimal impact on flame brightness compared to OF, even up to 90% HF, confirming that brightness is highly correlated with OF and AFT.

Four flames with a common Re $= 1600$ were chosen in order to analyze the flame macrostructure at constant Re because the current combustor was seen in Fig. 6.56c to blow out at approximately constant Re. The results are shown in Fig. 6.61. The flames are not identical, as can be observed; rather, they get shorter, brighter, and generally more stable with rising OF at constant Re, even with falling HF. This supports the result that OF has a stronger impact than HF, which was made earlier. It's interesting

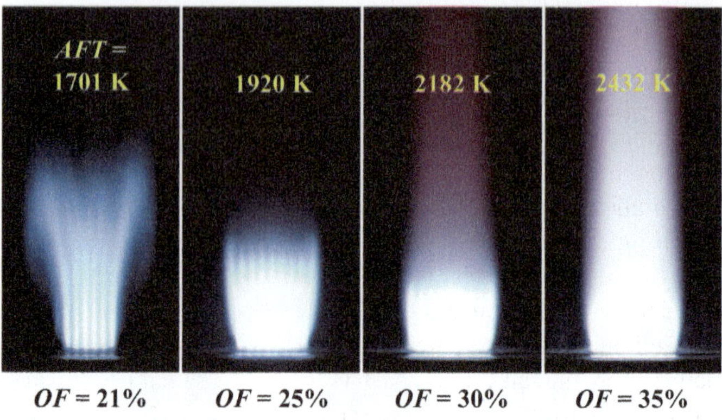

Fig. 6.60 Impact of OF on flame shape at a fixed HF = 30%

to note that the macrostructures of the three flames in the OF = 18–24% range are somewhat similar, suggesting that the effects of decreasing HF and increasing OF are almost balanced out at constant Re, leading to roughly constant burning velocity and similar macrostructure. Given that the blowout region is constrained by the reaction kinetics and Da, this offers an additional justification for why the blowout limit was found to occur at constant Re. However, it is important to note that boosting OF from 18 to 24% (i.e., by just 6%) is sufficient to balance decreasing HF from 52 to 0%.

Flame macrostructure was also examined at constant headend pressure drop, p, since the p map is comparable to that of Re, in order to throw more light on the opposing effects of HF and OF at constant Re (see Fig. 6.56c and d). Figure 6.62 shows the four typical p flames. As can be seen, despite the fact that HF is being reduced, the influence of OF sharply diminishes that of HF, as seen by substantially shorter, brighter, and more stable flames at higher OF. Therefore, unlike at constant

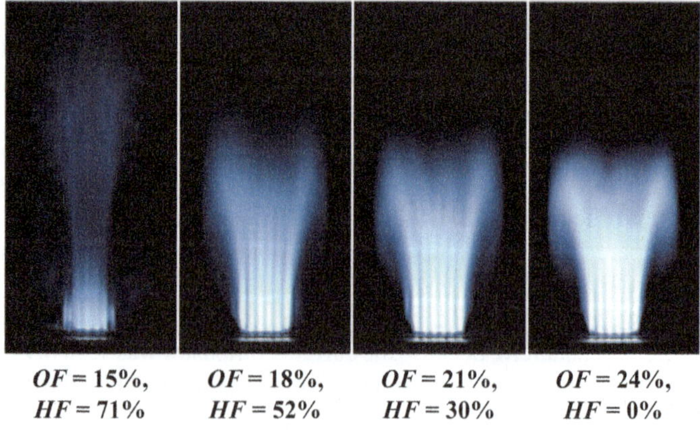

Fig. 6.61 Flame shape at a fixed Re = 1600 and different OF and HF

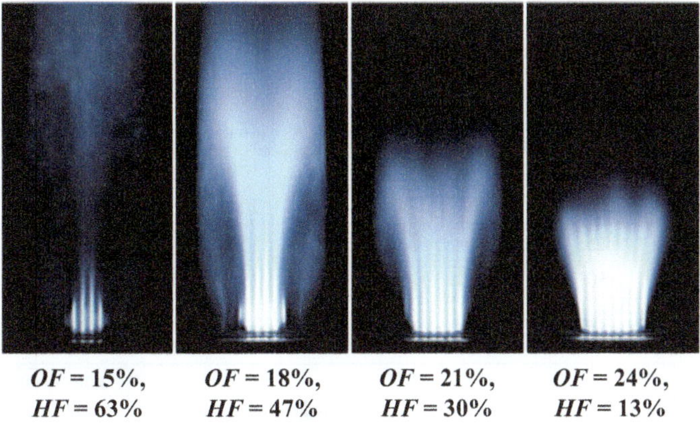

Fig. 6.62 Flame shape at a fixed $\Delta p = 20.7$ Pa and different OF and HF

Re, the effects of HF and OF are not balanced at constant p. Therefore, it can be concluded from the studies of flame shape in Figs. 6.61 and 6.62 that OF has a greater impact on oxy-flames than HF.

6.4.5 Profile Factor of the Combustor

In order to quantify the impacts of HF and OF on profile factor and peak profile temperature as well as to offer experimental validation for upcoming numerical simulations of the current flames, radial temperature profile measurements were made inside the quartz confinement close to the combustor exit. Table 6.4 lists the five flames that were chosen, where flames A through C share a common Re of 1600 and flames C through E have a set OF of 24%. For each flame, the measured peak profile temperature is also listed, along with the profile factor that was determined from the temperature data by applying the definition given in [144]:

Profile factor

$$= \frac{\text{Peak profile temperature} - \text{Mean profile temperature}}{\text{Mean profile temperature} - \text{Reactant temperature at combustor inlet}} \qquad (6.52)$$

In the gas turbine sector, a measurement called the profile factor is used to measure how far the peak profile temperature deviates from the profile mean. Thus, a near-zero profile factor is advised since it denotes a uniform-temperature flow entering the turbine without any hot areas, maintaining the health of the turbine.

The temperature profiles at constant Re are compared in Fig. 6.63. As would be predicted, the profiles are almost axisymmetric. Despite the fact that the HF has

Table 6.4 Operating conditions of the flames selected for temperature measurements

Flame	OF (%)	HF (%)	At constant	AFT (K)	Peak profile temperature (K)	Calculated profile factor
A	21.0	30	Re = 1600	1701	1513	0.118
B	22.7	15		1785	1566	0.127
C	24.0	0		1847	1613	0.149
D	24.0	15	OF = 24%	1855	1595	0.117
E	24.0	30		1866	1583	0.097

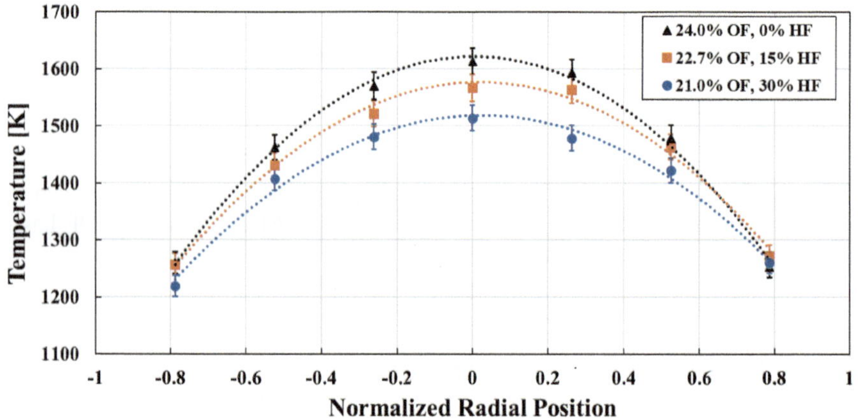

Fig. 6.63 Radial temperature plots at 28 cm at constant Re = 1600 and different OF and HF

decreased, it can also be shown that raising the OF generally raises radial temperatures. This supports the earlier findings, according to which OF has a stronger impact on AFT than HF. This assertion is supported by the AFT values provided in Table 6.4; AFT visibly rises with OF, which explains the hotter temperature profiles at higher OF. A surprising benefit of hydrogen enrichment is that the profile factor actually reduces as HF increases, making the profile less skewed. The temperature profiles at constant OF are compared in Fig. 6.64. Since HF has a little impact on AFT (see Table 6.4), none of the profiles are noticeably hotter than the others. This is expected at constant OF. However, lesser profile components at greater HF are where hydrogen enrichment's beneficial effects are once more visible.

6.5   Concluding Remarks

The discussion of fuel/oxidizer-flexible premixed combustion for clean energy generation in gas turbines is supported by experimental and numerical research in this chapter. Due to the growing need for this technology in combined cycle power

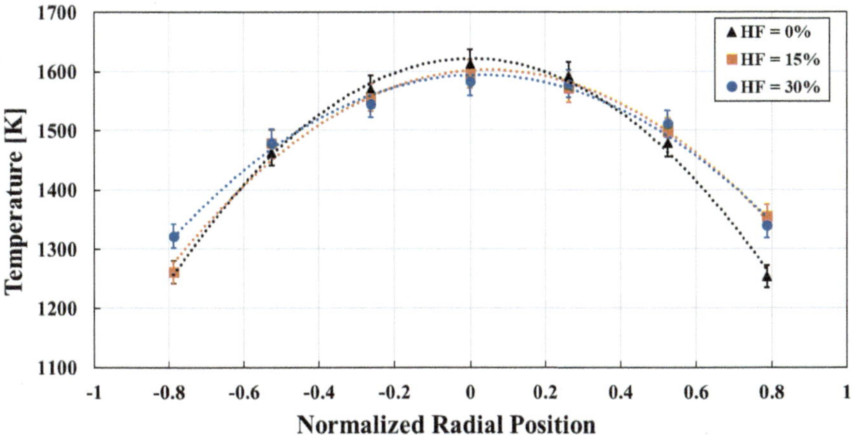

Fig. 6.64 Radial temperature plots at 28 cm at constant OF = 24% and different HF

plants, special focus has been placed on hydrogen-enrichment and synthetic gas combustion. This chapter presents a variety of research on oxidizer/fuel-flexible combustion for clean energy generation in gas turbines taking into account different burner designs and fuel and oxidizer compositions. The first study proposes investigating the impacts of stoichiometric premixed $CH_4/H_2/O_2/CO_2$ on flame features and stability, as well as the consequent emissions of hydrogen-enrichment, oxygen fraction, and bulk throat velocity. The findings demonstrate that for lower hydrogen fraction (HF) and oxygen fraction (OF) values, the outer recirculation zone (ORZ) contributes significantly to flame stability, while at larger levels, this contribution is reduced. The primary characteristic of the inner recirculation zone (IRZ) is the development of secondary eddies, whose number and magnitude largely depend on the OF and HF of the mixture. Due to the higher amounts of chemically active radicals, the rate of reaction increases noticeably as HF and OF are increased. As HF grows, the mole fraction of CO increases while the mole fraction of CO_2 declines. High temperatures cause CO_2 to split into CO and O atoms. The availability of more H species also results in an increase in OH species. As CO emissions rise at higher OF, CO_2 concentration similarly falls. More O atoms accessible to CO should boost its oxidation to CO_2, yet CO_2 dissociation (caused by higher temperatures) favors CO production. Increases in HF and OF cause a fall in the Damköhler number (Da), while increases in inlet velocity cause a rise. The higher values of Da imply that the compositional structure of the flame is primarily controlled by the overall stoichiometry, i.e., the propagation reactions dominate the flame-front region and largely maintain its microstructure, despite the macro-structure of the flame being somewhat affected by swirl-generated turbulence.

To evaluate the effects of equivalence ratio, Reynolds number, and swirl number on stability and macrostructure of various flames, the second study simply extends the first one. According to reports, corner stabilization and a single major eddy in the IRZ characterize flames at lower equivalence ratios. Higher equivalence ratios cause secondary smaller eddies and discrete ORZs, which stabilize swirling flames. As the combustion becomes leaner, the strain rate decreases. The model predicts the lean blowout at = 0.5, which is consistent with the experimental finding. Reynolds number did not appear to have a major impact on the flame structure, but it did have an impact on the IRZ downstream of the combustor. The Reynolds number rises the magnitude of the principal eddies in the IRZ though having no discernible impact on the ORZ. The higher swirl numbers result in a flow with a larger main eddy in the IRZ and a smaller ORZ. With a Karlovitz number of less than one and a high Damkohler number, all of the cases under investigation fall into the category of thin flame regimes, where the flame chunkiness is less than the minimum turbulent scale. When this occurs, the inner flame structure, which is still nearby a laminar flame and wrinkled by turbulence motions, is not susceptible to the turbulence.

The stability properties and combustion behavior of oxy-propane flames ($C_3H_8/O_2/CO_2$) and O2-enriched air-propane flames ($C_3H_8/O_2/N_2$) under identical oxygen fraction and equivalency ratio conditions were then compared in two investigations. For both CO_2 and N_2 flames at a fixed flame-base bulk throat velocity, the following novel findings were drawn. (a) Despite the fact that both flame sets have unique AFT maps, their stabilization zones can be readily identified by AFT by itself. As a result, at similar cold flow circumstances, the stable combustion zones are predominantly controlled by reaction kinetics. (b) In comparison to $C_3H_8/O_2/CO_2$ flames, the stability limits of the $C_3H_8/O_2/N_2$ flames are shifted toward leaner circumstances. In other words, N_2 flames hit their blowout and flashback extinction limits at a lower OF than CO_2 flames did for the identical equivalence ratio. (c) The macrostructure and stabilization mode undergoes comparable modifications as the AFT of CO_2 and N_2 flames is increased from blowout to flashback limit. (d) Flame shapes are similar between flames with similar AFT, even when the OF and are different, demonstrating the supremacy of AFT and response kinetics in flame behavior. Rising acoustic levels coincided with the improvement in reaction kinetics with increased AFT. It proves that, with constant bulk throat velocity, the knowledge from AFT alone is sufficient to accurately anticipate the macrostructure and stability of CO_2 and N_2 flames. Both types of flames' blowout limits essentially followed the same AFT contour, which was around 1580 K, while their flashback limits did the same at 2350 K. This suggests that when compared to $C_3H_8/O_2/CO_2$ flames, $C_3H_8/O_2/N_2$ flames blow out in leaner conditions. At increasing OF, the influence of diluent type on the flame blowout limit decreases; above 55%, both flames blow out at 0.2. The findings of this analysis indicate that AFT consideration, rather than percent exhaust recirculation or dilution ratio, should be taken into account in the design and operation of potential oxy-propane gas-turbine combustors. For any given operational scenario, AFT is easily estimable. Based solely on the data from AFT, it is possible to clearly infer the geometries and stabilization mechanisms of stable oxy-propane flames. As a result, AFT should be the basis for the design of heat shields at the combustor headend. In multi-premixer combustors, AFT can also be used to predict how nearby flames would interact. The effect of AFT on the stability of premixed $C_3H_8/O_2/N_2$ and $C_3H_8/O_2/CO_2$ flames has been identified by the current study. Gas-turbine manufacturers could benefit from knowing how AFT and O_2 enrichment affect these flames' emission characteristics in order to switch from air–fuel combustors to oxy-fuel-based systems.

The study's conclusion takes into account an experimental examination that was conducted to look at hydrogen-enriched oxy-methane flames in a micromixer gas-turbine combustor at a fixed equivalency ratio of 0.9 and inlet jet velocity of 5.2 m/s. The following cutting-edge results were discovered: The Allam oxy-fuel power cycle's primary reaction zone has nominal dilution requirements of 23%, but the micromixer technology's greater stability enables operation at oxygen fractions as low as 21% (by vol.) even without hydrogen enrichment. In contrast, swirl-based burners from earlier research were unable to maintain oxy-flames at oxygen fractions lower than 30%. Thus, the technology advised for oxy-fuel combustion in zero-emission power plants is micromixers. It was discovered that adding hydrogen to the fuel improved flame stability, enabling further oxygen reduction to a record-low 13% before blowout. A record low for oxy-fuel flames, the associated adiabatic flame temperature of 990 °C (1800 °F) even exceeds the Allam cycle's (1160 °C) dilution criteria. This provides excellent operating flexibility and a ground-breaking boost in the cycle's turndown and low-load capacities. The findings also demonstrated that hydrogen enrichment has little impact on combustor power density (MW/m^3/bar) and flame temperature, allowing for additional operational flexibility in adjusting the hydrogen fraction to maintain flame stability without affecting cycle peak temperature or having an adverse effect on the health of the turbine. Additionally, it was discovered that hydrogen enrichment decreased the combustor profile factor, or the skewness of the exit temperature profile, which is important for maintaining the health of the turbine. The opposite effects of increasing hydrogen fraction and lowering oxygen fraction on reaction kinetics and flame macrostructure were found to practically cancel each other out, causing flame blowout to occur at nearly constant input jet Reynolds numbers. Visualizing how a limited number of illuminated core jets serve as the pilot to stabilize the micromixer technology.

Acknowledgements The authors appreciate the support received for the preparation of this book from the Deanship of Research Oversight and Coordination (DROC) at King Fahd University of Petroleum & Minerals (KFUPM). The support provided by the Interdisciplinary Research Center for Hydrogen Technologies and Carbon Management (IRC-HTCM) on project number INHE2308 is highly appreciated. Also, the support received through the KFUPM Consortium for Hydrogen Future through the projects numbered H2FC2309 and H2FC2315 is highly appreciated.

References

1. B.J.P. Buhre, L.K. Elliott, C.D. Sheng, R.P. Gupta, T.F. Wall, Oxy-fuel combustion technology for coal-fired power generation. Prog. Energy Combust. Sci. **31**, 283–307 (2005). https://doi.org/10.1016/j.pecs.2005.07.001
2. H. Liu, R. Zailani, B.M. Gibbs, Pulverized coal combustion in air and in O_2/CO_2 mixtures with NOx recycle. Fuel **84**, 2109–2115 (2005). https://doi.org/10.1016/j.fuel.2005.04.028

3. L. Chen, S. Zheng Yong, A.F. Ghoniem, Oxy-fuel combustion of pulverized coal: characterization, fundamentals, stabilization and CFD modeling. Prog. Energy Combust. Sci. **38**, 156–214 (2012). https://doi.org/10.1016/j.pecs.2011.09.003

4. B.A. Imteyaz, M.A. Nemitallah, A.A. Abdelhafez, M.A. Habib, Combustion behavior and stability map of hydrogen-enriched oxy-methane premixed flames in a model gas turbine combustor. Int. J. Hydrogen Energy **43**, 16652–16666 (2018). https://doi.org/10.1016/j.ijhydene.2018.07.087

5. P. Jourdaine, C. Mirat, J. Caudal, A. Lo, T. Schuller, A comparison between the stabilization of premixed swirling CO_2-diluted methane oxy-flames and methane/air flames. Fuel **201**, 156–164 (2016). https://doi.org/10.1016/j.fuel.2016.11.017

6. T.C. Williams, C.R. Shaddix, R.W. Schefer, Effect of syngas composition and CO_2-diluted oxygen on performance of a premixed swirl-stabilized combustor. Combust. Sci. Technol. **180**, 64–88 (2008). https://doi.org/10.1080/00102200701487061

7. B. Imteyaz, M.A. Habib, Study of combustion characteristics of ethanol at different dilution with the carrier gas. J. Energy Resour. Technol. **137**, 032205 (2015). https://doi.org/10.1115/1.4028866

8. B. Imteyaz, M.A. Habib, M.A. Nemitallah, A. Jamal, Investigation of liquid ethanol evaporation and combustion in air and oxygen environments inside a 25 kW vertical reactor. Proc. Inst. Mech. Eng. Part A: J. Power Energy **229**, 647–661 (2015). https://doi.org/10.1177/0957650915588854

9. M.A. Nemitallah, B. Imteyaz, A. Abdelhafez, M.A. Habib, Experimental and computational study on stability characteristics of hydrogen-enriched oxy-methane premixed flames. Appl. Energy **250**, 433–443 (2019). https://doi.org/10.1016/j.apenergy.2019.05.087

10. M.A. Habib, B. Imteyaz, M.A. Nemitallah, Second law analysis of premixed and non-premixed oxy-fuel combustion cycles utilizing oxygen separation membranes. Appl. Energy **259**, 114213 (2020). https://doi.org/10.1016/j.apenergy.2019.114213

11. H.S. Kim, V.K. Arghode, A.K. Gupta, Flame characteristics of hydrogen-enriched methane-air premixed swirling flames. Int. J. Hydrogen Energy **34**, 1063–1073 (2009). https://doi.org/10.1016/j.ijhydene.2008.10.035

12. M. Gu, H. Chu, F. Liu, Effects of simultaneous hydrogen enrichment and carbon dioxide dilution of fuel on soot formation in an axisymmetric coflow laminar ethylene/air diffusion flame. Combust. Flame **166**, 216–228 (2016). https://doi.org/10.1016/j.combustflame.2016.01.023

13. S.H. Park, K.M. Lee, C.H. Hwang, Effects of hydrogen addition on soot formation and oxidation in laminar premixed C_2H_2/air flames. Int. J. Hydrogen Energy **36**, 9304–9311 (2011). https://doi.org/10.1016/j.ijhydene.2011.05.031

14. F. Halter, C. Chauveau, I. Gökalp, Characterization of the effects of hydrogen addition in premixed methane/air flames. Int. J. Hydrogen Energy **32**, 2585–2592 (2007). https://doi.org/10.1016/j.ijhydene.2006.11.033

15. R.W. Schefer, Hydrogen enrichment for improved lean flame stability. Int. J. Hydrogen Energy **28**, 1131–1141 (2003)

16. S. Gersen, N.B. Anikin, A.V. Mokhov, H.B. Levinsky, Ignition properties of methane/hydrogen mixtures in a rapid compression machine. Int. J. Hydrogen Energy **33**, 1957–1964 (2008). https://doi.org/10.1016/j.ijhydene.2008.01.017

17. R.K. Cheng, A.K. Oppenheim, Autoignition in methane hydrogen mixtures. Combust. Flame **58**, 125–139 (1984). https://doi.org/10.1016/0010-2180(84)90088-9

18. E. Hu, Z. Huang, J. He, C. Jin, J. Zheng, Experimental and numerical study on laminar burning characteristics of premixed methane–hydrogen–air flames. Int. J. Hydrogen Energy **34**, 4876–4888 (2009). https://doi.org/10.1016/j.ijhydene.2009.03.058

19. V. Di Sarli, A. Di Benedetto, Laminar burning velocity of hydrogen–methane/air premixed flames. Int. J. Hydrogen Energy **32**, 637–646 (2007). https://doi.org/10.1016/j.ijhydene.2006.05.016

20. C. Tang, Z. Huang, C. Jin, J. He, J. Wang, X. Wang et al., Laminar burning velocities and combustion characteristics of propane – hydrogen – air premixed flames. Int. J. Hydrogen Energy **33**, 4906–4914 (2008). https://doi.org/10.1016/j.ijhydene.2008.06.063

21. R. Sankaran, H.G. Im, Effects of hydrogen addition on the Markstein length and flammability limit of stretched methane/air premixed flames. Combust. Sci. Technol. **178**, 37–41 (2006). https://doi.org/10.1080/00102200500536217org/10.1080/00102200500536217

22. M. Nakahara, H. Kido, Study on the turbulent burning velocity of hydrogen mixtures including hydrocarbons. AIAA J. **46** (2008). https://doi.org/10.2514/1.23560

23. S. Daniele, P. Jansohn, J. Mantzaras, K. Boulouchos, Turbulent flame speed for syngas at gas turbine relevant conditions. Proc. Combust. Inst. **33**, 2937–2944 (2011). https://doi.org/10.1016/J.PROCI.2010.05.057

24. S. Daniele, P. Jansohn, K. Boulouchos, Experimental investigation of lean premixed syngas combustion at gas turbine relevant conditions: lean blow out limits, emissions and turbulent flame speed, in *Proceedings of 32nd Annual Meeting of Italian Section Combustion Colloquium* (2009)

25. Z. Abubakar, E.M.A. Mokheimer, Thermoacoustic combustion instability of propane-oxy-combustion with CO_2 dilution: experimental analysis. Int. J. Energy Res. **44**, 1031–1045 (2020). https://doi.org/10.1002/er.4980

26. N. Syred, M. Abdulsada, A. Griffiths, T. O'Doherty, P. Bowen, The effect of hydrogen containing fuel blends upon flashback in swirl burners. Appl. Energy **89**, 106–110 (2012). https://doi.org/10.1016/j.apenergy.2011.01.057

27. A.E.E. Khalil, A.K. Gupta, Fuel flexible distributed combustion for efficient and clean gas turbine engines. Appl. Energy **109**, 267–274 (2013). https://doi.org/10.1016/j.apenergy.2013.04.052

28. A. Ali, M.A. Nemitallah, A. Abdelhafez, B. Imteyaz, M.M. Kamal, M.A. Habib, Numerical and experimental study of swirl premixed $CH_4/H_2/O_2/CO_2$ flames for controlled-emissions gas turbines. Int. J. Hydrogen Energy (2020). https://doi.org/10.1016/j.ijhydene.2020.07.210

29. Q. Zhang, D.R. Noble, T. Lieuwen, Characterization of fuel composition effects in $H_2/CO/CH_4$ mixtures upon lean blowout. J. Eng. Gas Turbines Power **129**, 688 (2007). https://doi.org/10.1115/1.2718566

30. A. Abdelhafez, M.A. Nemitallah, S.S. Rashwan, M.A. Habib, Adiabatic flame temperature for controlling the macrostructures and stabilization modes of premixed methane flames in a model gas-turbine combustor. Energy Fuels **32**, 7868–7877 (2018). https://doi.org/10.1021/acs.energyfuels.8b01133

31. A. Abdelhafez, S.S. Rashwan, M.A. Nemitallah, M.A. Habib, Stability map and shape of premixed $CH_4/O_2/CO_2$ flames in a model gas- turbine combustor. Appl. Energy **215**, 63–74 (2018)

32. S. Abdelwahid, M. Nemitallah, B. Imteyaz, A. Abdelhafez, M. Habib, Effects of H_2 enrichment and inlet velocity on stability limits and shape of $CH_4/H_2-O_2/CO_2$ flames in a premixed swirl combustor. Energy Fuels **32**, 9916–9925 (2018)

33. M.A. Nemitallah, B. Imteyaz, A. Abdelhafez, M.A. Habib, Experimental and computational study on combustion and stability characteristics of CH_4-H_2-O_2-CO_2 premixed flames for gas turbine applications. Applied Energy **250**, 433–443 (2019)

34. B. Imteyaz, M.A. Nemitallah, A.A. Abdelhafez, M.A. Habib, Combustion behavior and stability map of hydrogen-enriched oxy-methane premixed flames in a model gas turbine combustor. Int. J. Hydrogen Energy **43**, 16652–16666 (2018)

35. J.M. Beer, *Combustion Aerodynamics* (1972)

36. A.K. Gupta, *Swirl Flows* (1984)

37. S. Brohez, C. Delvosalle, G. Marlair, A two-thermocouples probe for radiation corrections of measured temperatures in compartment fires. Fire Saf. J. **39**, 399–411 (2004)

38. R. Porter, F. Liu, M. Pourkashanian, S.D. Williams, Evaluation of solution methods for radiative heat transfer in gaseous oxy-fuel combustion environments. J. Quant. Spectrosc. Radiat. Transf. **111**, 2084–2094 (2010)

39. R. Johansson, B. Leckner, K. Andersson, F. Johnsson, Account for variations in the H_2O to CO_2 molar ratio when modelling gaseous radiative heat transfer with the weighted-sum-of-grey-gases model. Combust. Flame **158**, 893–901 (2011)

40. Z. Li, X. Cheng, W. Wei, L. Qiu, H. Wu, Effects of hydrogen addition on laminar flame speeds of methane, ethane and propane: experimental and numerical analysis. Int. J. Hydrogen Energy **42**, 24055–24066 (2017)

41. C.L. Tang, Z.H. Huang, C.K. Law, Determination, correlation, and mechanistic interpretation of effects of hydrogen addition on laminar flame speeds of hydrocarbon–air mixtures. Proc. Combust. Inst. **33**, 921–928 (2011)

42. B. Imteyaz, M.A. Habib, R. Ben-Mansour, The characteristics of oxycombustion of liquid fuel in a typical water-tube boiler. Energy Fuels **31**, 6305–6313 (2017). https://doi.org/10.1021/acs.energyfuels.7b00489

43. L.M. Bollinger, D.T. Williams, *Effect of Reynolds Number in Turbulent-Flow Range on Flame Speeds of Bunsen Burner Flames* (1949)

44. T.T. Vu, Y.-I. Lim, D. Song, T.-Y. Mun, J.-H. Moon, D. Sun, Y.-T. Hwang, J.-G. Lee, Y.C. Park, Techno-economic analysis of ultra-supercritical power plants using air- and oxy-combustion circulating fluidized bed with and without CO_2 capture. Energy **194**, 116855 (2020). https://doi.org/10.1016/j.energy.2019.116855

45. C. Gaber, C. Schluckner, P. Wachter, M. Demuth, C. Hochenauer, Experimental study on the influence of the nitrogen concentration in the oxidizer on NOx and CO emissions during the oxy-fuel combustion of natural gas. Energy **214**, 118905 (2021). https://doi.org/10.1016/j.energy.2020.118905

46. X. Liang, Q. Wang, Z. Luo, E. Eddings, T. Ring, S. Li, P. Yu, J. Yan, X. Yang, X. Jia, Experimental and numerical investigation on nitrogen transformation in pressurized oxy-fuel combustion of pulverized coal. J. Clean. Prod. **278**, 123240 (2021). https://doi.org/10.1016/j.jclepro.2020.123240

47. X. Liang, Q. Wang, Z. Luo, E. Eddings, T. Ring, S. Li, L. Han, J. Lin, G. Xie, Experimental study on sulfur-containing products in pressurised oxy-fuel pyrolysis of pulverised coal. J. Clean. Prod. **279**, 123818 (2021). https://doi.org/10.1016/j.jclepro.2020.123818

48. K. Kiriishi, T. Fujimine, A. Hayakawa, High efficiency furnace with oxy-fuel combustion and zero-emission by CO_2 recovery. Int. Gas Union World Gas Conf. Pap. **5**, 3618–3635 (2009)

49. J.H. Ahn, T.S. Kim, Effect of oxygen supply method on the performance of a micro gas turbine-based triple combined cycle with oxy-combustion carbon capture. Energy **211**, 119010 (2020). https://doi.org/10.1016/j.energy.2020.119010

50. H.K. Nguyen, J.-H. Moon, S.-H. Jo, S.J. Park, M.W. Seo, H.W. Ra, S.-J. Yoon, S.-M. Yoon, B. Song, U. Lee, C.W. Yang, T.-Y. Mun, J.-G. Lee, Oxy-combustion characteristics as a function of oxygen concentration and biomass co-firing ratio in a 0.1 MWth circulating fluidized bed combustion test-rig. Energy **196**, 117020 (2020). https://doi.org/10.1016/j.energy.2020.117020

51. M.A. Nemitallah, S.S. Rashwan, I.B. Mansir, A.A. Abdelhafez, M.A. Habib, Review of novel combustion techniques for clean power production in gas turbines. Energy Fuels **32**, 979–1004 (2018). https://doi.org/10.1021/acs.energyfuels.7b03607

52. E. Portillo, B. Alonso-Fariñas, F. Vega, M. Cano, B. Navarrete, Alternatives for oxygen-selective membrane systems and their integration into the oxy-fuel combustion process: a review. Sep. Purif. Technol. **229**, 115708 (2019). https://doi.org/10.1016/j.seppur.2019.115708

53. M.A. Nemitallah, A.A. Abdelhafez, A. Ali, I. Mansir, M.A. Habib, Frontiers in combustion techniques and burner designs for emissions control and CO_2 capture: a review. Int. J. Energy Res. 1–33 (2019). https://doi.org/10.1002/er.4730

54. N. Khallaghi, D.P. Hanak, V. Manovic, Techno-economic evaluation of near-zero CO_2 emission gas-fired power generation technologies: a review. J. Nat. Gas Sci. Eng. **74**, 103095 (2020). https://doi.org/10.1016/j.jngse.2019.103095

55. M.A. Haque, M.A. Nemitallah, A.A. Abdelhafez, I.B. Mansir, M.A.M. Habib, Review of fuel/oxidizer-flexible combustion in gas turbines, Energy Fuels (2020). https://doi.org/10.1021/acs.energyfuels.0c02097

56. M. Nemitallah, S. Alkhaldi, A. Abdelhafez, M. Habib, Effect analysis on the macrostructure and static stability limits of oxy-methane flames in a premixed swirl combustor. Energy **159**, 86–96 (2018). https://doi.org/10.1016/j.energy.2018.06.131

57. M.A. Habib, M.A. Nemitallah, P. Ahmed, M.H. Sharqawy, H.M. Badr, I. Muhammad, M. Yaqub, Experimental analysis of oxygen-methane combustion inside a gas turbine reactor under various operating conditions. Energy **86**, 105–114 (2015). https://doi.org/10.1016/j.energy.2015.03.120

58. A. Reihani, J. Hoard, S. Klinkert, C.K. Kuan, D. Styles, G. McConville, Experimental response surface study of the effects of low-pressure exhaust gas recirculation mixing on turbocharger compressor performance. Appl. Energy **261**, 114349 (2020). https://doi.org/10.1016/j.apenergy.2019.114349

59. M.A. Habib, S.S. Rashwan, M.A. Nemitallah, A. Abdelhafez, Stability maps of non-premixed methane flames in different oxidizing environments of a gas turbine model combustor. Appl. Energy **189**, 177–186 (2017). https://doi.org/10.1016/j.apenergy.2016.12.067

60. S.J. Shanbhogue, Y.S. Sanusi, S. Taamallah, M.A. Habib, E.M.A. Mokheimer, A.F. Ghoniem, Flame macrostructures, combustion instability and extinction strain scaling in swirl-stabilized premixed CH_4/H_2 combustion. Combust. Flame **163**, 494–507 (2016). https://doi.org/10.1016/j.combustflame.2015.10.026

61. M. Aliyu, M.A. Nemitallah, S.A. Said, M.A. Habib, Characteristics of H_2-enriched CH_4-O_2 diffusion flames in a swirl-stabilized gas turbine combustor: experimental and numerical study. Int. J. Hydrogen Energy **41**, 20418–20432 (2016). https://doi.org/10.1016/j.ijhydene.2016.08.144

62. S.A. Said, M. Aliyu, M.A. Nemitallah, M.A. Habib, I.B. Mansir, Experimental investigation of the stability of a turbulent diffusion flame in a gas turbine combustor. Energy **157**, 904–913 (2018). https://doi.org/10.1016/j.energy.2018.05.177

63. Y. Tu, S. Xu, M. Xu, H. Liu, W. Yang, Numerical study of methane combustion under moderate or intense low-oxygen dilution regime at elevated pressure conditions up to 8 atm. Energy **197**, 117158 (2020). https://doi.org/10.1016/j.energy.2020.117158

64. S. Karyeyen, J.S. Feser, E. Jahoda, A.K. Gupta, Development of distributed combustion index from a swirl-assisted burner. Appl. Energy **268**, 114967 (2020). https://doi.org/10.1016/j.apenergy.2020.114967

65. Z. Abubakar, S.Y. Sanusi, E.M.A. Mokheimer, Stability of Propane-air and oxyfuel diffusion flames in a swirl-stabilized combustor; an experimental study. Energy Procedia **142**, 1552–1557 (2017). https://doi.org/10.1016/j.egypro.2017.12.607

66. Z. Abubakar, S.Y. Sanusi, E.M.A. Mokheimer, Experimental analysis of the stability and combustion characteristics of propane-oxyfuel and propane-air flames in a non-premixed, swirl-stabilized combustor. Energy Fuels **32**, 8837–8844 (2018). https://doi.org/10.1021/acs.energyfuels.8b01819

67. A. Ali, M.A. Nemitallah, A. Abdelhafez, I.G. Alsakhawy, M.M. Kamal, M.A. Habib, Static stability and combustion characteristics of oxy-propane flames in a premixed fuel-flexible swirl combustor. Energy Fuels **33**, 11996–12007 (2019). https://doi.org/10.1021/acs.energyfuels.9b03157

68. Z. Abubakar, M.R. Shakeel, E.M.A. Mokheimer, Experimental and numerical analysis of non-premixed oxy-combustion of hydrogen-enriched propane in a swirl stabilized combustor. Energy **165**, 1401–1414 (2018). https://doi.org/10.1016/j.energy.2018.10.102

69. Z. Mansouri, M. Aouissi, T. Boushaki, Numerical computations of premixed propane flame in a swirl-stabilized burner: effects of hydrogen enrichment, swirl number and equivalence ratio on flame characteristics. Int. J. Hydrogen Energy **41**, 9664–9678 (2016). https://doi.org/10.1016/j.ijhydene.2016.04.023

70. A. Ali, M.A. Nemitallah, A. Abdelhafez, M. Hussain, M.M. Kamal, M.A. Habib, Comparative analysis of the stability and structure of premixed $C_3H_8/O_2/CO_2$ and $C_3H_8/O_2/N_2$ flames for clean flexible energy production. Energy **214**, 1–10 (2021). https://doi.org/10.1016/j.energy.2020.118887

71. F. Guo, L. Ding, Z. Gao, L. Yu, J. Ji, Effects of wind flow and sidewall restriction on the geometric characteristics of propane diffusion flames in tunnels. Energy **198**, 117332 (2020). https://doi.org/10.1016/j.energy.2020.117332

72. Y. Sun, Z. Rao, D. Zhao, B. Wang, D. Sun, X. Sun, Characterizing nonlinear dynamic features of self-sustained thermoacoustic oscillations in a premixed swirling combustor. Appl. Energy **264**, 114698 (2020). https://doi.org/10.1016/j.apenergy.2020.114698

73. J.M. Beér, Combustion aerodynamics, in *Combustion Technology* (Elsevier, 1974), pp. 61–89

74. A.K. Gupta, D.G. Lilley, N. Syred, *Swirl Flows*, Tw. (1984)

75. L. Chen, A.F. Ghoniem, Simulation of oxy-coal combustion in a 100 kW_{th} test facility using RANS and LES: a validation study. Energy Fuels **26**, 4783–4798 (2012). https://doi.org/10.1021/ef3006993

76. R. Johansson, B. Leckner, K. Andersson, F. Johnsson, Account for variations in the H_2O to CO_2 molar ratio when modelling gaseous radiative heat transfer with the weighted-sum-of-grey-gases model. Combust. Flame **158**, 893–901 (2011). https://doi.org/10.1016/j.combustflame.2011.02.001

77. G.P. Smith, D.M. Golden, M. Frenklach, N.W. Moriarty, B. Eiteneer, M. Goldenberg, C.T. Bowman, R.K. Hanson, S. Song, W.C. Gardiner Jr., V.V. Lissianski, Z. Qin, *GRI-MECH 3.0* (n.d.). http://www.me.berkeley.edu/gri_mech/

78. M. Aliyu, A. Abdelhafez, S.A.M. Said, M.A. Habib, M.A. Nemitallah, I.B. Mansir, Characteristics of oxyfuel combustion in lean-premixed multihole burners. Energy Fuels **33**, 11948–11958 (2019). https://doi.org/10.1021/acs.energyfuels.9b02821

79. M. Hussain, A. Abdelhafez, M.A. Nemitallah, A.A. Araoye, R. Ben-Mansour, M.A. Habib, A highly diluted oxy-fuel micromixer combustor with hydrogen enrichment for enhancing turndown in gas turbines. Appl. Energy **279**, 115818 (2020). https://doi.org/10.1016/j.apenergy.2020.115818

80. A. Abdelhafez, S.S. Rashwan, M.A. Nemitallah, M.A. Habib, Stability map and shape of premixed $CH_4/O_2/CO_2$ flames in a model gas-turbine combustor. Appl. Energy **215**, 63–74 (2018). https://doi.org/10.1016/j.apenergy.2018.01.097

81. M.A. Nemitallah, G. Kewlani, S. Hong, S.J. Shanbhogue, M.A. Habib, A.F. Ghoniem, Investigation of a turbulent premixed combustion flame in a backward-facing step combustor; effect of equivalence ratio. Energy **95**, 211–222 (2016). https://doi.org/10.1016/j.energy.2015.12.010

82. S. Taamallah, Z.A. LaBry, S.J. Shanbhogue, M.A.M. Habib, A.F. Ghoniem, Correspondence between "stable" flame macrostructure and thermo-acoustic instability in premixed swirl-stabilized turbulent combustion. J. Eng. Gas Turbines Power **137** (2015). https://doi.org/10.1115/1.4029173

83. A. Ali, M.A. Nemitallah, A. Abdelhafez, B. Imteyaz, M.M. Kamal, M.A. Habib, Numerical and experimental study of swirl premixed $CH_4/H_2/O_2/CO_2$ flames for controlled- emissions gas turbines. Int. J. Hydrogen Energy (2020). https://doi.org/10.1016/j.ijhydene.2020.07.210

84. M.R. Shakeel, Y.S. Sanusi, E.M.A. Mokheimer, Numerical modeling of oxy-methane combustion in a model gas turbine combustor. Appl. Energy **228**, 68–81 (2018). https://doi.org/10.1016/j.apenergy.2018.06.071

85. A. Abdelhafez, M.A. Nemitallah, S.S. Rashwan, M.A. Habib, Adiabatic flame temperature for controlling the macrostructures and stabilization modes of premixed methane flames in a model gas-turbine combustor. Energy Fuels **32**(7), 7868–7877 (2018)

86. N.L. Carr, R. Kobayashi, D.B. Burrows, Viscosity of hydrocarbon gases under pressure. J. Petrol. Technol. **6**(10), 47–55 (1954)

87. Y.A. Cengel, M.A. Boles, *Thermodynamics An Engineering Approach* (2015)

88. W. Jerzak, M. Kuźnia, Experimental study of impact of swirl number as well as oxygen and carbon dioxide content in natural gas combustion air on flame flashback and blow-off. J. Nat. Gas Sci. Eng. **29**, 46–54 (2016)

89. A. Abdelhafez, S.S. Rashwan, M.A. Nemitallah, M.A. Habib, Stability map and shape of premixed $CH_4/O_2/CO_2$ flames in a model gas-turbine combustor. Appl. Energy **215**, 63–74 (2018)

90. W.E. Kaskan, The dependence of flame temperature on mass burning velocity. Sympos. Combust. **6**(1), 134–143 (1957)

91. A. Van Maaren, D.S. Thung, L.P.H. De Goey, Measurement of flame temperature and adiabatic burning velocity of methane/air mixtures. Combust. Sci. Technol. **96**(4–6), 327–344 (1994)
92. B. Ademe, CO_2 capture and storage in the subsurface. *Geosci. Issues, Fr.* (2007)
93. IPCC, *Climate Change—The Physical Science Basis* (2007)
94. J.N. Armor, Addressing the CO_2 dilemma. Catal. Lett. **114**(3–4), 115–121 (2007)
95. B.J.P. Buhre, L.K. Elliott, C.D. Sheng, R.P. Gupta, T.F. Wall, Oxy-fuel combustion technology for coal-fired power generation. Prog. Energy Combust. Sci. **31**(4), 283–307 (2005)
96. N. Perrin, C. Paufique, M. Leclerc, Latest performances and improvement perspective of Oxycombustion for carbon capture on coal power plants. Energy Procedia **63**, 524–531 (2014)
97. M.A. Nemitallah, M.A. Habib, Experimental and numerical investigations of an atmospheric diffusion oxy-combustion flame in a gas turbine model combustor. Appl. Energy **111**, 401–415 (2013)
98. K. Lee, H. Kim, P. Park, S. Yang, Y. Ko, CO_2 radiation heat loss effects on NOx emissions and combustion instabilities in lean premixed flames. Fuel **106**, 682–689 (2013)
99. Y. Lafay, B. Taupin, G. Martins, G. Cabot, B. Renou, A. Boukhalfa, Experimental study of biogas combustion using a gas turbine configuration. Exp. Fluids **43**(2–3), 395–410 (2007)
100. K.K. Gupta, A. Rehman, R.M. Sarviya, Bio-fuels for the gas turbine: a review. Renew. Sustain. Energy Rev. **14**(9), 2946–2955 (2010)
101. T.C. Williams, C.R. Shaddix, R.W. Schefer, Effect of syngas composition and CO_2-diluted oxygen on performance of a premixed swirl-stabilized combustor. Combust. Sci. Technol. **180**(1), 64–88 (2007)
102. J. Zhang, J. Mi, P. Li, F. Wang, B.B. Dally, Moderate or intense low-oxygen dilution combustion of methane diluted by CO_2 and N_2. Energy Fuels **29**(7), 4576–4585 (2015)
103. M.A. Nemitallah, S.S. Rashwan, I.B. Mansir, A. Abdelhafez, M.A. Habib, Review of novel combustion techniques for clean power production in gas turbines. Energy Fuels **32**, 979 (2018)
104. I.A. Ramadan, A.H. Ibrahim, T.W. Abou-Arab, S.S. Rashwan, M.A. Nemitallah, M.A. Habib, Effects of oxidizer flexibility and bluff-body blockage ratio on flammability limits of diffusion flames. Appl. Energy **178**, 19–28 (2016)
105. F. Halter, F. Foucher, L. Landry, C. Mounaim-Rousselle, Effect of dilution by nitrogen and/or carbon dioxide on methane and iso-octane air flames. Combust. Sci. Technol. **181**(6), 813–827 (2009)
106. M. Ditaranto, J. Hals, Combustion instabilities in sudden expansion oxy-fuel flames. Combust. Flame **146**(3), 493–512 (2006)
107. P. Kutne, B.K. Kapadia, W. Meier, M. Aigner, Experimental analysis of the combustion behaviour of oxyfuel flames in a gas turbine model combustor. Proc. Combust. Inst. **33**(2), 3383–3390 (2011)
108. A. Abdelhafez, M.A. Nemitallah, S.S. Rashwan, M.A. Habib, Adiabatic flame temperature for controlling the macrostructures and stabilization modes of premixed methane flames in a model gas-turbine combustor. Energy Fuels **32**, 7868–7877 (2018)
109. C. Yin, J. Yan, Oxy-fuel combustion of pulverized fuels: combustion fundamentals and modeling. Appl. Energy **162**, 742–762 (2016)
110. M.A. Habib, M.A. Nemitallah, P. Ahmed, M.H. Sharqawy, H.M. Badr, I. Muhammad et al., Experimental analysis of oxygen-methane combustion inside a gas turbine reactor under various operating conditions. Energy **86**, 105–114 (2015)
111. Y. Song, C. Zou, Y. He, C. Zheng, The chemical mechanism of the effect of CO_2 on the temperature in methane oxy-fuel combustion. Int. J. Heat Mass Transf. **86**, 622–628 (2015)
112. https://www.mendeley.com/authors/15046461500/
113. https://en.wikipedia.org/wiki/Rodney_John_Allam
114. J.D. Laumb, M.J. Holmes, J.J. Stanislowski, X. Lu, B.A. Forrest, M. McGroddy, Supercritical CO_2 cycles for power production. Energy Procedia **114**, 573–580 (2017)
115. R.J. Allam, J.E. Fetvedt, B.A. Forrest, D.A. Freed, The oxy-fuel, supercritical CO_2 Allam cycle: new cycle developments to produce even lower-cost electricity from fossil fuels without atmospheric emissions. in *ASME Turbo Expo, GT2014-26952* (2014)

116. Y. Iwai, M. Itoh, Y. Morisawa, S. Suzuki, D. Cusano, M. Harris, Development approach to the combustor of gas turbine for oxy-fuel, supercritical CO_2 cycle, in *ASME Turbo Expo, GT2015-43160* (2015)
117. X. Lu, B.A. Forrest, S. Martin, J.E. Fetvedt, M. McGroddy, D.A. Freed, Integration and optimization of coal gasification systems with a near-zero emissions supercritical carbon dioxide power cycle, in *ASME Turbo Expo, GT2016-58066* (2016)
118. R.J. Allam, S. Martin, B.A. Forrest, J.E. Fetvedt, X. Lu, D.A. Freed et al., Demonstration of the Allam cycle: an update on the development status of a high efficiency supercritical carbon dioxide power process employing full carbon capture. Energy Procedia **114**, 5948–5966 (2017)
119. E. Hu, Z. Huang, J. He, C. Jin, J. Zheng, Experimental and numerical study on laminar burning characteristics of premixed methane-hydrogen-air flames. Int. J. Hydrogen Energy **34**, 4876–4888 (2009)
120. S. Daniele, P. Jansohn, J. Mantzaras, K. Boulouchos, Turbulent flame speed for syngas at gas turbine relevant conditions. Proc. Combust. Inst. **33**, 2937–2944 (2011)
121. A.F. Ghoniem, A. Annaswamy, S. Park, Z.C. Sobhani, Stability and emissions control using air injection and H_2 addition in premixed combustion. Proc. Combust. Inst. **30**(2), 1765–1773 (2005)
122. M. Gu, H. Chu, F. Liu, Effects of simultaneous hydrogen enrichment and carbon dioxide dilution of fuel on soot formation in an axisymmetric coflow laminar ethylene/air diffusion flame. Combust. Flame **166**, 216–228 (2016)
123. R.W. Schefer, Hydrogen enrichment for improved lean flame stability. Int. J. Hydrogen Energy **28**(10), 1131–1141 (2003)
124. S. de Ferrières, A. El Bakali, B. Lefort, M. Montero, J.F. Pauwels, Experimental and numerical investigation of low-pressure laminar premixed synthetic natural gas/O_2/N_2 and natural gas/H_2/O_2/N_2 flames. Combust. Flame **154**(3), 601–623 (2008)
125. B.A. Imteyaz, M.A. Nemitallah, A.A. Abdelhafez, M.A. Habib, Combustion behavior and stability map of hydrogen-enriched oxy-methane premixed flames in a model gas turbine combustor. Int. J. Hydrogen Energy **43**(34), 16652–16666 (2018)
126. S. Abdelwahid, M. Nemitallah, B. Imteyaz, A. Abdelhafez, M. Habib, Effects of H_2 enrichment and inlet velocity on stability limits and shape of CH_4/H_2-O_2/CO_2 flames in a premixed swirl combustor. Energy Fuels **32**(9), 9916–9925 (2018)
127. M.A. Nemitallah, B. Imteyaz, A. Abdelhafez, M.A. Habib, Experimental and computational study on stability characteristics of hydrogen-enriched oxy-methane premixed flames. Appl. Energy **250**, 433–443 (2019)
128. H.H.W. Funke, J. Keinz, K. Kusterer, A. Haj Ayed, M. Kazari, J. Kitajima, et al., Development and testing of a low NOx micromix combustion chamber for industrial gas turbines. Int. J. Gas Turbine Propuls. Power Syst. **9**(1), 27–36 (2017)
129. H.H.-W. Funke, N. Beckmann, J. Keinz, S. Abanteriba, Numerical and experimental evaluation of a dual-fuel dry-low-NOx micromix combustor for industrial gas turbine applications. J. Therm. Sci. Eng. Appl. **11**(1), 011015 (2019)
130. T. Asai, S. Dodo, M. Karishuku, N. Yagi, Y. Akiyama, A. Hayashi, Performance of multiple-injection dry low-NOx combustors on hydrogen-rich syngas fuel in an IGCC pilot plant. J. Eng. Gas Turbines Power **137**(9), 091504 (2015)
131. M.H. du Toit, A.V. Avdeenkov, D. Bessarabov, Reviewing H_2 combustion: a case study for non-fuel-cell power systems and safety in passive autocatalytic recombiners. Energy Fuels **32**, 6401–6422 (2018)
132. W.D. York, W.S. Ziminsky, E. Yilmaz, Development and testing of a low NOx hydrogen combustion system for heavy-duty gas turbines. J. Eng. Gas Turbines Power **135**(2), 022001 (2013)
133. T. Asai, S. Dodo, H. Koizumi, H. Takahashi, S. Yoshida, H. Inoue, Effects of multiple-injection-burner configurations on combustion characteristics for dry low-NOx combustion of hydrogen-rich fuels, in *Volume 2: Combustion, Fuels and Emissions, Parts A and B* (2011), pp. 311–320

134. H.H.W. Funke, N. Beckmann, S. Abanteriba, An overview on dry low NOx micromix combustor development for hydrogen-rich gas turbine applications. Int. J. Hydrogen Energy **44**(13), 6978–6990 (2019)
135. S. Dodo, T. Asai, H. Koizumi, H. Takahashi, S. Yoshida, H. Inoue, Combustion characteristics of a multiple-injection combustor for dry low-NOx combustion of hydrogen-rich fuels under medium pressure, in *Proceedings of ASME Turpo Expo 2011, GT2011-45459,* (2011), pp. 467–476
136. S. Brohez, C. Delvosalle, G. Marlair, A two-thermocouples probe for radiation corrections of measured temperatures in compartment fires. Fire Saf. J. **39**(5), 399–411 (2004)
137. S. Taamallah, N.W. Chakroun, H. Watanabe, S.J. Shanbhogue, A.F. Ghoniem, On the characteristic flow and flame times for scaling oxy and air flame stabilization modes in premixed swirl combustion. Proc. Combust. Inst. **36**(3), 3799–3807 (2017)
138. Y.A. Cengel, M.A. Boles, *Thermodynamics: An Engineering Approach*, 8th edn. (McGraw-Hill Education, New York, 2015)
139. N.L. Carr, R. Kobayashi, D.B. Burrows, J. Pet. Technol. **6**, 47–55 (1954)
140. H.K. Kayadelen, Effect of natural gas components on its flame temperature, equilibrium combustion products and thermodynamic properties. J. Nat. Gas Sci. Eng. **45**, 456–473 (2017)
141. M. Li, Y. Tong, M. Thern, J. Klingmann, Investigation of methane oxy-fuel combustion in a swirl-stabilised gas turbine model combustor. Energies **10**(5) (2017)
142. G. Beardsell, G. Blanquart, Fully compressible simulations of the impact of acoustic waves on the dynamics of laminar premixed flames for engine-relevant conditions. Proc. Combust. Inst. (in press). Available online 22 July 2020. https://doi.org/10.1016/j.proci.2020.06.003
143. E. Sullivan-Lewis, V. McDonell, Predicting flameholding for hydrogen and natural gas flames at gas turbine premixer conditions. J. Eng. Gas Turbines Power **138**(12), 1–9 (2016)
144. A.H. Lefebvre, D.R. Ballal, Gas Turbine Combustion: Alternative Fuels and Emissions, 3rd edn (CRC Press, 2010), p. 134. ISBN 9781420086058

Chapter 7
Energy and Hydrogen Production in Novel Membrane Reactors

7.1 Green Hydrogen Production

Although research and development of hydrogen production has been extensive in the last decades, the recent growing concerns with regard to the sustainability of energy and the deterrence or slowing of declining worldwide ecosystem has reintroduced the search for more economical, effective and simple procedures for green hydrogen production [1]. The great energy content in hydrogen, the production from renewable resources and environmental friendliness generate a growing international curiosity in its production and generation at large-scale. Relatively little capital venture is essential to produce hydrogen with inferior operating cost because of the easiness of the thermal production process. Since its technology has been currently established, the solar thermal energy can be the main energy foundation for the heating process. The application of the high temperature solar energy coupling is limited due to its present high cost. However, the recent success in more effective knowhow for harnessing the solar power suggests an expected lessening in the charge in the immediate future [2]. Currently, numerous technologies are available for relatively cheap generation of hydrogen. These include electrolysis using electricity from renewable as well as nuclear energy or photovoltaics, hydrogen from biomass, coal electrolysis to fossil fuels, and water splitting utilizing solar energy. It is also possible to obtain hydrogen from water through radiolysis, electrolysis, thermolysis, thermochemical cycles, bio-photolysis, bio-catalysis, photo-catalysis, etc., as well as steam reforming of methane and Plasma reforming [3–5]. Among the new methods for hydrogen production, the technology of water splitting by means of thermochemical or electrolysis cycles has been lately given exceptional consideration [6, 7]. It is possible to produce hydrogen more economically by implementing the concept of water splitting. Water splitting using electrolysis is characterized founded on the type of employed electrolyte. Currently, water electrolyzers satisfy just about 3.9% of the world's hydrogen mandate [8]. The encouraging ones comprise polymer electrolysis

© The Author(s), under exclusive license to Springer Nature Singapore Pte Ltd. 2024
M. A. Nemitallah et al., *Hydrogen for Clean Energy Production: Combustion Fundamentals and Applications*, https://doi.org/10.1007/978-981-97-7925-3_7

membrane (PEM) cells (using acidic ionomer), alkaline electrolysis cells (AEC), and great temperature steam electrolysis cells (utilizing solid oxides) [9–13].

Utilizing the technology of oxygen transport membranes (OTM) is considered to be an economical and reasonably a simple resources of manufacturing hydrogen from water splitting. The OTMs are manufactured from mixed ionic and electronic conductor membranes. These membranes have 100% oxygen selectivity from gas mixtures like N_2 and O_2 or H_2 and O_2. This is attributed to the existence of oxygen vacancies in their crystal lattice and the process of oxygen separation works as follows. At temperature high sufficient to overawed the activation energy of the process, water splits into hydrogen and oxygen. The oxygen is, then, permeated across the ion transport membrane (ITM) to the sweep side which is known as the oxygen permeation side. Thus, hydrogen is engendered from the steam at the feed side (hydrogen generation side). The process of permeation is derived by the partial pressure variance across the ITM. The reactions involved on both sides of the oxygen transport membrane (Feed and Sweep sides) are presented in the following equations:

Methane reduction step:

$$CH_4 + MO \rightarrow MO \text{ reduced} + CO + H_2$$

H_2O/CO_2 splitting step (oxidation step):

$$MO \text{ reduced} + H_2O(\text{or } CO_2) \rightarrow MO + H_2(\text{or } CO)$$

The production of hydrogen in the feed side and oxygen permeation to the sweep side can be enhanced significantly by reducing the oxygen concentration by a strong reductant fuel like CH_4. The methane reduction of the membrane metal oxide utilizes excellent reducing power of CH_4, and because methane is composed of two powerful reductants, C and H_2, a mole of CH_4 supplies a total of 3 mol of reducing agent. The work of Han et al., in 2018 revealed that CO and H_2 are the main products, instead of CO_2 and H_2O under conditions of depleted oxygen with spare CH_4 [14]. The exclusion of oxygen by the membrane shifts water splitting equilibrium reaction headed for dissociation into hydrogen and oxygen. Consequently, considerable quantities of oxygen and hydrogen are generated. The oxygen transport process is initiated when the oxygen vacancies are created from the methane reduction sites, allowing oxygen ions to drawn-out through the crystal lattice of ceramics, e.g., perovskite, by hopping through oxygen vacancy defect sites. The process comprises mass transfer of gaseous oxygen from the gas stream of the higher oxygen partial pressure in the feed side (feed or hydrogen generation side) to the membrane surface. This process is tracked by the adsorption of oxygen molecules, formerly, surface reaction at the feed side. Bulk diffusion, or the movement ent of oxygen ions crosswise of the membrane, occurs after this procedure. The lower O_2 partial pressure side (also known as the sweep side or oxygen permeation side) then experiences surface response. Ultimately, oxygen is moved into the stream of gas from the sweep side membrane surface [15–17]. The results that are available in previous works have revealed that oxygen transport through OTM depends on factors comprising of operational temperature, rates of gas flow, difference in partial pressures of oxygen through the membrane, as well as the membrane thickness [18–20]. Accordingly, those aspects greatly influence the production of hydrogen through water splitting with the use of ITMs.

The growing demand for clean hydrogen for petrochemical industry applications has led to extensive investigations towards improving the performance inorganic membranes for hydrogen production and purification [21–24]. The dense ceramic membranes have received more attention for the gas separation applications due to their capability to separate and transport oxygen [25–28]. The oxygen transport rate through a membrane is a vital parameter in determining hydrogen production via water splitting [29]. Moreover, there are several significant factors that affect the hydrogen production via water-splitting and oxygen permeation rate as well. These parameters include (1) gradient of partial pressure of oxygen across the membrane, (2) oxygen ion and electron conductivity, (3) membrane thickness, (4) temperature and (5) surface oxygen exchange kinetics [27, 30, 31]. Extra details can be found in the review on the mixed ionic-electronic conducting membranes for hydrogen production from water splitting by Wenping et al. [32]. The usage of oxygen transport membranes (OTMs) displayed that the equilibrium can be driven to the product side to increase the rate of hydrogen permeation as more oxygen is permeated across the membrane [33–36]. The dissociation process of this reaction produces a very low concentration of oxygen and hydrogen. Consequently, a new membrane has been introduced following the OTMs is the mixed ion and electron conducting (MIEC) OTMs which can deliver high oxygen permeation fluxes.

Numerous investigations have been conducted on various membranes for hydrogen generation from water splitting using the technology of oxygen permeation. Meng et al. [37] and Wang et al. [38] studied performances of membranes for hydrogen production and power generation. They reported that below 700 °C, the main parameter affecting the process is the proton conductivity. Above this temperature, it was shown that the electronic conductivity becomes more important. Using a thick and a thin membrane, they conducted experiments to produce hydrogen. The finding of their experiments displayed that, in the case of a thick membrane operating at higher temperatures, the main governing parameter for the oxygen permeation and, consequently, the hydrogen generation from water vapor, is the oxygen bulk diffusion. Conversely, it was found that, for thin membranes, the permeation process is jointly controlled by surface exchange reactions and the diffusion of the produced O_2 via the MIEC membrane. OTMs were also investigated by Park et al. [39]. A stable thin-film $La_{0.7}Sr_{0.3}Cu_{0.2}Fe_{0.8}O_{3-\delta}$ (LSCF) having a 2.2 cm thickness was used to generate significant quantity of hydrogen by flowing water vapor on the feed side of the membrane and syngas gas (CO_2/CO) on the permeate side of the membrane. The syngas consumes the permeated oxygen, hence, enhancing the chemical potential for oxygen permeation and hydrogen generation. The authors indicated that, based on the results, that substantial quantity of hydrogen can be generated by utilizing the product from coal gasification [12]. Results displaying the possibility of oxygen and hydrogen formation of OTMs from water splitting at moderate temperature (500–900 °C) were published by Balachandran et al. [27]. Prior studies have looked into the use of reactive gas on the membrane's oxygen permeation side. Coal gas was used with an LSCF membrane by Park et al. [39], methane was used with an LSCF membrane by Hong et al. [16] and Habib et al. [40], a BSCF membrane by Ben-Mansour et al. [41], a BCFZ membrane by Jiang et al. [42], and an LSTF membrane by Lee et al. [43]. It was acknowledged in each of these investigations that using reactive gas as sweep gas raises the oxygen penetration flux. This is attributed to the reaction of the gas with the permeated oxygen, in this manner, increasing the potential chemical gradient. This process leads, as well, to higher rates of hydrogen generation through shifting the water dissociation equilibrium to produce hydrogen after permeating the oxygen and its removal from the feed side by the membranes. The results of investigations employing reactive gases have shown that the inlet temperature plays an essential role in enhancing the oxygen permeation and, thus, hydrogen generation.

7.2　Production of Syngas Using Membranes for H_2O/CO_2 Splitting

Fossil fuels will continue to hold a major share in the ever-growing world energy requirements in the coming decades because of many reasons, including availability, ease of usage, transportation and the existing relevant infrastructure. This has necessitated the research groups around the globe to focus on the alarming emissions of greenhouse gasses and find efficient methodologies to capture carbon from the environment. One of the lucrative methods to mitigate CO_2 emissions is CO_2 splitting to produce CO and O_2. Carbon monoxide can be further used for various industrial purposes or the production of syngas, which is a mixture of CO and H_2 with varying ratios. Splitting of CO_2 is a highly-endothermic reaction but can be made more environmental-friendly by rendering the energy demand through solar or any renewable energy source. Oxyfuel combustion is another very promising carbon capturing technology, which distinguishes itself from the conventional combustion by the virtue that the combustion takes place in oxygen diluted environment with carbon dioxide. Thus, CO_2 can easily be separated from the combustion products, which are basically mixture of CO_2 and water vapor, by simple condensation process. The main concern associated with the oxyfuel combustion is the procurement of pure oxygen, which is an energy intensive process. Hence, an integration of the duo processes has the capacity to mutually enhance the efficiency of the system.

Nguyen and Blum [44] have provided a thorough analysis of the production of syngas from CO_2 and H_2O splitting as well as its application. Using the Fischer–Tropsch (F-T) process to transform syngas into synthetic liquid fuels is a significant use for it. The generated synfuel presents a viable substitute fuel for the transportation sector, given its safe and effective usage without requiring modifications to motor engine technology. Fu et al. [45] give an economic evaluation of the combined process using process modeling and sensitivity analysis. The most crucial economic constraint was assessed to be the electricity price as the combined cycle is energy intensive. Hence, they suggested that the energy supply to produce syngas should be met through cheap energy resources such as nuclear or wind energy, so that syngas can compete as a prospective fuel. For syngas production, the renewable energy source that can be effectively employed is the concentrated solar power (CSP). Agrafiotis et al. [46] provide a comprehensive account of the origins, progression, and present state of CSP-assisted syngas generation technology. An experimental study on CO_2 splitting increased by yttria stabilized zirconia (YSZ) membrane was conducted by Itoh et al. [47]. CO_2 thermal breakdown is described as a highly endothermic process. A favorable initial performance was seen in an experimental investigation [48–50] on the production of syngas by co-electrolysis of CO_2 and H_2O; it was somewhat lower than that of H_2O electrolysis but noticeably greater than that of CO_2 electrolysis. They discovered that at low current densities, the solid oxide cells' Ni/YSZ electrode degradation predominated; however, at higher current densities, the degradation of the LSM electrode and serial resistance demonstrated a significant role in the performance loss.

In an experimental investigation, the second order decomposition rate coefficient for the thermal decomposition of CO_2 was examined [51]. The computed rate coefficients in the experimental investigation were found to be within $\pm$ 20% of the previously reported values. They further stated that it was the first time that the incubation period before the thermal breakdown of CO_2 in a reflected shock wave experiment was detected. A two-step cycle-based metal oxide redox process was used to study CO_2 splitting by Galvez et al. [52]. The first stage involves the endothermic thermal breakdown of the metal oxide into metal, and the second is the exothermic reaction, in which the reduced metal combines with carbon dioxide to form carbon monoxide and regenerate the metal oxide. A concentrated sun beam was employed. It was possible to gain more understanding of the kinetics of the two-step cycle-based H_2O/CO_2 splitting reaction to create syngas [53]. They talked about the specific reaction mechanisms that would serve as guides for designing a reactor that would split CO_2 and H_2O to make syngas. A thermodynamic analysis of the second Zn/ZnO step of the thermochemical H_2O and CO_2 splitting cycle was carried out by Venstrom and Davidson [54]. They claimed that there are two main benefits to heterogeneous oxidation of zinc vapor over reactions using solid or liquid zinc: the process is quick and results in full zinc conversion. According to investigations [55], basic Cu(II)/Cu(0) electrodes in a two compartment electrochemical cell in a neutral aqueous solution can be used to effectively accomplish electrocatalytic splitting of water and CO_2. The simplicity of the catalysts, the state of the solution, and the structure of the cell are the primary benefits of employing this type of electrochemical cell.

Experimental studies on the H_2O/CO_2 splitting process using a ceria-based redox reaction cycle to produce syngas were conducted [56–58]. In eight hours, they ran ten consecutive cycles of H_2O/CO_2 splitting using a solar concentrator that used porous ceria kept at 1800 K inside the solar cavity receiver. As needed for a practical solar fuel application, the results displayed that the ceria-based redox reaction cycle could produce syngas in a repetitive and controlled volume. However, the prolonged deposition of ceria vapor on the parabolic concentrator affected the radiative power input, resulting in a lower temperature level and, as a result, the production yield. A thorough analysis of the photochemical and thermochemical splitting of H_2O/CO_2 using metal oxide catalysts was provided by other researchers [59–61]. According to their findings, TiO_2 is the most stable and practical metal oxide for H_2O/CO_2 photochemical splitting.

Membrane technology is becoming more and more important in the chemical and petrochemical industries for a variety of process engineering applications. It is also being used in biotechnology, water treatment, gas separation, syngas generation, methane-steam reforming, energy production, and other fields. These membranes' function can be expanded to include the separation of oxygen from ingested CO_2. The membrane permits carbon monoxide to be produced when oxygen moves from a CO_2 stream across it to the other side. This process produces CO, which can be used as fuel on its own or combined with H_2 or H_2O to create a variety of other hydrocarbon fuels. It is possible to create additional compounds such as fertilizers, syngas, methanol—a crucial fuel for automobiles—and other substances. This procedure may play a significant role in the technologies for carbon capture, use, and sequestration (CCUS). This strategy could lessen the effect that burning fossil fuels has on global warming if it is used in the electric power generation industry. The perovskite membrane only permits CO_2 atoms to pass since it is 100% selective for oxygen. Hot temperatures of up to 900 °C are what fuel the separation process. Maintaining a strong flow of the separated oxygen from CO_2 across the membrane is essential to the process's operation. A vacuum could be created on the membrane's sweep side to accomplish this. To keep the process going, a lot of energy would be needed. Additionally, the sweep side can employ a fuel stream—such as H_2 or CH_4—instead of a vacuum. Heat is the energy input needed to keep the process running, and it might come from waste heat or solar energy. The power plant itself or other sources may provide the waste heat. Compared to other energy storage methods, chemical energy storage is distinguished by its high energy density (energy per unit weight).

Wu [62] conducted a study on hydrogen/syngas production using oxygen permeable perovskite membrane by thermo-chemical splitting of H_2O/CO_2 as an energy storage technology. $La_{0.9}Ca_{0.1}FeO_{3-\delta}$ (LCF-91) perovskite membrane, pertaining to high chemical stability, was used to develop a button cell reactor with H_2O/CO_2 on the feed side and inert/reactive gas on the sweep side. Dense LCF-91 membranes have been found to have selective oxygen permeability [63, 64], however, sweep side reaction has been found to be the rate limiting step in the reactive cases. Adding porous layers and catalysts on the sweep side surface have been to improve the oxygen flux rate through LCF-91 membranes [65]. Nickle catalysts have been found to be very effective in the partial oxidation of methane in a membrane reactor [66–68]. So, another improvement was made by adding the catalyst to the porous layer on the sweep side. The result was a production rate more than double than that of without the catalyst. It was found that at lower fuel concentration (~1%), the surface reaction at the sweep side was the rate limiting step [69, 70].

7.3 Production of Syngas Using Membrane Reactors for Partial Oxidation of Methane

The need for natural gas is growing as more countries try to diversify their energy sources. Because natural gas has less carbon than other fossil fuels, it is advised as a good source for producing electric power. A new natural gas plant can be constructed in roughly two years, which is a short amount of time when compared to other plants. It is also simple to convert an existing plant to run on natural gas. The significance of natural gas, which is primarily composed of methane, has prompted scientists to concentrate on the direct and indirect conversion of hydrocarbons, like methane, to fuels that are easy to transport in recent years [71–73]. The literature primarily considers two conversion approaches. The first is the direct conversion approach through partial oxidation of methane (POM) to methanol. This is a challenging method since the reaction's byproduct, methanol, is far more reactive than methane, the original reactant, and could cause further oxidation [74, 75]. The second strategy is indirect and necessitates the oxidation of methane in the first stage in order to produce syngas that contains hydrogen (H_2) and carbon monoxide (CO). Several methods, including (1) steam reforming, (2) POM, and/or a mix of the two, can be used to accomplish this. Currently, the most common method for creating syngas from methane conversion is the indirect steam-reforming method [76, 77]. This strategy, however, has a number of disadvantages. For example, this reaction uses excessive amounts of energy, is not selective for CO, and produces a high H_2/CO generation ratio [78]. These days, a lot of researchers are instead concentrating on the POM method of producing syngas [79–81] by using air as the oxygen source. The oxygen separation unit has the largest monetary value in relation to POM.

Oxygen transport ceramic membranes can be used to separate oxygen from fed air and provide the pure oxygen needed for POM. Heating the membrane to a temperature (over 650 °C) that activates its surface for oxygen separation is necessary for this operation. The membrane for oxygen separation can be heated in the same unit using the heat produced by POM. This effort is focused on the partial oxidation of hydrocarbons, including methane, to create syngas by the use of ceramic membranes that can oxidize methane partially with air, saving the expense of an oxygen separation facility. Oxygen-permeable membranes provide a significant answer to a number of methane conversion issues [82]. Zhu et al. [83] have provided a brief overview of current research that take into account POM to syngas in addition to catalyst development. Ceramic material membranes have the benefit of being readily produced in a variety of shapes [84]. The reactor design that is most frequently used is the hollow-tube or concentric-tube reactor, in which methane is fed at the inner surface of the membrane and air is supplied in the annulus over the outer surface. Only at very high temperatures is the membrane accessible to oxygen; other gases, such nitrogen, are not. As a result, only airborne oxygen will be able to react with methane inside the inner tube. In his assessment of dense ceramic membranes for methane conversion, Bouwmeester [85] talked about the material characteristics that control the membrane's performance under real-world working circumstances. Other geometric configurations of the reactor were presented by Balachandran et al. [86]. Prior research has indicated that oxygen penetration rates from fed air across ceramic membranes can reach sufficiently high levels to support commercialization [87].

Dong et al. [82, 88] investigated the POM conversion to syngas using a different catalyst (LiLaNiO/γ-Al$_2$O$_3$). They used this catalyst in a disk-type BSCF (Ba$_{0.5}$Sr$_{0.5}$Co$_{0.8}$Fe$_{0.2}$O$_{3-\delta}$) membrane reactor for the sake of comparison. Methane conversion was greater than 98% and CO selectivity was greater than 93% at the same operating temperature of 1123 K. After 500 h of continuous operation, the membrane reactor reached an oxygen penetration flux of 10 ml/cm^2/min. Another study on the impact of utilizing ceramic membranes for POM to produce syngas was conducted by Balachandran et al. [89]. More than 99% methane conversion efficiencies were observed in tubular reactor geometries that ran for more than 1000 h. However, in the absence of a reforming catalyst, they recorded an oxygen permeation flow of roughly 0.3 s/cm^2/min, a methane conversion of roughly 35%, and a CO$_2$ selectivity of almost 90%. A review of the catalytic partial oxidation of natural gas to syngas was conducted by Bharadwaj and Schmidt [90]. The study found that POM's quick processing speed, great selectivity, and low energy usage represent potential benefits. The cost of capital and operating syngas production could be greatly reduced. Heitnes et al. [91] used quartz and steel reactors to store the catalyst during investigations in a traditional flow setup. They looked into POM for gas synthesis utilizing a variety of catalytic devices, including platinum gauze and fixed-bed, monolithic-type catalysts. The poll displayed that a good plan for observing quick reactions at extremely short times would be the gauze catalyst system. Actually, the best results come from using oxide catalysts during the partial oxidation of methane to syngas [92–94]. The H$_2$/CO ratio, CO selectivity percentage, and CH4 conversion for various catalytic compositions were reported for new selective and stable performance oxide catalysts without the presence of a noble metal by Sokolovskii et al. [95]. An asymmetric catalytic membrane reactor's POM to syngas was investigated both theoretically and empirically by Shelepova et al. [96]. As seen by the data, the membrane reactor operated well and steadily. The process temperature, gas flow rates, and operating parameter that guarantees the highest possible methane conversion to syngas were examined using the model they created.

The key factor influencing the POM process for syngas production is really increasing the membrane's performance in terms of a larger oxygen permeation flux, which is necessary to make this conversion process economically viable. A thorough numerical analysis of the features of methane partial oxidation and oxygen permeation in a catalytic membrane reactor for syngas generation is provided in the following subsections. The goal of the study is to optimize the membrane reactor's operational parameters for increased syngas generation rate and, consequently, higher oxygen permeation rate. A validation study was conducted using the literature's accessible data. To better understand the features of the POM process for producing syngas under variable operating conditions, the effects of sweep fuel concentration, feed air flow rate, and sweep gas mixture flow rate on oxygen permeation flux and reactor performance were studied numerically. Under varied operating conditions, the fuel conversion rate and syngas selectivity have a wealth of published data.

7.3.1   Reactor Design and Operating Conditions

The lab-scale membrane reactor shown in Fig. 7.1 was taken into consideration when conducting the investigation. As per the reactor architecture illustrated in Fig. 7.1, a mathematical model was created to solve for the flow field and reaction kinetics within the catalytic membrane reactor (CMR). The calculations were carried out taking into account half of the reactor in the whole three-dimensional (3-D) domain. According to the top schematic in Fig. 7.1, the CMR is made up of a multi-layer membrane with a diameter of 3.0 cm that divides the feed side header from the sweep side header, which has the same diameter and a length of 2.5 cm. The macroporous Ni–Al foam substrate used to create the membrane, as depicted in Fig. 7.2, is composed of three layers of perovskite-fluorite nanocomposites with graded (meso-micro) porosity, a thin, dense layer of $MnFe_2O_4$–$Ce0.9Gd0.1O_2$, and a porous catalytic layer of $LaNi0.9Pt0.1O3/Pr0.3Ce0.35Zr0.35O_2$-x […]. Air (21% O_2 and 79% N_2, by vol.) is delivered to the feed side header of the CMR's feed side using an inlet tube that measures 0.5 cm in diameter and 3.5 cm in length. Because the membrane is selective for oxygen, it isolates oxygen to the sweep side and allows oxygen-depleted air to exit the feed header through a 0.5 cm-diameter output tube. A helium and fuel (CH_4) mixture is fed to the sweep side header via a 3.5 cm long and 0.5 cm diameter inlet tube. The separated O_2 in the sweep side header is used to partially oxidize fuel, resulting in syngas mixture. This mixture exits the reactor sweep side through an outlet tube with a 0.5 cm diameter and 3.5 cm length.

Every simulation was run in atmospheric settings with a fixed inlet gas temperature of 1173 K, a fixed membrane thickness of 0.1 mm. The effects of feed air flow rate (2.0–7.0 L/h), sweep gas mixture flow rate (5–70 L/h), and sweep fuel concentration (10–20% by vol.) on oxygen permeation flux and CMR performance are investigated using the reference base case that was chosen. Table 7.1 provides an overview of all the study's operating circumstances.

7.3.2   Modelling Partial Oxidation of Methane with Oxygen Separation

Since the reactor is symmetric with respect to its axis, the mesh for numerical simulations was constructed taking into account only half of the domain. Several meshes with varying cell counts were constructed using commercial Gambit software in order to conduct a mesh independency analysis before to choosing the best mesh. Three distinct mesh sizes, measuring 0.5 mm × 1.0 mm (radial × axial), 0.5 mm × 0.5 mm, and 0.25 mm × 0.5 mm, were investigated. When comparing the oxygen permeation flux utilizing the second and third mesh to the first mesh, the results did not demonstrate a statistically significant difference. Based on that, a mesh consisting of 152,920 cells overall and measuring 0.5 mm by 1.0 mm was employed. The produced mesh was input into the commercial program ANSYS-2019-R2-ACADEMIC, which was

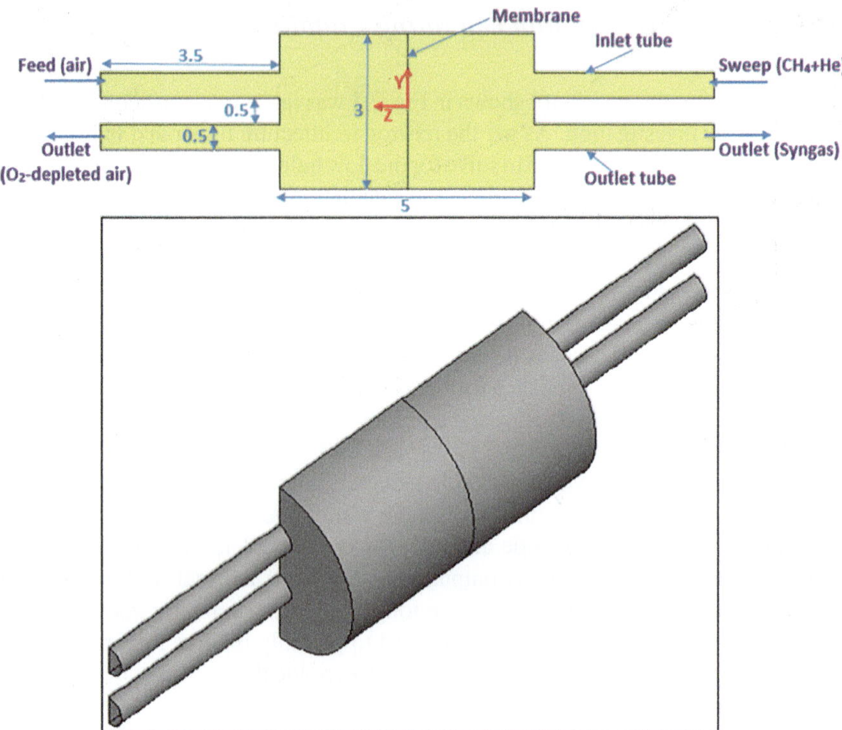

Fig. 7.1 Diagram showing the proposed catalytic membrane reactor: (top) 2-D reactor architecture with all dimensions in centimetres; (bottom) 3-D mesh of the reactor's studied half-domain

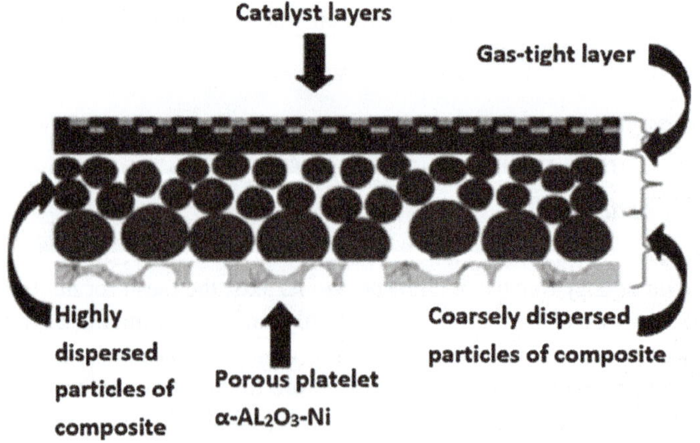

Fig. 7.2 Structure of the catalytic membrane [96]

Table 7.1 Summary of parameter variations in this study

Study	Parameter	Range
Base case	Membrane thickness	0.1 mm
	Temperature	1173 K
	Feed flow rate (Q_f°)	2 L/h
	Sweep flow rate (Q_s°)	5 L/h
	Feed air oxygen concentration	21% (by vol.)
	Sweep fuel concentration	10% (by vol.)
Effect of sweep fuel concentration	Membrane thickness	0.1 mm
	Temperature	1173 K
	Feed flow rate (Q_f°)	2 L/h
	Sweep flow rate (Q_s°)	5 L/h
	Feed air oxygen concentration	21% (by vol.)
	Sweep fuel concentration	10–20% (by vol.)
Effect of feed air flow rate	Membrane thickness	0.1 mm
	Temperature	1173 K
	Feed flow rate (Q_f°)	2–7 L/h
	Sweep flow rate (Q_s°)	5 L/h
	Feed air oxygen concentration	21% (by vol.)
	Sweep fuel concentration	10% (by vol.)
Effect of sweep (CH_4 + He) flow rate	Membrane thickness	0.1 mm
	Temperature	1173 K
	Feed flow rate (Q_f°)	2 L/h
	Sweep flow rate (Q_s°)	5–70 L/h
	Feed air oxygen concentration	21% (by vol.)
	Sweep fuel concentration	10%

used to set up the various models and run numerical simulations. While isothermal conditions, steady-state, well-mixed continuous flow reactor, and model setup were presumptive. The computed Peclet number for this reactor for the specified ranges of feed and sweep flow rates is less than unity, which could support the hypotheses.

ANSYS-2019-R2-ACADEMIC software was utilized to create the numerical model of the CMR, and the computations were carried out in the reactor's entire three-dimensional domain. Together with the species transport equation, the conservation equations of mass, energy, and momentum are also solved numerically for the reacting flow field. To account for the flow of oxygen from the feed side to the permeate side of the membrane, a source/sink component, Si, was added to the mass and momentum equations. The following is a presentation of the equations [97]:

$$\nabla \cdot (\rho U) = S_i \tag{7.1}$$

$$\nabla \cdot (\rho U U) = -\nabla p + \mu \nabla^2 U \tag{7.2}$$

$$(\rho C_p)_f \, U \cdot \nabla T = \nabla \cdot (k_f \nabla T) \tag{7.3}$$

$$\nabla \cdot (\rho U X_i) - \nabla \cdot (\rho D_{i,m} \nabla X_i) = S_i \tag{7.4}$$

For a given oxygen flux, J_{O2}, the source/sink term can be calculated as [98]:

$$S_i = \begin{cases} +\dfrac{J_{O_2,OTM} \cdot A_{cell} \cdot MW_{O_2}}{V_{cell}}, & \text{at the adjacent membrane cells in the sweep side} \\[2ex] -\dfrac{J_{O_2,OTM} \cdot A_{cell} \cdot MW_{O_2}}{V_{cell}}, & \text{at the adjacent membrane cells in the feed side} \end{cases} \tag{7.5}$$

The oxygen flux was estimated using [96]:

$$J_{O2} = Q_0 \frac{A_m}{\delta_m} \left[\sqrt{P_{O2}^{F.S}} - \sqrt{P_{O2}^{S.S}} \right] \tag{7.6}$$

With a membrane thickness of 0.1 mm, thermal conductivity of 4.0 W/m/K, specific heat of 450 J/kg/K, density of 6000 kg/m³, and emissivity of 0.8 [99], the membrane surface was regarded as a grey body in all computations. In order to account for heat release caused by partial methane oxidation in the sweep side of the CMR, a heat transfer connection was created amongst the feed and sweep sides of the membrane. The numerical model used the semi-implicit technique for pressure-linked equations (SIMPLE) approach to connect the pressure and velocity fields. Additionally, the second order upwind technique was used to discretize the governing equations' convective terms. Strict convergence criteria were followed, and the calculations were closely watched. A case was considered to have congregated as soon as the residuals of every variable fell beneath a rate of 10^{-6}. In every computation, the radiative transfer equation (RTE) was solved exhausting the discrete ordinate (DO) radiation model. When methane is partially oxidized using pure oxygen, the exponential wide band model (EWBM) is used to accurately represent the reaction temperature [100]. In order to do this, the vector gradient of radiation heat transfer was primary computed and, when the radiation intensity was known, substituted into the energy equation. Methane can be transformed into syngas using direct or indirect methods for fuel conversion modeling under POM circumstances [101]. Methane is directly transformed into syngas (CO + H_2) using the direct POM method. In the indirect POM approach, on the other hand, part of the methane is first reacted to custom H_2O and CO_2 in the sweep side's adjacent zone to the membrane surface. The remaining part of the methane then reacts with the H_2O and CO_2 produced in the core stream in a process known as dry and steam reforming to yield syngas. Constructed on the findings of Sazonova et al., the indirect technique was used in this investigation under such CPOM settings [102]. The following chemical processes illustrate the indirect approach:

Table 7.2 Kinetic parameters of the indirect approach POM reactions on LaNi$_{0.9}$Pt$_{0.1}$O$_3$ catalyst [102]

Reaction	k_{0i} (m/s)	E_j (kJ/mol)
$CH_4 + 2O_2 \rightarrow CO_2 + 2H_2O$	2.32×10^6	86
$CH_4 + H_2O \leftrightarrow CO + 3H_2$	1.76×10^3	46
$CH_4 + CO_2 \leftrightarrow 2CO + 2H_2$	1.12×10^5	70

$$CH_4 + 2O_2 \rightarrow CO_2 + 2H_2O$$

$$CH_4 + H_2O \leftrightarrow CO + 3H_2$$

$$CH_4 + CO_2 \leftrightarrow 2CO + 2H_2$$

The kinetic parameters for such reactions are given in Table 7.2. The reaction rate constants were estimated based on the Arrhenius equation [96]:

$$k_i = k_{0i} \exp\left(-E_j/RT\right) \tag{7.7}$$

7.3.3   Influence of Concentration of Sweep Fuel

The application was initially confirmed by contrasting the numerical consequences that were obtained through the experimental records of Shelepova et al. on CH_4 conversion [96]. To seize the CPOM process intimate the CMR, the model was run under the exact same operational circumstances as the tests. The CMR that has been utilized in this numerical analysis is the exact same one that Shelepova et al. used in their experimental inquiry [96]. To validate the model, the numerical model assumed the identical experimental flow circumstances contained by the CMR. In order to determine the conversion percentage of the sweep fuel for the validation research, the model was configured exactly like it was in the experiment. Figure 7.3 displays an assessment of the numerical and experimental figures. The comparison was carried out with the sweep fuel content in the entering sweep gas mixture varied, at an intake temperature of 1173 K, a flow rate of feed air of 2 L/hr, and a flow rate of sweep gas of 5 L/hr. With a mediocre percentage fault of roughly 13%, the numerical outcomes demonstrate respectable covenant with the experimental data, suggesting that the oxygen permeation and reaction mechanisms that have been implemented may successfully capture the CPOM process inside the CMR. A greater concentration of sweep fuel leads in a lower conversion rate, according to both experimental and numerical data. That could be explained by the reactor sweep side's inadequate oxygen availability, which prevents the fuel from being converted. As a result, the reactor exit tube on the sweep side has a larger mole proportion of CH_4.

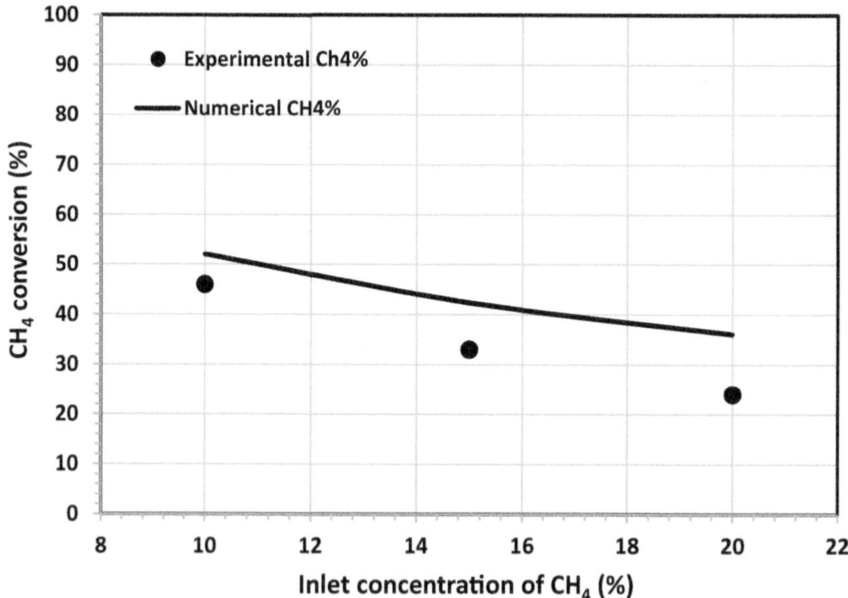

Fig. 7.3 Comparison of experimentally measured (by Shelepova et al. [96]) and numerically predicted (using the present model) CH_4 conversion as function of inlet CH_4 concentration at $T = 1173$ K, $Q_f^\circ = 2$ L/h, and $Q_s^\circ = 5$ L/h

Figure 7.4 displays the distributions of the CH_4 mole fraction for various sweep fuel concentrations inside the CMR's intake and exit tubes on the sweep side. When the sweep fuel concentration is lower, there are greater differences in the concentrations of CH_4 inside the inlet and output tubes, which suggests a higher fuel conversion. Figure 7.5 displays the velocity vectors mapped over the contour plots of temperature and CH_4 mole fraction over the CMR's symmetry plane under base case conditions to provide more insight into the fuel conversion process and the interplay between the flow and reaction fields. As seen in Fig. 7.5, the fuel starts to convert CH_4 as soon as it enters the sweep side header, and as it moves toward the membrane surface— where there is a lot of available O_2—its rate of conversion increases. The synthesis of syngas is dominated by the steam and dry reforming reactions, or reactions (2) and (3), as CH_4 approaches its minimal concentration near the membrane surface. The nearly constant concentration of CH_4 in the exit tube indicates that all reactions have been quenched due to the higher flow rate and deficiency of O_2 molecules. As soon as the reactions start, the temperature rises at the entrance of sweep side header's and eventually reaches its peak next to the membrane surface. The temperature within the outlet tube is reduced due to the quenching effect of all reactions.

Figure 7.6 displays the obtained oxygen permeation flux and the axial temperature distribution through the CMR's exit tube as a function of sweep fuel concentration. The oxygen permeation flux increased as the inlet CH_4 concentration rose from 15

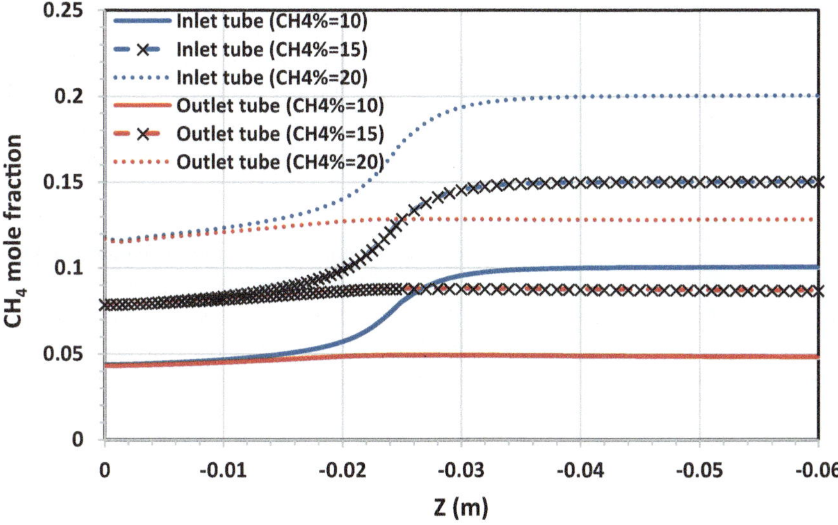

Fig. 7.4 Impact of sweep fuel concentration on axial distributions of the CH_4 mole fraction through the reactor sweep side's inlet and output tubes

to 20%, while it decreased when the concentration augmented from 10 to 15%. The oxygen permeation flow, as expressed by Eq. (7.6), is primarily determined by the membrane's surface temperature and the differential in partial pressure of oxygen between the feed and sweep sides. As seen in Fig. 7.6, growing the concentration of sweep fuel causes the reaction zone next to the membrane to become a fuel-rich zone, which lowers reaction temperatures by reducing reaction intensities. The oxygen permeation flux decreased from about 62 $\mu mol/cm^2/s$ at 10% inlet CH_4 concentration to approximately 23 $\mu mol/cm^2/s$ at 15% inlet CH_4 concentration due to the upsurge in membrane barrier for oxygen transport caused by the temperature drop. At 10% intake CH_4 concentration, the maximum temperature next to the membrane surface decreased to 1662 K, and at 15% inlet CH_4 concentration, it plummeted to 1215 K. The temperature rose to 1543 K and the oxygen permeation flow to approximately 37 $\mu mol/cm^2/s$ as a result of growing concentration of the inlet CH_4 to 20%. This could be explained by greater fuel diffusivity close to the membrane surface, which lowers the partial pressure of oxygen in the sweep side. This increases the oxygen penetration flux, which raises the temperature. Lower oxygen penetration flux and partial oxidation of the fuel occur in membrane reactors operating in fuel-rich environments. Any additional growth in oxygen permeation flux ought to accelerate the rate at which fuel is converted, raising the temperature.

Figure 7.7 displays the selectivity of syngas, which reflects the trend of oxygen penetration flux across the membrane depicted in Fig. 7.6. The selectivity for the production of syngas is decreased when the sweep fuel concentration is raised from 10 to 20%. The CO selectivity decreased from 62 to 32%, or nearly half of the value

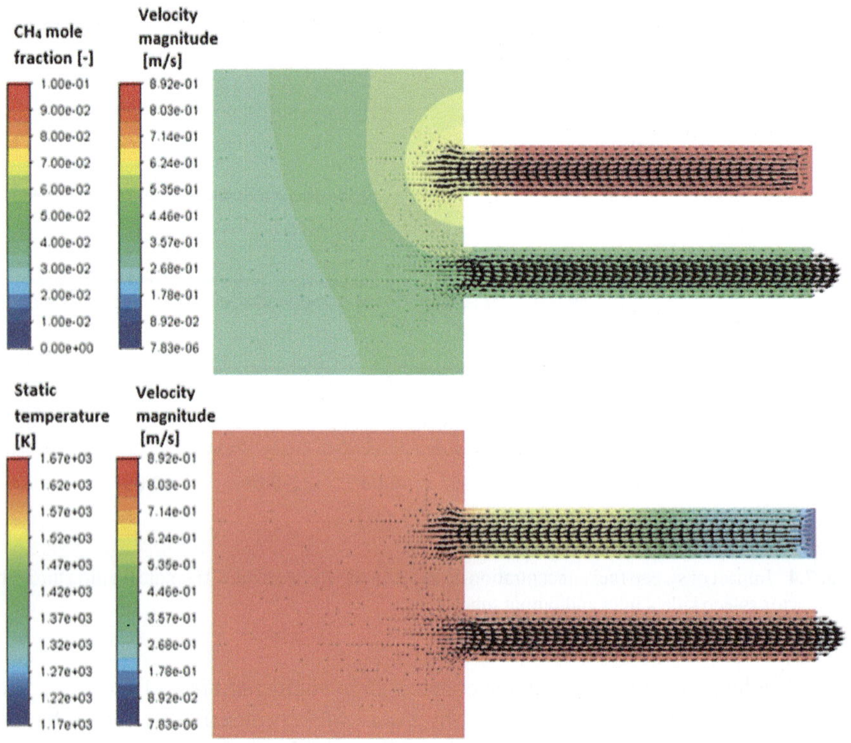

Fig. 7.5 Velocity vectors and CH₄ mole fraction and temperature for the base case at T = 1173 K, $Q_f^\circ = 2$ L/h, and $Q_s^\circ = 5$ L/h

drop. Similarly, with a similar increase in sweep fuel concentration, H_2's selectivity decreased from 37 to 30%. The selectivity for CO grew to 38% as a result of the oxygen permeation flux increasing at 20% inlet CH_4 concentration; on the other hand, the selectivity for H_2 continued to decrease, reaching 23%. Because H_2 molecules have a faster reaction rate and diffusivity than CO molecules, H_2 molecules always prevail in the fight for oxygen atoms when CO and H_2 molecules coexist in an environment where oxygen atoms are present. Better oxygen permeation, increased temperature, and enhanced fuel diffusivity were the outcomes of raising the sweep fuel concentration to 20%. Since H_2 has a higher diffusivity than CO at a suitable temperature, H_2 absorbs oxygen atoms first and transforms into H_2O in the homogenous mixture before exiting the reactor. This means that there is less H_2 at the reactor's exit, which further reduces H_2 selectivity. It is necessary to note that the current study's flow arrangement is co-current, with feed and seep flows connected by a flow type. Syngas selectivity was significantly reduced when the flow configuration was changed from counter-current to co-current, according to a comparison between the current results and those of Shelepova et al. [96], who used the same CMR but a counter-current flow configuration. The counter-current flow

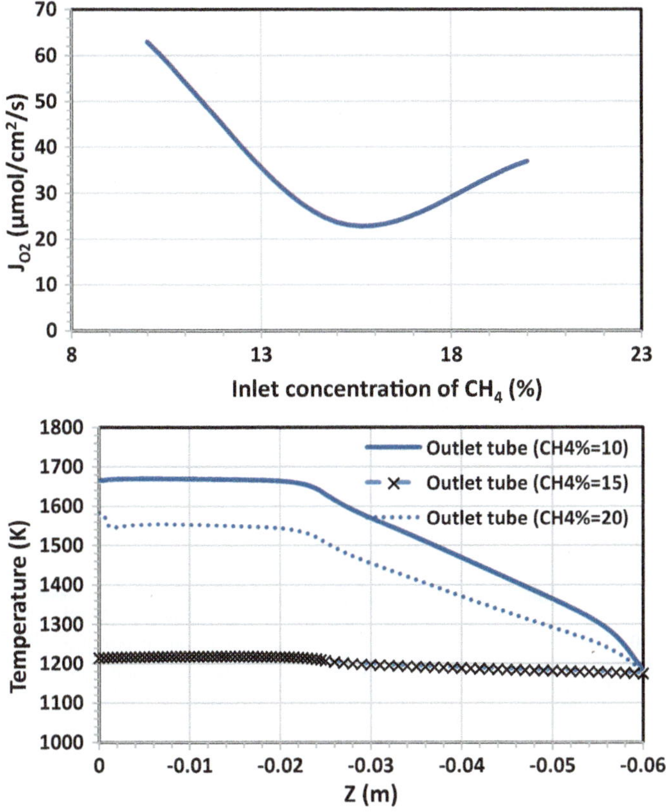

Fig. 7.6 Impact of oxygen permeation flux and axial temperature distributions through the reactor sweep side outflow tube as a function of sweep fuel concentration

arrangement produced selectivities of 89% and 73% for CO and H_2, respectively, at 20% intake CH_4 concentration, as opposed to the equivalent values of 38% and 23% obtained for the co-current flow configuration. This may be explained by the fact that the counter-current flow configuration of the reactor has superior heat transfer and flow conditions than the co-current one, which leads to a greater oxygen permeation flux and enhanced reactor performance in terms of a larger syngas production. This demonstrates how important flow arrangement is in regulating the rate at which syngas is produced in CMRs. Figure 7.7 also displays the axial distributions of the CO and H_2 mole fractions via the CMR's inlet and exit tubes. As a function of sweep fuel concentration, the trends of the acquired CO and H_2 selectivities are generally reflected on their axial distributions, as Fig. 7.7 illustrates. The CO and H_2 productions begin in the inflow tube close to the sweep header and peak close to the membrane, which is the location of the oxygen supply. Subsequently, their concentrations stay nearly constant in the exit tube as a result of all processes being quenched by the increased flow velocity and oxygen deficiency.

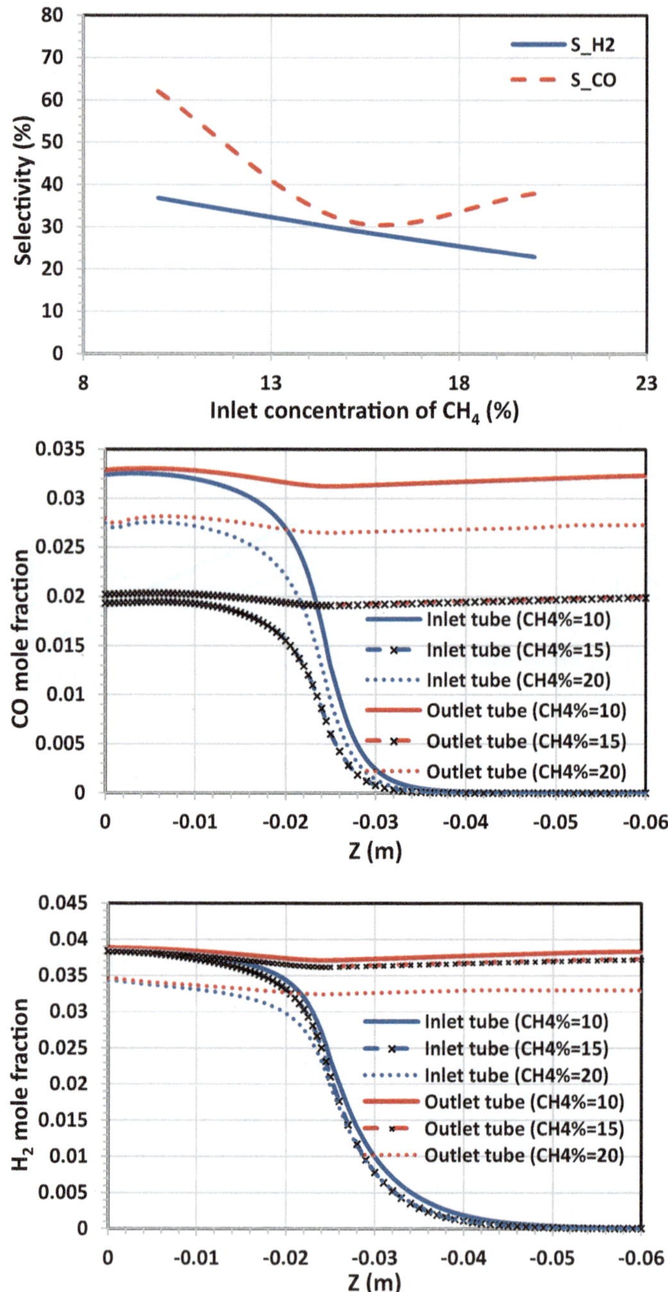

Fig. 7.7 Impact of sweep fuel concentration on axial mole fraction distributions through the reactor sweep side's intake and outflow tubes, as well as CO and H₂ selectivity

7.3.4 Effect of Feed Air Flow Rate

The literature has extensively demonstrated that an increase in feed air flow rate leads to an improvement in oxygen permeation flux [96, 98, 102, 103]. This was ascribed to an increase in the partial pressure differential between the feed and sweep sides of the membrane for oxygen, which is the primary factor causing oxygen to permeate across the membrane. Figure 7.10 in our study confirms these facts for CPOM in a CMR. Higher fuel conversion and decreased syngas selectivity are the outcomes of increased oxygen permeation flux as a function of feed air flow rate (Figs. 7.8, 7.9). Fuel conversion increases in tandem with a decrease in syngas selectivity due to an increase in oxygen content at the membrane's permeate side. The results of raising the feed air flow rate from 3 to 7 L/hr are displayed in Fig. 7.9. This led to a decrease in CO selectivity of 41.7% to 35.14% and a decrease in H_2 selectivity of 27.45–23.39%. As shown in Fig. 7.8, the fuel conversion increased from roughly 53% to roughly 61% in order to increase the feed air flow rate from 3 to 7 L/h. Figure 7.10 illustrates how the increase in feed air flow rate from 3 to 7 L/h resulted in an increase in the oxygen permeation flux from approximately 13 μmol/cm^2/ s to approximately 25.12 μmol/cm^2/s. Figure 7.10 illustrates how raising the feed flow rate and improving the oxygen permeation flux raised the reactor's temperature. When the feed flow rate exceeded 5 L/hr, the reactor's maximum temperature was around 1520 degrees Celsius next to the membrane surface. As shown in Fig. 7.8, all reactions were quenched, resulting in a fixed fuel concentration in the CMR's exit tube.

The radial distributions of the heat of reaction and the kinetic rate of reaction-1 ($CH_4 + 2O_2 \rightarrow CO_2 + 2H_2O$) extremely close to the membrane surface in the sweep side of the CMR are displayed in Fig. 7.11. As Fig. 7.11 illustrates, increasing the feed air flow rate increased the reaction rate, which in turn decreased the amount of heat released. This is directly related to the previously mentioned increase in oxygen penetration flow. The oxygen that has penetrated the membrane in front of the input tube is scavenged by the inlet flow's impingement, which lowers the partial pressure of the oxygen and improves the oxygen permeation flux in that area. As seen in Fig. 7.11, this improves the reaction rate in the area corresponding to the inlet tube. When the flow moves in the direction of the output tube, the rate of reaction-1 decreases because syngas generation processes predominate in this area. Figure 7.11 displays the similar patterns that were obtained for the reactor's heat of reaction as a function of feed air flow rate. The radial direction of the reaction heat distributions saw a series of variations. This could be explained by the sweep header's internal flow recirculation as a result of the flow's repeated impingements on the membrane surface and the sweep header's walls.

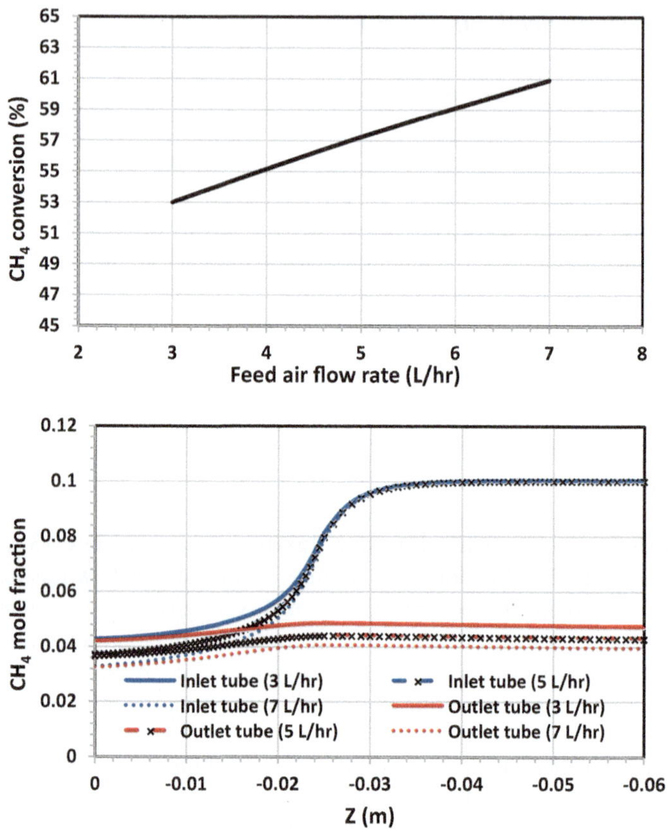

Fig. 7.8 Impact of feed air flow rate on axial CH_4 mole fraction distributions through the reactor sweep side's intake and outflow tubes and CH_4 conversion

7.3.5   *Effect of Sweep Gas Flow Rate*

In actuality, the oxygen permeation flux rate across the membrane governs this method of producing syngas utilizing oxygen permeable membranes. The primary factors influencing oxygen flux are partial pressure gradients across the membrane surfaces and the temperature of the membrane surface. The oxygen permeation flux across a specific membrane should be regulated to a certain maximum permissible value at a given fixed temperature and at limiting sweep and feed gas flow rates, under which the gradients of oxygen partial pressure cannot be further enhanced. Conversely, in the event that an excess of fuel is delivered to the sweep side of a CMR, the fuel's percentage conversion should decrease as the sweep gas flow rate increases. As shown in Fig. 7.12, that is precisely what was achieved for this CMR while increasing the sweep gas flow rate at a given sweep fuel concentration of 10%. When the sweep gas flow rate was increased from 5 L/hr to 30 L/hr, there was a

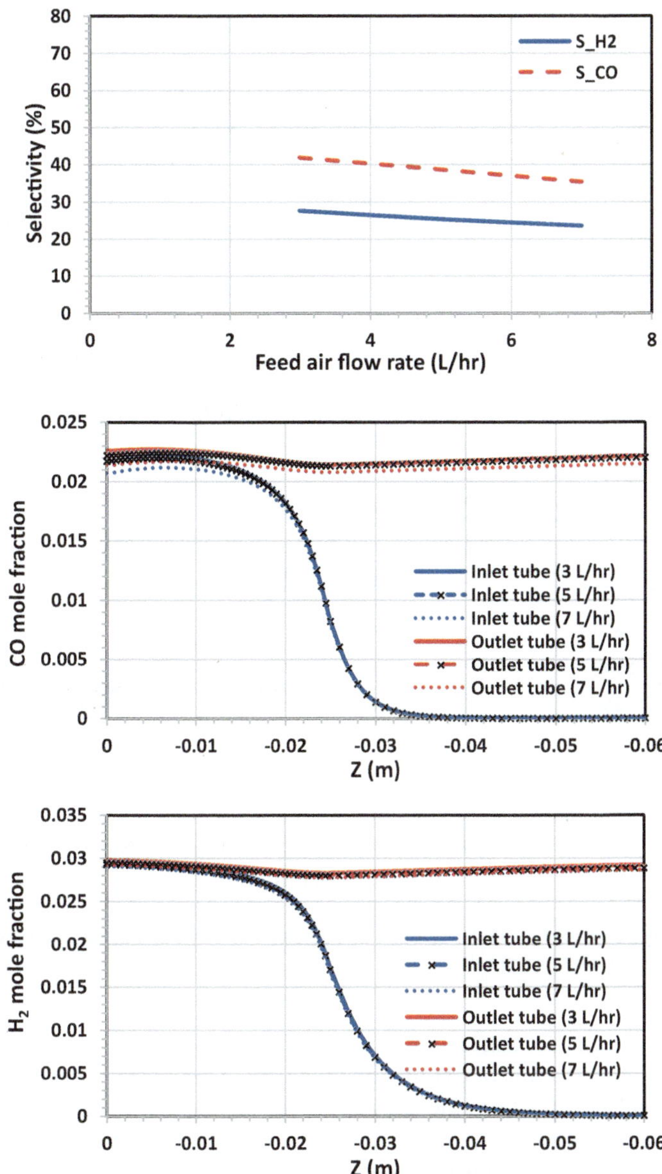

Fig. 7.9 Impact of feed air flow rate on axial mole fraction distributions across the reactor sweep side's input and output tubes, as well as CO and H$_2$ selectivity

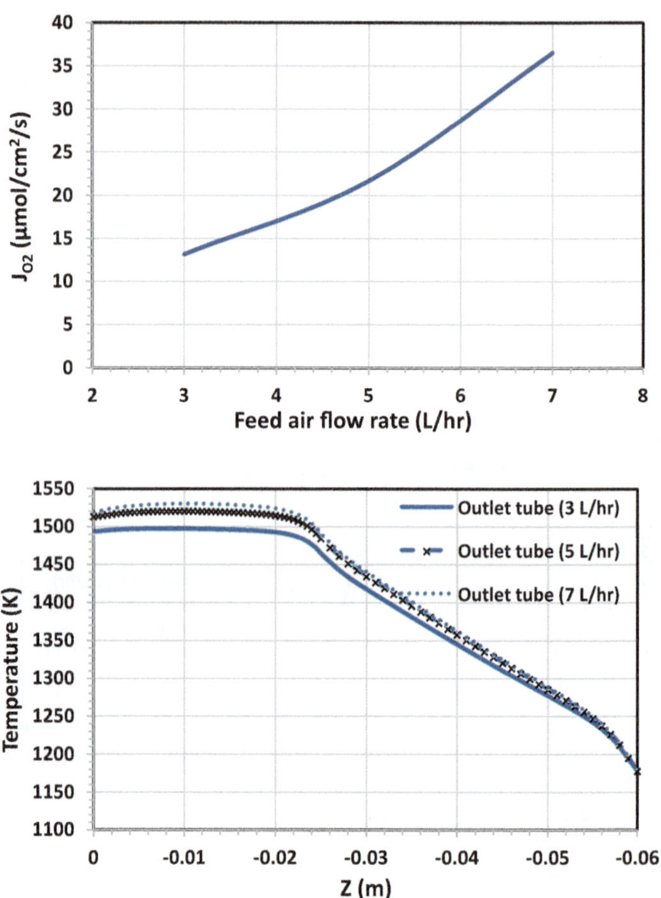

Fig. 7.10 Impact of feed air flow rate on axial temperature distributions and oxygen permeation flux through the reactor sweep side exit tube

noticeable drop in fuel conversion from roughly 53% to roughly 5%. Figure 7.12 displays the mole fraction of CH_4 distributions inside the reactor's intake and output tubes. The fuel mole fraction efficiently decreases approaching the membrane surface with lower sweep flow rates, indicating a higher conversion rate, with higher sweep gas flow rates, the situation is inverted.

The production rates of H_2O and CO_2 for steam and dry reforming reactions of the fuel for syngas generation were decreased by the drastic decrease in the fuel conversion % while increasing the sweep gas flow rate. Consequently, as shown in Fig. 7.13, the selectivities of CO and H_2 are likewise decreased as a function of sweep gas flow rate. Above 30 L/min, the selectivities of CO and H_2 decrease to less than 5%. As seen in Fig. 7.13, the concentrations of CO and H_2 demonstrated no build-up in the axial direction towards the membrane surface at sweep flow rates more than

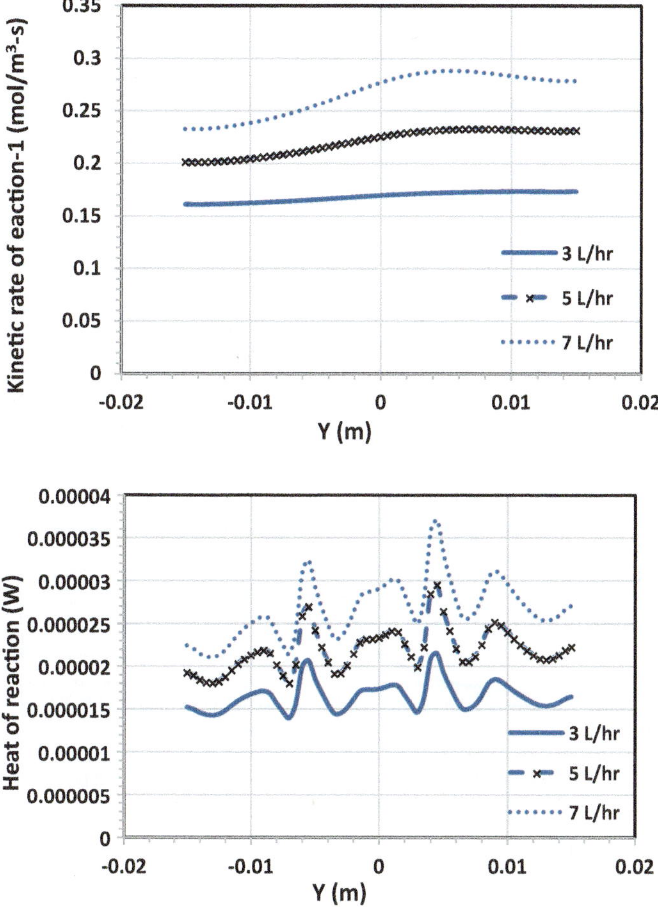

Fig. 7.11 Impact of feed air flow rate on the reactor sweep side membrane's radial distributions of the heat of reaction and the kinetic rate of reaction-1

30 L/hr. This suggests that there is not enough oxygen present in the CMR to convert fuel at a higher sweep gas flow rate. It's interesting to note that, as Fig. 7.14 illustrates, increasing the sweep gas flow rate above 5 L/hr reduced the oxygen penetration flux across the membrane. This is explained by the fact that with larger sweep flow rates, the membrane surface temperature significantly drops. As illustrated in Fig. 7.14, the gas temperature even decreases axially in the direction of the membrane surface at sweep flow rates more than 30 L/hr. The reason for this is that the reactor's sweep flow velocity increased in tandem, providing a very short residence time for the fuel to react with oxygen.

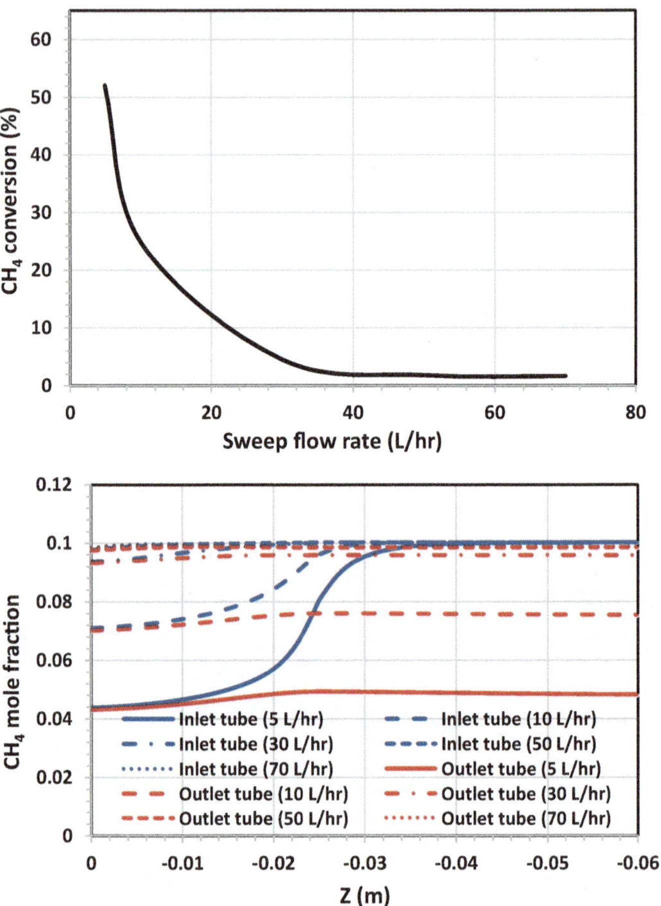

Fig. 7.12 Impact of sweep flow rate on axial CH$_4$ mole fraction distributions and CH$_4$ conversion through the reactor sweep side's inlet and output tubes

7.4 Oxy-syngas Combustion with Carbon Capture

Both spatial storage and carbon capture include a variety of methods. Three primary methods for achieving carbon capture have been found by Gibbins and Chalmers [104]: pre-combustion, post-combustion, and oxyfuel combustion capture. Pre-combustion capture requires less energy than post-combustion, although post-combustion has been shown to have a greater thermal efficiency when it comes to producing electricity. It is possible to perform pre-combustion carbon capture by partially oxidizing fuel and then shifting the process. As the name suggests, it entails decarbonizing the fuel, which prevents CO$_2$ from being produced during the combustion stage by removing carbon from the fuel before it is burned. After the gas combination (CO and H$_2$) is created, it undergoes a shift reaction, which produces H$_2$ and

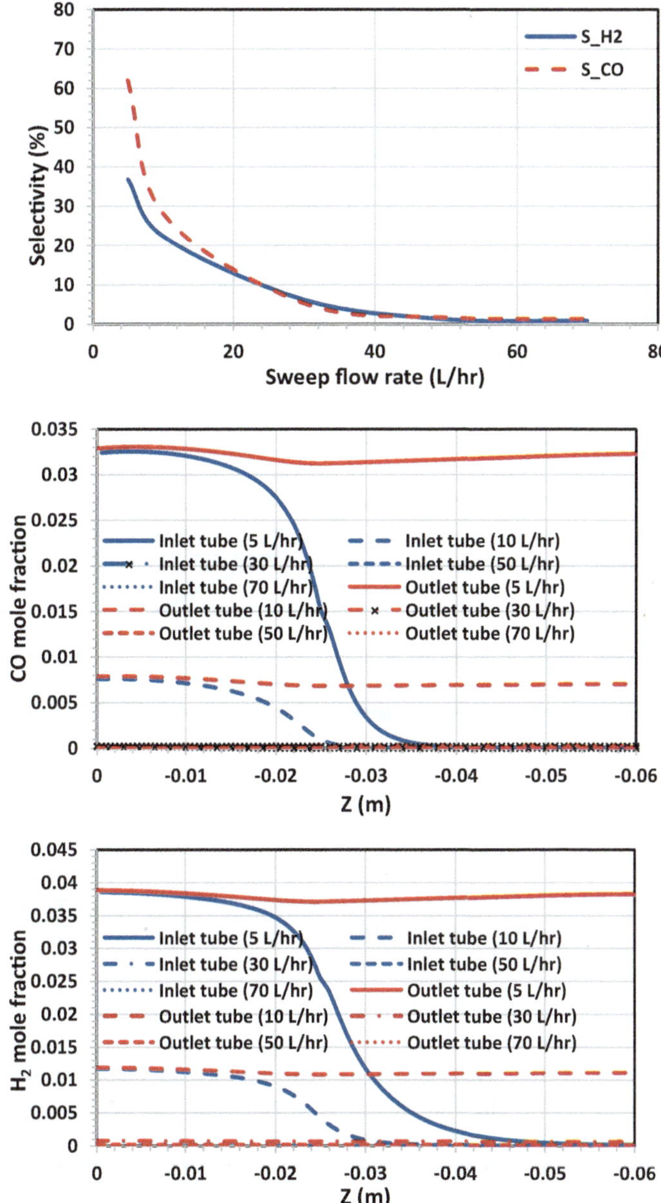

Fig. 7.13 Effect of sweep air flow rate on CO and H_2 selectivity and axial profiles through the reactor sweep side

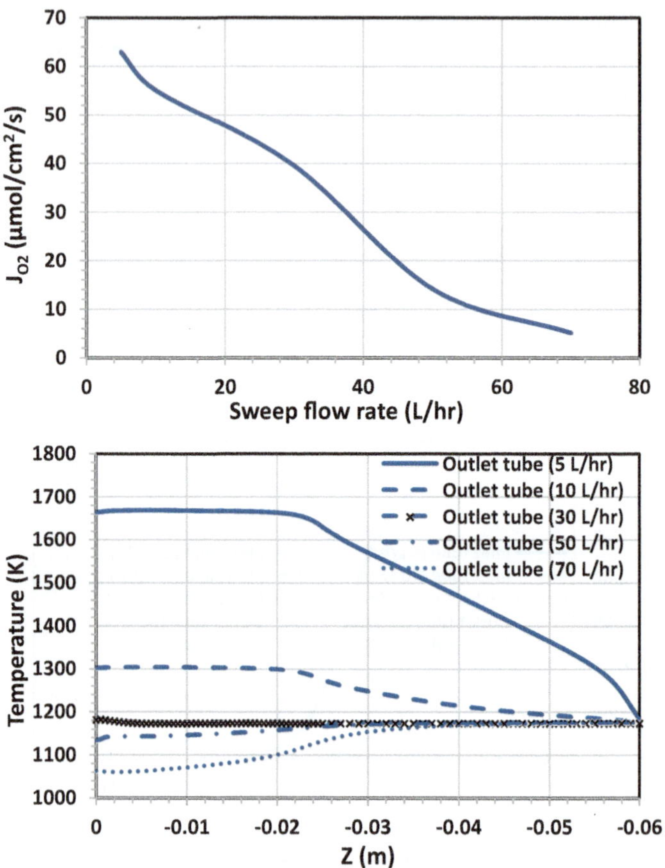

Fig. 7.14 Effect of sweep flow rate on oxygen permeation flux and axial temperature profiles through the outlet tube of the reactor permeate side

CO_2 at high temperatures and pressures, allowing CO_2 to be extracted. Figure 7.15 illustrates the pre-combustion steps.

After the combustion process, CO_2 is removed through chemical absorption in post-combustion [105]. The flue gas is scrubbed using an amine solution (alkanolamine) as part of the absorption process. Thus, an amine-CO_2 complex is created. After that, the complex is heated to release only CO_2. On the other hand, oxygen fuel combustion capture works by first removing oxygen from the air, after which combustion occurs to produce CO_2 and H_2O. Oxyfuel combustion has a capture efficiency of over 90% [106]. It is suggested that more research will still be required to fully explore the potential of oxy-combustion. These factors and effects include the degree of corrosion at different temperatures, oxygen concentration levels, ash quality, and the ability to predict the amount of sulphur and nitrogen oxides that will be released. In order to reduce carbon emissions into the environment, storage of the

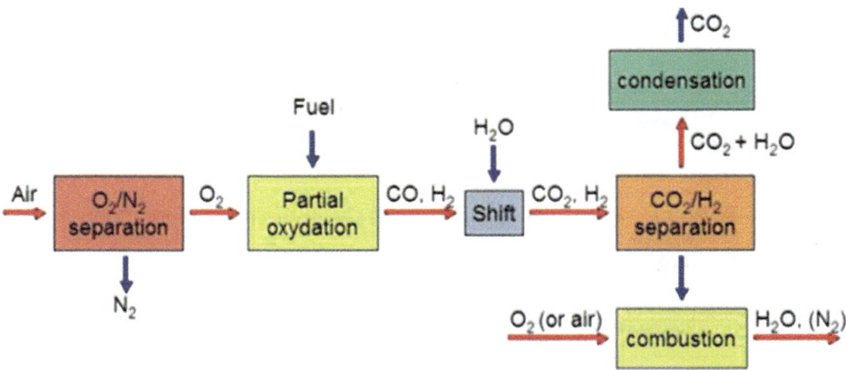

Fig. 7.15 Pre-combustion processes

captured carbon is equally crucial. It is possible to store carbon in saline aquifers, abandoned gas fields, and oil fields [104]. Another effective method of storing CO_2 is to employ the mineral serpentinite, which interacts with CO_2 to produce a magnesium compound [107]. Figure 7.16 provides a summary of the three primary CO_2 collection methods' operating principles.

7.4.1 Oxy-fuel Combustion

Oxy-combustion is a practical method for capturing CO_2. In contrast to air combustion, oxy-combustion produces CO_2 and H_2O only when the fuel burns in the presence of pure oxygen. This raises the CO_2 concentration to facilitate easy capture following the condensation-induced separation of H_2O. Additionally, because nitrogen is not present during combustion, it permits a decrease in the generation of nitrogen oxides (NOx). Both NOx and SOx emissions are reduced by approximately one-third (1/3) and thirty percent, respectively, as compared to air combustion [106]. Since O_2 is a necessary feed gas for practically all renewable energy technologies, oxygen fuel combustion is a crucial carbon capture process [15]. Figure 7.17 depicts the essential concepts of oxyfuel combustion.

Numerous researchers have studied the properties and combustion behavior of oxyfuel fires. In order to investigate the flame structure and stability of oxyfuel combustion in a gas turbine, Kutne et al. [109] conducted experiments using oxygen at mole fractions of 20% to 40%, equivalency ratios of 0.5 to 1.0, and powers ranging from 10 to 30 kW. The results of Nemitallah and Habib [110] also indicated extinction of the flame if the oxygen in the mixture is less than 21%. The authors pointed out that if the mole fraction of oxygen is less than 22%, the burner will be impossible to operate, as it will result in unstable operation and blow out even if the equivalency ratio is 1.0 and at powers of 20 and 30kW. Therefore, flame stability can only occur when the oxygen mole fraction exceeds 21%. Additionally, while using the

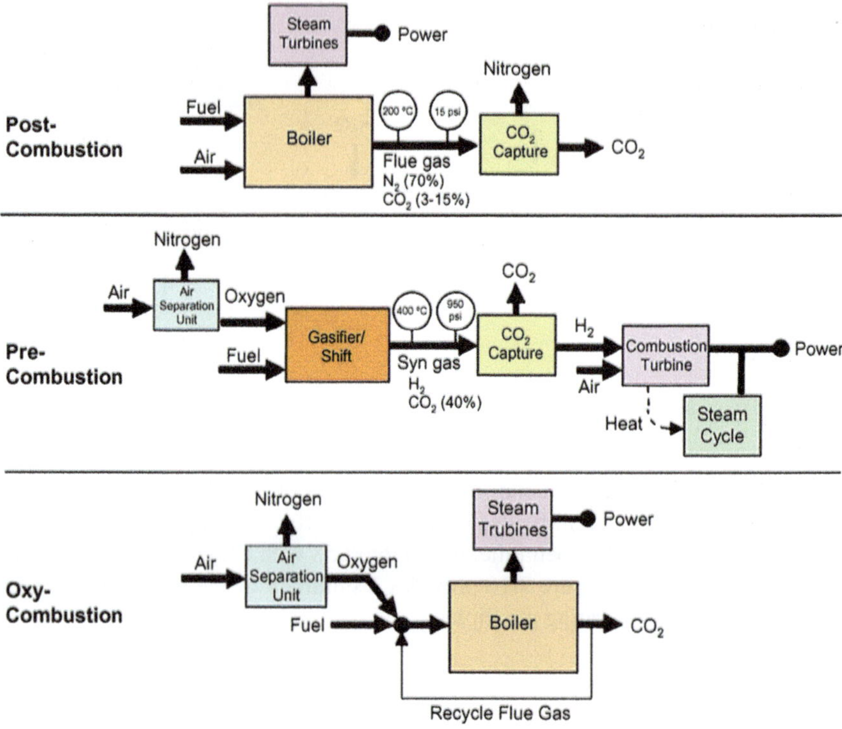

Fig. 7.16 CO_2 capture techniques [108]

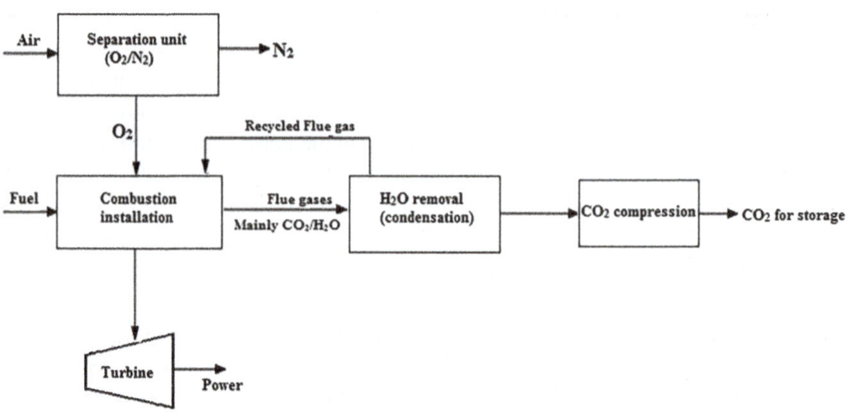

Fig. 7.17 Basic principles of oxyfuel combustion

same burner, the authors discovered that there was greater heat loss in the CH_4/oxy-combustion flames than in the CH_4/air flames. In all likelihood, inadequate supply of oxygen, fuel, and diluent will prevent oxy-combustion [111]. Enhancing the oxidizer mixture's initial O_2 percentage (implying a decrease in CO_2 concentration) enhances combustion stability and raises the exhaust gas's H_2O concentration [110]. Additionally, consistent flames and minimal pollutants will result from oxyfuel burning at a specific oxygen/diluent ratio [111]. Equivalency ratio has a major impact on both the oxidizers and the H_2O content. While the availability of oxygen decreases, the concentration of H_2O increases when the equivalency ratio is raised while maintaining a constant ratio of $O_2/CO_2\%$ and fuel volume flow rate [110]. Additionally, important are the effects of pressure on the combustion products. It was discovered that the flue gases' O_2 content and CO emissions were greater under high pressure than they were at low pressure [111]. Lower combustion temperature is caused by incomplete combustion, which is the cause of this.

7.4.2 Characteristics of Syngas Combustion

Mostly made up of H_2 and CO, syngas also contains trace amounts of CO_2, H_2O, CH_4, and N_2. The fuel sources used to generate syngas and the processing parameters affect its composition [112]. In essence, it is produced through the reforming or gasification of carbon-containing materials, such as wastes or natural gas, coal, and biomass. Syngas created from biological sources, such as organic waste, is regarded as a renewable fuel, and the energy released during combustion is thus regarded as renewable. Carbonaceous materials can be gasified or reformatted to yield syngas. Rostrup-Nielsen et al. [113] demonstrated production by steam and CO_2 reforming in the presence of catalysts such as nickel, cobalt, and the group 8 metals; however, the group 8 metals are highly expensive despite their activity. Syngas generation is better with nickel-based catalysts than with mineral-based catalysts [114]. According to the results of Rostrup-Nielsen et al. [113], steam reforming is a useful method for creating syngas, with the ratio of H_2/CO being extremely close to 3. A high temperature is necessary for the reforming process to maximize the conversion of CH_4. It was also mentioned that CO_2 may be used in place of steam, however the result would be a reduction in the H_2/CO ratio to 1. The authors also said that the creation of H_2 is dependent on the water gas shift reaction, which proceeds after the reforming process with copper acting as a catalyst and results in the total conversion of carbon monoxide. Syngas can also be produced by partial oxidation. In order to fully convert and reduce CH_4 in the generation of soot, the non-catalytic variant of this process uses high temperatures. The result is typically syngas with an H_2/CO ratio of 1.7–1.8 [113]. Using a fixed catalyst bed and a burner, the auto-thermal reforming (ATR) type combines partial oxidation with steam reforming. This leads to reduced oxygen use, enhanced efficiency, and a lower maximum temperature. With all reactants premixed and all reactions occurring in a catalytic reactor, catalytic partial oxidation (CPO) yields H_2/CO with a ratio of 2 and low reaction heat [113]. It also occurs without the need for a burner [115]. The main reactions involved in the generation of syngas are shown in Table 7.3.

Gasification of coal, which involves gasifying the coal at a high temperature using steam and oxygen or air as indicated in Table 7.4, is another method of generating syngas. Another method for creating syngas is biomass gasification. Unlike fossil fuels, biomass is a carbon-containing renewable resource that emits no greenhouse gas because it only emits the carbon it absorbs during growth and is further absorbed by the growth of new biomass, meaning that it does not contribute to the emission of CO_2. Lv et al. [116] used direct catalytic gasification to create syngas from biomass in a single step. Two reactors were used in their experiments: a downstream fixed bed reactor using nickel-based catalyst and a fluidized bed gasifier using dolomite-based catalyst. In their experiments, steam and air were employed as gasifying agents. The results of their research displayed that adding more catalyst increased the amount of H_2 produced by roughly 52.47% by volume, resulting in a high H_2/CO ratio between 1.87 and 4.45, with the fixed bed reactor's optimal temperature happening at 750oC. Additionally, they observed that whereas the generation of CO initially decreases but eventually increases with temperature, the creation of hydrogen increases as the temperature rises. The significant CO_2 and CH_4 levels of the generated syngas were also discovered by the authors. They think that increasing the use of catalysts and providing ideal operating conditions will fix this. Catalytic pyrolysis of the biomass combined with gasification is another method for producing syngas from biomass; the resulting syngas has a yield of around 3.29 Nm3/kg. Xie et al. [114] developed a two-stage method that applies catalytic pyrolysis and gasification procedures to produce syngas. The scientists examined how the temperatures of the two processes affected the amount of syngas produced, and their findings indicated that a specific temperature is necessary for syngas production to reach its maximum. At 750 °C during the pyrolysis stage, the maximum syngas output was reached. In contrast,

Table 7.3 The major reactions taking place in syngas production [113, 116]

Process	ΔH (kJ/mol)
Steam reforming	
$CH_4 + H_2O \rightleftharpoons CO + 3H_2$	−206
$C_nH_m + nH_2O \rightleftharpoons nCO + (n + 0.5m)H_2$	−1179 (for n-C_7H_{16})
$CO + H_2O \rightleftharpoons CO_2 + H_2$	41
CO_2 Reforming	
$CH_4 + CO_2 \rightleftharpoons 2CO + 2H_2$	−247
Autothermal Reforming (ATR)	
$CH_4 + 1.5O_2 \rightleftharpoons CO + 2H_2O$	520
$CH_4 + H_2O \rightleftharpoons CO + 3H_2$	−206
$CO + H_2O \rightleftharpoons CO_2 + H_2$	41
Catalytic Partial Oxidation (CPO)	
$CH_4 + 0.5O_2 \rightleftharpoons CO + 2H_2$	38

the gasification stage's highest syngas generation happened at 850 °C following a reaction between steam and the pyrolysis char. The gas yield decreased as the temperature was attempted to be raised further higher. Ji et al.'s investigation into the production of syngas using ethanol catalytic breakdown [117]. They discovered that while the CO concentration drops with an increase in the ethanol flow rate, the hydrogen content of the syngas increases as the ethanol flow rate rises. This lowers emissions, increases system efficiency, and creates syngas with a high H_2/CO ratio.

The properties of syngas, such as its flame speed, extinction limits, and emission characteristics during combustion, have been researched in relation to the fuel's H_2/CO ratio, the oxidizers used, the quantity of diluents used, the burning methods used, and other burning conditions. In combustion studies, diluents such as CO_2, water vapor, and N_2 are utilized; however, dilution with CO_2 and water vapor is more successful than that with N_2 in reducing NOx production [118]. The flame extinction temperature of dry syngas with CO_2 addition is lower than that of dry syngas with H_2O dilution, according to results published by Ding et al. [119]. Ding et al. [114] also discovered that syngas containing methane impurities has a high extinction limit and increases flame temperature because of an increase in H_2 level, which promotes burning. When CO_2 is added to impure syngas, the temperature and flame extinction limit rise relative to dry syngas [120, 121], but the extinction temperature in moist syngas is nearly unchanged. Extinction limits and temperature of moist syngas diluted with CO_2 or N_2 at high pressure have been found to be higher than those at atmospheric pressure [119]. This is because a boost in pressure causes the rate of burning to increase, raising the temperature of the flame as a result.

Fuels' volumetric heating value is also influenced by CO_2 dilution [118]. Fuels with high CO_2 dilution typically have low volumetric heating values, which leads to high CO emissions, and vice versa. This is due to the fact that CO_2 causes incomplete combustion by lowering reaction intensity. More CO emissions result as a result of this. Nonetheless, by lowering the fuel's equivalency ratio, the CO emission caused by CO_2 dilution can be decreased. According to Lieuwen et al. [121], the flame speed decreases when synthetic gas is diluted with any of the diluents listed above. At the same dilution percentage and equivalency ratio, nitrogen has the maximum flame speed and temperature, followed by H_2O and then CO_2.

The proportion of H_2 and CO in the mixture affects the laminar burning speed of the syngas flame. The upstretched laminar burning speed of $H_2/CO/O_2/N_2/CO_2$ was assessed by Di Sarli et al. [120] using the Davis reaction scheme in conjunction with the CHEMKIN program [122]. According to Wang et al.'s findings [112], adding CO_2 was found to decrease burning velocity to flame extinction [110, 121]. A rise in the fuel mixture's CO_2 concentration was also discovered to be an inhibitor of the reaction process. Liu et al. [123] investigated the chemical impacts of CO_2 on

Table 7.4 Syngas production through coal gasification

Gasification of Coal in the presence of steam and air or oxygen	ΔH (kJ/mol)
$C + H_2O \rightleftharpoons CO + H_2$	-131

premixed flames of methane and hydrogen. They found that the effect of decreasing burning velocity was more pronounced in the hydrogen flames than in the methane flames, and that the effect diminished as the equivalency ratio decreased. On the other hand, increasing the mixture's H_2 content improves the flame structure and speeds up laminar burning [112, 124–126]. According to research by Bouvet et al. [125], when the hydrogen concentration of the gas mixture increased, so did the laminar flame speed of syngas. They discovered that a small amount of H_2 increases the CO/air mixture's laminar flame speed. Around 10 cm/s was the flame speed with 1/99% H_2/CO at an equivalency ratio of 0.6, and 20 cm/s was the flame speed with 10/90% H_2/CO [125]. In contrast to fuels with low hydrogen content, which have low flame temperature and high NO concentration, fuels with high hydrogen constituent—such as 45% CO, 50% H_2, and 5% CH_4 in [119]—have very high flame temperatures and low NO concentrations.

Singh et al. [124] investigated the effect of adding water to fuel mixtures on flame speed for two distinct syngas fuel combinations (5% H_2–95% CO and 50% H_2–50% CO). According to the findings of their research, the flame speed of the 5:95 fuel mixtures increases as the fuel's H_2O percentage rises to 20% of the mixture and then decreases as the fuel's H_2O content falls. Any small addition of H_2O to the 50:50 ($CO:H_2$) fuel combination lowers the flame speed. Singh et al. [124] reported on the impacts of preheat temperature and equivalency ratio on syngas combustion. Their findings demonstrated that, in comparison to fuel with a larger hydrogen component, fuel with a lower hydrogen content reached its maximum flame speed at a higher equivalency ratio. Through studies with different preheat temperatures, they also investigated how temperature affects flame speed. They discovered that the flame speed increases with the preheat temperature at all equivalency ratios. The richness of the fuel determines how the equivalency ratio affects the fuel's burning properties. According to Xie et al.'s experimental and numerical research [126], the laminar burning velocity of CH_4/CO_2/O_2 mixes for fuel-rich mixtures falls as the equivalent ratio rises, but it increases for fuel-lean mixtures.

7.4.3 Emissions and Reduction Techniques of Syngas Combustion

Syngas can release carbon monoxide (CO), nitrogen oxides (NOx), and sulfur dioxide, much like any other fuel. The kind, composition, and qualities of the syngas, as well as the combustor's operating parameters, all affect the emissions produced. When compared to traditional fossil fuels, syngas is recognized for producing incredibly low emissions. The primary components of syngas, H_2 and CO, aid in raising the fuel's temperature, particularly in syngas with a high H_2/CO ratio. This high temperature lowers emissions while improving thermal efficiency and full combustion. Therefore, the production of electricity from syngas fuels is more environmentally benign than that of fossil fuels. The formation of NOx occurs through three primary mechanisms: heat formation, fuel NOx formation, and rapid NOx formation. When syngas is burned, NO is the primary NOx species [121]. Both heat formation and fuels NO formation are mostly to blame for this. NOx emissions can be managed via post-combustion control, which is implemented after combustion, or in-furnace control, which takes place inside the combustion device and includes techniques like flue gas recirculation. The first approach works best for controlling NO production during syngas combustion. The primary cause of CO emissions from syngas combustion is incomplete combustion.

Experiments were conducted by Alavandi and Agrawal [127] on the burning of an $H_2/CO/CH_4$ combination in a porous burner. Their findings demonstrated that while the adiabatic flame temperature rises, the stoichiometric air/fuel ratio drops as the H_2/CO ratio in the mixture increases. According to experimental research, fuels containing syngas mixes have lower emissions than pure gasoline because syngas combustion in a porous burner results in reduced pollutant emissions and fuel flexibility. Moreover, the addition of syngas to CH_4 fuel reduces NOx and CO emissions. For instance, pure CH_4 fuel emits roughly 12 and 25 parts per million of CO and NOx at an adiabatic flame temperature of Tad $= 1600$ °C. The fuel's emissions of CO and NOx were lowered to 5 ppm and 7 ppm, respectively, by adding 35% H_2 and 35% CO. Emissions of NOx and HC are reduced when syngas is added to gasoline and CH_4 fuel [117, 128]. Zhang et al. [118] also investigated the NOx emissions from syngas combustion and found that the micro-mixing technique, which uses miniature diffusive combustion, reduced NOx emissions but increased the wall temperature at the nozzle outlet. However, this can be decreased with the right addition. Thus, in addition to lowering the NOx emission, CO_2 dilution was added to the experiment to lower the wall temperature at the nozzle outlet. Ji et al.'s study [117] examined how adding syngas—produced by the catalytic breakdown of ethanol—affected the gasoline engine's lean performance. The amount of unburned hydrocarbon-related HC and NOx emissions decreases as the syngas volume fraction rises. A small drop in cylinder temperature as a result of an increase in ethanol and a decrease in gasoline eventually lowers the NOx emission.

In comparison to CH_4 air combustion, syngas oxy-combustion with CO_2 mixture exhibits a larger radiation heat loss and syngas-air flame [106, 112]. When developing a combustor, this needs to be taken into account. When syngas is introduced to a gasoline engine running low, the engine's NOx and CO emissions increase somewhat [128]. Syngas addition lowers HC emissions with surplus air ratios less than 1.21, according to research by Dai et al. [128]. However, if the fuel is leaner, HC emissions tend to rise with syngas addition. In the studies conducted by Dam et al. [129], the flame stabilities of syngas oxyfuel ($H_2/CO-O_2$) and methane-oxyfuel (CH_4-O_2) were compared with the addition of CO_2 and steam as diluents. The impact of H_2 concentration and dilution on flame stability were investigated. Because of the greater adiabatic flame temperature, CH_4-O_2 flames did not stable in the absence of diluent. They also understood that a larger burner diameter would result in a shorter stability period. Their findings also displayed that, due to a lower flame temperature than CH_4-O_2 flames, syngas Oxy-fuel (10% H_2–90% CO–O_2) without CO_2 addition will still be stable with a burner diameter of 3mm. At larger burner diameters, CO_2 will be necessary for both flames to remain stable. Additionally, they discovered that the syngas oxyfuel (10% H_2–90% CO–O_2) flame experienced a longer time of stability than the methane oxyfuel (CH_4–O_2) flame.

7.5   Energy and Hydrogen Production in Water Splitting Membrane Reactor

Significant study into hydrogen production was spurred by the 1973 oil embargo, which had previously been overlooked during the heyday of fossil fuel production [1]. The search for simpler, more cost-effective methods for producing hydrogen on a wide scale has resumed in light of growing concerns over energy security and sustainability as well as the preservation of the world ecology [1]. There is growing interest in the large-scale development and production of hydrogen due to its high energy content, environmental friendliness, and use of renewable resources worldwide. Because the thermal production method is straightforward, producing hydrogen with lower operating costs requires relatively little capital investment. Since solar thermal energy has established technology, it can serve as the main energy source for the heating process. The implementation of high temperature solar energy harvesting may be limited by its current cost, however recent advancements in more effective solar power harvesting technology indicate a potential cost reduction in the near future [2].

There are numerous options available right now for producing hydrogen. These include the production of hydrogen from biomass, the electrolysis of coal to produce fossil fuels, the splitting of water using direct solar energy, and the use of electricity from nuclear and renewable sources. Moreover, thermolysis, electrolysis, radiolysis, thermochemical cycles, photo-catalysis, bio-photolysis, bio-catalysis, steam reforming of methane, and plasma reforming can all be used to extract hydrogen from water [4, 130]. The photo-biological method of decomposing water uses light and certain microorganisms, such as green algae and cyanobacteria. Similar to how plants make oxygen through photosynthesis, the microorganisms take in water and use their metabolic processes to produce hydrogen as a byproduct. A bright future for bio-photolysis in the production of hydrogen is ensured by this photolytic water conversion [5]. Among these methods, water splitting by electrolysis or thermochemical cycles has garnered particular attention recently [131, 132]. By using the direct water splitting concept, hydrogen can be produced economically from water vapor at a moderate temperature, provided that an effective technique for separating hydrogen and oxygen is established and materials that can withstand the temperatures are viable and readily available [132]. Electrolysis of water is classified according to the kind of electrolyte used. Currently, water electrolyzers meet around 3.9% of the global hydrogen requirement [8]. The most promising ones are the high temperature steam electrolysis cells (using solid oxides), alkaline electrolysis cells (AEC), and polymer electrolysis membrane (PEM) cells (using acidic ionomer) [9, 12, 13, 133, 134].

By using ion transport membrane (ITM) technology, hydrogen can be produced from water splitting in a straightforward and affordable manner. The ITMs have 100% selectivity for oxygen from gas mixtures and are composed on mixed ionic and electronic conductors (MIEC) membranes. Their crystal lattice's oxygen vacancies are the cause of this. Water splits into hydrogen and oxygen at high temperatures. With the help of the ITM, the oxygen reaches the sweep side (also known as the oxygen permeation side), and at the feed side (also known as the hydrogen production side), hydrogen is produced from the steam. With ITM, it is possible to dramatically increase the generation of both hydrogen and oxygen by eliminating the oxygen following water splitting. Significant amounts of hydrogen and oxygen are produced when the membrane's elimination of oxygen causes the water's equilibrium reaction to move toward dissociation into hydrogen and oxygen. Mass transfer of gaseous oxygen from the gas stream to the membrane surface at the higher oxygen partial pressure side (feed or hydrogen production side) is the process of oxygen permeation. The adsorption of oxygen molecules comes next, and the feed side surface reaction comes next. The transfer of oxygen from the membrane surface at the sweep side to the gas stream occurs next, which is followed by surface response at the lower pressure side (also known as the oxygen permeation side) and bulk diffusion, which is the process of moving oxygen ions across the membrane [15–17]. According to findings published in the literature, the operating temperature, variations in the partial pressures of oxygen across the membrane, gas flow rates, and membrane thickness all affect the amount of oxygen that permeates through ITM [19, 135, 136]. Thus, these variables also have an impact on the production of hydrogen via water splitting using an ITM.

Due to the growing need for clean hydrogen for petrochemical applications, inorganic membranes have demonstrated extremely strong performance for hydrogen synthesis and purification [21–23, 137]. Because of its capacity to move or separate oxygen, dense ceramic membranes have drawn interest for use in gas separation applications [27, 28, 138, 139]. The oxygen penetration rate is a crucial factor in determining hydrogen synthesis by water splitting, provided that membrane water-splitting for hydrogen production (MWSHP) is a direct outcome of oxygen permeation via membranes [29]. In addition, a multitude of significant factors influence the rate of oxygen permeation and water-splitting hydrogen production. These factors include oxygen ion and electron conductivity, temperature, membrane thickness, gradient of partial pressure of oxygen across the membrane, and surface oxygen exchange kinetics [27, 30, 31]. Additional information about mixed ionic-electronic conducting (MIEC) membranes for hydrogen production from water splitting may be found in the review conducted by Wenping et al. [32].

When more oxygen permeates across the membrane as per reaction (A), the equilibrium can be driven to the product side to improve the hydrogen penetration rate, according to the use of oxygen transport membranes (OTMs) [33, 34]:

$$H_2O \leftrightarrow 0.5O_{2(g)} + H_{2(g)}$$

This reaction dissociates into extremely low concentrations of hydrogen and oxygen. As a result, a novel membrane known as mixed ion and electron conducting (MIEC) OTMs, which can offer high oxygen penetration fluxes, has been developed in place of OTMs. Table 7.5 presents a comparative analysis of hydrogen production rates across several membrane configurations.

Numerous researchers have researched on different membranes for hydrogen generation from water splitting and oxygen permeation. The bi-functional performances of $BaCe0.95Tb0.05O3-\delta$ (BCTb) membranes for hydrogen synthesis and power generation were investigated by Meng et al. [141]. They found that the BCTb exhibits dominating proton conductivity below 700 oC, and that electronic conductivity becomes more prominent above this temperature. $Gd0.2Ce0.8O1.9 - \delta-Gd0.08Sr0.88Ti0.95Al0.05O3-\delta$, GDC–GSTA) mixed electronic conducting and oxygen ionic membranes were used by Wang et al. [142] to extract a sizable amount of hydrogen through water splitting and oxygen removal. They experimented with both thick and thin membranes to produce hydrogen. Their experimental results displayed that, for the thick membrane, oxygen bulk diffusion is the primary controlling parameter for oxygen permeation and hydrogen generation from steam (water) at higher temperatures, whereas for the thin membrane, surface exchange reactions and O_2 bulk diffusion via the MIEC membrane jointly control the permeation process. Park et al. [39] also investigated oxygen transport membranes (OTMs). A substantial amount of hydrogen was produced using a stable thin-film $La0.7Sr0.3Cu0.2Fe0.8O3-d$ (LSCF7328) with a thickness of 22 μm. Steam and coal gas (CO_2/CO) were passed through the membrane on the sides that produce hydrogen and oxygen, respectively. The penetrated oxygen was used by the coal gas, which enhanced the chemical potential for hydrogen production and oxygen permeation. It was claimed that when a 99.5% pure carbon monoxide was passed through the oxygen permeation side, the rate of hydrogen creation was measured to be approximately 4.7 $cm^3 min^{-1} cm^{-2}$ at

Table 7.5 Comparison of hydrogen production rates considering different membranes

Type	Membrane configuration	Membrane temperature range	References	Hydrogen production
$La_{0.6} Sr_{0.4} Co_{0.2} Fe_{0.8} O_{3-\delta}$	Micro tubular	Study performed at 900 °C	Franca et al. [35]	
0.13 mm thick CGO/NI cermet membrane with coarse microstructure	MIEC cermet	Study temperature range from 700 to 900 °C	Balachandran et al. [28]	6.0 $cm^3/$ min/cm^2
0.09 mm thick CGO/Ni membrane with porous surface layers and 1 mm thick SFC_2	MIEC cermet	Study temperature range from 500 to 900 °C	Balachandran et al. [140]	10.0 $cm^3/$ min/cm^2
$0.9Zro0.1Y_2O_3$ and $0.8ZrO_2-0.1TiO_2 0.1Y_2O_3$	MIEC	1873 – 1956 K	Naito and Arashi [33]	4.0 E-07 $mol/s/cm^2$

900 °C. This demonstrated that using the byproduct of coal gasification, a significant amount of hydrogen may be produced [13]. The results of Balachandran et al.'s study [27] also displayed that ITM has the capacity to produce hydrogen and oxygen from water splitting at moderate temperatures (500–900 °C). Previous researchers have looked into the use of reactive gas on the oxygen permeation side of the membrane. Park et al. [39] used coal gas with an LSCF membrane; Hong et al. [16] and Habib et al. [143] used methane along with an LSCF membrane; Ben-Mansour et al. [41] used a BSCF (Ba0.5Sr0.5Co0.8Fe0.2Ox) membrane; Jiang et al. [42] used a BCFZ (BaCoxFeyZr1-x-yO3-δ) membrane; Lee et al. [144] used an LSTF (La0.6Sr0.4Ti0.2Fe0.8O3 $-$ δ) membrane. It was discovered that using reactive gas as sweep gas increases the oxygen penetration flux in every instance. This is because the gas reacts with the oxygen that has permeated it, raising the chemical potential gradient. By altering the water dissociation equilibrium for the generation of hydrogen following the membranes' removal of oxygen, this also results in a faster rate of hydrogen synthesis. Findings involving reactive gases also demonstrated that partial pressure differences are not as important in promoting hydrogen production and oxygen permeation as the inlet temperature.

7.5.1 Modelling Membrane Reactors for Energy and Hydrogen Production

A thorough numerical analysis of membrane reactor modeling for the production of energy and hydrogen is provided in this section. The application of the dense ceramic membranes for oxygen penetration and water splitting to produce hydrogen is examined in the study. Additionally, an investigation is conducted into the features of syngas oxy-combustion inside the water splitting oxygen transport reactor (OTR). The study took into account various water splitting operating circumstances, and the rate of hydrogen production for each scenario has been provided. The effects of temperature, membrane thickness, and fuel composition are examined. The effects of CO_2 circulation and sweep gas flow rate are examined. The lowest and maximum working temperatures for the oxygen separation membrane apply to the usage of membranes for water splitting. Since most dense ceramic membranes are not activated for oxygen separation below 600 °C, they are unable to separate oxygen below this temperature. Also, because of the corresponding thermal stresses that lead to membrane collapse, such membranes cannot function at temperatures higher than 1000 °C. As a result, the parameters for the hydrogen production process utilizing ceramic membranes and water splitting are restricted to the membrane's operational properties.

The study looks into the possibility of using a reactive gas, syngas, in an oxygen transport reactor to produce hydrogen and oxygen from water splitting, as well as the use of helium, an inert gas, as the sweep gas. This paper also investigates the oxygen-permeated syngas's oxy-combustion. Because of its high thermal and electrical efficiency, low carbon emissions, high stability and predictability, and capacity to be produced from renewable sources, syngas is regarded as the reactive gas in this work. Anticipate a reaction between the syngas and the penetrated oxygen, which will cause the water's balance to shift in favor of oxygen and hydrogen dissociation. With the aid of ANSYS meshing software, a mesh including 13,000 grid points was created, and the CFD commercial code ANSYS 15.0 was utilized for the computations. Since this sort of membrane has a great potential for oxygen permeability and is chemically stable at high temperatures, it was used in this instance to separate oxygen from water splitting. User defined function (UDF) was used to model the flow of oxygen through the membrane. In this instance, CO_2 and synthetic gas are combined to create the sweep gas. According to reports, CO_2 is a very useful diluent for syngas combustion [118]. As a result, it serves as a diluent and a way to moderate the reactor temperature in this work.

7.5.1.1   Reactor Specifications and Boundary Conditions

The purpose of this work is to examine the LSCF membrane's capacity for hydrogen generation and oxygen permeation through water splitting. For Newtonian fluids, the calculations were carried out under steady state and laminar conditions. An annulus of water vapor encircling a reactor in the shape of a tubule is created. The reactor has a length of 100 mm and two tubes that share a radius of 10 mm. Water is added from the hydrogen generating side at a volumetric proportion of 66.67% and 33.33% of hydrogen and oxygen, respectively. To start, helium, an inert gas, is injected into the oxygen permeation side in order to sweep the oxygen that has infiltrated. Later, a reactive gas called syngas is added to the inert gas to investigate the membrane's capacity for hydrogen production and oxygen permeability. When dealing with reactive gas, a blend of CO_2 and synthetic gas (CO/H_2) is added, both of which function as sweep gas. After splitting, oxygen is extracted from the water with the use of an LSCF membrane. Oxy-combustion of syngas in the combustion chamber is caused by the separated oxygen reacting with the fuel mixture as it passes through the membrane wall to the oxygen permeation side. At the membrane's oxygen permeation side, the separated oxygen's reaction with the syngas should result in a decrease in the partial pressure of oxygen. In an oxygen transport reactor, these are the forces that propel oxygen permeation and hydrogen creation in addition to temperature, which rises as a result of burning. The combustor's other end is used to release the combustion products, which are primarily CO_2 and H_2O. To learn more about how fuel combustion in the oxygen transport reactor affects hydrogen generation and oxygen penetration, researchers are looking into the oxy-combustion of syngas in these reactors. Two numerical models are used in this study: the kinetics of the synthetic gas process and the oxygen permeation/hydrogen production model. To determine whether the applied models are valid, validations using experimental data are carried out. Structural mesh is used in this study's computational geometry to minimize numerical mistakes. Two flow zones—one for the side of hydrogen

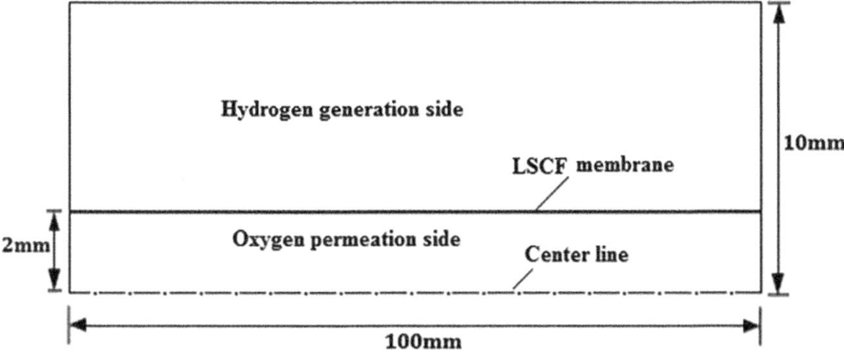

Fig. 7.18 Schematic diagram representing the geometry of the present membrane reactor

generation and the other for the side of oxygen permeation—are represented by the geometry in Fig. 7.18. A wall that acts as a membrane divides the two zones. Two mass flow inlets are provided by the geometry to introduce fuel and water vapor, while two pressure outlets are provided to release reaction products and retentate.

7.5.1.2 Governing Equations

The models produced in this paper are solved using the finite volume approach of CFD. The mixture's pressure and velocity fields are obtained by solving the continuity, momentum, and energy equations. The ANSYS Fluent 15.0 algorithm solves the conservation equations of species by using the convection–diffusion equation for the ith species and forecasting Yi, the local mass fraction of each species. The following is an expression for the conservation equations of mass, momentum, energy, and species [145]:

$$\nabla(\rho U) = S_i \tag{7.8}$$

$$(\rho UU) + \nabla p = \mu \nabla^2 U \tag{7.9}$$

$$(\rho C_p) U . \nabla T = \nabla .(k \nabla T) \tag{7.10}$$

$$\nabla .(\rho U Y_i) - \nabla .\left(\rho D_{i,m} \nabla Y_i\right) = S_i \tag{7.11}$$

The term S_i can be obtained using the Eq. (7.12) as follows [41]:

$$S_i = \begin{cases} \frac{+J_{O_2} A_{cell} MW_{O_2}}{V_{cell}} & \text{at the permeate side} \\ \frac{-J_{O_2} A_{cell} MW_{O_2}}{V_{cell}} & \text{at the feed side} \end{cases} \tag{7.12}$$

Computation of $D_{i,m}$, can be done using Eq. (7.13) [143]:

$$D_{i,m} = \frac{1 - X_i}{\sum_{j,j \neq 1} \left(\frac{X_i}{D_{i,j}} \right)} \tag{7.13}$$

The binary mass diffusion coefficient, $D_{i,j}$, can be calculated by utilizing Chapman-Enskog formula using the kinetic theory [146].

In this work, the oxygen permeation model based on the work done by Xu and Thomson [19] on a LSCF membrane is applied. The oxygen permeation equation is presented as follows [19]:

$$J_{O_2} = \frac{D_V K_r \left(P_{O_2}^{'\left(\frac{1}{2}\right)} - P_{O_2}^{''\left(\frac{1}{2}\right)} \right)}{2LK_f \left(P_{O_2}' P_{O_2}'' \right)^{\frac{1}{2}} + D_V \left(P_{O_2}^{'\left(\frac{1}{2}\right)} + P_{O_2}^{''\left(\frac{1}{2}\right)} \right)} \tag{7.14}$$

According to findings published by earlier researchers [142, 144], the rate at which hydrogen is produced by splitting water is twice that of oxygen penetration. This is predicated on the processes of oxygen permeation and steam dissociation reaction. As a result, the rate of hydrogen creation can be stated as follows [144] using Eq. (7.16):

$$H_2O \leftrightarrow H_2 + \frac{1}{2}O_2 \tag{7.15}$$

$$J_{H_2} = \frac{2D_V K_r \left(P_{O_2}^{'\left(\frac{1}{2}\right)} - P_{O_2}^{''\left(\frac{1}{2}\right)} \right)}{2LK_f \left(P_{O_2}' P_{O_2}'' \right)^{\frac{1}{2}} + D_V \left(P_{O_2}^{'\left(\frac{1}{2}\right)} + P_{O_2}^{''\left(\frac{1}{2}\right)} \right)} \tag{7.16}$$

The parameters in the above equation are specially modelled for hydrogen generation through water splitting and their values are presented in Table 7.6. The three resistances to oxygen permeation shown in Eq. (7.17a) can also be used to define the permeation flux model. Expressions to quantify the resistances are given in Eqs. (7.17b-7.17e) [144]:

$$J_{O_2} = \frac{\frac{K_r}{K_f} \left(P_{O_2}^{''-\left(\frac{1}{2}\right)} - P_{O_2}^{'-\left(\frac{1}{2}\right)} \right)}{R_t} \tag{7.17a}$$

$$R_t = R_{ex}' + R_{diff} + R_{ex}'' \tag{7.17b}$$

$$R_{ex}' = \frac{1}{K_f P_{O_2}^{'\left(\frac{1}{2}\right)}} \tag{7.17c}$$

$$R_{diff} = \frac{2L}{D_v} \qquad (7.17d)$$

$$R''_{ex} = \frac{1}{K_f P''_{O_2}{}^{(\frac{1}{2})}} \qquad (7.17e)$$

Prior studies [19, 136, 142] have demonstrated that the rate-limiting step at lower temperatures is surface exchange at the membrane's permeate side. Therefore, at lower temperatures, $[\![R]\!]\_ex^{,,}$ has a greater influence on oxygen permeation and hydrogen generation rate. On the other hand, resistance to bulk diffusion is the governing factor at higher temperatures.

The radiation model used in this study is the discrete ordinates (DO) model. This model solves the radiative transfer equation (RTE) for emitting, scattering and absorbing medium. The RTE equation in the direction $\vec{S}$ is given as follows [147]:

$$\frac{dI(\vec{r}, \vec{s})}{ds} + (a + \sigma_s)I(\vec{r}, \vec{s}) = \frac{an^2(\sigma T^4)}{\pi} + \frac{\sigma_s}{4\pi} \int_0^{4\pi} I(\vec{r}, \vec{s\prime})\phi(\vec{s}.\vec{s\prime})d\Omega' \qquad (7.18)$$

Expression (7.18) is transformed into an expression for radiation intensity in the spatial coordinates (x, y, z) by the DO model [147]. Variable absorption is the absorption coefficient that is employed; this is useful for simulating radiation in combustion problems. This study uses the weighted-sum-of-grey-gases-model (WSGGM) of domain based to take radiation's effect on gas mixtures into account. The overall emissivity in this model can be written as Eq. (7.19) [100]. In the current work, all of the walls have an internal emissivity of 0.8 [148].

$$\epsilon = \sum_{i=0}^{i=I} a_{\epsilon_i}(T)\left[1 - e^{-k_i PL}\right] \qquad (7.19)$$

The emissivity weighting factors a_{ϵ_i} depend on temperature T. k_i is the absorption coefficient and L in this case is the gas layer thickness.

Table 7.6 The values of the coefficients, D_v, k_f and k_r, used in the present model

Expressions	Pre-exponential coefficients		Activation energy (kJ/mol)
	Units	Values	
$D_V = D_V^o \exp(-E_D/RT)$	cm²/s	1.58	73.59
$k_f = k_f^0 \exp(-E_f/RT)$	cm/atm$^{0.5}$s	6.23×10^{10}	226.89
$k_r = k_r^0 \exp(-E_r/RT)$	mol/cm²s	2.91×10^6	241.29

7.5.1.3　Solution Procedures

This computational work makes use of two numerical models. The model for hydrogen generation is the first one. The hydrogen production model is constructed from oxygen permeability across the membrane because hydrogen generation from water splitting is directly dependent on this rate. Equation (7.16), which presents the hydrogen generation model for the LSCF, is applied. Table 7.6 displays the diffusion coefficient as well as the forward and reversed rates constants. The model incorporates bulk diffusion in terms of oxygen partial pressures at steady state, as well as surface exchange effects on the side of the membrane that produces hydrogen and permits oxygen to pass through. Many C++-written UDFs that take into account the LSCF membrane's requirements numerically achieve the oxygen permeation via the insert transfer membrane. The UDF takes into account three macros: "Define Initialize," "Define Adjust," and "Define Source." They allow the source term in the continuity and species transport equations to be added or subtracted. The UDF adds new parameters to the solver data at the membrane walls from both sides after every iteration. The kinetics of the synthetic gas reaction is the subject of the second model in this paper. This work uses the model reported by Cuoci et al. [149]. After optimizing the global mechanism's kinetic parameters, as reported by Westbrook and Dryer [150], the authors created the model. A streamlined synthetic gas mechanism is used in the model. The model works well for CFD applications because of this simplification. At the membrane's side where oxygen permeates, the syngas was added. Here, the hydrogen generation and oxygen permeation side's inflow section is subject to restrictions on the fuel's composition, temperature, flow rate, and quantity of diluents. Equations 7.20–7.22 [149] give the syngas reaction equations taking into account the combustion of syngas in a CO_2 medium with the reaction rates.

$$CO + 0.5O_2 \rightarrow CO_2 \quad r_1 = 2.30 \times 10^{11} \exp\left(-\frac{31700}{RT}\right)[CO][H_2O] \quad (7.20)$$

$$CO_2 \rightarrow CO + 0.5O_2 \quad r_2 = 4.45 \times 10^{9} \exp\left(-\frac{41300}{RT}\right)[CO_2] \quad (7.21)$$

$$H_2 + 0.5O_2 \rightarrow H_2O \quad r_3 = 1.35 * 10^{8} \exp\left(-\frac{6900}{RT}\right)[H_2]^{0.87}[O_2]^{1.1} \quad (7.22)$$

The species thermochemical and transport data files, along with the synthetic gas reaction mechanism, were imported in the CHEMKIN format into the ANSYS Fluent software. The two applied models are tested against the available experimental data in the literature to ensure that they are highly accurate. Based on the uncertainty of the applicable models, the computational work's uncertainty can be computed. The most precise models from earlier literature publications are used in these numerical simulations. Higher precision is ensured by using second order upwind discretization. In order to prevent interpolation mistakes and pressure gradient assumptions on borders, the pressure staggering option (PRESTO) scheme is employed for pressure computations. This technique improves the computational work's stability and rate of convergence. The pressure–velocity coupling is implemented using the semi-implicit technique for pressure-linked equations (SIMPLE) algorithm [151]. The numerical problem solves each of the separate governing equations in turn using a segregated algorithm [147]. To avoid divergence, the scalars and variables were thought to be under-relaxing. Using the aggressive advanced multi-grid (AAMG) technique, the computational work's convergence is further improved. For the pressure, conservation, and species equations, the scheme is set to "F-cycle," and the bi-conjugate gradient stabilized method (BCGSM) is used to prevent irregular convergence. The residual values for each species and conservation equation were adjusted to 10–9 in order to obtain a completely converged solution. Moreover, at the oxygen permeation (reaction) zone's outflow, the mass fraction of different species and static temperature are tracked.

A grid independence test was performed in order to lower discretization errors in the current CFD simulations. The test makes sure that the number of grids created has no effect on the answer and that there is no misleading diffusion in the numerical solution. ANSYS meshing software was used to create the grid, which is finer near the membrane on both sides. The experiment was conducted using a variety of grids, spanning from 1000 to 36,000 grid points, and the impact of grid refinement on the oxygen removal rate was computed. It was discovered that there were more than 30% variations in the oxygen penetration rates between grid point meshes of 1000 and 4000. The discrepancies were reduced to around 25% with grid refinement to 16,000 grid points, and to about 8% with grid refinement to 25,000 grid points. Ultimately, the difference was less than 1% when the grid was improved to 36,000. It was found that further refining had no benefit but could need more time and computation.

7.5.1.4 Model Validation

Equation (7.16) illustrates how the oxygen permeation model was used to create the hydrogen generation model. In order to validate the created hydrogen generation model, Park et al. [39] published experimental records, which were compared with the current numerical results. The pre-exponential coefficient values were fitted to the LSCF (La0.7Sr0.3Co0.2Fe0.8O3-δ) membrane experimental records published by Park et al. [152]. Table 7.6 displays the fitted values for Dv, kf, and kr. The influences of temperature on the rate of hydrogen generation were studied in the reported experimental investigations. There was a temperature range of 600 to 900 degrees Celsius. On the hydrogen generation side, humidified nitrogen (N_2/H_2O: 0.51/0.49) was added as feed gas, while on the oxygen permeation side, H_2/He: 0.8/0.2 was added as sweep gas. To accurately depict the experimental setup in the numerical model, a new mesh was created. The reported experimental circumstances were used to make numerical computations, and the outcomes were compared to the experimental records. Figure 7.19 present the similarities. As the operational temperature rises, hydrogen generation is observed to be enhanced. This may attributed to the fact that raising the operational temperature causes the equilibrium reaction of water

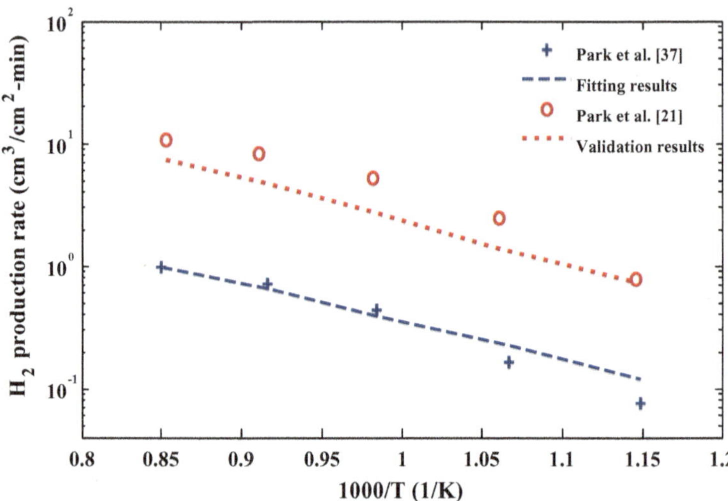

Fig. 7.19 Hydrogen production rate assessment with experimental data of Park et al. [39, 152] (symbols for experimental records, dotted line for validation results, and dash line for fitting results)

splitting to shift in favor of dissociation into hydrogen and oxygen, which lowers the membrane's resistance to oxygen permeability and, as a result, increases the amount of hydrogen produced. The figure's results demonstrate how well the proposed model can forecast the pace at which hydrogen is generated from water splitting.

In the current investigation, the synthetic gas reaction kinetics model created by Cuoci et al. [149] is utilized. Equations 7.20–7.22 present the simplified reactions. Barlow et al.'s [153] experimental results are used to validate the model. Boundary conditions, the geometry, the solution algorithm, and the experimental setup were all covered in [154]. As seen in Fig. 7.20, the estimated mass fractions of CO, H_2, and O_2 from the computational work are compared with the observed experimental values by Barlow et al. [153]. Plots, both experimental and numerical, are displayed from the intake to the exit sections along the reactor's center line. Because of the jet flow, the O_2 concentrations in the coflow are low in the inlet region. It is noted that as O_2 diffuses through the reactor, its concentration rises away from the jet. Due to their complete burning during the combustion process, the concentrations of CO and H_2 are lowered in the axial direction, as indicated by the figure, at an axial location of 0.17 m. The current numerical model and the experimental data exhibit good agreement, according to the findings. The two generated models' results agreeing with published experimental data indicates how reliable the models are for producing hydrogen and synthetic gas reactions.

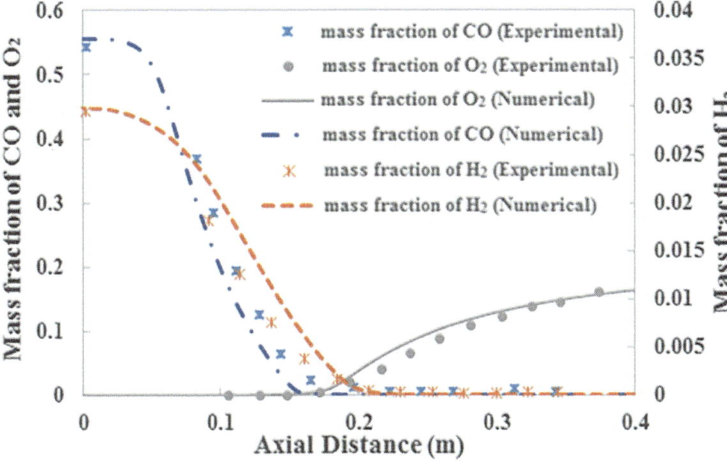

Fig. 7.20 Assessment between the current model results and the recorded experimental records by Barlow et al. [153] in terms of species mass fractions along the center line of the reactor

7.5.2   Non-reactive Flow Operation (Hydrogen Production Only)

This section's results are shown in terms of the rate at which hydrogen is created while helium is used as an inert sweep gas. Following water splitting, oxygen is transferred from the hydrogen generation side to the oxygen permeation side by the LSCF membrane, where helium is added as the sweep gas. It goes without saying that the pace at which oxygen permeates the water dissociation zone affects hydrogen creation. Therefore, increasing the oxygen penetration flux will increase the production of hydrogen immediately. Many parameters, such as feed and sweep gas flow rates, partial pressure of oxygen/partial pressure of water, operating temperature, membrane thickness, and combustion (in reactive situations), have been studied in relation to oxygen permeation across a perovskite membrane. These elements undoubtedly have an impact on the rate at which hydrogen is produced utilizing ITMs. Since the reactor in the non-reactive instance is isothermal, the membrane wall and both inlet sections were forced to maintain the same temperature. As a result, in the non-reactive scenario, oxygen partial pressure variations through the membrane are the primary factor influencing oxygen permeability and hydrogen creation. This section looks into how the inlet temperature, membrane thickness, and helium flow rate affect the rate at which hydrogen is generated The gases on either side of the membrane have intake temperatures ranging from 975 to 1175 K. While the flow rate of helium varied from 7.45E-8 to 3.45E-7 kg/s, the gas flow rate in the hydrogen generating side remained constant at 3.45E-7 kg/s. The water supply in the feed zone is solely composed of hydrogen and oxygen, with respective moles of 66.67 and 33.33%. Additionally, the thickness of the membrane was adjusted from 0.02 mm to 0.45 mm in order to examine its impact on the rate of hydrogen production.

Figures 7.21 and 7.22 display how the inlet temperature influences the system response. The rate of hydrogen generation rises with the inlet temperature. This is due to the membrane's decreased resistance to oxygen diffusion at higher temperatures. Bulk diffusion is the primary mechanism responsible for oxygen permeation and hydrogen formation at higher temperatures, according to numerous studies. In this instance, Eq. (7.17d) becomes the step that limits the rate and increases the membrane's potential for oxygen penetration. Therefore, at higher temperatures, oxygen penetration is increased. As a result, when the working temperature rises, so does the rate of hydrogen creation. The impacts of helium flow rate on hydrogen generation from the simulations are displayed in Fig. 7.21. The partial pressure of oxygen at the oxygen permeation side decreases when the sweep gas flow rate is increased. This increases the potential for oxygen penetration by increasing the oxygen partial pressure differences. Since the rate of oxygen permeation is directly correlated with this, more hydrogen is produced as a result. It is also observed that at higher temperatures, the effects of rising flow rates are more pronounced. This demonstrates that sweep flow rates are not as important in hydrogen creation as inlet temperature is. Figure 7.22 displays the effect of membrane thickness on the rate of hydrogen production. It is observed that the rate of hydrogen production decreases with increasing membrane thickness. This could be explained by a thicker membrane's greater diffusion resistance. It is evident from Eq. (7.17d) that a thicker membrane resists oxygen diffusion more strongly than a thinner one. Therefore, decreasing oxygen permeability and hydrogen generation rate results from thickening the membrane.

7.5.3　Reactive Flow Operation (Combined Hydrogen Production and Oxy-combustion of Syngas)

Research has shown that using inert or non-reactive gas as sweep gas in a membrane reactor results in less hydrogen being produced and less oxygen penetrating the reactor [41, 42, 155, 156]. The use of inert sweep gas, which produces small partial pressure variations across the membrane, may be the cause of this. This study looks into reactive flow employing syngas in the permeate side of the membrane to improve rates of hydrogen generation and oxygen permeation. It is anticipated that the syngas will consume the oxygen that has permeated and increase the chemical potential gradient of oxygen. The water equilibrium reaction is shifted toward the synthesis of more hydrogen as a result of the quick consumption of the permeated oxygen during the permeate side combustion process, which in turn increases the rates of hydrogen generation and oxygen permeation. Because of the oxy-combustion of the syngas with the penetrated oxygen, the addition of a reactive gas will also increase the temperature of the reactor. On the other hand, an excessively high reactor temperature

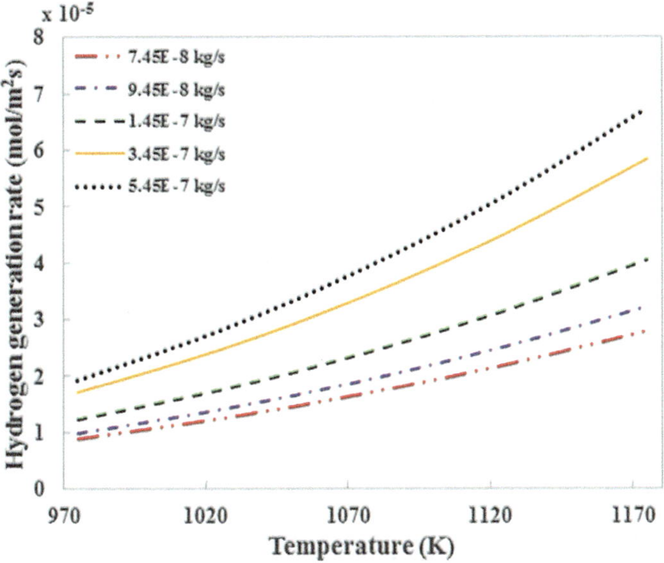

Fig. 7.21 Effects of helium flow rate on hydrogen generation rate for non-reactive case

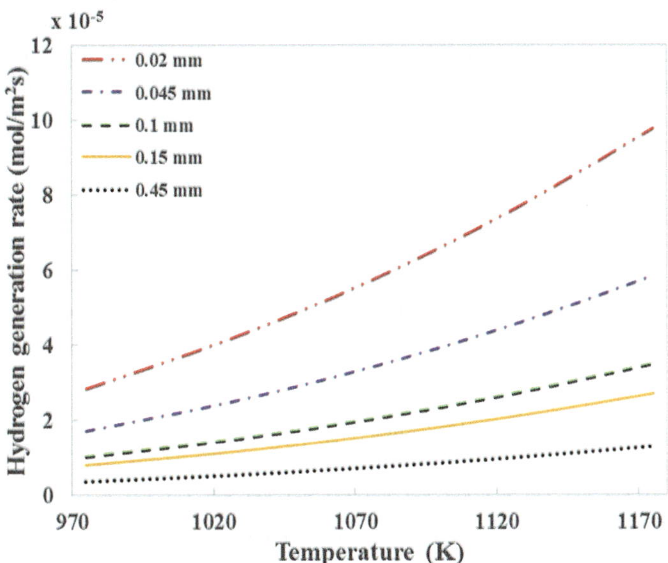

Fig. 7.22 Effects of membrane thickness on the rate of non-reactive hydrogen production

could harm the membrane. CO_2 is injected to the sweep gas to act as a diluent in the combustion zone and reduce the reactor temperature, preventing the membrane from overheating. In order to address the reactive situation in this study, the ANSYS Fluent software's volumetric reaction under species transport was turned on for the interaction between the synthetic gas and penetrated oxygen in accordance with the reaction mechanisms shown in Eqs. 7.22–7.22. Table 7.7 presents the model parameters and their values and acts as the foundational reference example for.

Figure 7.23 displays the axial distributions of oxygen permeation flux for the reactive and non-reactive cases using syngas under the same operating conditions as listed in Table 7.7. This illustrates how reactions affect the membrane's permeation potential and, in turn, the rate at which hydrogen is generated from water splitting. It is clear that the reactive situation has a higher oxygen flux than the non-reactive condition. This could be explained by syngas using up the oxygen that has infiltrated during combustion, which lowers the partial pressure of oxygen in the reaction zone and ultimately raises the variations in partial pressure of oxygen across the membrane. This suggests that the reactive situation generates hydrogen at a faster pace than the non-reactive case. At an axial distance of 0.9 cm from the reactor intake, the reactive case's oxygen permeation flux approaches a maximum value of approximately 6.04 $\times$ 10^{-5} mol/m^2/s, and the hydrogen generation rate approaches a value of approximately 12.07 $\times$ 10^{-5} mol/m^2/s. The region of the flame front, where the temperature peaks, also coincides with this point. This demonstrates that temperature has a dominant effect in the reactive scenario. Conversely, in the region of the reactor's input section, the oxygen permeation flux and hydrogen generation rate for the non-reactive scenario approach maximum values of around 1.4 $\times$ 10^{-5} mol/m^2/s and 2.8 $\times$ 10^{-5} mol/m^2/s, respectively. Due to the greatest partial pressure differential, the maximum values for the non-reactive situation are found close to the reactor's input section.

The impacts of several significant parameters on oxygen permeation and hydrogen generation rate are examined using a parametric research in the sections that follow. In a parametric research, one parameter is changed while the values of the other parameters are maintained at the levels specified in Table 7.7. Five distinct parameters, each with five levels, were examined in this parametric investigation. Table 7.8

Table 7.7 Model parameters and their values

Parameters	Values
Feed gas flow rate	3.45E-6 kg/s
Sweep gas flow rate	3.45E-7 kg/s
Fuel and CO_2 composition	CO/H$_2$/CO$_2$: 0.05/0.05/0.9
Feed side inlet temperature	1175 K
Sweep side inlet temperature	1175 K
Feed gas composition (by mole)	H$_2$/O$_2$: 0.67/0.33
Density of membrane	6000 kg/m^3
Thermal conductivity of membrane	4 W/m–K

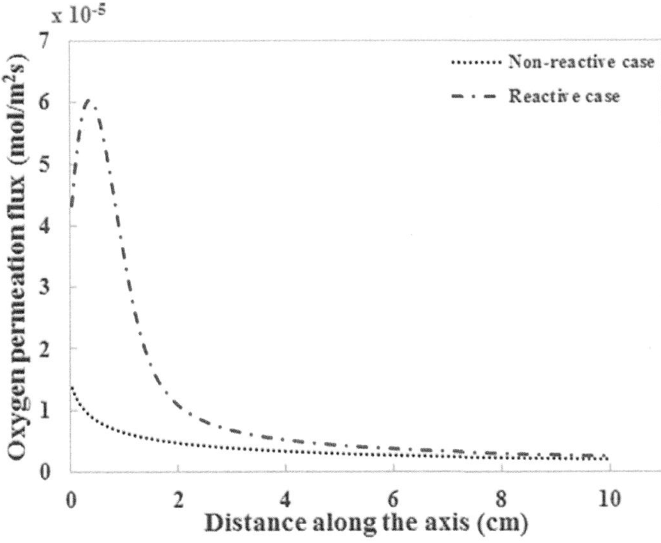

Fig. 7.23 Comparison of oxygen permeation flux; reactive versus non-reactive cases

Table 7.8 Levels of parameters used in the present parametric study

Parameters	Fuel composition CO/H_2 by %	Membrane thickness (mm)	Sweep flow rates (kg/s)	Inlet temperature (K)	CO_2 circulation syngas/CO_2 (%)
Levels	90/10	0.02	7.45E-8	975	6/94
	75/25	0.045	9.45E-8	1025	10/90
	50/50	0.1	1.45E-7	1075	16/84
	25/75	0.15	3.45E-7	1125	20/80
	10/90	0.45	5.45E-7	1175	24/76

summarizes the parameters, levels, and related values. In this instance, the sweep gas consists of CO and H_2, in addition to CO_2, which regulates the reactor's temperature. Analysis of syngas oxy-combustion characteristics using oxygen from water splitting is part of the work, along with consideration of the impacts of sweep gas, fuel composition, input temperature, CO_2 circulation, and membrane thickness.

7.5.3.1 Oxy-combustion Characteristics of Syngas

The findings and comments regarding the oxy-combustion of syngas with the help of permeating oxygen from water splitting are presented in this part. For the combustion process, the streamlined mechanism shown in Eqs. 7.20–7.22 is employed. The mass fraction of species, kinetic rate of reactions, and temperature distribution are all analyzed. This study makes use of the base case parameters and their values listed in Table 7.7. As illustrated in Fig. 7.24a, the temperature rises due to combustion from the inlet value of 1175 K to the peak value of 1241 K. At a distance of roughly 0.9 cm from the reactor input portion, this temperature reaches its highest. The kinetic rate of reactions is displayed in Fig. 7.24b; the kinetic. The location of the strong flame coincides with the maximal values of the kinetic rates for reactions 1 and 2. It is also discovered that the rate-determining step is reaction 3, which entails the consumption of H_2 for the creation of H_2O. This is because the kinetic rate for reaction 3 drops to a very low level of 0.008 kgmol/m3s as the fuel's hydrogen component is completely consumed, as seen in Fig. 7.24c. This happens at an axial distance of about 1.03 cm from the reactor's entrance portion, and as soon as the hydrogen is completely consumed, the temperature begins to drop. Reaction rates for reactions 1 and 2 nearly equalize at this point. The two reactions' potential equilibrium could be the cause of this. As shown in Fig. 7.24c, this is also reflected in the consumption of CO and the creation of CO_2. Figure 7.24c displays the mass fractions of several species to help explore the oxy-combustion features in further detail. Along the reactor's length, the concentration of inflow hydrogen in the syngas mixture in the permeation zone decreases as predicted. This could be explained by the fact that it is used up in the combustion process that produces H_2O. There is an increase in temperature as a result of burning, which increases the permeation potential. As a result, the concentration of oxygen increases along the reactor's axis. Figure 7.24c displays the distribution of CO and CO_2 concentrations, which exhibit distinct trends in comparison to H_2 and H_2O. This might be the result of the two CO and CO_2 reactions shown in Eq. 7.21. There is an increase in CO concentration at the expense of CO_2 in the area around the reactor's inlet section; this is consistent with the reaction rate curve displayed in Fig. 7.24b. In comparison to the generation of CO_2, the reaction leading to the formation of CO is now more prevalent (see Eq. 7.21). After some time, the situation is reversed, demonstrating the dominance of the alternative reaction that produces CO_2 (see Eq. 7.20). After that, balance is reached along the reactor's length. The CO and CO_2 trends match the findings published by Fogler [157].

7.5.3.2   Effects of Fuel Composition

The rate at which oxygen permeates and hydrogen is generated in an OTR is dependent on numerous factors. They also affect the syngas's oxy-combustion properties in a membrane reactor. The fuel composition, membrane thickness, operating temperature, sweep gas flow rate, and the amount of CO_2 circulation are a few of these characteristics. A parametric research is conducted in order to comprehend the consequences of certain parameters. It is noteworthy to mention that an increase in temperature and/or a partial pressure differential through the membrane can improve the rates of hydrogen production and oxygen permeation. Therefore, any parameter that increases the partial pressure difference or the reactor temperature will benefit the rates of hydrogen generation and oxygen permeation. It will also enhance syngas combustion since the penetrated oxygen will form a better combustible combination.

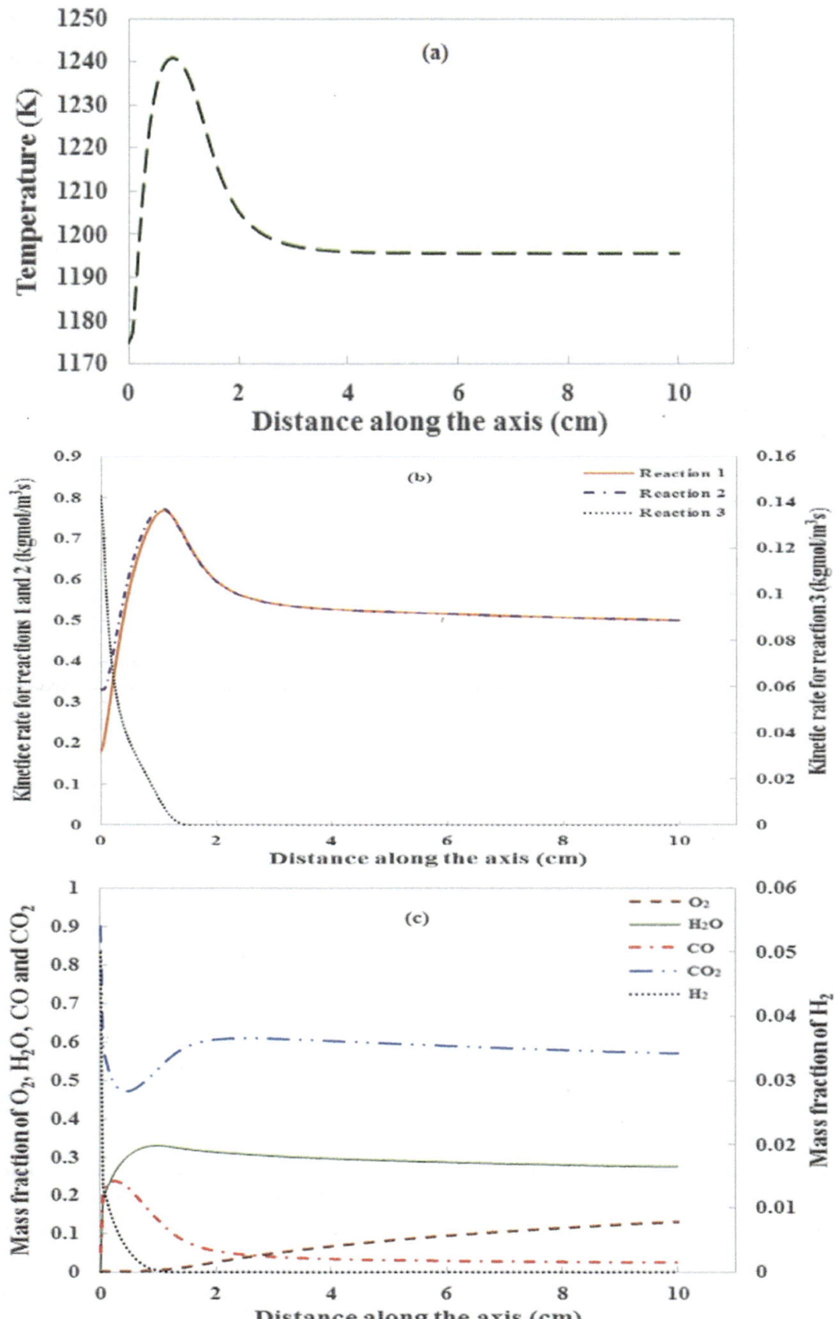

Fig. 7.24 Circulations along the center line of the reactor of: **a** temperature, **b** kinetic rate of reactions and **c** species mass fractions

The investigation focuses on the impact of fuel composition (CO and H_2 fractions) on reactor temperature as well as the rates of hydrogen generation and oxygen permeation. The concentration of CO in the fuel mixture was changed at the expense of hydrogen in order to investigate the impact of the CO/H_2 ratio. All other parameters are maintained at their base case values, as shown in Table 7.7, while this is being done. The fuel's CO and hydrogen concentrations were adjusted from 10/90% to 90/10% by mass. The effects of the modifications on the oxygen permeation flow and hydrogen production rate are shown in Fig. 7.25a and b. The image makes it quite evident that there is a direct correlation between the amount of hydrogen generated and the quantity of oxygen that permeates the syngas. This can be explained by raising the inflow hydrogen concentration in the syngas mixture and raising the combustion temperature, as seen in Fig. 7.25c. An increase in temperature accelerates processes and increases oxygen penetration. Additionally, raising the fuel's CO concentration causes the fuel's molecular weight to climb overall. As a result, there is a decrease in the volumetric flow rate, which causes the sweep gas to sweep less oxygen and, ultimately, less oxygen permeation flux. As a result, the quantity of hydrogen produced decreases, as seen in Fig. 7.25b. As a result, raising the fuel's CO/H_2 ratio lowers the rate at which hydrogen is generated and the oxygen permeability. The opposite occurs as the fuel's hydrogen content rises, increasing the ratio of CO to H_2. As seen in Fig. 7.25c, the fuel mix also affects the total reactor temperature. The figure indicates that raising the quantity of hydrogen in the syngas composition causes the CO/H_2 ratio to decrease, which raises the reactor temperature. This could be explained by hydrogen's superior igniting and fast flame speed properties. These outcomes concur with those Ding et al. [158] reported. Additionally, because more oxygen is needed for burning when the CO/H_2 ratio is lowered, the development of a combustible combination is delayed. This could explain why, as the hydrogen content in the fuel mixture rises, the temperature and oxygen flux peak sites are shifting away from the reactor inlet section.

7.5.3.3 Effects of Membrane Thickness

Figure 7.26a and b illustrate how membrane thickness affects the rates of hydrogen production and oxygen permeation. The thickness of the membrane is adjusted between 0.02 and 0.45 mm, while the remaining parameters remain constant at the levels indicated in Table 7.7. According to the investigation, there is less oxygen permeability and thus less hydrogen creation when the membrane is thicker. This could be explained by the fact that the membrane's resistance to oxygen diffusion increases with increasing membrane thickness. Equation (7.17d) states that as membrane thickness grows, so does the diffusion resistance, or R-diff. Consequently, as shown by Eqs. 7.14–7.16, this lowers oxygen permeation and hydrogen generation. Hence, there is less oxygen permeability via a thicker membrane. Figure 7.26c further illustrates how the thickness of the membrane affects the temperature of combustion. Reactor temperature drops with increasing membrane thickness. This could be explained by a thinner membrane resulting in a decreased heat transfer rate, which

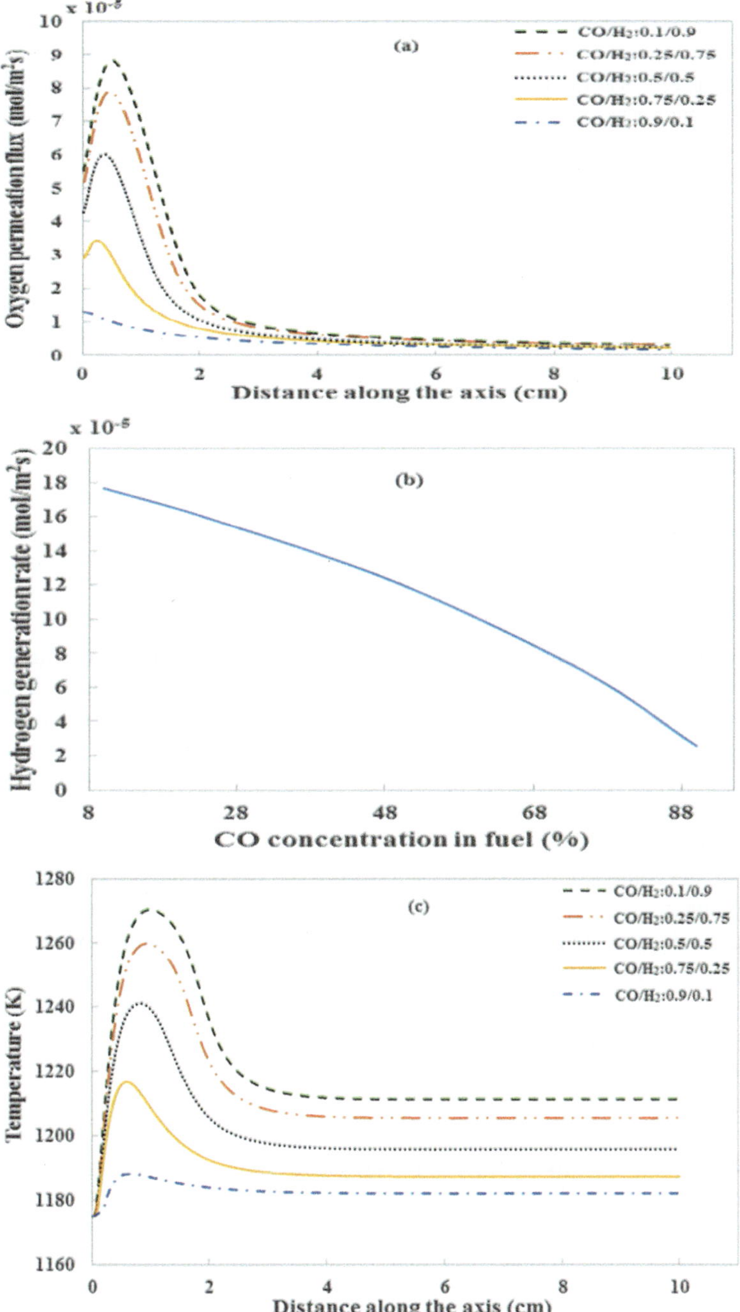

Fig. 7.25 Special effects of fuel composition on: **a** oxygen permeation flux, **b** hydrogen generation rate and **c** axial temperature distribution

lowers the reactor's temperature overall. Additionally, Fig. 7.26c demonstrates how, as thickness grows, the temperature peak is shifted toward the reactor exit section. As seen in Fig. 7.26a, this is explained by the decrease in oxygen penetration flow at greater membrane thickness. As a result, the fuel and oxygen mixture's ability to ignite is delayed. These results' tendencies correspond with the studies published by [27, 28, 135, 152].

7.5.3.4 Effects of Sweep Gas Flow Rate

Figure 7.27a and b display how the flow rate of sweep gas (syngas with CO_2) affects the amounts of hydrogen production and oxygen permeation. As demonstrated in Fig. 7.27b, increasing the sweep gas flow rate increases oxygen permeation and, in turn, increases the rate at which hydrogen is generated. This could be clarified by the fact that more fuel is implied on the oxygen permeation side when the sweep gas flow rate is increased. It is anticipated that the extra fuel will consume the oxygen that has penetrated and lower the partial pressure of oxygen at the oxygen-permeated side. This ultimately raises the rate of oxygen removal by creating a greater oxygen chemical potential gradient through the membrane. Accordingly, more hydrogen is produced as the water equilibrium reaction in the feed side is further shifted in favor of the production of oxygen and hydrogen. The pattern of the findings in this section is consistent with findings from earlier studies [143, 159]. The axial distribution of the combustion temperature as a function of the sweep flow rate is displayed in Fig. 7.27c. As the graphic illustrates, the temperature of the reactor rises in direct proportion to the upsurge in sweep flow rate. This could be brought on by the augmented oxygen penetration rate (Fig. 7.27a), which is brought on by the addition of fuel and ultimately leads to an improvement in both the reaction rates and the combustion efficiency. This ultimately causes the reactor's combustion temperature to rise. Figure 7.27c further displays that as the sweep flow rate increases, there are delays in the combustion process. This could be explained by the upsurge in fuel due to the upsurge in the sweep gas flow rate. This results in a delay in the combustion process and the displacement of the peaks of the oxygen flux and combustion temperature because more oxygen is needed to generate a combustible mixture.

7.5.3.5 Effects of Operating Temperature

The current work aims to investigate the impact of operating temperature by varying the inlet temperature between 975 and 1175 K, while maintaining the remaining parameters at their base case values. According to research, this temperature variety is perfect for oxygen permeability from water splitting. The findings of earlier studies have demonstrated that bulk diffusion is the regulating factor at higher temperatures, whereas surface exchange is primarily responsible for the rates of hydrogen generation and oxygen permeation at lower temperatures [19]. Figure 7.28a and b

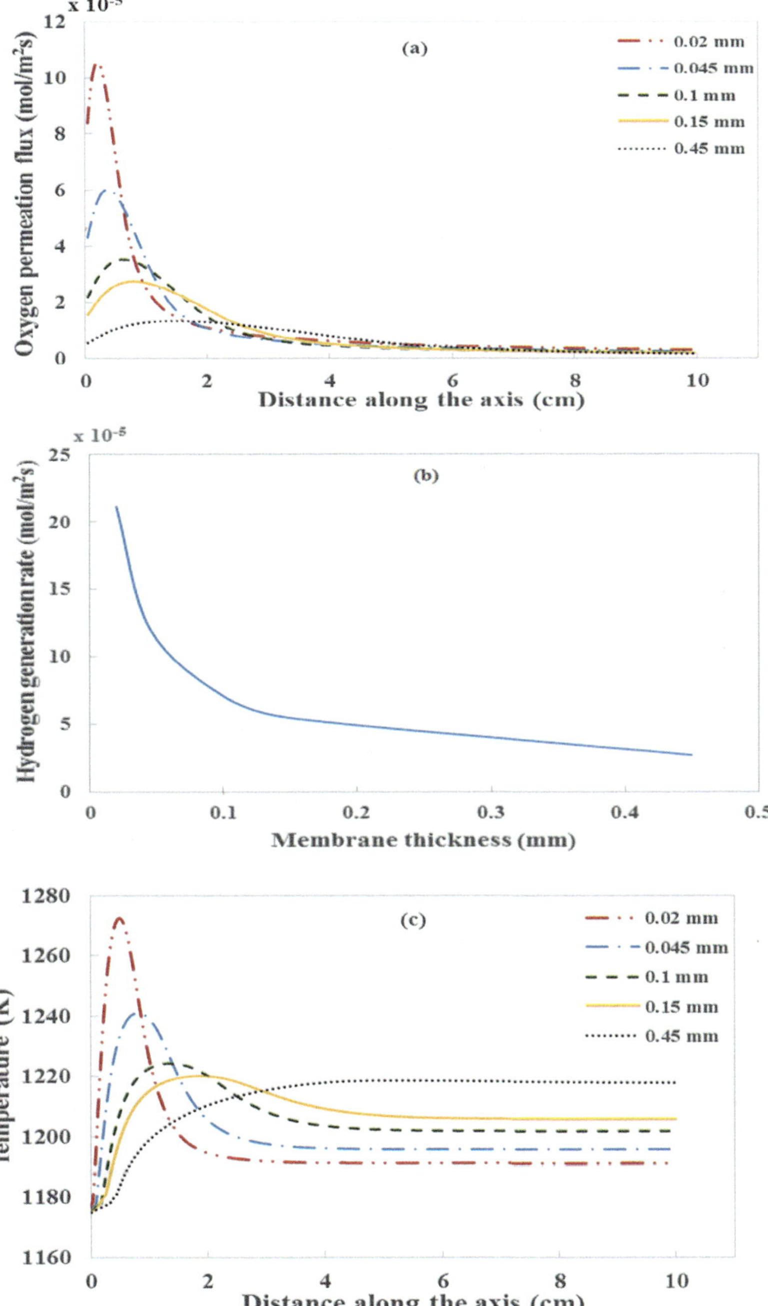

Fig. 7.26 Effects of membrane thickness on the axial temperature distribution, hydrogen generation rate, and oxygen permeation flux

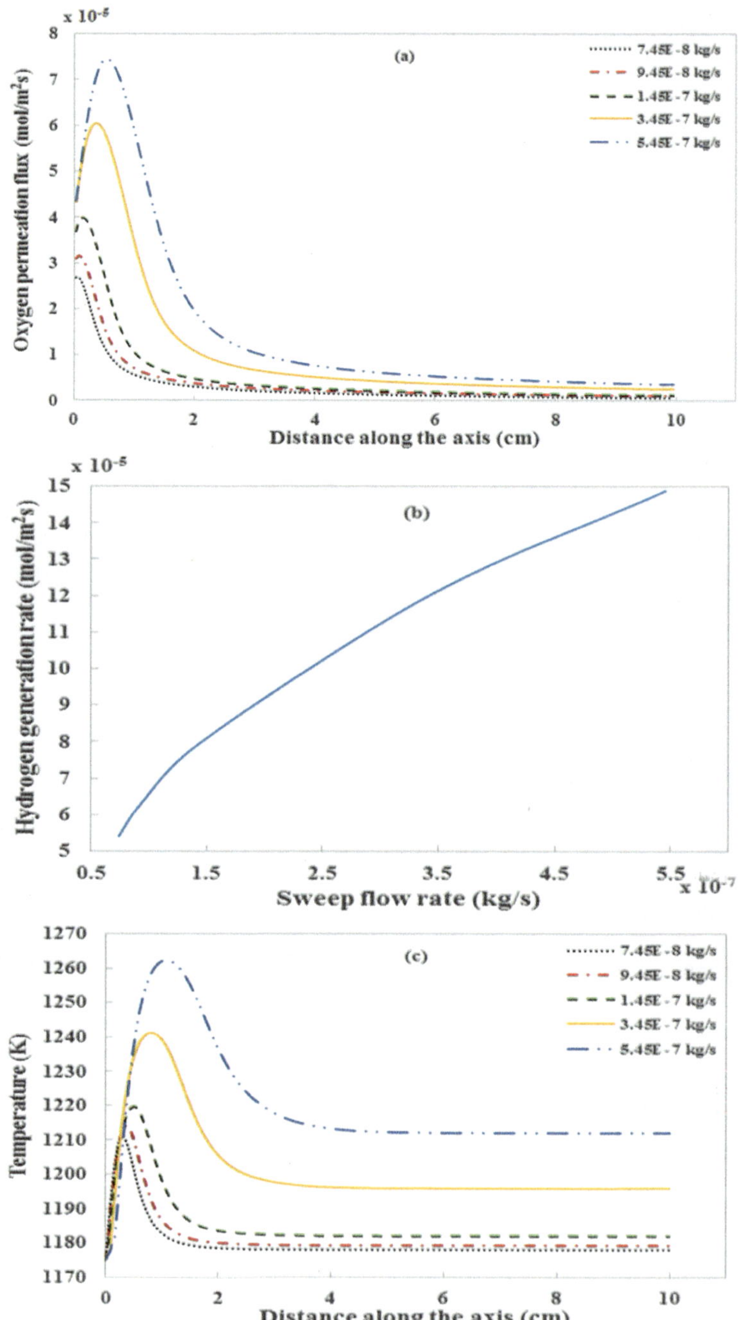

Fig. 7.27 Impacts of sweep flow rate on: **a** oxygen flux, **b** hydrogen production rate and **c** axial temperature profile

illustrate how the inlet temperature affects the rates of hydrogen production and oxygen permeation. It should be mentioned that an upsurge in the inlet temperature causes the reaction rate constants to riseKf and Kr. As seen in Fig. 7.28a, this has a direct impact on the amount of oxygen taken from the feed side and causes the water equilibrium reaction to change in favor of the creation of hydrogen and oxygen. As a result, as Fig. 7.28b illustrates, the rate of hydrogen creation is enhanced. This agrees with the parameters of the model as shown in Eqs. 7.14 and 7.16. Additionally, compared to the reverse rate constant, Kr, the forward rate constant, Kf, is extra sensitive to temperature increases. The rates of oxygen adsorption and desorption are determined, respectively, by these two rate constants. Thus, as seen in Fig. 7.28b, Kf raises the rate at which oxygen is detached from the hydrogen generating side, which ultimately results in a higher rate of hydrogen creation. It is also seen that the values of the flux peak of oxygen permeation shift toward the direction of the reactor's input section. The findings in this section are consistent with earlier research, such as [27, 28, 42, 144, 154]. The impact of the reactor's input temperature on its internal combustion temperature is depicted in Fig. 7.28c. The combustion temperature rises as the intake temperature rises, as the figure illustrates. This could be because of an increase of oxygen that has entered the atmosphere, which improves combustion by raising reaction rates. Additionally, it has been noted that a higher input temperature causes the peak combustion temperature to occur sooner. This is consistent with the outcomes of Singh et al. [124] and is triggered by an increase in the reaction rate. As seen in Fig. 7.28a, c, the location of the oxygen permeation peak and the combustion temperature peak are in alignment. As the intake temperature rises, the reactor's temperature peak shifts toward the inlet part of the reactor. This may be explained by the higher oxygen permeation flux at higher temperatures, which causes a combustible fuel and oxygen mixture to develop more quickly.

7.5.3.6   Effects of CO_2 Circulation

Diluent is ideally present when combustion occurs. The main diluents used in the majority of combustion applications are CO_2, H_2O, and N_2. Because the oxygen-combustion process burns fuel with only oxygen, it produces a high combustion temperature. For this reason, CO_2 should be employed to regulate the combustion temperature. The reactor temperature can be moderated by using CO_2 in addition to radiation-induced heat transfer. Figure 7.29a, b, and c, respectively, display an investigation of the impacts of the amount of CO_2 circulating on the rates of oxygen permeation and hydrogen formation as well as the combustion temperature. While the other portion is balanced with syngas, which has an equal amount of CO and H_2, the CO_2 content in the sweep gas is adjusted between 76 and 94%. Reducing the more reactive species (CO and H_2) in the sweep gas leads to an increase in the fraction of CO_2, which in turn lowers the syngas's reactivity and, ultimately, the reaction rates. As was previously mentioned, it is evident that reaction rates directly affect combustion temperature. As a result, there is less oxygen removed from the air, which lowers the rate at which hydrogen is produced. An increase in CO_2 concentration

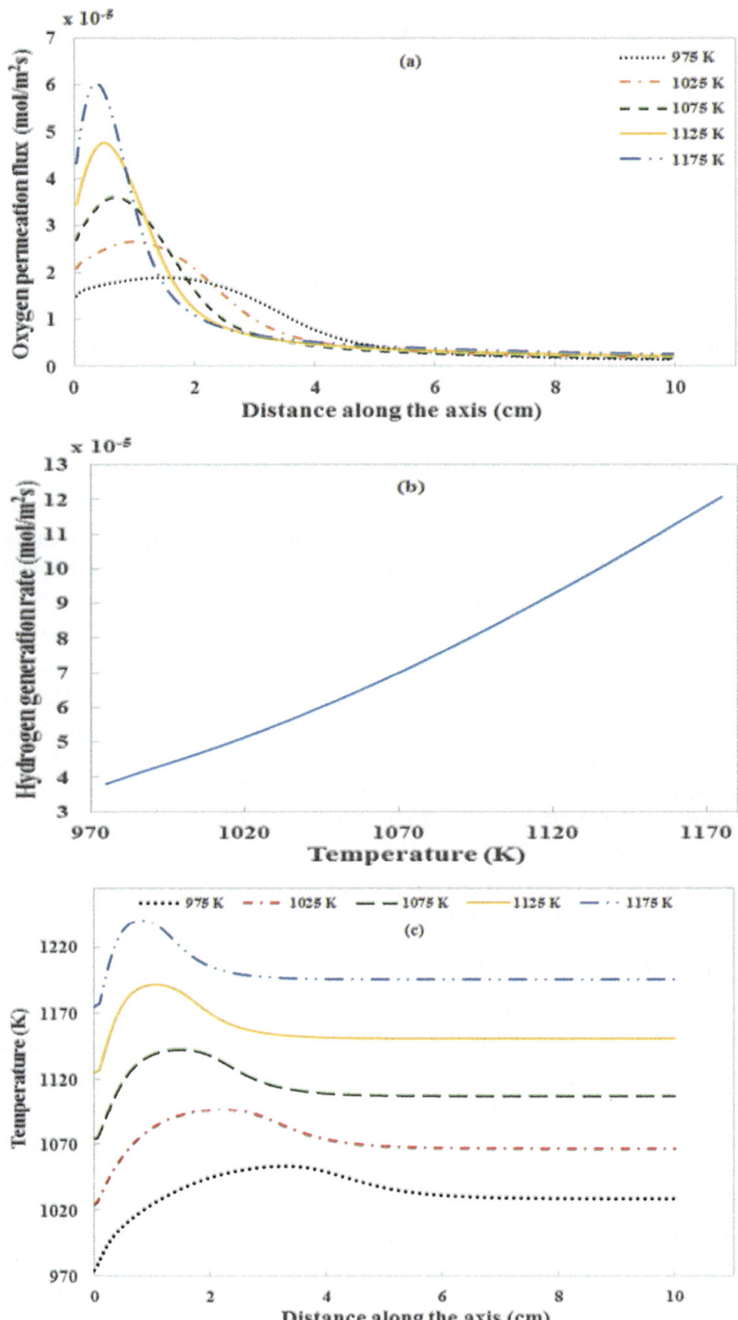

Fig. 7.28 Impacts of inlet temperature on: **a** oxygen flux, **b** hydrogen production rate and **c** axial temperature profile

may also induce a decrease in the amount of oxygen that permeates the sweep gas, as this increases the molecular weight of the sweep gas overall. Because a denser sweep gas has a lower volumetric flow rate than a less dense gas, it is less able to sweep the oxygen that has permeated, which lowers both the rate of hydrogen creation and the amount of oxygen permeation. As a result, as Fig. 7.29a illustrates, adding more CO_2 to the sweep gas lowers the oxygen permeation flux. This is in line with the findings of Farooqui et al. [160] and Habib et al. [143]. As a result, as Fig. 7.29b illustrates, the rate of hydrogen generation is reduced when using sweep gas with a greater CO_2 concentration. Additionally, it is discovered that, as Figure illustrates, the reactor temperature decreases as CO_2 circulation increases. This could be because of the decreased reaction rate brought on by the higher CO_2 concentration in the reaction zone, which in turn reduced flame speed. These in turn cause the intensity of the reactions to decrease, which lowers the reactor's overall combustion temperature. Additionally, as the fuel concentration decreases at the reactor's intake section, it is noted that the temperature and oxygen flux peak sites are moving in the direction of the reactor inlet. This could be explained by a decreased fuel composition and increased CO_2 circulation, which reduce the quantity of oxygen needed to generate a flammable mixture.

7.6 Concluding Remarks

In this chapter, techniques for hydrogen production have been discussed considering electrolysis, hydrogen from biomass, coal electrolysis to fossil fuels, and water splitting utilizing solar energy. Approaches for hydrogen and syngas production have been discussed supported by detailed dedicated studies on such topics. The use of oxygen separation membranes for clean oxy-combustion applications with carbon capture, conversion/splitting of carbon-dioxide/water into syngas and pure hydrogen, and partial oxidation of hydrocarbons for syngas production have been discussed in detail.

A detailed numerical investigation has been presented in this chapter to characterize the processes of oxygen permeation and partial oxidation of methane in a catalytic membrane reactor (CMR). The goal of the study was to enhance the CMR's performance for increased syngas output. Based on the indirect method for catalytic partial oxidation of methane (CPOM) for syngas production, a comprehensive numerical model has been created. Good agreements were found when the model was evaluated using experimental data that was available in the literature. Numerical research has been done on the impact of feed air flow rate, sweep gas flow rate, and sweep fuel concentration on the CMR's performance. The findings demonstrated that while sweep fuel concentration increased, syngas selectivity increased, and fuel conversion decreased. The arrangement of the flow between the feed and sweep flows plays a significant influence in regulating the syngas production performance of the CMR; the counter-current flow configuration performs significantly better. Fuel conversion increases and syngas selectivity decreases when the oxygen

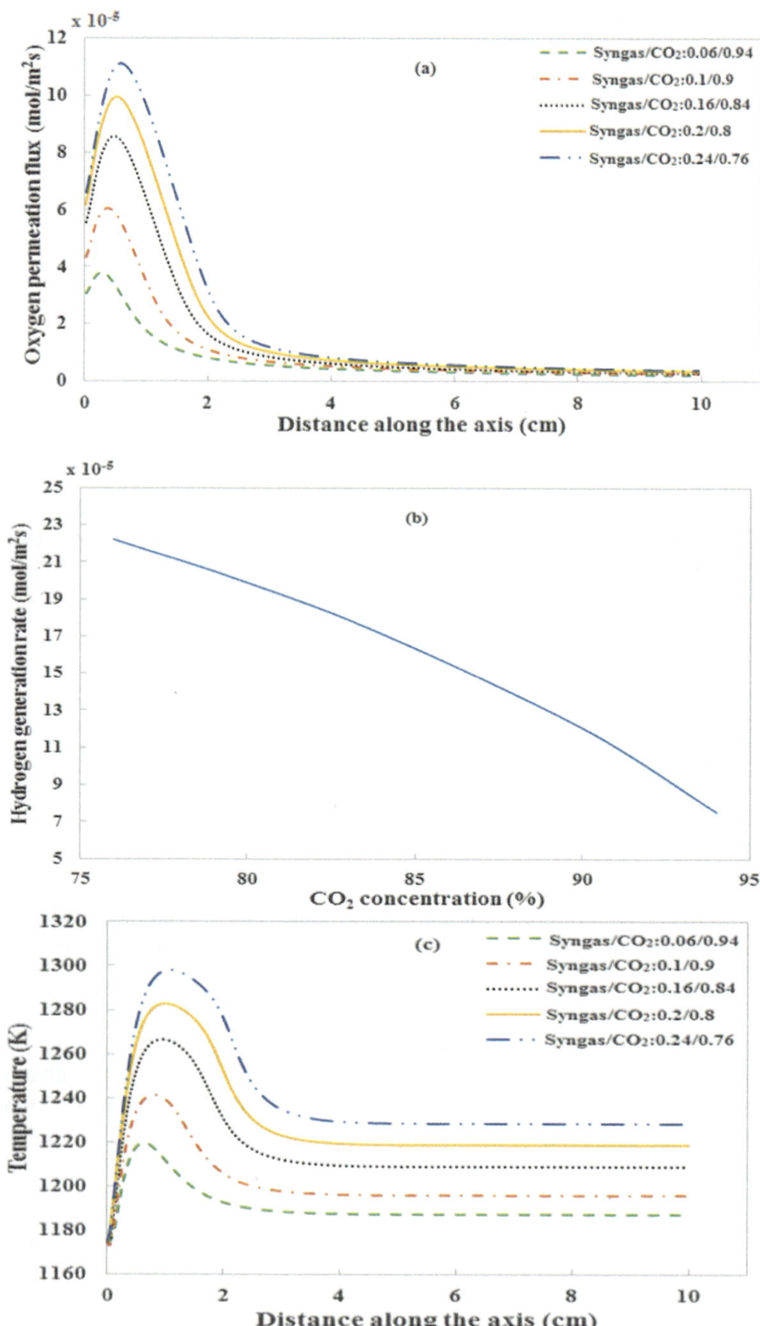

Fig. 7.29 Impacts of CO_2 circulation on: **a** oxygen flux, **b** hydrogen production rate and **c** axial temperature profile

content at the permeate side of the membrane rises in tandem with an increase in feed air flow rate. Elevating the sweep gas flow rate beyond a particular threshold led to a decline in the CMR's efficiency and a notable decrease in oxygen penetration flux, fuel conversion, and syngas selectivity. Even when the sweep gas flow rate was increased further, there was no increase in temperature, and the high velocity of the gases inside the reactor caused them to cool down, preventing any dwell time for reactions to initiate. This chapter also discusses the latest advancements in syngas clean combustion with carbon collected. It concludes with a detailed numerical study that addresses hydrogen production from water splitting via an LSCF membrane and oxygen permeation. With the use of CFD, the researchers also examined the oxy-combustion properties of synthetic gas in an oxygen transport reactor utilizing oxygen extracted from water splitting. Models for oxygen permation and reaction kinetics have both been validated using accessible experimental data from published works. The work required calculations first using helium, an inert sweep gas, and later syngas, a reactive sweep gas. The results of the investigations demonstrate that the water equilibrium reaction is shifted toward the formation of hydrogen and oxygen when oxygen is consumed with syngas across the membrane. This made it possible to produce substantial volumes of hydrogen and oxygen. In the current experiment, the combustion properties of synthetic gas—a mixture of CO and H_2—have also been examined. Studies have been done on the effects of fuel composition on the CO/H_2 ratio, membrane thickness, sweep gas flow rate, operating temperature, and amount of CO_2 circulation. It was evident from comparisons of the reactive and non-reactive cases that the reactive case has greater potential for hydrogen creation and oxygen removal. The rates of oxygen penetration, hydrogen production, and combustion temperature have all increased when the H_2 proportion in the syngas fuel has increased. A thinner membrane allows for more oxygen to pass through and produce hydrogen, which raises the combustion temperature. It also has a lower bulk diffusion resistance for oxygen permeation. Elevating the temperature at the input enhanced the combustion process, resulting in increased formation of hydrogen and oxygen permeation. In order to create a combustible combination, increasing the sweep gas flow rates means using more fuel for combustion and more oxygen. Low combustion temperature was the outcome of increasing the CO_2 concentration in the sweep gas since it decreased the rates of hydrogen production and oxygen permeation.

Acknowledgements The authors appreciate the support received for the preparation of this book from the Deanship of Research Oversight and Coordination (DROC) at King Fahd University of Petroleum & Minerals (KFUPM). The support provided by the Interdisciplinary Research Center for Hydrogen Technologies and Carbon Management (IRC-HTCM) on project number INHE2308 is highly appreciated. Also, the support received through the KFUPM Consortium for Hydrogen Future through the projects numbered H2FC2309 and H2FC2315 is highly appreciated.

References

1. J.E. Funk, Thermochemical hydrogen production: past and present. Int. J. Hydrogen Energy **26**(3), 185–190 (2001)
2. S. Ihara, Feasibility of hydrogen production by direct water splitting at high temperature. Int. J. Hydrogen Energy **3**(3), 287–296 (1978)
3. G. Lauermann, P. Haussinger, R. Lohmuller, A.M. Watson, *Hydrogen 1. Properties and Occurrence, Ullmann's Encyclopaedia of Industrial Chemistry* (2013), pp. 1–15
4. W.J. Lee, Y.K. Lee, Internal gas pressure characteristics generated during coal carbonization in a coke oven. Energy Fuels **15**(3), 618–623 (2001)
5. K. Ghasemzadeh, A.A. Babaluo, A. Aghaeinejad-Meybodi, Membrane reactors for the decomposition of H_2O, NOx, and CO_2, in *Membrane Reactors for Energy Applications and Basic Chemical Production* (2015), pp. 211–247
6. National Academy of Sciences, National Research Council, *The Hydrogen Economy: Opportunities, Costs, Barriers, and R&D Needs* (February 2004).
7. B. Yildiz, M.S. Kazimi, *Nuclear Energy Options for Hydrogen and Hydrogen Based Liquid Fuels Production, MIT-NES-TR-001*, September (2003)
8. B.C.R. Ewan, R.W.K. Allen, A figure of merit assessment of the routes to hydrogen. Int. J. Hydrogen Energy **30**(8), 809–819 (2005)
9. A. Hauch, S.D. Ebbesen, S.H. Jensen, M. Mogensen, Highly efficient high temperature electrolysis. J. Mater. Chem. **18**, 2331–2340 (2008)
10. S.P.S. Badwal, S. Giddey, C. Munnings, Hydrogen production via solid electrolytic routes. Wiley Interdiscip. Rev. Energy Environ. **2**(5), 473–487 (2013)
11. C.A. Marcelo, L.F. David, J. Mergel, D. Stolten, A comprehensive review on PEM water electrolysis. Int. J. Hydrogen Energy **38**(12), 4901–4934 (2013)
12. K. Zeng, D. Zhang, Recent progress in alkaline water electrolysis for hydrogen production and applications. Prog. Energy Combust. Sci. **36**(3), 307–326 (2010)
13. F. Barbir, PEM electrolysis for production of hydrogen from renewable energy sources. Sol. Energy **78**(5), 661–669 (2005)
14. T. Han, Y. Sohn, H. Ha, M. Yoo, S. Kim, S.R. Lee, H.Y Kim. J.H. Han, Modification of the amount of CH_4 supplied for the efficient CH_4 Reduction of SnO_2. Kor. J. Metals Mater. **56**(5), 384–391, (2018)
15. A. Leo, S. Liu, J.C.D. Da Costa, Development of mixed conducting membranes for clean coal energy delivery. Int. J. Greenhouse Gas Control **3**(4), 357–367 (2009)
16. J. Hong, P. Kirchen, A.F. Ghoniem, Numerical simulation of ion transport membrane reactors: oxygen permeation and transport and fuel conversion. J. Membr. Sci. **407–408**, 71–85 (2012)
17. X. Tan, K. Li, Modeling of air separation in a LSCF hollow-fiber membrane module. Am. Inst. Chem. Eng. **48**, 1469–1477 (2002)
18. A. Behrouzifar, A.A. Asadi, T. Mohammadi, A. Pak, Experimental investigation and mathematical modeling of oxygen permeation through dense $Ba_{0.5}Sr_{0.5\text{-}}Co_{0.8}Fe_{0.2}O_{3\text{-}d}$ (BSCF) perovskite-type ceramic membranes, Ceram. Int. **38**, 4797–4811 (2012)
19. S.J. Xu, W.J. Thomson, Oxygen permeation rates through ion-conducting perovskite membranes. Chem. Eng. Sci. **54**, 3839–3850 (1999)
20. A.A. Asadi, A. Behrouzifar, M. Iravaninia, T. Mohammadi, A. Pak, Preparation and oxygen permeation of $La_{0.6}Sr_{0.4}Co_{0.2}Fe_{0.8}O_{3\text{-}d}$ (LSCF) perovskite-type membranes: experimental study and mathematical modeling, Ind. Eng. Chem. Res. **51**, 3069–3080 (2012)
21. C.Y. Park, T.H. Lee, S.E. Dorris, Y. Lu, U. Balachandran, Oxygen permeation and coal-gas-assisted hydrogen production using oxygen transport membranes. Int. J. Hydrogen Energy **36**, 9345–9354 (2011)
22. C.Y. Park, T.H. Lee, S.E. Dorris, U. Balachandran, Hydrogen production from fossil and renewable sources using an oxygen transport membrane. Int. J. Hydrogen Energy **35**, 4103–4110 (2010)

23. M.V. Mundschau, X. Xie, I.V.C.R. Evenson, A.F. Sammells, Dense inorganic membranes for production of hydrogen from methane and coal with carbon dioxide sequestration. Catal. Today **118**, 12–23 (2006)

24. M.R. Rahimpour, Z. Arab Aboosadi, A.H. Jahanmiri, Synthesis gas production in a novel hydrogen and oxygen perm-selective membranes tri-reformer for methanol production. J. Nat. Gas. Sci. Eng. **9**, 149–159 (2012)

25. S.Y. Jeon, M.B. Choi, C.N. Park, E.D. Wachsman, S.J. Song, High sulfur tolerance dual-functional cermet hydrogen separation membranes. J. Membr. Sci. **382**, 323–327 (2011)

26. S.J. Song, J.H. Moon, T.H. Lee, S.E. Dorris, U. Balachandran, Thickness dependence of hydrogen permeability for Ni-BaCe0.8Y0.2O3-?? Solid State Ionics **179**, 1854–1857 (2008)

27. U. Balachandran, T.H. Lee, S.E. Dorris, Hydrogen production by water dissociation using mixed conducting dense ceramic membranes. Int. J. Hydrogen Energy **32**, 451–456 (2007)

28. U. Balachandran, T.H. Lee, S. Wang, S.E. Dorris, Use of mixed conducting membranes to produce hydrogen by water dissociation. Int. J. Hydrogen Energy **29**, 291–296 (2004)

29. S.Y. Jeon, H.N. Im, B. Singh, J.H. Hwang, S.J. Song, A thermodynamically stable La_2NiO_4/$Gd_{0.1}Ce_{0.9}O_{1.95}$ bilayer oxygen transport membrane in membrane-assisted water splitting for hydrogen production. Ceram. Int. **39**, 3893–3899 (2013)

30. S.J. Xu, W.J. Thomson, Perovskite-type oxide membranes for the oxidative coupling of methane. AIChE J. **43**, 2731–2740 (1997)

31. V.V. Kharton, A.A. Yaremchenko, A.V. Kovalevsky, A.P. Viskup, E.N. Naumovich, P.F. Kerko, Perovskite-type oxides for high-temperature oxygen separation membranes. J. Membr. Sci. **163**, 307–317 (1999)

32. W. Li, X. Zhu, Z. Cao, W. Wang, W. Yang, Mixed ionic-electronic conducting (MIEC) membranes for hydrogen production from water splitting. Int. J. Hydrogen Energy **40**, 3452–3461 (2015)

33. H. Naito, H. Arashi, Hydrogen production from direct water splitting at high temperatures using a ZrO_2-TiO_2-Y_2O_3 membrane. Solid State Ionics **79**, 366–370 (1995)

34. J. Lede, F. Lapicque, J. Villermaux, Production of hydrogen by direct thermal decomposition of water. Int. J. Hydrogen Energy **8**, 675–679 (1983)

35. R.V. Franca, A. Thursfield, I.S. Metcalfe, $La_{0.6}Sr_{0.4}Co_{0.2}Fe_{0.8}O_3$ microtubular membranes for hydrogen production from water splitting. J. Membr. Sci. **389**, 173–181 (2012)

36. U. Balachandran (Balu), T.H. Lee, S.E. Dorris, Hydrogen production by water dissociation using mixed conducting dense ceramic membranes. Int. J. Hydrogen Energy **32**, 451–456 (2007)

37. X. Meng, Y. Shang, B. Meng, N. Yang, X. Tan, J. Sunarso, et al., Bi-functional performances of $BaCe_{0.95}Tb_{0.05}O_{3-d}$-based hollow fiber membranes for power generation and hydrogen permeation. J. Eur. Ceram. Soc. **36**(16), 4123–4129 (2016)

38. H. Wang, S. Gopalan, U.B. Pal, Hydrogen generation and separation using Gd0.2Ce0.8O1.9_d(GDC)eGd0.08Sr0.88Ti0.95Al0.05O3±d mixed ionic and electronic conducting membranes, Electrochim. Acta **56**, 6989–6996 (2011)

39. C.Y. Park, T.H. Lee, S.E. Dorris, U. Balachandran, Hydrogen production from fossil and renewable sources using and oxygen transport membrane. Int. J. Hydrogen Energy **35**, 4103–4110 (2010)

40. M.A. Habib, P. Ahmed, R. Ben-Mansour, H.M. Badr, P. Kirchen, A.F. Ghoniem, Modeling of a combined ion transport and porous membrane reactor for oxycombustion. J. Membr. Sci. **446**, 230–243 (2013)

41. R. Ben-Mansour, M.A. Habib, H.M. Badr, E. Azharuddin, M. Nemitallah, Characteristics of oxy-fuel combustion in an oxygen transport reactor. Energy Fuels **26**, 4599–4606 (2012)

42. H. Jiang, F. Liang, O. Czuprat, K. Efimov, A. Feldhoff, S. Schirrmeister, T. Schiestel, H. Wang, J. Caro, Hydrogen production by water dissociation in surface-modified $BaCo_xFe_yZr_{1-x-y}O_{3-\delta}$ hollow-fiber membrane reactor with improved oxygen permeation. Chem. Eur. J. **16**, 7898–7903 (2010)

43. K.J. Lee, Y.J. Choe, J.S. Lee, H.J. Hwang, Fabrication of a microtubular production, Adv. Mater. Sci. Eng. **2015**, 6 (2015). Article ID 505989

44. V.N. Nguyen, L. Blum, Syngas and synfuels from H_2O and CO_2: current status. Chem.-Ing.-Tech. **87**, 354–375 (2015)

45. Q. Fu, C. Mabilat, M. Zahid, A. Brisse, L. Gautier, Syngas production via high-temperature steam/CO_2 co-electrolysis: an economic assessment. Energy Environ. Sci. **3**, 1382 (2010)

46. C. Agrafiotis, M. Roeb, C. Sattler, A review on solar thermal syngas production via redox pair-based water/carbon dioxide splitting thermochemical cycles. Renew. Sustain. Energy Rev. **42**, 254–285 (2015)

47. N. Itoh, M.A. Sanchez, W.-C. Xu, K. Haraya, M. Hongo, Application of a membrane reactor system to thermal decomposition of CO_2. J. Memb. Sci. **77**, 245–253 (1993)

48. C. Graves, S.D. Ebbesen, M. Mogensen, Co-electrolysis of CO_2 and H_2O in solid oxide cells: performance and durability. Solid State Ionics **192**, 398–403 (2011)

49. Z. Zhan et al., Syngas production by coelectrolysis of CO_2/H_2O: the basis for a renewable energy cycle. Energy Fuels **23**, 3089–3096 (2009)

50. Y. Nigara, B. Cales, Production of carbon monoxide by direct thermal splitting of carbon dioxide at high temperature. Bull. Chem. Soc. Jpn **59**, 1997–2002 (1986)

51. M.A. Oehlschlaeger, D.F. Davidson, J.B. Jeffries, R.K. Hanson, Carbon dioxide thermal decomposition: observation of incubation. Zeitsch. Phys. Chem. **219**, 555–567 (2005)

52. M.E. Galvez, P.G. Loutzenhiser, I. Hischier, A. Steinfeld, CO_2 splitting via two-step solar thermochemical cycles with Zn/ZnO and FeO/Fe_3O_4 redox reactions: thermodynamic analysis. Energy Fuels **22**, 3544–3550 (2008)

53. A. Stamatiou, P.G. Loutzenhiser, A. Steinfeld, Solar syngas production from H_2O and CO_2 via two-step thermochemical cycles based on Zn/ZnO and FeO/Fe_3O_4 redox reactions: kinetic analysis. Energy Fuels **24**, 2716–2722 (2010)

54. L.J. Venstrom, J.H. Davidson, Splitting water and carbon dioxide via the heterogeneous oxidation of zinc vapor: thermodynamic considerations. J. Sol. Energy Eng. **133**, 011017, (8 pp.) (2011)

55. Z. Chen, P. Kang, M.-T. Zhang, B.R. Stoner, T.J. Meyer, Cu(ii)/Cu(0) electrocatalyzed CO_2 and H_2O splitting. Energy Environ. Sci. **6**, 813 (2013)

56. P. Furler, J.R. Scheffe, A. Steinfeld, Syngas production by simultaneous splitting of H_2O and CO_2 via ceria redox reactions in a high-temperature solar reactor. Energy Environ. Sci. **5**, 6098 (2012)

57. P. Furler et al., Solar thermochemical CO_2 splitting utilizing a reticulated porous ceria redox system. Energy Fuels **26**, 7051–7059 (2012)

58. S. Lorentzou, G. Karagiannakis, C. Pagkoura, A. Zygogianni, CO_2 and H_2O splitting for thermochemical production of solar fuels using nonstoichiometric ceria and ceria/zirconia solid solutions. Energy Fuels **25**, 4836–4845 (2011)

59. G.P. Smestad, A. Steinfeld, Review: photochemical and thermochemical production of solar fuels from H_2O and CO_2 using metal oxide catalysts. Ind. Eng. Chem. Res. **51**, 11828–11840 (2012)

60. P. Lahijani, Z.A. Zainal, M. Mohammadi, A.R. Mohamed, Conversion of the greenhouse gas CO_2 to the fuel gas CO via the Boudouard reaction: a review. Renew. Sustain. Energy Rev. **41**, 615–632 (2015)

61. S. Rayne, Thermal carbon dioxide splitting: a summary of the peer-reviewed scientific literature. Nat. Preced. **1**, 1–17 (2008)

62. X. Wu, *Membrane-Supported Hydrogen/Syngas Production Using Reactive H_2O/CO_2 Splitting for Energy Storage* (Massachusetts Institute of Technology, 2017)

63. A. Hunt, G. Dimitrakopoulos, A.F. Ghoniem, Surface oxygen vacancy and oxygen permeation flux limits of perovskite ion transport membranes. J Memb Sci **489**, 248–257 (2015). https://doi.org/10.1016/j.memsci.2015.03.095

64. A. Hunt, G. Dimitrakopoulos, P. Kirchen, A.F. Ghoniem, Measuring the oxygen profile and permeation flux across an ion transport ($La_{0.9}Ca_{0.1}FeO_{3-\delta}$) membrane and the development and validation of a multistep surface exchange model. J. Membr. Sci. **468**, 62–72 (2014). https://doi.org/10.1016/j.memsci.2014.05.043

65. A.S. Yu, J. Kim, T.S. Oh, G. Kim, R.J. Gorte, J.M. Vohs, Decreasing interfacial losses with catalysts in $La_{0.9}Ca_{0.1}FeO_{3-\delta}$ membranes for syngas production. Appl. Catal. A Gen. **486**, 259–265 (2014). https://doi.org/10.1016/j.apcata.2014.08.028

66. R. Yuan, Z. He, Y. Zhang, W. Wang, C. Chen, H. Wu et al., Partial oxidation of methane to syngas in a packed bed catalyst membrane reactor. AIChE J. **62**, 2170–2176 (2016). https://doi.org/10.1002/aic.15202

67. A. Torkkeli, Droplet microfluidics on a planar surface. VTT Publ. **7**, 3–194 (2003). https://doi.org/10.1002/aic

68. B.C. Enger, R. Lodeng, A. Holmen, A review of catalytic partial oxidation of methane to synthesis gas with emphasis on reaction mechanisms over transition metal catalysts. Appl. Catal. A Gen. **346**, 1–27 (2008). https://doi.org/10.1016/j.apcata.2008.05.018

69. X.Y. Wu, A.F. Ghoniem, M. Uddi, Enhancing co-production of H2and syngas via water splitting and POM on surface-modified oxygen permeable membranes. AIChE J. **62**, 4427–4435 (2016). https://doi.org/10.1002/aic.15518

70. C. Park, T. Lee, S. Dorris, *Hydrogen Production from Fossil and Renewable Sources Using an Oxygen Transport Membrane* (Elsevier, 2010)

71. N. Spencer, C.J. Pereira, $V_2O_5SiO_2$-catalyzed methane partial oxidation with molecular oxygen. J. Catal. **116**, 399–406 (1989)

72. H. Gesser, N.R. Hunter, C.B. Prakash, The direct conversion of methane to methanol by controlled oxidation. Chem. Rev. **85**, 235–244 (1985)

73. K. Jabbour, Tuning combined steam and dry reforming of methane for metgas production: a thermodynamic approach and state-of-the-art catalysts. J. Energy Chem. (2020, In press). https://doi.org/10.1016/j.jechem.2019.12.017

74. K. Takanabe, Catalytic conversion of methane: carbon dioxide reforming and oxidative coupling. J. Jpn. Pet. Inst. **55**, 1–12 (2012)

75. K. Skutil, M. Taniewski, Some technological aspects of methane aromatization (direct and via oxidative coupling). Fuel Process. Technol. **87**, 511–521 (2006)

76. J.R. Rostrup-Nielsen, Production of synthesis gas. Catal. Today **18**, 305–324 (1993)

77. A. Alipour-Dehkordi, M. Khademi, Use of a micro-porous membrane multi-tubular fixed-bed reactor for tri-reforming of methane to syngas: CO_2, H_2O or O_2 side-feeding. Int. J. Hydrogen Energy **44**, 32066–32079 (2019)

78. H. Wang, Y. Cong, W. Yang, Investigation on the partial oxidation of methane to syngas in a tubular $Ba_{0.5}Sr_{0.5}Co_{0.8}Fe_{0.2}O_{3-\delta}$ membrane reactor. Catal. Today **82**, 157–166 (2003)

79. D. Dissanayake, M.P. Rosynek, K.C.C. Kharas, J.H. Lunsford, Partial oxidation of methane to carbon monoxide and hydrogen over a Ni/AI_2O_3 catalyst. J. Catal. **132**, 117–127 (1991)

80. D.A. Hickman, L.D. Schmidt, Production of syngas by direct catalytic oxidation of methane. Science **259**, 343–346 (1993)

81. A. Elbadawi, L. Ge, J. Zhang, L. Zhuang, S. Liu, X. Tan, S. Wang, Z. Zhu, Partial oxidation of methane to syngas in catalytic membrane reactor: role of catalyst oxygen vacancies. Chem. Eng. J. (2020, In press). Article 123739. https://doi.org/10.1016/j.cej.2019.123739

82. H. Dong, Z. Shao, G. Xiong, J. Tong, S. Sheng, W. Yang, Investigation on POM reaction in a new perovskite membrane reactor. Catal. Today **67**, 3–13 (2001)

83. Q. Zhu, X. Zhao, Y. Deng, Advances in the partial oxidation of methane to synthesis gas. J. Nat. Gas Chem. **13**, 191–203 (2004)

84. J. Fabián-Anguiano, C. Mendoza-Serrato, C. Gómez-Yáñez, B. Zeifert, X. Ma, J. Ortiz-Landeros, Simultaneous CO_2 and O_2 separation coupled to oxy-dry reforming of CH_4 by means of a ceramic-carbonate membrane reactor for in situ syngas production. Chem. Eng. Sci. **210** (2019). Article 115250

85. H.J.M. Bouwmeester, Dense ceramic membranes for methane conversion. Catal. Today **82**, 141–150 (2003)

86. U. Balachandran, J.T. Dusek, P.S. Maiya, B. Ma, R.L. Mieville, M.S. Kleefisch et al., Ceramic membrane reactor for converting methane to syngas. Catal. Today **36**, 265–272 (1997)

87. S. Kato, M. Ogasawara, M. Sugai, S. Nakata, Crystal structure and property of perovskite-type oxides containing ion vacancy. Catal. Surv. Asia **8**, 27–34 (2004)

88. Z. Shao, H. Dong, G. Xiong, Y. Cong, W. Yang, Performance of a mixed-conducting ceramic membrane reactor with high oxygen permeability for methane conversion. J. Membr. Sci. **183**, 181–192 (2001)
89. U. Balachandran, J.T. Dusek, R.L. Mieville, R.B. Poeppel, M.S. Kleefisch, S. Pei et al., Dense ceramic membranes for partial oxidation of methane to syngas. Appl. Catal. A, Gen. **133**, 19–29 (1995)
90. S.S. Bharadwaj, L.D. Schmidt, Catalytic partial oxidation of natural gas to syngas. Fuel Process. Technol. **42**, 109–127 (1995)
91. H.K. Heitnes, S. Lindberg, O.A. Rokstad, A. Holmen, Catalytic partial oxidation of methane to synthesis gas. Catal. Today **24**, 211–216 (1995)
92. P.M. Torniainen, X. Chu, L.D. Schmidt, Comparison of monolith-supported metals for the direct oxidation of methane to syngas. J. Catal. **146**, 1–10 (1994)
93. H. Nourbakhsh, J. Shahrouzi, H. Ebrahimi, A. Zamaniyan, M. Nasr, Experimental and numerical study of syngas production during premixed and ultra-rich partial oxidation of methane in a porous reactor. Int. J. Hydrogen Energy **44**, 31757–31771 (2019)
94. M. Sakbodin, E. Schulman, S. Oh, Y. Pan, E. Wachsman, D. Liu, Dual utilization of greenhouse gases to produce C_{2+} hydrocarbons and syngas in a hydrogen-permeable membrane reactor. J. Membr. Sci. **595** (2020). Article 117557
95. V.D. Sokolovskii, N.J. Coville, A. Parmaliana, I. Eskendirov, M. Makoa, Methane partial oxidation. Challenge and perspective. Catal. Today **42**, 191–195 (1998)
96. E. Shelepova, A. Vedyagin, V. Sadykov, N. Mezentseva, Y. Fedorova, O. Smorygo et al., Theoretical and experimental study of methane partial oxidation to syngas in catalytic membrane reactor with asymmetric oxygen-permeable membrane. Catal. Today **268**, 103–110 (2016)
97. M. Habib, M. Nemitallah, D. Afaneh, Numerical investigation of a hybrid polymeric-ceramic membrane unit for carbon-free oxy-combustion applications. Energy **147**, 362–376 (2018)
98. M. Nemitallah, A study of methane oxy-combustion characteristics inside a modified design button-cell membrane reactor utilizing a modified oxygen permeation model for reacting flows. J. Nat. Gas Sci. Eng. **28**, 61–73 (2016)
99. M. Nemitallah, M. Habib, K. Mezghani, Experimental and numerical study of oxygen separation and oxy-combustion characteristics inside a button-cell LNO-ITM reactor. Energy **84**, 600–611 (2015)
100. M. Rajhi, R. Ben-Mansour, M. Habib, M. Nemitallah, K. Andersson, Evaluation of gas radiation models in CFD modeling of oxy-combustion. Energy Convers. Manage. **81**, 83–97 (2014)
101. A. York, T. Xiao, M. Green, Brief overview of the partial oxidation of methane to synthesis gas. Top. Catal. **22**, 345–358 (2003)
102. N. Sazonova, S. Pavlova, S. Pokrovskaya, N. Chumakova, V. Sadykov, Structured reactor with a monolith catalyst fragment for kinetic studies: the case of CH_4 partial oxidation on LaNiPt-catalyst. Chem. Eng. J. **154**, 17–24 (2009)
103. R. Ben-Mansour, M. Nemitallah, M. Habib, Numerical investigation of oxygen permeation and methane oxy-combustion in a stagnation flow ion transport membrane reactor. Energy **54**, 322–332 (2013)
104. J. Gibbins, H. Chalmers, Carbon capture and storage. Energy Policy **36**(12), 4317–4322 (2008)
105. T. Fujimori, T. Yamada, Realization of oxyfuel combustion for near zero emission power generation. Proc. Combust. Inst. **34**(2), 2111–2130 (2013)
106. M.B. Toftegaard, J. Brix, P.A. Jensen, P. Glarborg, A.D. Jensen, Oxy-fuel combustion of solid fuels. Prog. Energy Combust. Sci. **36**(5), 581–625 (2010)
107. K. Cobb, http://www.resilience.org/stories/2007-02-01/best-method-carbon-sequestration. (2007)
108. J.D. Figueroa, T. Fout, S. Plasynski, H. McIlvried, R.D. Srivastava, Advances in CO_2 capture technology—The U.S. Department of Energy's carbon sequestration program. Int. J. Greenhouse Gas Control **2**(1), 9–20 (2008)

109. P. Kutne, B.K. Kapadia, W. Meier, M. Aigner, Experimental analysis of the combustion behaviour of oxyfuel flames in a gas turbine model combustor. Proc. Combust. Inst. **33**(2), 3383–3390 (2011)

110. M.A. Nemitallah, M.A. Habib, Experimental and numerical investigations of an atmospheric diffusion oxy-combustion flame in a gas turbine model combustor. Appl. Energy **111**, 401–415 (2013)

111. C.Y. Liu, G. Chen, N. Sipöcz, M. Assadi, X.S. Bai, Characteristics of oxy-fuel combustion in gas turbines. Appl. Energy **89**(1), 387–394 (2012)

112. J. Wang, Z. Huang, H. Kobayashi, Y. Ogami, Laminar burning velocities and flame characteristics of $CO–H_2–CO_2–O_2$ mixtures. Int. J. Hydrogen Energy **37**(24), 19158–19167 (2012)

113. J. Rostrup-Nielsen, J. Sehested, J. Norskov, Hydrogen and synthesis gas by steam- and CO_2 reforming. Adv. Catal. **47**, 65–139 (2002)

114. Q. Xie, S. Kong, Y. Liu, H. Zeng, Syngas production by two-stage method of biomass catalytic pyrolysis and gasification. Bioresour. Technol. **110**, 603–609 (2012)

115. A.S. Bodke, S. Bharadwaj, L. Schmidt, The effect of ceramic supports on partial oxidation of hydrocarbons over noble metal coated monoliths. J. Catal. **179**(1), 138–149 (1998)

116. P. Lv, Z. Yuan, C. Wu, L. Ma, Y. Chen, N. Tsubaki, Bio-syngas production from biomass catalytic gasification. Energy Convers. Manag. **48**(4), 1132–1139 (2007)

117. C. Ji, X. Dai, S. Wang, C. Liang, B. Ju, X. Liu, Experimental study on combustion and emissions performance of a hybrid syngas–gasoline engine. Int. J. Hydrogen Energy **38**(25), 11169–11173 (2013)

118. Y. Zhang, T. Yang, X. Liu, L. Tian, Z. Fu, K. Zhang, Reduction of emissions from a syngas flame using micromixing and dilution with CO_2. Energy Fuels **26**, 6595–6601 (2012)

119. N. Ding, R. Arora, M. Norconk, S.Y. Lee, Numerical investigation of diluent influence on flame extinction limits and emission characteristic of lean-premixed $H_2–CO$ (syngas) flames. Int. J. Hydrogen Energy **36**(4), 3222–3231 (2011)

120. V. Di Sarli, A. Basco, F. Cammarota, A. Di Benedetto, E. Salzano, V. Diocleziano, I. Chimica, N. Federico, Flammability of syngas/CO_2 mixtures in oxygen-enriched air, in *XXXV Meeting of the Italian Section of the Combustion Institute* (2012), pp. 1–6

121. T.C. Lieuwen, V. Yang, R. Yetter, *Synthesis Gas Combustion: Fundamentals and Applications* (CRC Press, Taylor & Francis Group, Boca Raton, 2010)

122. S.G. Davis, A.V. Joshi, H. Wang, F. Egolfopoulos, An optimized kinetic model of H_2/CO combustion. Proc. Combust. Inst. **30**(1), 1283–1292 (2005)

123. F. Liu, H. Guo, G.J. Smallwood, The chemical effect of CO_2 replacement of N_2 in air on the burning velocity of CH_4 and H_2 premixed flames. Combust. Flame **133**(4), 495–497 (2003)

124. D. Singh, T. Nishiie, S. Tanvir, L. Qiao, An experimental and kinetic study of syngas/air combustion at elevated temperatures and the effect of water addition. Fuel **94**, 448–456 (2012)

125. N. Bouvet, S. Lee, I. Gökalp, R.J. Santoro, Flame speed characteristics of syngas (H_2-CO) with straight burners for laminar premixed flames, in *Proceedings of the European Combustion Meeting* (2007), pp. 1–6

126. Y. Xie, J. Wang, M. Zhang, J. Gong, W. Jin, Z. Huang, Experimental and numerical study on laminar flame characteristics of methane oxy-fuel mixtures highly diluted with CO_2. Energy Fuels **27**(10), 6231–6237 (2013)

127. S.K. Alavandi, A.K. Agrawal, Experimental study of combustion of hydrogen–syngas/methane fuel mixtures in a porous burner. Int. J. Hydrogen Energy **33**(4), 1407–1415 (2008)

128. X. Dai, C. Ji, S. Wang, C. Liang, X. Liu, B. Ju, Effect of syngas addition on performance of a spark-ignited gasoline engine at lean conditions. Int. J. Hydrogen Energy **37**(19), 14624–14631 (2012)

129. B.K. Dam, N.D. Love, A.R. Choudhuri, Flame stability of methane and syngas oxy-fuel steam flames. Energy Fuels **27**, 523–529 (2012)

130. G. Lauermann, P. Haussinger, R. Lohmuller, A.M. Watson, *Hydrogen, 1. Properties and Occurrence. Ullmann's Encyclopedia of Industrial Chemistry* (2013), pp. 1–15

131. National Academy of Sciences, National Research Council. *The Hydrogen Economy: Opportunities, Costs, Barriers, and R&D Needs* (February 2004)

132. B. Yildiz, M.S. Kazimi, Nuclear energy options for hydrogen and hydrogen based liquid fuels production. MIT-NES-TR-001 (September 2003)

133. S.P.S. Badwal, S. Giddey, C. Munnings, Hydrogen production via solid electrolytic routes. Wiley Interdiscipl. Rev. Energy Environ. **2**(5), 473–487 (2013)

134. C.A. Marcelo, L.F. David, J. Mergel, D. Stolten, A comprehensive review on PEM water electrolysis". Int. J. Hydrogen Energy **38**(12), 4901–4934 (2013)

135. A. Behrouzifar, A.A. Asadi, T. Mohammadi, A. Pak, Experimental investigation and mathematical modeling of oxygen permeation through dense $Ba_{0.5}Sr_{0.5}Co_{0.8}Fe_{0.2}O_{3-\delta}$ (BSCF) perovskite-type ceramic membranes. Ceram. Int. **38**, 4797–4811 (2012)

136. A.A. Asadi, A. Behrouzifar, M. Iravaninia, T. Mohammadi, A. Pak, Preparation and oxygen permeation of $La_{0.6}Sr_{0.4}Co_{0.2}Fe_{0.8}O_{3-\delta}$ (LSCF) perovskite-type membranes: experimental study and mathematical modeling. Ind. Eng. Chem. Res. **51**, 3069–3080 (2012)

137. M.R. Rahimpour, Z. Arab Aboosadi, A.H. Jahanmiri, Synthesis gas production in a novel hydrogen and oxygen perm-selective membranes tri-reformer for methanol production. J Nat Gas Sci Eng **9**, 149–159 (2012)

138. S.J. Song, J.H. Moon, T.H. Lee, S.E. Dorris, U. Balachandran, Thickness dependence of hydrogen permeability for Ni-$BaCe_{0.8}Y_{0.2}O_{3-\delta}$. Solid State Ionics **179**, 1854–1857 (2008)

139. S.Y. Jeon, M.B. Choi, C.N. Park, E.D. Wachsman, S.J. Song, High sulfur tolerance dual-functional cermet hydrogen separation membranes. J. Memb. Sci. **382**, 323–327 (2011)

140. U. Balachandran, T.H. Lee, S.E. Dorris, Hydrogen production by water dissociation using mixed conducting dense ceramic membranes. Int. J. Hydrogen Energy 32:451–456 (2007)

141. X. Meng, Y. Shang, B. Meng, N. Yang, X. Tan, J. Sunarso, et al., Bi-functional performances of $BaCe_{0.95}Tb_{0.05}O_{3-\delta}$-based hollow fiber membranes for power generation and hydrogen permeation. J. Eur. Ceram. Soc. (2016)

142. H. Wang, S. Gopalan, U.B. Pal, Hydrogen generation and separation using $Gd_{0.2}Ce_{0.8}O_{1.9-\delta}$(GDC)–$Gd_{0.08}Sr_{0.88}Ti_{0.95}Al_{0.05}O_{3\pm\delta}$ mixed ionic and electronic conducting membranes, Electrochim. Acta **56**, 6989–6996 (2011)

143. M.A. Habib, P. Ahmed, R. Ben-Mansour, H.M. Badr, P. Kirchen, A.F. Ghoniem, Modeling of a combined ion transport and porous membrane reactor for oxy-combustion. J. Membr. Sci. **446**, 230–243 (2013)

144. K.J. Lee, Y.J. Choe, J.S. Lee, H.J. Hwang, Fabrication of a microtubular $La_{0.6}Sr_{0.4}Ti_{0.2}Fe_{0.8}O_{3-\delta}$ membrane by electrophoretic deposition for hydrogen production. Adv. Mater. Sci. Eng. (2015). Article ID 505989

145. L. Anetor, E. Osakue, C. Odetunde, Reduced mechanism approach of modeling premixed propane-air mixture using Ansys Fluent. Eng. J. **16** (2012)

146. H.A. McGee, *Molecular Engineering* (McGraw-Hill, New York, 1991)

147. *User's Guide, Fluent 6.3 Documentation* (Fluent Inc., Lebanon, NH 2006)

148. P. Ahmed, M.A. Habib, R. Ben-Mansour, P. Kirchen, A.F. Ghoniem, CFD (computational fluid dynamics) analysis of a novel reactor design using ion transport membranes for oxy-fuel combustion. Energy **77**, 932–944 (2014)

149. A. Cuoci, A. Frassoldati, T. Faravelli, E. Ranzi, Accuracy and flexibility of simplified kinetic models for CFD applications, in *Combustion Colloquai: Italian Section of the Combustion Institute* (2009)

150. C.K. Westbrook, F.L. Dryer, Chemical kinetic modeling of hydrocarbon combustion. Prog. Energy Combust. Sci. **10**, 1–57 (1984)

151. S.V. Patankar, *Numerical Heat Transfer and Fluid Flow* (Hemisphere Publishing Corporation, Washington D.C., 1980)

152. C.Y. Park, T.H. Lee, S.E. Dorris, U. Balachandran, $La_{0.7}Sr_{0.3}Co_{0.2}Fe_{0.8}O_{3-\delta}$ as oxygen transport membrane for producing hydrogen via water splitting. ECS Trans. **13**(26), 393–403 (2008)

153. R.S. Barlow, G.J. Fiechtner, C.D. Carter, J.Y. Chen, Experiments on the scalar structure of turbulent $CO/H_2/N_2$ jet flames. Combust. Flame **120**, 549–569 (2000)

154. M.A. Habib, S.A. Salaudeen, M.A. Nemitallah, R. Ben-Mansour, E.M.A. Mokheimer, Numerical investigation of syngas oxy-combustion inside a LSCF-6428 oxygen transport membrane reactor. Energy (2016, just accepted)
155. H. Jiang, Z. Cao, S. Schirrmeister, T. Schiestel, J. Caro, A coupling strategy to produce hydrogen and ethylene in a membrane reactor. Angew. Chem. Int. Ed. **49**(33), 5656–5660 (2010)
156. H. Jiang, H. Wang, S. Werth, T. Schiestel, J. Caro, Simultaneous production of hydrogen and synthesis gas by combining water splitting with partial oxidation of methane in a hollow-fiber membrane reactor. Angew. Chem. Int. Ed. **47**, 9341–9344 (2008)
157. H.S. Fogler, *Elements of Chemical Reaction Engineering*, 4th edn., (Prentice Hall International Series, 2006), pp. 340–343
158. N. Ding, R. Arora, M. Norconk, S.Y. Lee, Numerical investigation of diluent influence on flame extinction limits and emission characteristic of lean-premixed H_2–CO (syngas) flames. Int. J. Hydrogen Energy **36**, 3222–3231 (2011)
159. K. Li, X. Tan, Y. Liu, Single-step fabrication of ceramic hollow fibers for oxygen permeation. J. Membr. Sci. **272**, 1–5 (2006)
160. A.E. Farooqui, H.M. Badr, M.A. Habib, R. Ben-mansour, Numerical investigation of combustion characteristics in an oxygen transport reactor. Int. J. Energy Res. **38**(5), 638–651 (2013)